Nancy Drickey

Mathematics Methods and Modeling for Today's Mathematics Classroom

A Contemporary Approach to Teaching Grades 7–12

Mathematics Methods and Modeling for Today's Mathematics Classroom

A Contemporary Approach to Teaching Grades 7–12

John A. Dossey
Illinois State University, Emeritus

Sharon McCrone
Illinois State University

Frank R. Giordano

Maurice D. Weir
U.S. Naval Postgraduate School

BROOKS/COLE
THOMSON LEARNING™

Australia • Canada • Mexico • Singapore • Spain • United Kingdom • United States

BROOKS/COLE
THOMSON LEARNING™

Sponsoring Editors: *Gary W. Ostedt, Robert W. Pirtle*
Marketing Team: *Karin Sandberg, Samantha Cabaluna*
Editorial Assistant: *Molly Nance*
Production Editor: *Tom Novack*
Production Service: *Luana Richards*
Manuscript Editor: *Luana Richards*
Permissions Editor: *Roxane Buck-Ezcurra*
Interior Design: *John Edeen*
Cover Design: Cheryl Carrington
Cover Photo: *Wides and Holl/FPG International*
Interior Illustration: *Brian Betsill*
Photo Researcher: *Kathleen Olson*
Print Buyer: *Jessica Reed*
Typesetting: *Integre Technical Publishing Company, Inc.*
Cover Printing: *Transcontinental*
Printing and Binding: *Transcontinental*

For more information about this or any other Brooks/Cole products, contact:
BROOKS/COLE
511 Forest Lodge Road
Pacific Grove, CA 93950 USA
www.brookscole.com
1-800-423-0563 (Thomson Learning Academic Resource Center)

Printed in Canada

10 9 8 7 6 5 4 3 2 1

Library of Congress Cataloging-in-Publication Data

Mathematics modeling for today's mathematics classroom : a contemporary approach to teaching grades 7–12 / John A. Dossey ... [et al.].
p. cm.
Includes bibliographical references and index.
ISBN 0-534-36604-X (alk. paper)
1. Mathematics—Study and teaching (Middle school) 2. Mathematics—Study and teaching (Secondary) I. Dossey, John A.

QA11 .M37577 2002
510′.71′2—dc21

2001035730

Preface

Building on the recommendations of NCTM's *Principles and Standards for School Mathematics* (2000), this text provides guidance for teaching mathematics content and for developing mathematics processes from middle school through high school. Its emphasis on mathematical modeling provides future teachers with an approach to teaching mathematics that supports the approach taken by many contemporary curricular materials for mathematics instruction.

Mathematical modeling is one of the richest forms of representation in mathematics. It requires students to work with and apply a variety of mathematical concepts, processes, and relationships. Students must be able to connect, through creative problem solving, their understanding of specific content to the modeling situation. For example, predicting the growth of a wildlife population involves exponential or logistic functions. A model for using a shipping facility efficiently involves probability and matrix algebra. Models such as these encourage students to deeply investigate problem situations and to use their investigations to understand, predict, and control aspects of the situation.

The *Principles and Standards* (p. 40) states that "one of the most powerful uses of mathematics is the mathematical modeling of phenomena. Students at all levels should have opportunities to model a wide variety of phenomena mathematically in ways that are appropriate to their level." If we are to realize this goal, teachers of mathematics must be able to select and direct activities that will forge their students' capabilities to form, interpret, and apply mathematical models in a variety of settings.

This text integrates mathematical modeling content with various methods of teaching mathematics. This was done to support discussions that will help pre- and in-service teachers move toward a *Standards*-like approach to the teaching and learning of mathematics. Central to this objective is encouraging future teachers to merge their knowledge of mathematics with their knowledge of teaching in ways that help them:

1. Understand the middle and high school curriculum in mathematics from a perspective of the NCTM *Principles and Standards for School Mathematics*.
2. Develop methods for teaching the content areas of the curriculum as outlined in the *Standards*.
3. Promote the processes of problem solving, reasoning, communicating, representing, and connecting mathematical concepts and principles using modeling as a medium.
4. Relate modeling concepts to the content of contemporary middle school and secondary programs developed with support from the National Science Foundation and that are now being used in schools.
5. Assess and evaluate students in realistic and authentic ways.
6. Model and solve significant problems in our society and in our everyday lives.

7. Apply technology (calculators and computers) to develop major mathematical concepts and relationships with students.
8. Understand student differences and needs in the classroom while promoting quality mathematics for all students.

Student Background and Course Content

In writing the text, we assumed that readers would have a solid mathematical preparation similar to that found for teacher certification at the middle school or high school level in most states. In many instances, the mathematics employed does not assume a technical level beyond that expected of juniors or seniors in an undergraduate mathematics education program without developing the mathematics in the text itself. In the optional chapter on Modeling Using Calculus, we assume that readers have completed a calculus course sequence. We hope that readers will approach the text with a spirit of investigation and interest. Our methods move to connect their understanding of mathematics with its myriad uses. The modeling serves to focus on the fact that the understanding and ability to use important mathematics grows out of rich and authentic opportunities to learn that mathematics.

Organization of the Text

The text is organized into three main parts. Part I includes Chapters 1–3 and offers an overview of the contemporary context for the teaching of mathematics and the role that mathematical modeling plays in that teaching. Part II, Chapters 4–11, deals with the content and processes found in the major content areas of the curriculum. These chapters also focus on the teaching and learning of mathematics in the individual content areas. Part III, Chapters 12–16, deals with preparing to teach, organizing the classroom, assessing students, and other pedagogical issues.

The chapters are organized in a manner that allows the book to be used in a number of ways. It can be used in a one-term mathematics methods course with the instructor choosing what content to cover. Or, it can be used over a number of terms with the instructor working through the material in greater depth in a sequence of courses. From a methods standpoint, one would most likely want to address the chapters in Part I before launching into the chapters in Part II. The mathematical content chapters in Part II could be covered in a different order than ours, or some could be selected for special emphasis, depending on the nature of the course. The chapters in Part III give future teachers "nuts and bolts" ways to organize their classrooms, and to provide their students successful experiences with mathematics.

The text provides a well-grounded introduction to the entire modeling process. The student has the opportunity, on several occasions, to experience the following facets of modeling:

1. *Creative and Empirical Model Construction:* Given a real-world scenario, the student learns to identify a problem, make assumptions and collect data, propose a model, test the assumptions, refine the model as necessary, fit the model

to data if appropriate, and analyze the underlying mathematical structure of the model to appraise the sensitivity of the conclusions when the assumptions are not met precisely.

2. *Model Analysis:* Given a model, the student learns to work backward to uncover the implicit assumptions, critically assess how well those assumptions fit the scenario at hand, and estimate the sensitivity of the conclusions when the assumptions are not met precisely.
3. *Model Research:* The student investigates a specific area to gain a deeper understanding of some behavior and learns to use what has already been created or discovered.

The modeling sections in the chapters are designed to enhance a student's modeling capabilities. We identify the following phases of this process:

1. Problem identification
2. Model construction or selection
3. Identification and collection of data
4. Model validation
5. Calculation of solutions to the model
6. Model implementation and maintenance

Unfortunately, many teachers have only experienced calculating model solutions—they have never built a model. As a result, they feel anxious about materials that call on them to lead student investigations of real-world problem situations or modeling projects, especially those for which no model is provided. With this knowledge in hand, we have attempted to slowly introduce students to modeling and support their growth in the ability to carry out all phases of the modeling process. We have found that the early introduction of these processes, beginning with short projects, facilitates their progressive development and confidence in both teaching mathematics and modeling important situations.

The study of modeling begins in Chapter 1 with the consideration of models and the use of discrete dynamical systems to model change with difference equations. This is extended in Chapter 3 to the consideration of the mathematical behavior of dynamical systems and systems of difference equations. In Chapter 5, the focus shifts to using functions to "linearize" data to understand the proportional relations underlying the set of data. These skills are extended to modeling with geometric similarity and measures in Chapter 6. The teaching of probability and statistics is considered in Chapter 7 using simulations to model random events. Chapter 8 examines models drawn from graph theory. Chapter 9 fits models to empirical data using different criteria for deciding what model might be optimal. The modeling in Chapter 10 relates topics in secondary school algebra and geometry to linear programming and numerical search methods. Chapter 11 extends the modeling process by using calculus to model change in continuous settings. This content closes the loop on many of the topics considered across these chapters in that readers see the close ties between discrete and continuous methods and between empirical and theory-based approaches to modeling.

While the modeling content in these chapters has been arranged in the order in which we prefer to teach our classes and also to link this content to the 7–12 school curriculum, the presentation of the material can be varied to fit different situations. In general, however, the modeling content in Chapters 1–6 will need to be covered before tackling the modeling topics in the remaining chapters. This will give students a solid foundation in the essence of change, using difference equations, sensing the nature of proportionality in modeling situations, and fitting models.

Features of the Text

Various features enhance the text's approach to the subject matter and provide a broad range of experiences for pre-service mathematics teachers. These include

- Overview/Focus on the *Standards* and Classroom—Each chapter begins with a brief description of the chapter content and special issues raised. It connects the chapter's content to the *Principles and Standards* and, at the same time, relates the chapter content to the reality of the classroom.
- Technology—Technology is emphasized throughout the text. A variety of technological aids are suggested, where appropriate, for the teaching of particular content, for enhancing problem-solving abilities, and for preparing middle school and high school students for today's technological society.
- Section Exercises—Each chapter section contains several exercises that continue and extend investigations similar to those presented in the section narrative. Many of the exercises are appropriate for use at the high school level. Most sets of exercises end with a Projects section.
- "Find Out For Yourself" Projects—These projects direct the student to investigate a topic in the teaching and learning of mathematics. These involve problems or classroom situations the student is likely to encounter. In other cases, student teachers are asked to observe in actual classrooms, interview students, and build a real sense of a teacher's day in class. These projects challenge students to find out what it's like to teach. They also offer ideas for instructional activities and give students opportunities to experience activities that promote problem solving and modeling approaches.
- "Modeling Exploration" Projects—These projects provide additional real-world applications of the modeling methods considered in the text. In some sections, these modeling projects extend the examples considered in the text. In others, they acquaint students with other resources drawn from the accompanying CD, UMAP modules, and other modeling resources. UMAP modules are self-contained modeling materials developed and distributed by the Consortium for Mathematics and Its Applications (COMAP). They are available through COMAP at *www.comap.com*.

Technology

We have assumed that instructors and students using the text will be familiar with graphing calculators and the basics of computer-based spreadsheets, statistics programs, and geometry software packages. While we seldom make special use of these in the text, we have included the output from such technology use in examples and in some exercises and lesson plans. We believe that expertise in these areas should be developed as part of the full sequence of courses that pre- and in-service mathematics teachers take and that this text simply builds on that knowledge from a methods standpoint.

Resources

The text has is accompanied by a compact disk, which contains additional modeling information, resources from COMAP's activity files on modeling, and problems from the collegiate Mathematical Contest in Modeling (MCM) and the High School Mathematical Contest in Modeling (HiMCM) contests on mathematical modeling. An instructor's manual is also available.

Acknowledgments

It is always a pleasure to acknowledge individuals who have played a role in the development of a book. We are indebted to the many colleagues and teachers who have stimulated our interest in teaching methods and modeling and for the support and guidance they have given to our careers. We are also indebted to the many individuals who have authored or co-authored curriculum or modeling materials we have referred to across the text. Without such materials, the teaching of mathematics from a modeling standpoint would be impossible. We give special thanks to the participants of the summer 2000 NSF workshop on teaching secondary school methods that examined preliminary versions of these chapters and offered us feedback on their content and tone.

We would like to thank the National Science Foundation for their support of this project, especially Dr. Elizabeth Teles who provided sage counsel throughout the duration of the project. We would also like to thank the folks at COMAP who were both supportive and cheerful when we needed help. We would especially like to thank Rick Jennings who helped us with the workshops, Roland Cheyney and Gary Froelich who designed the companion software, Jan Beebe who coordinated arangements for workshops and materials, Clarice Callahan and Laurie Aragon for their help with administering the project and workshops. Finally, a special thanks to Sol Garfunkel, the Executive Director of COMAP, for inspiring the project and encouraging us throughout its completion.

The production of any mathematics text is a complex process, and we have been especially fortunate in having a superb and creative production staff at Brooks/Cole. In particular, we express our thanks to our editors Gary Ostedt and Bob Pirtle; Tom Novack, our production editor; Luana Richards, our production service; Karin Sandberg, our marketing manager; and the many unnamed others who assisted in bringing this text to completion.

Finally, we are grateful to our spouses—Anne Dossey, Judi Giordano, Matt McCrone, and Gale Weir—for their support and understanding. We additionally thank our families for their interest in, and support of, our continued study of the teaching and learning of mathematics.

John A. Dossey
Frank R. Giordano
Sharon S. McCrone
Maurice D. Weir

Contents

Part III Decisions in Organizing for Mathematics Instruction 483

Part I

The Context for Teaching Mathematics

1
What Is Mathematics and Why Do We Teach It?

"I venture to suggest that if one were to ask for that single attribute of the human intellect which would most clearly indicate the degree of civilization of a [society], the answer would be, the power of close reasoning, and that this power could best be determined in a general way by the mathematical skill which members of [that society] displayed."

A. B. Chace, 1927

Overview

What is mathematics? Why do we have to learn it? These questions often arise in the classroom from students—students who want to know. Such questions are not as easy to answer as they are to ask. The names and contributions of individuals involved in the history of mathematics are easy to cite. The wide variety of its applications in our society are easy to list. But, the nature of mathematics itself is hard to capture. This results from a lack of consensus, even among mathematicians, as to what constitutes "mathematics" and what "doing mathematics" means.

In this chapter we answer these two questions by briefly examining the history of mathematics and looking at what mathematics is and at some of its uses. We also focus on the learning of mathematics—who should learn it, how they should learn it, and why they should learn it.

Focus on the NCTM *Standards*

In 1989, the National Council of Teachers of Mathematics (NCTM) issued its *Curriculum and Evaluation Standards for School Mathematics*. In 2000, the Council updated these *Standards* to maintain their currency and reiterated the fundamental principles and goals that guide quality mathematics programs. These *Standards* list characteristics that define quality programs and describe valued outcomes for individual students. The characteristics, or *principles* as NCTM calls them, evolved from a vision of mathematics and how it should be portrayed in schools. They lay out the conditions that define quality teaching and learning of mathematics in these six principles:

- *Equity.* Excellence in mathematics education requires equally high expectations and strong support for all students.
- *Curriculum.* A curriculum is more than a collection of activities; it must be coherent, focused on important mathematics, and well articulated across the grades.
- *Teaching.* Effective mathematics teaching requires understanding what students know and need to learn and then challenging and supporting them to learn it well.
- *Learning.* Students must learn mathematics with understanding, actively building new knowledge from experience and prior knowledge.
- *Assessment.* Assessment should support the learning of important mathematics and furnish useful information to both teachers and students.
- *Technology.* Technology is essential in teaching and learning mathematics; it influences the mathematics that is taught and enhances students' learning (NCTM, 2000, p. 11).

Examine these principles. How do they speak to the issues of learning and teaching mathematics? How would the absence of even one affect the quality of a mathematics program at the classroom level?

Focus on the Classroom

Math teachers, operating at the classroom level in daily contact with students, bring instructional programs in mathematics to life. Mathematics learning takes place as teachers and students jointly search for meaning in the situations they encounter. The information and suggestions in this text provide a basis for teachers and future teachers to develop and assess the degree to which such mathematical encounters help achieve the principles and goals established for teaching mathematics.

The NCTM has developed its *Principles and Standards* with the goal of changing mathematics education so that each student has the opportunity to learn in a classroom that

- demonstrates the value of mathematics
- develops confidence to participate in mathematics
- nurtures capability for solving problems
- builds skills for communicating in mathematical contexts
- strengthens insight into the patterns of mathematical reasoning

With these same goals and principles in mind, we designed this text to increase your knowledge of methods for teaching mathematics and of how mathematics helps model events in our world. Along the way, the readings, activities, and content of this text will help you develop a personal style that embodies the NCTM principles and goals.

1.1 What Is Mathematics?

What is mathematics? This is an easy question to ask, but a very difficult one to answer. Many books address this question, and each book answers the question differently. In the 1940s, the English number theorist G. H. Hardy wrote

> A mathematician, like a painter or a poet, is a maker of patterns. If his patterns are more permanent than theirs, it is because they are made with *ideas*. A painter makes patterns with shapes and colours, a poet with words. A painting may embody an idea, but the idea is usually commonplace and unimportant. In poetry, ideas count for a good deal more. A mathematician, on the other hand, has no material to work with but ideas, and so his patterns are likely to last longer, since ideas wear less with time than words.
>
> The mathematician's patterns, like the painter's or the poet's, must be *beautiful*; the ideas, like the colours or the words, must fit together in a harmonious way. Beauty is the first test: there is no permanent place in the world for ugly mathematics.... The best mathematics is *serious* as well as beautiful... I am not thinking about the "practical" consequences of mathematics.... The "seriousness" of a mathematical theorem lies, not in its practical consequences, which are usually negligible, but in the *significance* of the mathematics ideas which it connects. We may say, roughly, that a mathematical idea is "significant"

> if it can be connected, in a natural and illuminating way, with a large complex of other mathematical ideas (Hardy, 1940).

In this statement, Hardy links mathematics with ideas that capture the relationships inherent in the study of quantity, shape, form, change, growth, chance, and measure. At the same time, Hardy downplays the role of applications as a measure of the worth of mathematics. Hardy said that "it is not possible to justify the life of any genuine professional mathematician on the ground of the utility of his work" (Hardy, 1992, 119–120). Here we see the tension between the development of "beautiful abstract ideas and relationships" and the use of mathematics to solve real-world problems. This tension plays throughout the history of mathematics.

One of the oldest documents we have detailing human views about mathematics is the Rhind papyrus, a document written by the Egyptian scribe Ahmes in approximately 1650 B.C. This papyrus roll, measuring 1 foot by 18 feet, starts with a bold claim to provide an "Accurate Reckoning: The entrance into the knowledge of all existing things and all obscure secrets" (Chace, 1927, p. 50). In reality, the document was a practical handbook of mathematical exercises encompassing rules for operating with fractions, finding solutions to problems by a method known as "false position," and procedures for adjusting estimates to quantities to exact answers. The contents deal with exercises about dividing measures of grain, bread, and beer among groups of individuals. The work portrays mathematics as a tool for solving problems in commerce and governmental affairs.

The 24th problem in the Rhind papyrus is a good example of this. It asks the reader to find the value of a heap if a heap and a seventh of a heap is 19. Using the method of false position, the Egyptians began by assuming that the value of a heap was 7. Hence, if x is the value of a heap, $x + x/7 = 8$. Clearly this was a "false position." However, if one notes that $8(2 \text{ and } \frac{3}{8}) = 19$, one can multiply 7 by 2 and $\frac{3}{8}$, getting the correct value of a heap as 16 and $\frac{5}{8}$. Will this process always work? If so, why? (Gillings, 1982, pp. 154–55.)

Greek discussions of mathematics from the fourth century B.C. contain exchanges between Plato and his student Aristotle. Plato took the position that the objects of mathematics have an existence of their own, beyond the human mind. Plato drew clear distinctions between the ideas of the mind and their representations in the world of the senses. He also drew clear distinctions between the world of the arithmetic theory of numbers and that of logistics, the application of computation to commerce. In *The Republic*, Plato argued that the study of arithmetic has a positive effect on individuals, compelling them to reason about abstract numbers, but at the same time bristled at the use of physical arguments by technicians to "prove" results involving numbers in applied settings. For Plato, mathematics was an abstract mental exercise for thinkers of this time. This is documented in his writings about the five regular polyhedra, since called the Platonic solids and his support and encouragement of the mathematical development of Athens (Boyer, 1968).

Aristotle, Plato's student, viewed mathematics as one of the three categories into which knowledge might be divided, the others being physical and theological knowledge. He noted that mathematics illustrates "quality with respect to forms and local motions, seeking figure, number, and magnitude, and also place, time, and similar things. . . . Such an essence falls, as it were, between the other two [forms of knowledge], because it can be conceived both through the senses and without the senses" (Ptolemy, 1952, p. 5).

Aristotle's affirmation of the role of the senses as a source for abstracting and generalizing ideas related to quantity, form, and shape is vastly different from that of his teacher. Aristotle based his view of mathematical knowledge on experience and reality—where knowledge is obtained through experimentation, observation, and abstraction. His view supports the conception that we construct the relations inherent in a given mathematical context.

The Three Views of Mathematics

Near the beginning of the twentieth century, the German mathematician Gottlob Frege promoted logicism as a way of conceiving mathematics. This approach was an outgrowth of the Platonic school of thought. Logicists argued that the ideas of mathematics can be viewed as a subset of the ideas of formal logic. The proponents of this school of thought attempted to show that all mathematical propositions can be expressed as completely general propositions whose truth follows from their form rather than from their interpretation in any applied setting. This approach ran into trouble in the early 1900s with the discovery of Russell's paradox by the Englishmen Bertrand Russell and Albert North Whitehead. The logicists thought that mathematics consists of objects and an external (outside the senses) structure of generalized statements. This structure is composed of sentences about the objects and their mathematical relationships to one another. The logicists used classical arguments, starting with the axioms for the natural numbers and set theory, to prove the validity of those relationships. Russell's paradox arose as they attempted to extend and develop such a system. In such a structure, it is natural to consider a set of all sets that do not have the property of being elements of themselves. That is, the set $S = \{A \mid A \text{ is a set and } A \notin A\}$. Since S itself is a set, it must be the case that $S \in S$ or $S \notin S$. However, if $S \in S$, then S is not one of the sets A and cannot be an element of S. On the other hand, if $S \notin S$, then S satisfies the property described and must be an element of S, again an impossibility. Since both cases lead to a contradiction, the statement about the set S leads to a paradox. This finding shocked the mathematical world, especially the logicists, who thought that all of mathematics could be placed in a nice, consistent logical structure.

Another school of mathematical thought arose in the Netherlands at nearly the same time as logicism. This school, referred to as intuitionism, followed the works of the Dutch mathematician L. E. J. Brouwer. He argued that no idea of classical mathematics can be accepted until it can be constructed via a combination of clear inductive steps from basic axioms. The intuitionists rejected all theorems of the form $p \rightarrow q$ that were developed by assuming $\tilde{q}$ and then showing that this leads to a contradiction (thus arriving at the conclusion that q must be true). Such indirect proofs depend on the law of the excluded middle; that is, either q or $\tilde{q}$ must be true. In contrast to the logicists, the intuitionists portrayed mathematics as an object consisting of results derived from the "close introspection" of relationships and events.

The third school of thought to originate in the early 1900s was the formalist school headed by the German mathematician David Hilbert. Hilbert's views, like those of Brouwer, are more in line with Aristotelian tradition than with the traditions of Plato or Frege. However, Hilbert did not believe that arithmetic and geometry exist as descriptions of *a priori* knowledge to the same degree that Brouwer did. Hilbert saw mathematics arising from intuition based on objects that have concrete representations in the mind. His teaching was punctuated with the use of geometric models for contemplation and experimentation. Formalists attempted to characterize observed and imagined relationships in terms of formal axiomatic systems. They aimed to free mathematics from the bounds and contradictions that plagued Russell and Whitehead's attempt to develop mathematics as a formal stystem based on a small axiom set and logic. Hilbert freed geometry from many of the oversights left by Euclid. Nevertheless, Kurt Gödel in 1931, with his development of a firm foundation for geometry, showed that the formalist's approach to developing mathematics suffered from major problems as well.

Thus, the three major schools of thought in the early 1900s failed to provide a widely accepted foundation for the nature of mathematics. Yet, all three views characterize the contents of mathematics as products or objects. The logicists saw the contents of mathematics as the elements of classical mathematics, its definitions, its postulates, and its theorems. The intuitionists saw the contents as theorems constructed from basic axioms using "valid" patterns of reasoning. The formalists saw mathematics as formal axiomatic structures developed to rid classical mathematics of its shortcomings. Platonic and Artistotelian notions run as strong undercurrents in these theories. The origin of the "product" as a preexisting external object, or as an object created through experience from sense perceptions or experimentation, remained an issue.

Current Views of Mathematics

Today's professional mathematicians carry a strong belief in the existence of mathematical concepts outside the human mind. When pushed to clarify their conceptions of mathematics, most retreat to a formalist, or Aristotelian-based, position of mathematics as a game played with symbol systems according to a fixed set of socially accepted rules (Davis & Hersh, 1980). However, in reality, most professional mathematicians think little about the fundamental nature of their subject as they "do" mathematics.

The conception of mathematics held by teachers significantly influences how they teach it. A person holding strong Platonic-based beliefs tends to present mathematics as a structure existing outside the mind and experiences of the student. The focus is on the development of generalized rules and concepts, devoid of real-world contexts. A teacher holding to Aristotelian-based notions relies heavily on experiences, perhaps to the detriment of developing relations in a more abstract and general setting.

Hersh (1986) argues that a new philosophy of mathematics is needed, one that will serve both the working mathematician and the mathematics educator. According to Hersh, the working mathematician is not controlled by constant attention to validating every step with an accepted formal argument. Rather, guided by intuition, the mathematician proceeds by exploring concepts and their interactions. That is, they focus on understanding rather than on long formal derivations of carefully quantified results.

This mode of operation calls for a major change in our conceptions of mathematics. Mathematics must be accepted as a human activity, an activity not strictly governed by any one school of thought (logicist, formalist, or intuitionist). Such a conception answers the question of what mathematics is by saying that "Mathematics deals with ideas. Not pencil marks or chalk marks, not physical triangles or physical sets, but ideas (which may be represented or suggested by physical objects)" (Hersh, 1986, p. 22).

Hersh continues by asking

> What are the main properties of mathematical activity or mathematical knowledge, as known to all of us from daily experience?

Hersh then answers his own question by setting up the following conditions:

1. Mathematical objects are invented or created by humans.
2. They are created, not arbitrarily, but arise from activity with already existing mathematical objects, and from the needs of science and daily life.
3. Once created, mathematical objects have properties which are well determined, which we may have great difficulty in discovering, but which are possessed independently of our knowledge of them.

Hersh's call for a new statement on the nature of mathematics parallels the view put forth in the NCTM *Standards*: Mathematics is something that we do, experience, and connect to our previous knowledge. Lynn Steen shared the following metaphors:

> Many educated persons, especially scientists and engineers, harbor an image of mathematics as akin to a tree of knowledge: formulas, theorems, and results hanging like ripe fruits to be plucked by passing scientists to nourish their theories. Mathematicians, in contrast, see their field as a rapidly growing rain forest, nourished and shaped by forces outside mathematics while contributing to human civilization a rich and ever-changing variety of intellectual flora and fauna. These differences in perception are due primarily to the steep and harsh terrain of abstract language that separates the mathematical rain forest from the domain of ordinary human activity. (Steen, 1988, p. 611)

Today, we talk of "pure mathematics" when we discuss the formal structures of mathematics in terms of undefined terms, defined terms, postulates, valid forms of logical argument, and deduced theorems. We talk of "applied mathematics" when we use algebra to devise security codes, partial differential equations to develop pricing models for options in the financial world, and statistics to determine the reliability of a manufacturing process. Today, school curricula are bringing the applications of mathematics to the classroom rather than delaying them until students master arithmetic, algebra, and the basic proofs of geometry. In many cases, these applications justify the validity of the mathematical ideas studied.

Mathematical modeling is central to understanding the real world, while simultaneously developing worthwhile mathematics. Modeling generally develops significant mathematics from basic concepts and principles, and uses it to understand, predict, and control events in the real world. Section 1.2 demonstrates this approach by modeling change in discrete settings.

1.2 Mathematical Literacy and Modeling with Discrete Systems

Literacy is generally thought about in terms of reading, but contemporary events require that an individual also have a mathematical literacy defined in terms of mathematical knowledge and skills. In addition to the traditional **prose literacy** that describes an individual's ability to read and interpret written prose, today's world requires competence in **document literacy** and **quantitative literacy**. Document literacy is one's ability to read tables, charts, and graphs of varied complexity. Quantitative literacy is one's ability to deal with numerical, measurement, geometrical, probabilistic, statistical, and algebraic relationships in meaningful ways. Combined, these three forms of literacy speak to an individual's ability to function in a world that requires shifting between verbal, symbolic, numerical, and graphical ways of viewing situations we encounter.

A **mathematically literate** person is one who can interpret and apply these various types of mathematical knowledge to understand, predict, and control factors in contexts important to his or her life. Such an individual can reason in numerical, data, spatial, and probabilistic settings; can integrate and apply mathematical concepts and procedural skills; and can develop and interpret models related to the problems encountered.

Applying mathematics to the real world is often a tentative and intuitive process. The clarity of the problem, the nature and adequacy of the information available, the form of the solution needed, and the support available for executing the data collection, calculations, and verification of results are often questionable at best. The skills required to apply mathematics in such a situation differs greatly from those traditionally taught in the classroom. Most applications problems in textbooks present all of the data and ask all of the important questions. They are merely opportunities to practice skills, not to develop them. In this section we begin to examine how mathematical modeling assists in "making sense" of our world. We do this by examining the role models play in understanding events around us.

Modeling Change

In this section we build mathematical models that describe change in an observed behavior. When we observe change, we are often interested in understanding why the change occurs in the way it does, perhaps to analyze the effects of different conditions or to predict the future.

A powerful modeling paradigm is

$$\textit{future value} = \textit{present value} + \textit{change}$$

If the behavior is taking place over *discrete time periods*, this paradigm leads to a **difference equation**, studied in this chapter. If the behavior takes place *continuously* over time, then the paradigm leads to a **differential equation**, studied in Chapter 11. Both give powerful models for studying change to explain and predict behavior.

Let's begin our study of discrete change by examining behavior modeled *exactly* by difference equations. In Section 1.3, we use proportion to *approximate change* we have observed. In Section 3.3, we construct *numerical solutions* to the difference equations we have built to determine the types of *long-term behaviors* they predict. In Section 3.4, we model **interactive systems**, such as ecological systems involving predators and prey, using systems of difference equations. Again we examine the long-term behaviors of these systems by constructing numerical solutions.

Example 1 *A Savings Certificate*

Consider the value of a savings certificate initially worth \$1000 that accumulates interest paid each month at 1% per month. The following *sequence* of numbers represents the value of the certificate for consecutive months:

$$A = (1000, 1010, 1020.10, 1030.30, \ldots)$$

■

Definition 1 For a sequence of numbers $A = (a_0, a_1, a_2, a_3, \ldots)$ the **first differences** are defined as follows:

$$\begin{aligned}
\Delta a_0 &= a_1 - a_0 \\
\Delta a_1 &= a_2 - a_1 \\
\Delta a_2 &= a_3 - a_2 \\
\Delta a_3 &= a_4 - a_3 \\
&\vdots \quad \vdots
\end{aligned}$$

In general, the ***n*th first difference** is defined as

$$\Delta a_n = a_{n+1} - a_n$$

■

Note from Figure 1.1 that the first difference represents the rise or fall, that is, the *change*, in the graph of the sequence during one time period.

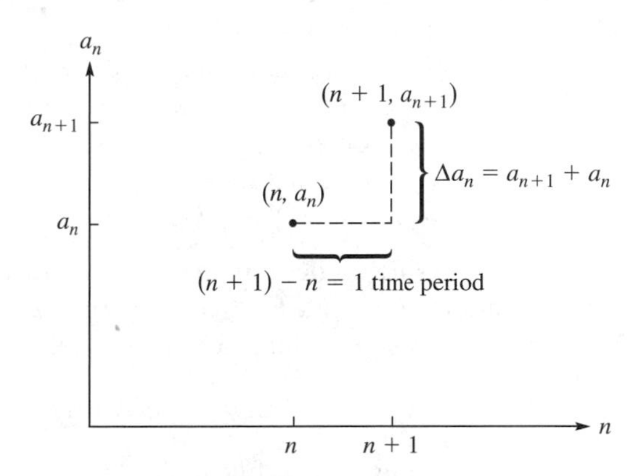

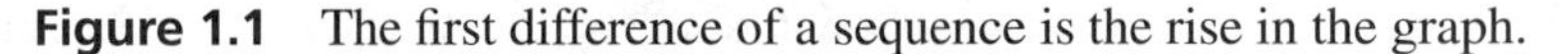
Figure 1.1 The first difference of a sequence is the rise in the graph.

For example, several of the first differences for the sequence representing the value of the savings certificate are as follows:

$$\Delta a_0 = a_1 - a_0 = 1010 - 1000 = 10$$
$$\Delta a_1 = a_2 - a_1 = 1020.10 - 1010 = 10.10$$
$$\Delta a_2 = a_3 - a_2 = 1030.30 - 1020.10 = 10.20$$

Note that the first differences represent the *change in the sequence* during the period, or the *interest earned* in the case of the savings certificate example. The symbol Δ is a Greek capital letter "dee," pronounced delta, and the notation does not mean something that multiplies something else.

The first difference is useful for modeling change that occurs in discrete intervals. In the current example we know that the change in the value of the certificate from one period to the next is the interest paid during that period. If n is the number of months and a_n the value of the certificate after n months, then the change, or interest growth, for the next month is the nth difference

$$\Delta a_n = a_{n+1} - a_n = 0.01a_n$$

This expression can be rewritten as the difference equation

$$a_{n+1} = a_n + 0.01a_n$$

which gives the value of the certificate next month. We also know the initial deposit (initial value), which allows us to write the complete **initial value model**:

$$a_{n+1} = 1.01a_n \qquad n = 0, 1, 2, 3, \ldots \tag{1}$$
$$a_0 = 1000$$

where a_n is the amount accrued after n months. Since n represents the nonnegative integers $\{0, 1, 2, 3, \ldots\}$, equation (1) is an *infinite set* of algebraic equations. The difference equation allows us to compute the next term if we know the preceding term in the sequence, but it does not allow us to compute a specific term directly (say the certificate value a_{100} after 100 months).

Let's modify our example and withdraw \$50 from the account each month. Now the change during a single period is the interest earned during that period minus the monthly withdrawal, or

$$\Delta a_n = a_{n+1} - a_n = 0.01a_n - 50$$

In most examples, describing the change mathematically will not be as precise as the procedure just illustrated. Often we need to *plot the change*, *observe a pattern*, and then *describe the change* in mathematical terms. That is, we try to find

$$\textit{change} = \Delta a_n = \textit{some function } f$$

The change may be a function of previous terms in the sequence (as was the case with no monthly withdrawals), and may even involve external terms (such as with a monthly withdrawal or some expression involving n). Thus, in constructing models representing change, we will **model change in discrete intervals**, where

$$change = \Delta a_n = a_{n+1} - a_n = f(\textit{terms in the sequence, external terms})$$

Modeling change in this way becomes the art of determining or approximating a function f that captures or satisfactorily reflects the change.

Example 2 ***Mortgaging a Home***

Six years ago your parents purchased a home by financing \$80,000 for 20 years, and making monthly payments of \$880.87 with a monthly interest of 1%. They have now made 72 payments and wish to know how much they owe on the mortgage, which they are considering paying off with money from an inheritance. Or they may refinance the mortgage with one of several interest rate options, depending on the length of the payback period. The change in the amount owed each period increases by the amount of interest and decreases by the amount of the payment:

$$\Delta b_n = b_{n+1} - b_n = 0.01 b_n - 880.87$$

Solving for b_{n+1} and incorporating the initial condition gives the **dynamical system model**

$$b_{n+1} = b_n + 0.01 b_n - 880.87$$
$$b_0 = 80{,}000$$

where b_n represents the amount owed after n months. Thus

$$b_1 = 80{,}000 + 0.01(80{,}000) - 880.87 = 79{,}919.13$$
$$b_2 = 79{,}919.13 + 0.01(79{,}919.13) - 880.87 = 79{,}837.45$$

yielding the sequence or *numerical solution*

$$B = (80{,}000, 79{,}919.13, 79{,}837.45, \ldots)$$

The sequence is graphed in Figure 1.2. ■

In the following problem set you will encounter other real-world behaviors modeled exactly by difference equations. In the next section we use difference equations to *approximate* some observed change. After collecting data and discerning patterns of the behavior, we use proportion to test and fit models that we propose. First, let's summarize the important definitions and get some practice using difference equations to model real-world behavior exactly.

Definition 2 A **sequence** is a function whose domain is the set of all nonnegative integers and whose range is a subset of the real numbers. ■

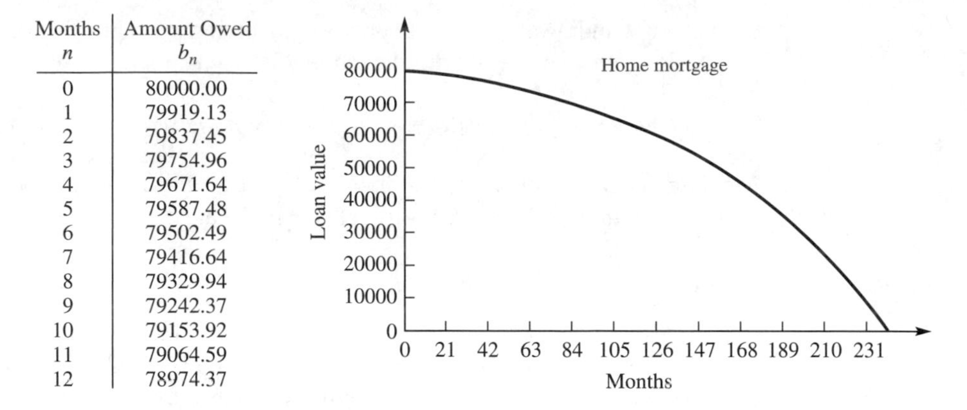

Months n	Amount Owed b_n
0	80000.00
1	79919.13
2	79837.45
3	79754.96
4	79671.64
5	79587.48
6	79502.49
7	79416.64
8	79329.94
9	79242.37
10	79153.92
11	79064.59
12	78974.37

Figure 1.2 Numerical solution and graph for the home mortgage in Example 2.

Definition 3 A **difference equation** is a relationship among terms in a sequence. ■

Definition 4 A **numerical solution** is a table of values satisfying the difference equation. ■

Definition 5 A **recurrence relation** is an equation of the form $a_{n+1} = f(a_n, a_{n-1}, a_{n-2}, \ldots)$ where f is a function. For example, $a_{n+1} = 2a_n + a_{n-1} + 5$ for $n > 1$. ■

Definition 6 A **discrete dynamical system** is a sequence defined by a recurrence relation. ■

Exercises 1.2

Sequences

1. Write out the first five terms of the following sequences.

a. $a_{n+1} = 3a_n, \quad a_0 = 1$
b. $a_{n+1} = 2a_n + 6, \quad a_0 = 0$
c. $a_{n+1} = a_n^2, \quad a_0 = 1$
d. $a_{n+1} = 2a_n(a_n + 3), \quad a_0 = 4$

Difference Equations

2. Examine the following sequences and write a difference equation that represents the change during the nth interval as a function of the previous term in the sequence.

a. $2, 4, 6, 8, 10, \ldots$
b. $2, 4, 16, 256, \ldots$
c. $1, 3, 7, 15, 31, \ldots$
d. $1, 7, 25, 79, \ldots$

3. Substitute $n = 0, 1, 2, 3$ and write out the first four algebraic equations represented by the following **initial value problems**.

a. $a_{n+1} = 3a_n, \quad a_0 = 1$
b. $a_{n+1} = 2a_n + 6, \quad a_0 = 0$
c. $a_{n+1} = a_n^2, \quad a_0 = 1$
d. $a_{n+1} = 2a_n(a_n + 3), \quad a_0 = 4$

Modeling Change Exactly

For Exercises 4–7, formulate a difference equation that models the situations described.

4. You currently have $5000 in a savings account that pays 0.5% interest each month. You add another $200 each month.

5. You owe $500 on a credit card that charges 1.5% interest each month. You can pay $50 each month and you make no new charges.

6. Your parents are considering a 30-year $100,000 mortgage that charges 0.5% interest each month. Formulate a model in terms of a monthly payment p that allows the mortgage (loan) to be paid off after 360 payments. [*Hint:* If a_n represents the amount owed after n months, what are a_0 and a_{360}?]

7. Your grandparents have an annuity. The value of the annuity increases each month by 1% interest on the previous month's balance. Your grandparents withdraw $1000 each month for living expenses. Presently, they have $50,000 in the annuity. Model the annuity with a dynamical system. Will the annuity run out of money? When? [*Hint:* What value will a_n have when the annuity is depleted?]

8. Name several behaviors you think can be modeled by dynamical systems.

PROJECT Modeling Exploration

You want to buy a new car. You narrow your choices to Saturn SL, Cavalier, and Tercel. Each company offers you their prime deal:

Saturn SL	$11,990	$1,000 down	3.5% interest for up to 60 months
Cavalier	$11,550	$1,500 down	4.5% interest for up to 60 months
Tercel	$10,900	$500 down	6.5% interest for up to 48 months

You are able to spend at most $475 a month on a car payment. Use a difference equation to determine which car you should buy.

1.3 Approximating Change with Difference Equations

In most situations, describing the observed change mathematically is not as precise a procedure as it was for the savings certificate and mortgage examples in the previous section. In such cases, we plot the change, observe a pattern, and then approximate the change mathematically.

Modeling change in this way is the "art" of determining or approximating some appropriate function f that represents the change. We begin by distinguishing between change that takes place continuously and that which occurs in discrete time intervals.

Discrete Versus Continuous Change

When constructing models, it is important to distinguish between change that takes place in **discrete** time intervals (such as the depositing of interest in an account) and change that occurs **continuously** over time (such as the change in the temperature of a cold can of soda on a warm day). Difference equations represent change in the case of discrete time intervals. But we can also approximate continuous change by examining data taken at discrete time increments.

Approximating Change

Few models represent the real world exactly. Generally, mathematical models simply *approximate* real-world behaviors. This happens because some *simplification* is required in order to model the real-world behavior in mathematical terms. For example, to model the spotted owl population in a habitat in order to predict the effects of changes in environmental policy (such as intense logging operations), we might omit such variables as competition for food, predators, and natural disasters all affecting the owl population over time.

In a mathematical construct, we cannot hope to capture every detail, but we can eventually build models that consider more variables. First, let's construct some very simple models that allow us to focus on the process.

Example 1 *Growth of a Yeast Culture*

Consider the data in Figure 1.3[1] from an experiment that measures the growth of a yeast culture. A plot of the data suggests that the change in biomass is proportional to the

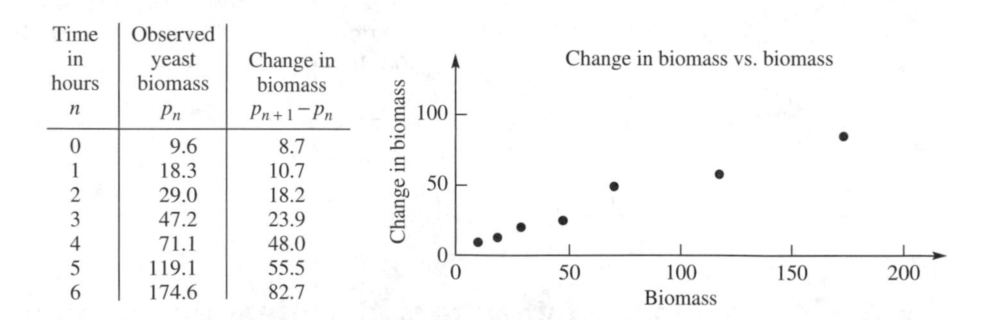

Time in hours n	Observed yeast biomass p_n	Change in biomass $p_{n+1} - p_n$
0	9.6	8.7
1	18.3	10.7
2	29.0	18.2
3	47.2	23.9
4	71.1	48.0
5	119.1	55.5
6	174.6	82.7

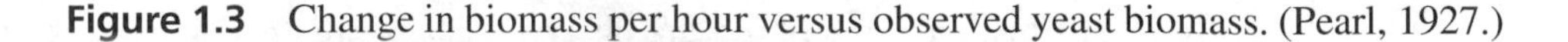

Figure 1.3 Change in biomass per hour versus observed yeast biomass. (Pearl, 1927.)

[1] Data in Figures 1.3 through 1.6 from R. Pearl, The growth of population, *Quart. Rev. Biol.* **2**(1927):532–548.

current size of the yeast population. That is, in the equation $\Delta p_n = p_{n+1} - p_n = kp_n$, p_n represents the size of the yeast population (measured in units of biomass) after n hours where k is a positive constant.

Although the data do not appear to lie precisely along a straight line passing through the origin, we can *approximate* the data by a straight line. Placing a ruler over the data to approximate a straight line through the origin, we estimate the slope of the line to be about 0.6. Using the estimate $k = 0.6$, we propose the proportionality model

$$\Delta p_n = p_{n+1} - p_n = 0.6p_n$$

yielding the prediction $\Delta p_n = 0.6p_n$. Thus every hour the yeast population increases to 1.6 times its size the preceding hour. Notice that this model predicts a population that increases forever, which is of course unrealistic. ■

Model Refinement: Modeling Births, Deaths, and Resources

If both births and deaths during a time period are proportional to the existing population, then the change in population itself should be proportional to the size of population, as illustrated above. However, certain finite resources (food, for instance) can support only a maximum population level rather than one that increases indefinitely. As the maximum level is approached, the growth rate should slow. The data in Figure 1.4 show what happens to the yeast culture growing in a restricted area as time increases beyond the eight observations given in the table in Figure 1.3.

Notice from the third column of the table in Figure 1.4 that the change in population per hour decreases as the resources become more limited or constrained. From the

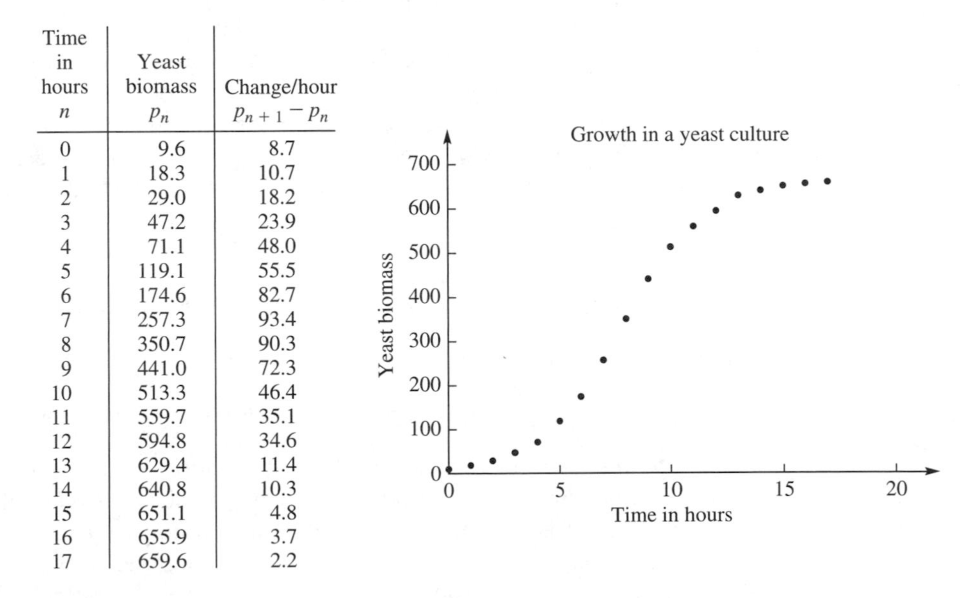

Time in hours n	Yeast biomass p_n	Change/hour $p_{n+1} - p_n$
0	9.6	8.7
1	18.3	10.7
2	29.0	18.2
3	47.2	23.9
4	71.1	48.0
5	119.1	55.5
6	174.6	82.7
7	257.3	93.4
8	350.7	90.3
9	441.0	72.3
10	513.3	46.4
11	559.7	35.1
12	594.8	34.6
13	629.4	11.4
14	640.8	10.3
15	651.1	4.8
16	655.9	3.7
17	659.6	2.2

Figure 1.4 Yeast biomass approaches a limiting population level. (Pearl, 1927.)

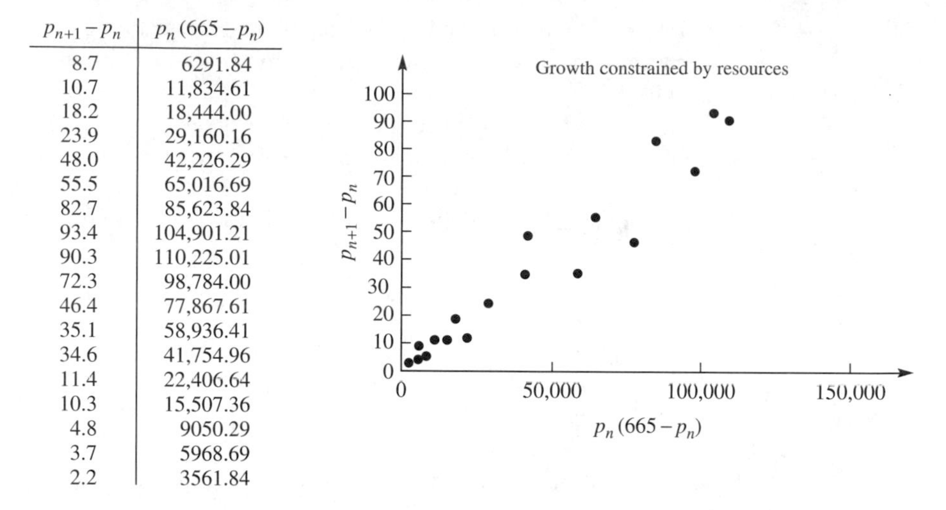

$p_{n+1} - p_n$	$p_n(665 - p_n)$
8.7	6291.84
10.7	11,834.61
18.2	18,444.00
23.9	29,160.16
48.0	42,226.29
55.5	65,016.69
82.7	85,623.84
93.4	104,901.21
90.3	110,225.01
72.3	98,784.00
46.4	77,867.61
35.1	58,936.41
34.6	41,754.96
11.4	22,406.64
10.3	15,507.36
4.8	9050.29
3.7	5968.69
2.2	3561.84

Figure 1.5 Testing the constrained growth model. (Pearl, 1927.)

graph of population versus time, the population appears to approach a limiting value or **carrying capacity**. Suppose that based on our graph we estimate the carrying capacity to be 665. Note that the graph doesn't precisely tell us the correct number is 665, and not 664 or 666, for example. As p_n approaches 665, the change slows considerably. Because $665 - p_n$ gets smaller as p_n approaches 665, our new model

$$\Delta p_n = p_{n+1} - p_n = k(665 - p_n)p_n$$

shows that Δp_n becomes increasingly small as p_n approaches 665.

Let's test this hypothesized model against the data. To do this, we plot $(p_{n+1} - p_n)$ versus $(665 - p_n)p_n$ to see if there appears to be a proportional relationship. Then we estimate the proportionality constant k.

Examining Figure 1.5, we see that the plot reasonably approximates a straight line projected through the origin. Accepting the proportionality argument, we estimate the slope of the line approximating the data to be $k \approx 0.00082$, yielding the model

$$p_{n+1} - p_n = 0.00082(665 - p_n)p_n$$

Solve the Model Numerically and Verify the Results

Solving for p_{n+1} gives

$$p_{n+1} = p_n + 0.00082(665 - p_n)p_n$$

Note that the right-hand side of this last equation is a quadratic in p_n. Such dynamical systems are classified as **nonlinear** and generally cannot be solved to give analytical solutions. That is, we usually cannot find a formula expressing p_n in terms of n. Nevertheless, given that $p_0 = 9.6$, we can substitute in the expression to compute p_1:

$$p_1 = p_0 + 0.00082(665 - p_0)p_0 = 9.6 + 0.00082(665 - 9.6)9.6 = 14.76$$

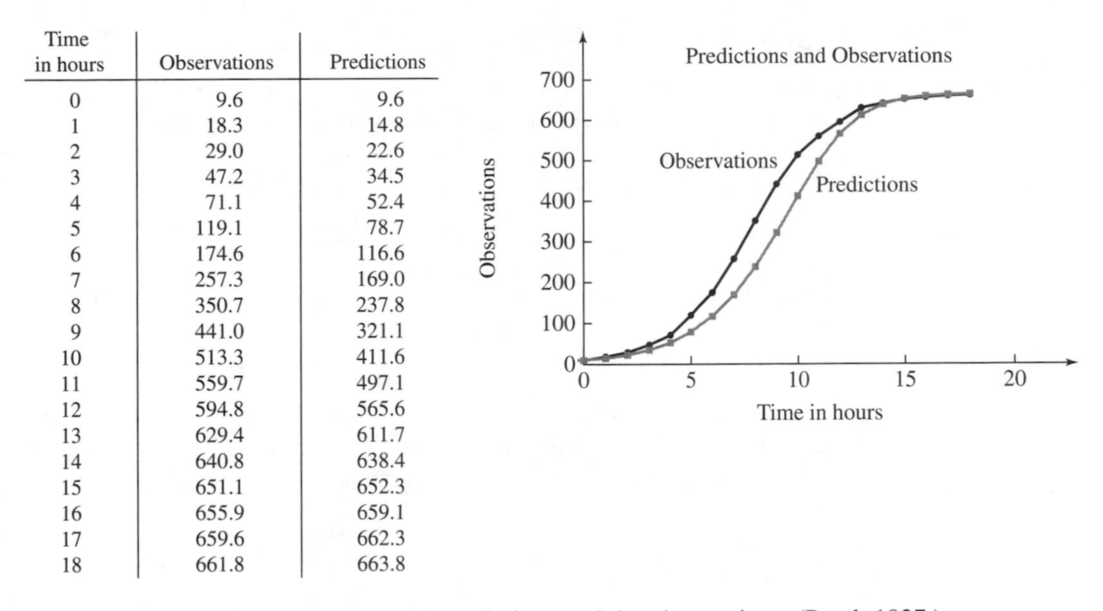

Time in hours	Observations	Predictions
0	9.6	9.6
1	18.3	14.8
2	29.0	22.6
3	47.2	34.5
4	71.1	52.4
5	119.1	78.7
6	174.6	116.6
7	257.3	169.0
8	350.7	237.8
9	441.0	321.1
10	513.3	411.6
11	559.7	497.1
12	594.8	565.6
13	629.4	611.7
14	640.8	638.4
15	651.1	652.3
16	655.9	659.1
17	659.6	662.3
18	661.8	663.8

Figure 1.6 Plotting the model predictions and the observations. (Pearl, 1927.)

In a similar manner, we substitute $p_1 = 14.76$ to compute $p_2 = 22.63$. **Iterating** in this way, we can construct a table of values providing a **numerical solution** to the model. This numerical solution is presented in Figure 1.6. The predictions and observations are both plotted versus time on the same graph. Note that the model captures the *trend* of the observed data very well.

We next construct several models that we will solve in subsequent sections.

Example 2 ***The Spotted Owl Population***

Here we build four models of the spotted owl population, each incorporating different assumptions. We begin with a very simple model that assumes an abundance of resources. The second model adds the assumption that resources are restricted. In the third and fourth models we assume that another species also lives in the ecosystem. More specifically, in the third model we assume the two species compete for the same scarce resources, and in the fourth model we assume the two species have a predator-prey relationship.

Unconstrained Growth Model

Assume the population changes by only births and deaths, and that Δp_n is proportional to p_n (as in Example 1). Then

$$\Delta p_n = p_{n+1} - p_n = kp_n$$

where $k > 0$ represents the growth constant. (For most populations k is a positive number.)

Constrained Growth Model

Now suppose the habitat only supports an owl population of size M, where M is the carrying capacity of the environment. That is, if there are more than M spotted owls, the growth rate becomes negative. Assume too that the growth rate slows as p nears M. We have seen that one model capturing these assumptions is

$$\Delta p_n = p_{n+1} - p_n = k(M - p_n)p_n$$

where k is a positive constant.

Competing Species Model

Next suppose a second species lives in the habitat. We denote the population of the competing species after n periods by c_n. Let's also suppose that in the absence of the other species each individual species exhibits unconstrained growth:

$$\Delta c_n = c_{n+1} - c_n = k_1 c_n$$

and

$$\Delta p_n = p_{n+1} - p_n = k_2 p_n$$

where k_1 and k_2 are the positive constant growth rates. The effect of the presence of a second species is to diminish the growth rate of the first species, and vice versa. While there are many ways to model the mutually detrimental interaction of the two species, we assume that this decrease is roughly proportional to the number of possible interactions between the two species. So, one submodel assumes that the decrease is proportional to the product of c_n and p_n. These considerations lead to the model

$$\Delta c_n = c_{n+1} - c_n = k_1 c_n - k_3 c_n p_n$$

and

$$\Delta p_n = p_{n+1} - p_n = k_2 p_n - k_4 c_n p_n$$

The positive constants k_3 and k_4 are the *relative intensities* of the competitive interactions.

Predator-Prey Model

Finally, we assume that the spotted owl's only food source is a single prey, say mice. We denote the size of the mouse population after n periods by m_n. In the absence of the predatory spotted owl, the mouse population prospers. Then we model the detrimental effect on the mouse growth rate by the spotted owl (predator) population in a manner similar to our model for detrimental competition:

$$\Delta m_n = m_{n+1} - m_n = k_1 m_n - k_2 p_n m_n$$

where k_1 and k_2 are positive constants. On the other hand, if the spotted owl's single food source is the mouse, then in the absence of mice the spotted owl population will diminish to zero, say at a rate proportional to p_n (or $-k_3 p_n$ for k_3 a positive constant). The presence of more mice increases the growth rate of owls, and we assume the increase in the growth rate of the predatory owl is proportional to the product of p_n and m_n, yielding the model

$$\Delta p_n = p_{n+1} - p_n = -k_3 p_n + k_4 p_n m_n$$

Summarizing, this **predator-prey model** is the system

$$\Delta m_n = m_{n+1} - m_n = k_1 m_n - k_2 p_n m_n$$

$$\Delta p_n = p_{n+1} - p_n = -k_3 p_n + k_4 p_n m_n$$

Notice the similarities, but sign differences, between the predator-prey and the competitive species models. ■

Example 3 *Spread of a Contagious Disease*

In a college dormitory of 400 students, at least one student has a contagious flu. Let i_n represent the number of students infected after n time periods. Assume that some interaction between those infected and those not infected is required to pass the disease. If all students are susceptible to the disease, $(400 - i_n)$ is the number of susceptible but not yet infected students. If those infected remain contagious, we can model the change in the infected as being proportional to the product of those infected and those susceptible but not yet infected. That is,

$$\Delta i_n = i_{n+1} - i_n = k i_n (400 - i_n)$$

In this model, $i_n(400 - i_n)$ represents possible interactions between those infected and those not infected at time n. A fraction k of $i_n(400 - i_n)$ would now become infected at this stage. This model is identical in form to the constrained growth model in Example 1.

A number of refinements can be made to our model. For example, we can assume that a segment of the population is not susceptible to the disease, that the infection period is limited, or that infected students are removed from the dorm to prevent their interaction with uninfected students. More sophisticated models treat the infected and susceptible populations separately, similar to the predator-prey model. ■

Example 4 *Warming of a Cooled Object*

Here is an example in which the behavior takes place continuously. A cold can of soda is taken from a refrigerator and placed in a warm classroom, and the temperature is measured periodically. The temperature of the soda is initially 40°F and the room temperature is 72°F. Since the volume of soda is small relative to the volume of the room, we can reasonably assume the room temperature to remain constant. Further, we assume the entire can of soda has the same temperature, neglecting any variation within the contents of the can. We might expect the change in temperature per time period to be greater when the difference in temperatures between the soda and room is large, and the change in temperature per unit time to be less when the difference in the temperatures is small.

Letting t_n represent the temperature of the soda after n time periods and k a positive constant of proportionality, we propose

$$\Delta t_n = t_{n+1} - t_n = k(72 - t_n)$$

This model represents a simplification: The air in the room will cool a little, the temperature within the can will vary a little, the rate of cooling depends on the shape and conductivity of the can, and other simplifications.

Again, many refinements are possible for this model. While we have assumed k to be constant, it actually depends on the shape and conductivity of the container, the time period between temperature measurements, and so forth. Also, the temperature of the environment may not be constant and the temperature of the soda may not be uniform throughout. The temperature may vary in one dimension (as in the case of a thin wire), two dimensions (such as for a flat plate), or three dimensions (as in the case of a space capsule reentering the earth's atmosphere). ■

Exercises 1.3

Approximating Change Using Proportionality

1. Digoxin is used to treat heart disease. In the table, y represents the amount of digoxin in the bloodstream and t represents the time in days after taking a single dose. The initial dosage is 0.5 mg.

t	0	1	2	3	4	5	6	7	8
y	0.500	0.345	0.238	0.164	0.113	0.078	0.054	0.037	0.026

   a. Formulate a model using a difference equation in which the *change* in concentration per day is proportional to the amount of digoxin present. Test your model by plotting the change versus the amount present at the beginning of the period.
   b. Assume that after the "initial dose" of 0.5 mg, each day a "maintenance dose" of 0.1 mg is taken. Formulate the refined model.
   c. Find a numerical solution by constructing a table of values for part b for 15 days.

2. A certain drug is effective in treating a disease if the concentration remains above 100 mg/L. The initial concentration is 640 mg/L. It is known from laboratory experiments that the drug decays at the rate of 20% of the amount present each hour.
   a. Formulate a model representing the concentration at each hour.
   b. Construct a table of values and determine when the concentration reaches 100 mg/L.

3. A humanoid skull is discovered near the remains of an ancient campfire. Archaeologists are convinced the skull is the same age as the original campfire. It is determined from laboratory testing that only 1% of the original amount of carbon-14

remains in the burned wood taken from the campfire. It is known that carbon-14 decays at a rate proportional to the amount remaining and that carbon-14 decays 50% over 5700 years. Formulate a model for carbon-14 dating. How old is the humanoid skull?

4. Sociologists recognize a phenomenon called *social diffusion*, which is the spreading of a piece of information, a technological innovation, or a cultural fad among a population. The members of the population can be divided into two classes: those who have the information and those who do not. In a fixed population whose size is known, it is reasonable to assume that the rate of diffusion is proportional to the number who have the information times the number yet to receive it. If a_n denotes the number of people who have the information in a population of N people after n days, formulate a dynamical system to approximate the change in the number of people in the population who have the information.

5. Consider the spreading of a highly communicable disease on an isolated island with population size N. A portion of the population travels abroad and returns to the island infected with the disease. Formulate a dynamical system to approximate the change in the number of people in the population who have the disease.

6. The issue is the survival of whales, and you are to assume that if the number of whales falls below a minimum survival level m the species will become extinct. Assume also that the population is limited by the carrying capacity M of the environment. That is, if the whale population is above M, then it will experience a decline because the environment cannot sustain that large a population. In the following model, a_n represents the whale population after n years. Discuss the model:

$$\Delta a_n = a_{n+1} - a_n = k(M - a_n)(a_n - m)$$

7. The following data were obtained for the growth of a sheep population introduced into a new environment on the island of Tasmania.[2]

Year	1814	1824	1834	1844	1854	1864
Population (in thousands)	125	275	830	1200	1750	1650

Plot the data. Is there a trend? Plot the change in population versus years elapsed after 1814. Formulate a discrete dynamical system that reasonably approximates the change you have observed.

8. The following data represent the U.S. population from 1790 to 1990. Find a difference equation that fits the data fairly well. Test your model by plotting the predictions of the model against the data.

[2] Adapted from J. Davidson, On the growth of the sheep population in Tasmania, *Trans. Roy. Soc. S. Australia* **62**(1938):342–346.

Year	U.S. Population	Year	U.S. Population	Year	U.S. Population
1790	3,929,000	1860	31,443,000	1930	122,755,000
1800	5,308,000	1870	38,558,000	1940	131,669,000
1810	7,240,000	1880	50,156,000	1950	150,697,000
1820	9,638,000	1890	62,948,000	1960	179,323,000
1830	12,866,000	1900	75,995,000	1970	203,212,000
1840	17,069,000	1910	91,972,000	1980	226,505,000
1850	23,192,000	1920	105,711,000	1990	248,710,000

9. The gross national product (GNP) represents the sum of consumer purchases of goods and services, government purchases of goods and services, and gross private investment (which is the increase in inventories plus buildings constructed and equipment acquired). Assume that the GNP is increasing at the rate of 3% per year, and that the national debt is increasing at a rate proportional to the GNP. Let a_n represent the GNP after n years and b_n the national debt after n years. Construct a dynamical system to model the change in the GNP and national debt.

10. In 1868, the accidental introduction into the United States of the cottony cushion insect (*Icerya purchasi*) from Australia threatened to destroy the American citrus industry. To counteract this situation, a natural Australian predator, a ladybird beetle (*Novius cardinalis*) was imported. The beetles kept the insects at relatively low numbers. When it was discovered that DDT (an insecticide) killed scale insects, farmers applied it in the hopes of reducing the scale insect population ever further. However, DDT turned out to be lethal to the beetle as well, and the overall effect of the insecticide was to increase the numbers of the scale insect. Modify the predator-prey model to reflect a predator-prey system where farmers apply (on a regular basis) an insecticide that destroys both the insect predator and the insect prey at a rate proportional to the numbers present.

11. Consider two species whose survival depends upon their mutual cooperation. An example would be a species of bee that feeds primarily on the nectar of one plant species and simultaneously pollinates that plant. Letting a_n and b_n represent the bee and plant population levels after n days, we have the following model:

$$a_{n+1} = a_n - k_1 a_n + k_2 a_n b_n$$

$$b_{n+1} = b_n - k_3 b_n + k_4 a_n b_n$$

where the k_i are positive constants.

a. Discuss the meaning of each k_i in terms of mutual cooperation.

b. What assumptions are being made about growth of each species in the absence of cooperation?

12. Place a cold can of soda in a room. Measure the temperature of the room and periodically measure the temperature of the soda. Formulate a model to predict the change in the temperature of the soda. Estimate any constants of proportionality from your data. What are some of the sources of error in your model?

PROJECTS Modeling Explorations

1. Complete the UMAP module "The Diffusion of Innovation in Family Planning," by Kathryn N. Harmon, UMAP 303. This module gives an interesting application of finite difference equations to study the process through which public policies are diffused in order to understand how national governments might adopt family planning policies.

2. Complete the UMAP module "Difference Equations with Applications," by Donald R. Sherbert, UMAP 322. This is a good introduction to first- and second-order linear difference equations, including undetermined coefficients for nonhomogeneous equations. Applications to problems in population and economic modeling are presented.

Further Reading

Frauenthal, J. C. *Introduction to population modeling*. Lexington, MA: COMAP, 1979.

Hutchinson, G. E. *An introduction to population ecology*. New Haven, CT: Yale University Press, 1978.

Levins, R. The strategy of model building in population biology. *American Scientist* **54**(1966):421–431.

Lotka, A. J. *Elements of mathematical biology*. New York: Dover, 1956.

Odum, E. P. *Fundamentals of ecology*. Philadelphia: Saunders, 1971.

Pearl, R., & Reed, L. J., On the rate of growth of the population of the United States since 1790. *Proceedings of the National Academy of Science* **6**(1920):275–288.

1.4 Why Do We Study Mathematics?

Why do we study mathematics? In three words—to understand, predict, and control. We study math in order *to understand* things in our environment, things that are important to us from a daily-living, career, or interest standpoint. We study math in order *to predict*, through some form of modeling, what might happen in a given setting under certain circumstances. And, we study math in order *to control* the outcome of a given process or procedure.

Understanding, predicting, and controlling encompass a broad range of meanings. The British pure mathematician Hardy would view them as understanding a situation in pure mathematics, predicting the direction its investigation might most fruitfully take, and controlling the number of cases needed in a proof setting via combinatorial reasoning. Archimedes, the famous Greek mathematician (ca. A.D. 200), might have conceived of these three reasons as understanding geometrically the principles of buoyancy relative to ship design, predicting the stability of a ship by examining a cross section of its hull, and controlling its stability in rough seas through geometric modifications of that hull design.

The degree to which individuals develop these skills of understanding, predicting, and controlling is directly related to their economic success in life. Mathematical knowledge is the intellectual currency of the technological age. Knowing and being able to use math productively creates economic opportunity for individuals. However, we do not want to argue for mathematics strictly from a utilitarian standpoint. Mathematics plays a key role in our history. As such, it deserves our study as part of the central core of subjects. It has been one of the core liberal arts throughout the history of academia. Individuals representing all cultures have contributed to our knowledge of number, measure, geometry, chance, statistics, algebra, and the study of change.

Another argument for developing mathematical knowledge is that of enlightened citizenship. Students need math skills so that they can understand issues that arise in the local newspaper, in local and national elections, in project designs or plans, and in interpreting the results of a survey. Being able to envision problems and to use known and acquired knowledge in their solution is empowering. Many students finish their study of mathematics without ever experiencing the kind of conjecturing, hypothesizing, and nonroutine applications of basic concepts and principles that help them understand real-world problems. To be mathematically literate, a person must be able to use the power of mathematics in realistic settings.

Students who are mathematically literate are able to think and reason mathematically, actively using concepts, principles, and skills to make sense of the world around them. This requires that they not only integrate the content, but also use cognitive processes of mathematics to probe, interpret, suggest, validate, and communicate what they know about a situation. For example, a student may be asked to determine the height at which a type of fireworks explodes. The student must find information and relevant data about fireworks in general and then determine what aspects of this information apply directly to the problem at hand. Or, a student may be asked to estimate a distance, determine the magnitude of a force, or find the probability of an event occurring. Or, a student might wonder how the north-south orientation of a house, the width of its eaves, and the placement and size of a window relates to the amount of sunlight that falls on the carpet in summer and winter. Problems such as these require students to model the situation, bring to bear their mathematical knowledge, and methodically work toward a solution.

Mathematically literate people are capable of understanding and applying their mathematical knowledge to interpret, predict, and control factors important to their life. Such individuals can integrate and apply mathematical concepts and procedural skills, and then develop and interpret models related to the problems they encounter.

How do we develop a mathematics curriculum that addresses these needs? How do we create programs that bring mathematics to all students and help them "make sense" of mathematics? These efforts directly relate to the five goals listed earlier in this chapter for individual growth in school mathematics. Improving the confidence of each learner in mathematical contexts and developing the learner's value of mathematics is imperative. Learning to reason, make connections, and communicate in mathematical contexts requires that all students have rich opportunities to learn.

Exercises 1.4

1. Write your own definition of "mathematics."
2. List five applications of mathematics that make your life easier.

PROJECTS — Find Out for Yourself

1. Collect clippings from your local newspaper that exemplify applications of mathematics. (This file will be useful in your future classroom.) Write a brief summary.
2. Use the Internet to explore current discussions about mathematics and its role in contemporary society. Write a brief summary.
3. Interview your grandparents, parents, and siblings, if possible, about their views of what mathematics is and what purposes it serves in our current world. Compare and contrast their answers.

References

Boyer, C. B. (1968). *A history of mathematics*. New York: Wiley.

Chace, A. B. (1927). *The Rhind mathematical papyrus*, Volume I. Oberlin, OH: The Mathematical Association of America.

Davidson, J. (1938). On the growth of the sheep population in Tasmania. *Trans. Roy. Soc. S. Australia* **62**:342–346.

Davis, P., & Hersh, R. (1980). *The mathematics experience*. Boston: Birkhäuser.

Gillings, R. J. (1982). *Mathematics in the time of the pharaohs*. New York: Dover.

Hardy, G. (1940). *A mathematician's apology*. Cambridge, England: Cambridge University Press.

Hersh, R. (1986). Some proposals for reviving the philosophy of mathematics. *In* T. Tymoczko (ed.), *New directions in the philosophy of mathematics* (pp. 9–28). Boston: Birkhäuser.

National Council of Teachers of Mathematics (1989). *Curriculum and evaluation standards for school mathematics*. Reston, VA: NCTM.

——— (2000). *Principles and standards for school mathematics*. Reston, VA: NCTM.

Ptolemy (1952). The almagest. *In* R. M. Hutchins (ed.), *Great books of the western world: vol. 16. Ptolemy, Copernicus, and Kepler* (p. 1478). Chicago: Encyclopaedia Britannica.

Snapper, E. (1979). Three crises in mathematics: Logicism, intuitionism, and formalism. *Mathematics Magazine*, **52**:207–216.

Steen, L. (1988). The science of patterns. *Science*, **240**:611–616.

2
A Model for Mathematics Education

Changing the nature of both the mathematical tasks posed and the discourse norms in the classroom creates new classroom roles for teachers. If students are to have opportunities to explore rich problems within which the mathematics will be confronted, the teacher has to learn how to be effective in at least four new roles: (1) engaging the students in the problem, (2) pushing student thinking while the exploration is proceeding, (3) helping the students to make the mathematics more explicit during whole-class and group interaction and synthesis, (4) using and responding to the diversity of the classroom to create an environment in which all students feel empowered to learn mathematics.

Glenda Lappan and Diane Briars

Overview

What are the goals of mathematics education? Teachers rarely consider this question until they are confronted with a view opposing theirs. Generally, this is too late to be considering why we teach what we teach. While we don't answer this question in this chapter, we begin to consider the basis for answering this and other questions. In particular, we consider where we are now and how we have come to be here.

Mathematics education in the United States is, at present, a disorganized enterprise. Each of the nearly 16,000 different school districts in the United States essentially has the legal right to set its own mathematics goals for its students. Further, each has the right to select its own textbooks and hire and assign its own teachers. These decisions may be guided by state laws regarding financial support for school textbook programs and teacher certification laws; however, local schools have tremendous decision making authority over the design and administration of their programs. This results in a myriad of different curricular programs and goals for students in U.S. schools. In many other countries, there are tight guidelines for what students are to learn in a given year, a single textbook used throughout the country, and suggested ways of teaching that material to students.

In this chapter we examine the current data on student achievement in mathematics in the United States and other countries, the essence of the recommendations for change, and the relationship of these recommendations to current school practices. We also look at research findings on how schools' mathematics programs function. Finally, we consider what students think about school mathematics. What are their beliefs about, and attitudes toward, mathematics? What role do these attitudes play in changing school mathematics to better serve the students' future needs for mathematical knowledge and skills?

Focus on the NCTM *Standards*

Central to the changes taking place in mathematics education today is the National Council of Teachers of Mathematics' standards movement. Beginning in 1989 with the publication of the *Curriculum and Evaluation Standards for School Mathematics* and continuing in 2000 with the release of *Principles and Standards for School Mathematics*, the Council addresses what mathematics educators feel is important and needed by students in today's world. The *Standards* outline a course of study for students at all ages that includes both process and content goals.

The process goals speak to the ways in which students cognitively approach mathematical content: deriving meaning, creating models, testing ideas, applying algorithms, and solving problems. These processes describe the ways in which individuals learn mathematics. The content goals for school mathematics address what students should know and be able to do. These goals detail the mathematical content students should acquire to be mathematically literate, culturally literate, prepared for the workplace, and prepared for further study.

The mathematical content of the mathematics curriculum falls into the general categories of number and operation, geometry, measurement, probability and statistics, and algebra and functions. However, the intended learning goes beyond this content to include the development of students' abilities to understand, represent, and apply this content-based knowledge in meaningful ways. Students need to be able to apply their mathematics knowledge and skills to real-world situations and represent these situations in graphical, symbolical, numerical, and verbal formats. They need to be able to reason, solve problems, and communicate their thoughts clearly and concisely.

Students' future needs for mathematics are large. If we believe that what one knows is related to how it was learned, these requirements speak loudly to the need to change how we teach math. These processes, content, and instructional goals, coupled with the programs, are the focus of this chapter.

Focus on the Classroom

School mathematics, and especially secondary school mathematics, can be examined at several levels. One level is the intended curriculum—that is, the content of the program and its delivery as envisioned by specialists in math education. Let's call this the perfect curriculum, encapsulating research findings, worthwhile content, and the wisdom of master teachers. At a second level, we can look at the implemented curriculum—that is, the curriculum that results when teachers close their classroom doors and teach math. A third level is the achieved curriculum—what students take with them in terms of what they know and can do.

In some instances, the three differ very little. In such settings, teachers deliver the "goods," and students meet the outcome goals set for them. In other settings, the implemented curriculum closely resembles the intended one, but student outcomes fall far short of the mark. Perhaps the intended curriculum is a mismatch for students' backgrounds or the delivery design is flawed. In yet other settings, the implemented curriculum varies greatly from the intended one, but student outcomes come close to achievement goals anyway.

Optimally, we of course hope for congruence among the three levels. In this chapter we examine the issues surrounding curriculum development and implementation, as well as student attitudes and beliefs as they approach learning mathematics at the middle and high school levels. How do our programs function, and what policies affect the success or failure of math programs? What can teachers do in terms of curricular design and delivery?

2.1 The Intended Curriculum

Prior to 1989, curricular guidance in mathematics for U.S. schools came either from local or state statements of objectives or from the contents of the textbooks themselves. Across the roughly 16,000 school districts in the United States, a great variety of math programs were in place. Even in those states that adopted statewide learning outcomes for mathematics, individual school districts opted to achieve those objectives in numerous ways. With the release in 1989 of the NCTM's *Curriculum and Evaluation Standards for School Mathematics*, the picture began to change. The majority of states either wrote standards for school mathematics for the first time or rewrote existing standards. Thus, bringing local goals and objectives into line with state or NCTM *Standards* became a major focus for many districts.

The *Standards* and the *Principles and Standards*

The NCTM *Curriculum and Evaluation Standards for School Mathematics* (1989) and the *Principles and Standards for School Mathematics* (2000) are based on practice, professional opinions of best practice, comparisons with other nations' mathematics programs, and research on the teaching and learning of mathematics. The NCTM recommends what the profession feels is a solid and defensible program for students in mathematics. It suggests how mathematics curricula might be best structured. The *Principles and Standards* stress that what students learn is a product of how they encounter it in the classroom. In particular, concepts, relations, and processes are often best learned in situations in which students construct the knowledge themselves under the watchful eye of their teacher. Such teachers carefully create conditions that encourage students to focus on the important aspects of the content and to make connections between those aspects and other important information.

The *Standards* resulted from a century of ferment and change centered on school mathematics curricula and what an ideal program should entail. Various groups, dating back to the Committee of Ten in 1896, made numerous suggestions, but these suggestions never outlined goals for a K–12 program in terms of content and process learnings. Neither did they systematically list (by grade level bands—K–4, 5–8, 9–12) what students should know about problem solving, reasoning, communication, connections, and other aspects of mathematics. Further, none of these groups circulated drafts of their recommendations so that teachers and parents could give them feedback.

Background

The 1989 NCTM *Standards* grew out of practice, research on the teaching and learning of mathematics, and comparisons of U.S. programs with themselves over time and with those in other countries. Since 1973, our government has collected data on the mathematical achievement of U.S. students through the National Assessment of Educational Progress (NAEP). From 1973 to 1990, the data were collected from students of 9, 13, and 17 years of age. In 1990, this program was altered not only to continue to collect long-term trend data from the age groups, but also to change the main data focus on grade level groups using grades 4, 8, and 12. During this same period, the United States also participated in international comparisons of student achievement in mathematics.

Figure 2.1a compares the performance of U.S. students from 1973 to 1999. The scores for the NAEP Long-Term Trend Assessment ranges from 0 to 500. The data show that students in all three age groups made significant growth over time on this assessment, which uses the same tests as those used in 1973. Hence basic skills have not slipped from those held in 1973. Figure 2.1b compares student performance on the national NAEP examination. This assessment employs a test, different from the NAEP Long-Term Trend Assessment, which is also scaled from 0 to 500. The NAEP Assessment measures students' grasp of current curricula content. An examination of the data

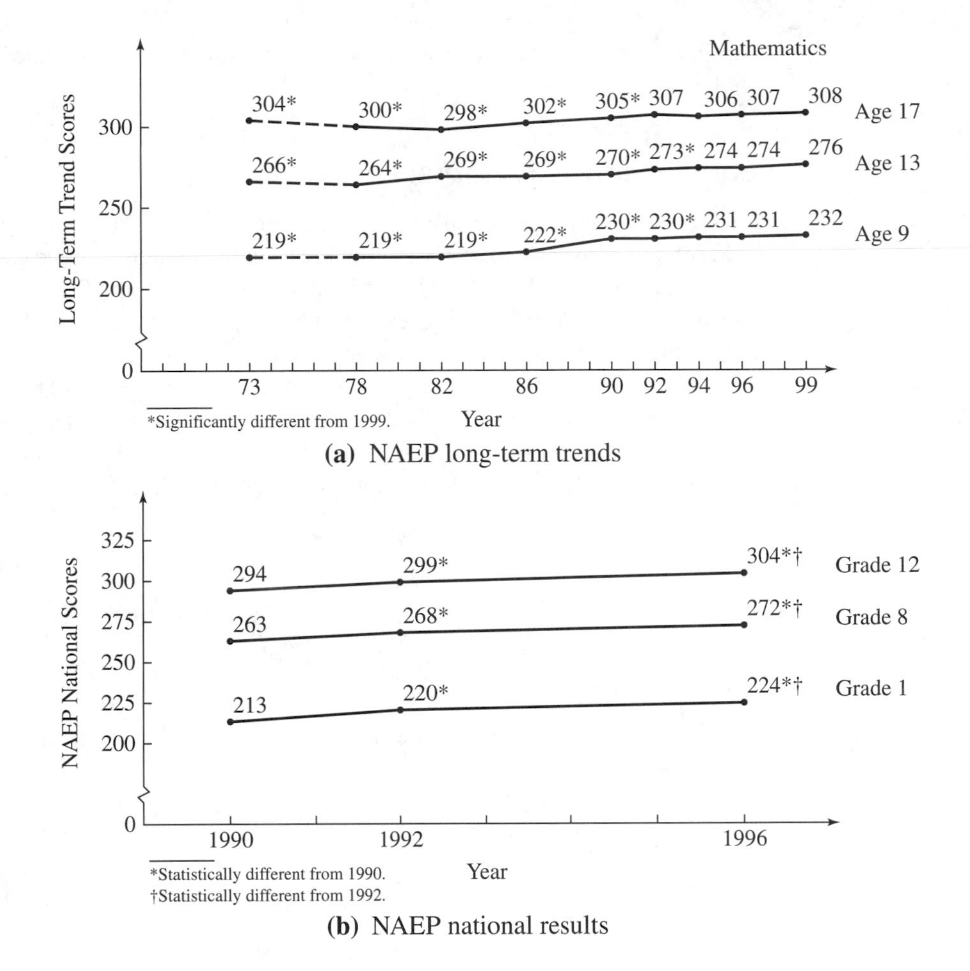

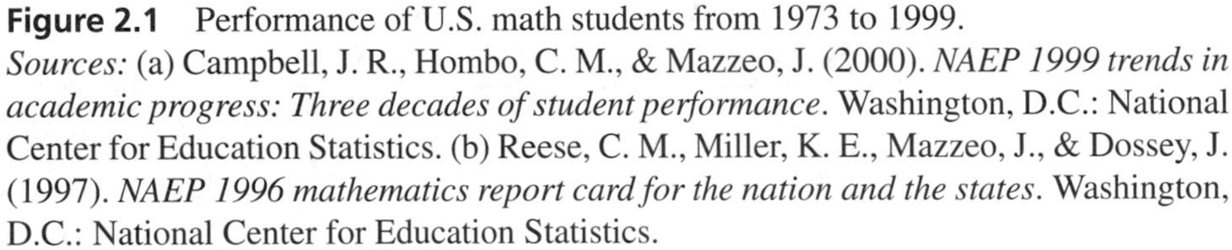

Figure 2.1 Performance of U.S. math students from 1973 to 1999.
Sources: (a) Campbell, J. R., Hombo, C. M., & Mazzeo, J. (2000). *NAEP 1999 trends in academic progress: Three decades of student performance*. Washington, D.C.: National Center for Education Statistics. (b) Reese, C. M., Miller, K. E., Mazzeo, J., & Dossey, J. (1997). *NAEP 1996 mathematics report card for the nation and the states*. Washington, D.C.: National Center for Education Statistics.

in Figure 2.1b shows that at all three grade levels in 1996 students were performing at approximately one grade level—that is, 10 scale score points—higher than their counterparts in 1990. This reflects a considerable improvement in math skills over the past decade for U.S. students (Reese, Miller, Mazzeo, & Dossey, 1997).

While we would like to claim that a new focus in schools brought about this change, it appears that this growth resulted from improving the scores of the lowest performing students. Other data show that great differences exist in students' opportunities to learn. A major reason the NCTM *Standards* were developed was to close this gap in opportunity and increase the focus on worthwhile mathematics important to all students.

Comparisons with Others

At the same time as American educators gathered data to compare national performance, U.S. schools also participated in international studies centered on the teaching and learning of mathematics. The Third International Mathematics and Science Study (TIMSS), carried out in 1995–1996, provides a wealth of information on school organization, curricular intention, classroom practice, and student learning (Robitaille, 1997).

Figure 2.2 compares the organization of U.S. schools with that of the Netherlands. The height of the graphs represents 100% of the age group at each point on the graph. The numbers along the top represent the grade structure of the schools and the numbers along the bottom student ages. As shown in Figure 2.2a, U.S. students enter kindergarten

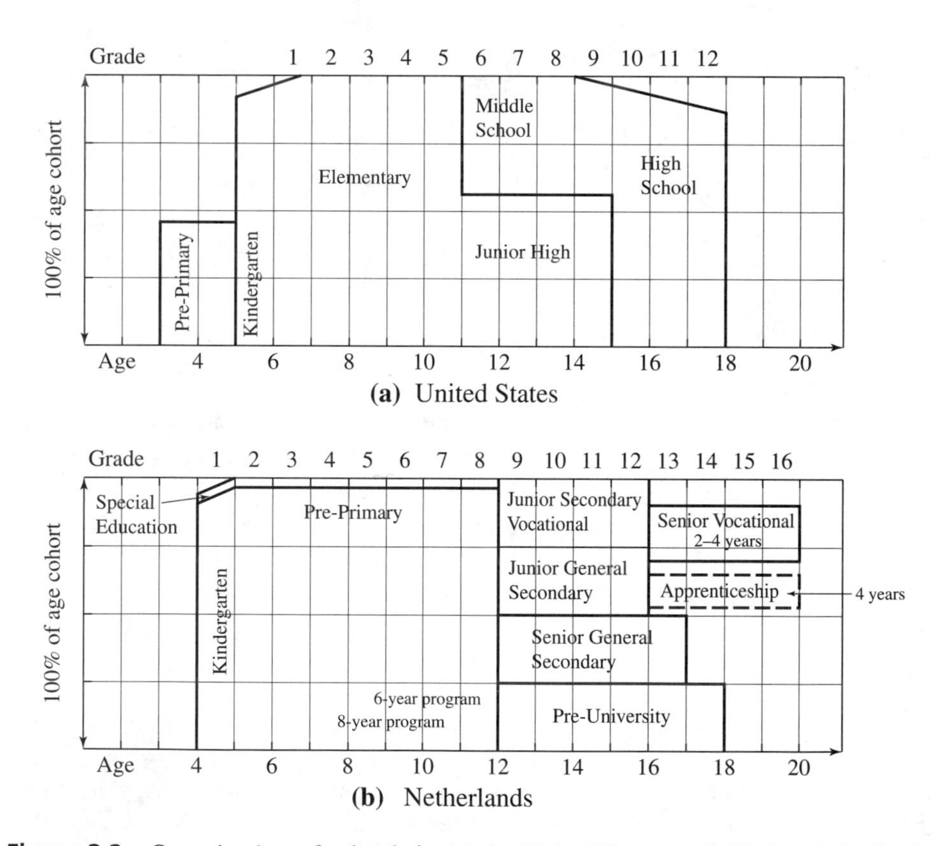

Figure 2.2 Organization of schools in (a) the United States and (b) the Netherlands. *Source:* Dossey, J. A. (1999). How should U.S. performance in school mathematics be interpreted? *In* Z. Usiskin (ed.). *Developments in school mathematics education around the world* (pp. 228–239). Reston, VA: National Council of Teachers of Mathematics.

at age 5 and continue until they graduate or drop out of school, as illustrated by the slanted line at the top of the graph beginning at grade 9. The only differentiation between U.S. school systems is whether students attend a middle school or junior high school in their early adolescent years.

Schools in the Netherlands differ from ours in structure, even though like the United States, schooling is compulsory from ages 5 to 16. About 95% of students enter school at age 4 and move through a self-contained classroom system until the end of grade 8. At that point, students follow one of four tracks. Junior secondary vocational schools prepare students for jobs in technology, commerce, agriculture, and so on. Some students who opt for this track continue in two- or four-year vocational programs that combine classwork and apprenticeships. Other students complete junior general secondary education. This track empties more directly into the labor market, but some students on this track also go on to complete apprenticeships. Senior general secondary education is a five-year program that prepares students for higher vocational education. Students on this track pursue advanced technical training, akin to our community college and college-based apprenticeship programs. Finally, the pre-university track prepares students for college-level work. In the first three years of the secondary tracks, all students complete a core curriculum of 15 subjects, including mathematics, combined physics/chemistry, biology, geography, earth science, social studies, and languages.

The core curriculum of the Netherlands challenges students to link mathematical concepts and principles to their real-life experiences and knowledge of other subjects. There is a far greater reliance on reasoning, problem-solving, and inquiry than in U.S. curricula and a greater coherence between mathematics and other subjects in the curriculum. The final years of the secondary tracks are divided between the study of Math A, which focuses on applications, and Math B, which focuses on formal and abstract concepts. The latter course prepares students for advanced mathematical, scientific, and technical studies.

The TIMSS study tested math skills of students at 9 years (29 countries), 13 years (41 countries), and the terminal year of secondary education (21 countries). The results were not especially pleasing to U.S. educators. While U.S. students at grade 4 performed among the best in the world, at grade 8, U.S. students tested below the international median. At grade 12, two separate samples of students were drawn. In the first sample, a random cross section of all U.S. students still enrolled in school at grade 12 were tested. These students ranked 19th out of the 21 nations tested. The second sample tested only those students enrolled in precalculus or calculus. These U.S. students ranked 15th out of the 16 nations tested.

Needless to say, these results are disquieting. Unfortunately, they confirm results from other national and international studies. Further, these studies found that our programs contain too much review, cover too many topics, and fail to provide a study focus.

In the TIMSS study, Japanese and U.S. teachers were asked what topics they teach during eighth grade. Figure 2.3 shows their responses. As the data shows, the number of topics available to teach are about the same, 26 in Japan and 27 in the United States. However, fewer than 25% of Japanese teachers teach 50% or more topics, while approximately 98% of U.S. teachers do. Factor into this that U.S. teachers do so in a significantly shorter school year.

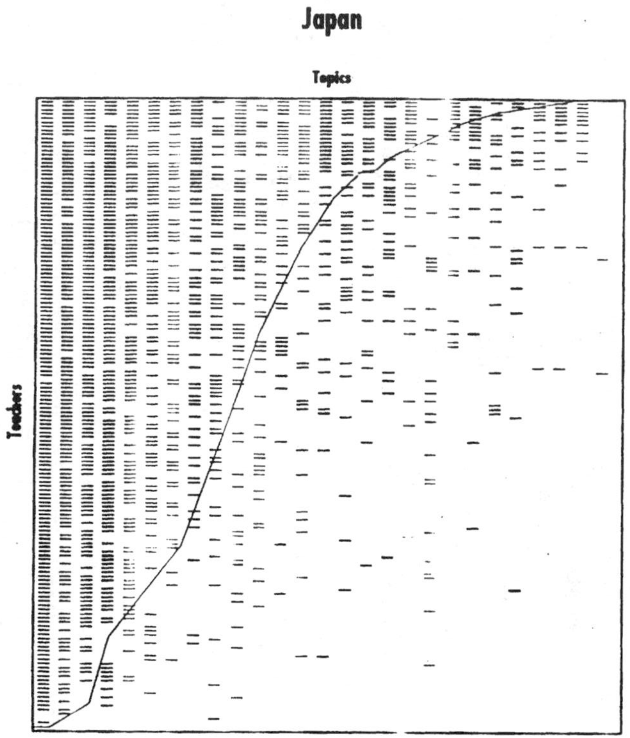

Japan

Topics

Teachers

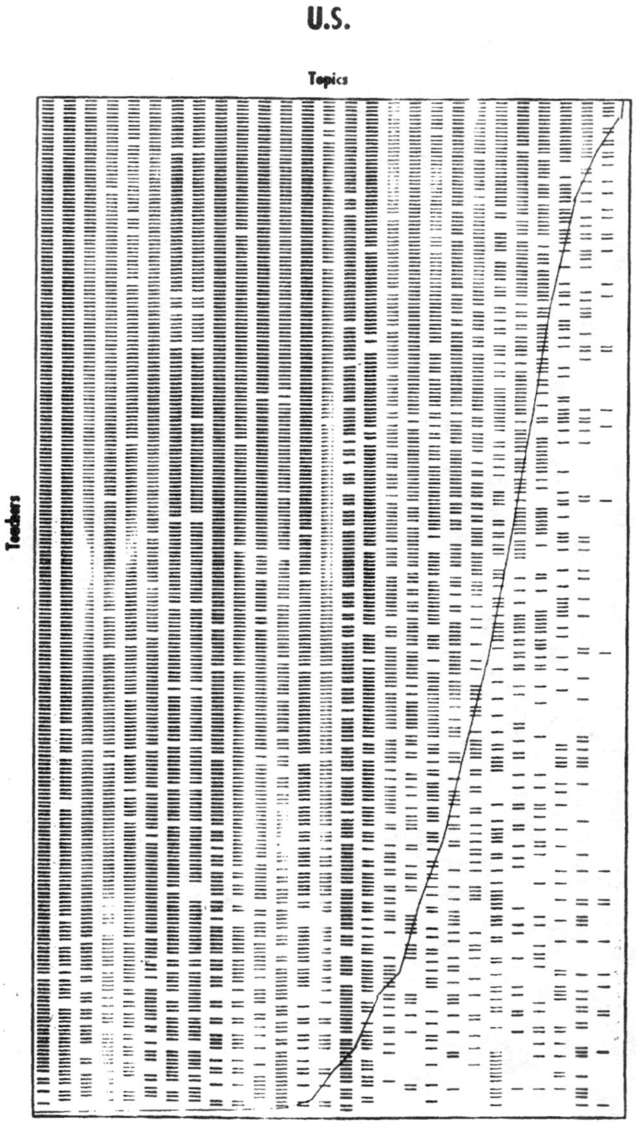

U.S.

Topics

Teachers

Figure 2.3 Topics taught in Japanese and U.S. schools at grade 8. Columns represent number of topics taught (26 for Japan and 27 for the United States). Teacher responses (the horizonal bars) are sorted by topic so that the topic most often taught is the leftmost column. The curved lines represent the average number of topics taught by one teacher during the school year. *Source:* Schmidt, W. H., McKnight, C. C., & Raizen, S. A. (1997). *A splintered vision: An investigation of U.S. science and mathematics education*. Dordrecht, The Netherlands: Kluwer Academic Publishers.

An analysis of textbook content reveals the need for greater focus in the curriculum. Figure 2.4 depicts the contents of two eighth grade mathematics textbooks schematically. The vertical axes of the two profiles represent the number of content topics covered in the two texts. The horizontal axes represent the length of the textbooks. Note that the Japanese textbook is shorter and covers fewer topics. The shaded bars represent consecutive segments of the text spent on the same content topic. It is here that we see the major difference in the structure of what the textbook provides the teacher. The Japanese textbook focuses on a single topic at the beginning of the year, followed by shorter, but still discernible, extended coverage of other topics. The U.S. textbook flits from one topic to another with few periods of extended coverage.

It is this lack of focus and our predilection to include massive amounts of review materials in our K–8 books that Jim Flanders (1987) describes (Figure 2.5). After 8 years of a review-style approach, in ninth grade students are suddenly confronted with Algebra I, where little review is provided. Many students encounter difficulty with the shift in the pace of coverage and the lack of review work.

Our predisposition to review later, rather than teach for mastery now, results in mathematics curricula in a state of confusion. Topics are continually added and all previously taught topics are continually reviewed, so less and less time is given to the teaching of new content. Targets are not set to drop topics once they are mastered. The NCTM *Standards*, and other moves to reform mathematics education, reflect an attempt to address this problem through a more focused curriculum.

Other studies, such as Skip Kifer's (1992), indicate that opportunities to learn math are not equally distributed across students in the United States. Ethnic and racial groupings, as well as gender, affect who gets to study what as differentiated programs begin to appear in the higher grades. Over the 1980s, educators strived to eliminate gender discrimination in math programs. However, unequal access still exists both in the breadth of mathematics curricula offered in different schools and who gets into the classes when they are offered.

Learning in Math Classes

The third thrust in math classes, besides content and structure, is the delivery of that instruction. The historical pesentation of mathematics instruction followed a typical format. Teachers presented the material, worked a few examples, had the students try a few examples, and then assigned students a set of similar problems to work for the next day. Students were not asked to reason through the development of the concept. Neither were they asked to work extended problems or projects where they constructed a solution, and communicated, either in writing or verbally, their solution to the problem.

In many countries, students actively participate in the development of the concepts they are studying. At each grade level, teachers pose problems and help students move through the development of the major concepts and principles. That is not to say that they do not lecture at some points. However, students are held accountable, to a greater degree, for understanding the "whys" of the subject.

Teachers in classrooms such as these are much more a guide and active partner in the work of learning math. Their role shifts from being that of sole source of new

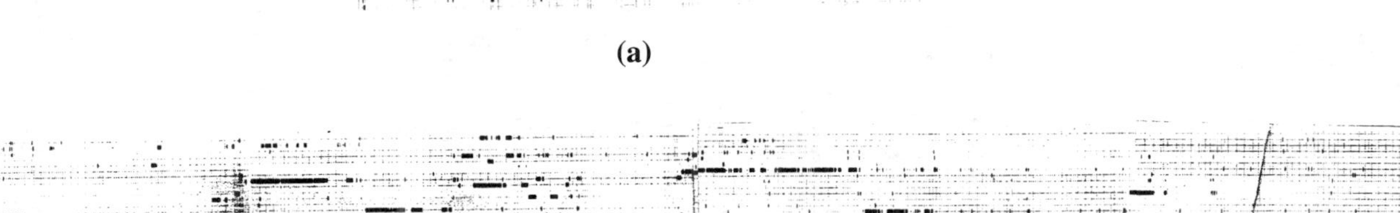

(a)

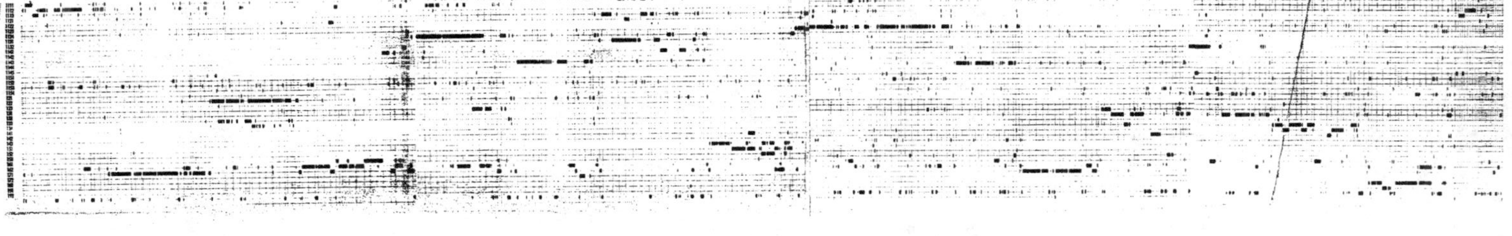

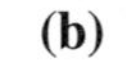

(b)

Figure 2.4 Schematic profiles of (a) Japanese and (b) U.S. Population 2 (eighth grade) textbooks. The long, dense row segments indicate coverage that focuses on a single topic. Each column corresponds to one teaching block. The Japanese textbook covers a limited number of topics and, as shown by the row segments, gives extended coverage to certain topics. The U.S. textbook covers far more topics and extended coverage is severely limited. *Source:* Schmidt, W. H., McKnight, C. C., & Raizen, S. A. (1997). *A Splintered Vision: An Investigation of U.S. Science and Mathematics Education*. Dordrecht, The Netherlands: Kluwer Academic Publishers.

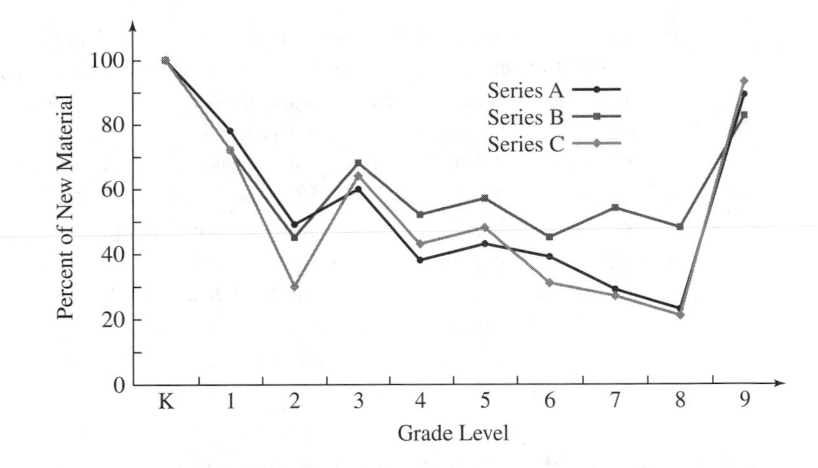

Figure 2.5 Percentage of new content in three U.S. math textbook series. Note the spike in new material in ninth grade, when students first encounter algebra. *Source:* Flanders, J. (1987). How much of the content in mathematics textbooks is new? *Arithmetic Teacher* **35**(1):18–23.

knowledge to one who structures learning situations, effectively guides and assesses students, and works alongside students as they learn. The focus of the classroom shifts from a one-way transmission of content. Rather, students build new knowledge through engagement with mathematical situations. Students and teachers are partners in these investigations, where teachers observe, guide, question, and assess as students actively participate in learning activities. This form of teaching probably places a greater load on the teacher than simply lecturing. However, the outcomes in terms of student learning and growth are much greater.

Problems and tasks play a central role in programs in which students are called on to help in the construction of the math education taking place. Instructional activities in such classrooms pivot around problems or statements of situations that promote the learning of a concept. Such problems can and should vary greatly in their nature because students vary greatly in their backgrounds, knowledge, and experiences. It is the task of the teacher to find and structure situations that make sense to students and are appropriately difficult so that they actively engage the students. Central to students constructing knowledge is the opportunity for social interaction with others as part of the problem-solving activity. This differs from the past when learning mathematics was viewed as a solitary activity.

In a problem-centered classroom, students frequently participate in either teacher-led whole-group activities or small-group work. Such classes often open with a brief introduction of the problem or investigation, followed by students working in small groups. The introduction aims not to provide direction in working the problem, but rather to ensure that students understand the nature of the problem and any special terminology or constraints that exist. Students then work in small groups while the teacher circulates to observe, question and guide, assess, and motivate. As this type of problem solving is taking place, the teacher notes where the students are in learning the material and in making the appropriate generalizations concerning the content goals. Throughout these activities, the focus is on students sharpening their problem-solving abilities, as well as their abilities to reason, communicate, connect ideas, and shift among representations of mathematical concepts and ideas.

To ensure that high school teachers have curricular programs in place that model the approaches set out in the NCTM *Standards*, the National Science Foundation (NSF) funded five middle school and five high school curriculum-development projects in the early 1990s. These programs not only developed curricular materials, but also included pilot testing, evaluation, re-design, and field testing. Each program was developed by teams that included teachers, university mathematics educators, mathematicians, and professional evaluators. The following descriptions highlight some of the features in these programs.

Middle School Mathematics Programs

Connected Mathematics Project (CMP) (Prentice Hall)

Materials in the Connected Mathematics Project reflect an interest in developing student math knowledge that is rich in connections. These connections exist between the various topics, between mathematics and its applications in other disciplines, between the planned learning activities and the special aptitudes and interests of middle school students, and between student skills developed in elementary schools and the goals of secondary schools. These materials were developed by a group of researchers, teachers, and mathematicians at Michigan State University.

The curriculum is organized around a selected number of important mathematical concepts and process goals. Each concept is studied in depth within a general theme. Instruction emphasizes inquiry and discovery of mathematical ideas through investigation of structurally rich problem situations. Students grow in their ability to reason effectively with information presented in graphic, numeric, symbolic, and verbal forms and in their ability to move flexibly among these representations.

Mathematics in Context (Encyclopaedia Britannica)

Mathematics in Context was developed by research and development teams at the Freudenthal Institute at the University of Utrecht, the Netherlands, and at the University of Wisconsin, along with a group of middle school teachers. A total of 40 units were developed for grades 5–8. These units make extensive use of realistic contexts. From the context of tiling a floor, for example, flow a wealth of mathematical applications such as similarity, ratio and proportion, and scaling. Units emphasize the interrelationships of mathematical content areas such as number, algebra, geometry, and statistics. Mathematics is not presented as a set of disjointed facts and rules. Rather, students come to view math as an interesting, powerful tool that enables them to better understand their world. Activities have multiple levels so that able students can go into more depth while students having trouble can still make sense of the activity. Teachers are rewarded by seeing students excited by mathematical inquiry. The curriculum redefines the teacher's role as guide and facilitator of inquiry. In addition, the curriculum advocates collaboration with other teachers that can result in innovative approaches to instruction, in increased enthusiasm for teaching, and in improved student attitudes about mathematics.

MathScape (Creative Publications)

The MathScape curriculum, designed by Education Development Center (EDC), builds on the central theme of mathematics as a human experience. Throughout the curriculum, students experience math as it is used to plan, predict, design, explore, explain, coordinate, compare, make decisions, and other activities fundamental to human endeavors, both currently and historically. The materials develop proportional reasoning, multiple representations, patterns and functional relationships, and modeling, as well as specific concepts, skills, and language in the areas of algebra, estimation/computation, discrete mathematics, functions, geometry/visual reasoning, measurement, number, probability, and statistics.

The pedagogy of this curriculum views learning as a process of constructing one's own knowledge and emphasizes the social context of learning for middle school students. Technology is integrated throughout the curriculum. In addition to working with calculators, students use spreadsheets and dynamic geometry software.

Middle Grades MathThematics (STEM) (McDougal Littell)

The Middle Grades MathThematics (STEM) materials were designed by a consortium of teachers, researchers, and mathematicians from Montana. In designing the curriculum, the staff worked with IBM, McDougal Littell, Texas Instruments, and Microsoft. The materials provide middle school teachers with curricular materials that are mathematically accurate, use technology, and bridge to science and other fields. The materials stress communication in mathematics by using reading, writing, and speaking as tools for learning mathematics. MathThematics materials are problem-centered, application-based, and use technology where appropriate. Many lessons are project-oriented and require students to work cooperatively. A major goal of the materials is to present math as exciting and useful. Topics such as quantitative literacy and discrete mathematics receive much more emphasis than in more traditional curricula.

The Middle School Mathematics Through Applications Project (MMAP) (Voyager Expanded Learning)

The MMAP curriculum provides a comprehensive middle school mathematics program centered around applications projects that engage students in real-world problem solving. The materials assume that no matter what students are doing, they are learning. As a result, the materials involve students as if they are professionals who use math in their work. In the guppies unit, for example, students assume the role of population biologists. The students role-play and a "community" of population biologists is formed. Through group work, shared goals, and discourse, mathematical problems emerge. Because understanding the mathematics helps them solve a real problem, student math skills and reasoning improve. Discussion about the problems takes place, and new understandings emerge.

The MMAP curriculum has three components from which teachers and curriculum decision makers construct a complete and balanced middle school curriculum. Application units support 4–8 weeks of activity, use specially designed software, and plunge students into an extended role play in which they learn and use math. Two shorter units fill out the curriculum. Extensions bridge the understanding students attain in the projects to standard notations and skills. Investigations introduce students to pure mathematics topics and methods. In addition, application projects are accompanied by specially created computer design tools, needed for about half the project. Students use the software together to solve design problems. This creates opportunities for students to learn mathematics through experimentation, model-building, representation, and argumentation.

High School Math Programs

Application Reform in Secondary Education (ARISE): Mathematics: Modeling Our World (Southwestern Educational Publishing)

The Consortium for Mathematics and Its Applications (COMAP, Inc.) has been developing mathematics education materials that feature contemporary applications since 1980. In creating *Mathematics: Modeling Our World*, COMAP worked with a team of over 20 authors to produce a curriculum that presents math concepts in the contexts in which they are actually used. The word "modeling" is key. The curriculum is founded on the principle that math is a necessary tool for understanding our physical and social worlds. Questions about the real world are presented first and motivate the development of the mathematics. Thus, the contextual questions drive the math. As students discover various ways to solve problems, they not only learn mathematics and content in other subjects, but they also learn how to reason mathematically, organize and analyze data, make predictions, prepare and present reports, and revise their predictions based on new information.

In *Mathematics: Modeling Our World*, each unit uses engaging, real-life situations and the problems and conditions associated with them. For example, students analyze various voting methods used throughout the world, predict changes in the Florida manatee population relative to powerboat use, and analyze the effectiveness of polling samples in medical testing. In the modeling process, students identify key features of the context being studied, build a simple model, test it against various criteria, modify the model in an effort to improve its description of the real context, and use the model to make predictions or to solve the problem. Both graphing calculators and computers are used extensively to enhance concept development.

Core-Plus Mathematics Project (CPMP): Contemporary Mathematics in Context (Everyday Learning Corporation)

The Core-Plus Mathematics Project (CPMP) builds on the theme that mathematics is sense-making. The materials for each grade level feature algebra and functions, geometry and trigonometry, statistics and probability, and discrete mathematics. These topics are connected within and across units by fundamental ideas such as symmetry, function, matrices, data analysis, and curve fitting. Effective mathematical habits such as visual thinking, recursive thinking, searching for and describing patterns, making and checking conjectures, reasoning with multiple representations, "inventing" mathematics, and providing convincing arguments are emphasized. The topics are linked further by the fundamental themes of understanding data representations, shapes, and change. Important ideas are continually revisited through these connections so students can develop a robust understanding of mathematics. Numerical, graphical, and programming capabilities of graphing calculators are capitalized on to encourage students to develop versatile ways of dealing with realistic situations.

Core topics are designed to be accessible to all students. Differences in student performance and interest are accommodated by the depth and level of abstraction to which topics are pursued, by the nature and degree of difficulty of applications, and by opportunities for student choices in homework tasks and projects. Instructional practices promote mathematical thinking through rich problem situations in which students, both in collaborative groups and individually, investigate, conjecture, verify, apply, evaluate, and communicate mathematical ideas. Comprehensive assessment of student understanding and progress is achieved through both curriculum-embedded assessment opportunities and supplementary assessment tasks. This allows teachers to monitor and evaluate each student's performance in terms of mathematical processes, content, and dispositions.

Interactive Mathematics Project (IMP): Interactive Mathematics Program (Key Curriculum Press)

The Interactive Mathematics Project (IMP) created a four-year program of problem-based mathematics that replaces the traditional Algebra I, Geometry, Algebra II/Trigonometry, and Precalculus sequence of secondary school courses. The IMP curriculum integrates traditional material with additional topics recommended by the NCTM *Standards*, such as statistics, probability, curve fitting, and matrix algebra. IMP units are generally structured around a complex central problem. Although each unit has a specific focus, other topics are brought in as needed to solve the central problem, rather than narrowly restricting the math content. Ideas that are developed in one unit are usually revisited and deepened in later units.

MATH Connections Project: MATH Connections: A Secondary Mathematics Core Curriculum Initiative (It's About Time Publishing)

The development of *MATH Connections: A Secondary Mathematics Core Curriculum* was guided by the Connecticut Business and Industry Association (CBIA) Education Foundation. The program materials were developed by an experienced and diverse team of curriculum developers—mathematicians, scientists, educators in mathematics, science, and technology, and business people.

MATH Connections blends algebra, geometry, probability, statistics, trigonometry, and discrete mathematics into a meaningful package that is interesting and accessible to all students. The text materials provide students with mathematical experiences that excite their curiosity, stimulate their imagination, and challenge their skills. All the while, the primary concern is the conceptual development of the learner while focusing on these goals: (1) math as problem solving; (2) math as communication; (3) math as reasoning; and (4) math as making connections. MATH Connections is based on topical (rather than problem) themes. That is, it is concept-driven. A common thread connects and blends many mathematical topics that traditionally have been taught separately and independently. This approach emphasizes the unity and interconnectedness of mathematical ideas.

Technology is integrated into MATH Connections through the use of graphing calculators and computers, which students use to investigate concepts in greater depth and breadth, make conjectures, and validate findings. Real-world applications and problem situations from the sciences, the humanities, and business and industry prepare students for post-secondary education and the demands of the twenty-first century. Alternative assessment methods such as written, oral, and demonstration formats assess higher-order thinking skills.

Systemic Initiative for Montana Mathematics and Science Project (SIMMS): Integrated Mathematics: A Modeling Approach Using Technology (Pearson Custom Publishing)

The SIMMS curriculum incorporates a modeling approach using technology. The curriculum is designed to replace all grade 9–12 mathematics courses, with the possible exception of advanced placement classes. The materials include work in algebra, geometry, trigonometry, analysis, statistics, probability, and matrices, as well as less traditional high school topics such as graph theory, game theory, and chaos theory. Students are expected to have ready access to technology including a graphing utility, spreadsheets, a geometry utility, a statistics program, a symbolic manipulator, and a word processor. The materials were written by secondary school teachers and university personnel. They consist of approximately 90 modules in which mathematics evolves from real-world problems. Each module is taught over a 2- to 3-week period in classes of 45–60 minutes and can be readily adapted to block scheduling.

These overviews of the NSF-sponsored curricula for middle school and high school mathematics reflect the central role that modeling, technology, and problem solving have taken in math programs designed to meet the future needs of our nation's youth. They represent much of the intended curriculum supported by the NCTM *Standards* documents.

The following exercises give you an opportunity to explore the *Standards*, their origins in comparative studies, and the new curricula from different perspectives.

Exercises 2.1

For Exercises 1 and 2, examine the results from the most recent NAEP mathematics assessment at the web site for the National Center for Education Statistics (http://nces.ed.gov).

1. Write a short report on middle school student performance since 1989.

2. Write a short report on high school student performance since 1989.

3. Examine the findings of the Second International Mathematics Study as detailed in C. C. McKnight et al., *The Underachieving Curriculum: U.S. Mathematics from an International Perspective*. Compare and contrast these findings with those of the TIMSS study, as presented through the Pursuing Excellence volumes, which can be found at (http://nces.ed.gov).

For Exercises 4 and 5, write an in-depth report on one of the middle school or high school programs developed with NSF funding. In your report, describe the project, outline the curriculum, and give examples of curriculum, evaluation data, teacher support and resources, and publisher information.

4. Middle school program of your choice (data can be found at http://showmecenter.missouri.edu/).

5. High school program of your choice (data can be found at http://www.ithaca.edu/compass/main.htm).

PROJECTS Find Out for Yourself

1. Visit a local middle school and examine their textbooks for general mathematics classes.
   a. How does the textbook compare with one from NSF middle school projects?
   b. Does the traditional textbook cover the same concepts and topics as the NSF curricula you chose in part (a)?
2. Visit a local high school Algebra I class and examine the textbook used.
   a. How does the Algebra I text compare with an NSF curricula first-year textbook?
   b. Does the NSF curriculum series cover as many Algebra I concepts as the traditional textbook?

2.2 Implemented Curriculum

While the NCTM *Standards* outline what students ought to receive and how the associated instruction might take place, the real question is, What happens when classroom doors close and math instruction begins? In other words, how is the curriculum implemented? Numerous indicators give us a picture of mathematics instruction in the United States. In addition to the data from the National Assessment of Educational Progress, annual composite reports from the Council of Chief State School Officers (Blank & Langesen, 1999) detail reports made by each state's Office of Public Instruction.

Trends in Student Performance

These sources indicate a positive trend in mathematics instruction over the years. The data show that through the 1990s almost all states revised their state learning outcomes for mathematics according to the NCTM *Standards*, and many were updating these recommendations at 2000. By 1998, 23 states required three years of math credit for graduation from high school, and 25 states required at least two years. Studies examining student performance in math indicate that correlated increases in student performance occurred across the same period of time (Blank & Langesen, 1999). However, at the same time, other studies show that while this growth is highly correlated with instructional time and courses taken (Wilson & Blank, 1999), it is also highly correlated with student socioeconomic status (Lee & Smith, 1993; and Weiss, 1994). Hence, we are making progress, but questions of equal access to quality instruction remain.

Perhaps the most interesting data from a curricular standpoint is the percentage of students who reach various checkpoints in the mathematics curriculum. While almost all students complete mathematics through grades 8 or 9 due to state laws on compulsory education, a different picture exists for advanced coursework. NAEP data from 1990, 1992, 1996, and 1999 (Campbell, Voelkl, & Donahue, 1997; Mitchell et al., 1999) suggest the patterns shown in Figure 2.6.

Here we see that over the recent past, fewer students are dropping out after General Mathematics or Algebra I and more students are completing courses in Geometry

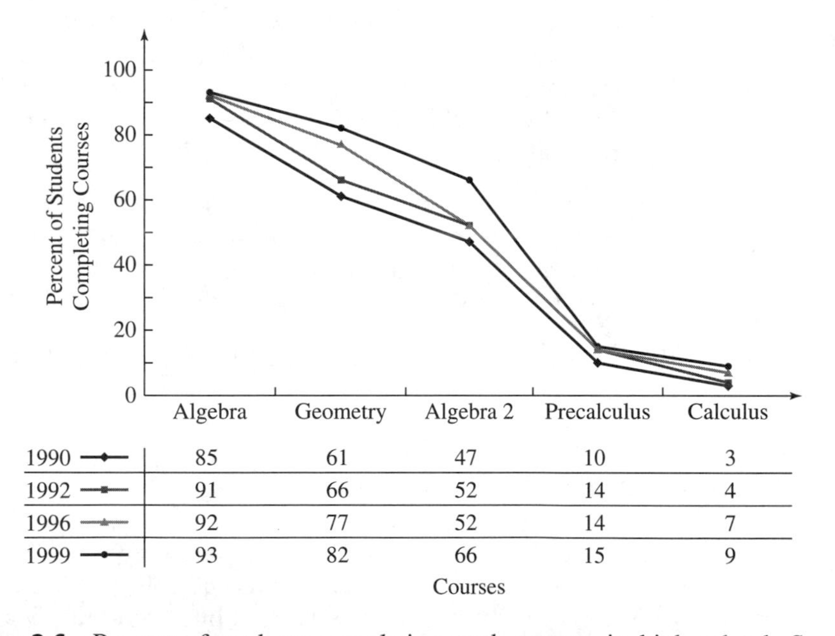

	Algebra	Geometry	Algebra 2	Precalculus	Calculus
1990	85	61	47	10	3
1992	91	66	52	14	4
1996	92	77	52	14	7
1999	93	82	66	15	9

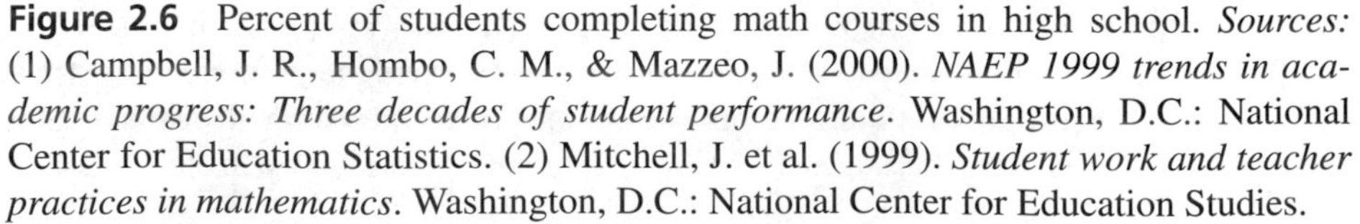

Figure 2.6 Percent of students completing math courses in high school. *Sources:* (1) Campbell, J. R., Hombo, C. M., & Mazzeo, J. (2000). *NAEP 1999 trends in academic progress: Three decades of student performance.* Washington, D.C.: National Center for Education Statistics. (2) Mitchell, J. et al. (1999). *Student work and teacher practices in mathematics.* Washington, D.C.: National Center for Education Studies.

or Algebra II. However, the percentages of students completing coursework in Precalculus or Calculus remains exceedingly low, both in comparison to the percentage who completed coursework through Algebra II and to the percentage of students reaching that level in countries that are our economic peers.

In an interesting tangent, one study asks what percent of students take the equivalent of Algebra I as middle school or junior high school students. NAEP data from 1996 indicate that approximately 25% of the nation's students take the equivalent of Algebra I by grade 8 (Hawkins, Stancavage, & Dossey, 1998).

Instructional Programs

As with the trends in student coursework, small changes in U.S. instructional programs reflect movement toward a more process-oriented mathematics curriculum. In 1996, teachers indicated that they placed "a lot" of emphasis on these areas of mathematics curricula: 79% on learning facts and concepts, 79% on learning skills and procedures, 52% on developing reasoning and the ability to solve unique problems, and 43% on helping students learn how to communicate ideas in mathematics effectively (Mitchell et al., 1999). These percentages are marginally up from 74%, 79%, 49%, and 40% reported, respectively, for the same areas in 1992 (Dossey et al., 1994). Similar growth occurred in data reporting teachers' use of manipulatives, small groups, writing to learn, projects, portfolios for assessment, and tasks that reflect real-life situations involving mathematics.

Calculator Usage

An area that grew significantly occurred in calculator use in class. Thus, 55% of eighth graders use calculators almost every day, while 91% use them at least once or twice a month. In high school, 78% of twelfth grade math students use them almost every day, while 95% use them at least once or twice a month. Further, eighth grade teachers report that 47% of students have generally unrestricted use of calculators in their math classes and 67% of the students are allowed to use calculators on tests (Mitchell et al., 1999).

These data on coursework trends, the program features teachers emphasize, and calculator usage are both encouraging and discouraging. If we look at the data in terms of the low emphasis given some of the recommended processes, the data are discouraging. If we regard the data from the standpoint of change, then it is apparent that U.S. math programs are changing. Still, change is often a slow process. Whether the nation's schools choose overnight reform or slow evolution is a choice that individual school districts must make.

Exercises 2.2

1. What are the most frequently used mathematics textbooks in the region where you are planning to teach? Design a survey, collect some data, and write a report detailing your findings.

2. What criteria would you use to evaluate math textbooks? What role should the *Standards* play in such an activity?

3. Consider research on calculator usage and student achievement in mathematics. What do the NAEP reports say? What information has research provided? You might want to begin by looking at the NAEP data at (http://www.nces.ed.gov) or in the March, 1986 *Journal for Research in Mathematics Education.*

4. Describe the various ways technology might be used in the classroom to support the goals of the NCTM *Principles and Standards*. Be brief, but respond to both the process and content goals.

PROJECTS Find Out for Yourself

1. Determine the nature of the curriculum in the high school you attended. Talk to teachers and find out whether a written curriculum exists, what are the school's graduation requirements, what percent of students reach each level of study, the various instructional emphases, and calculator usage. Use data accurate for your former school. Write a brief report detailing your findings.

2. Interview a student at a local high school to find out what the student knows about the school's offerings in mathematics and how they fit together. Does the student know what courses are prerequisites for other courses and which courses are required for college admission?

3. Interview the mathematics department chairperson at a high school. What percentage of students take each course in the curriculum? What percentage of students complete the series of Algebra I, Geometry, Algebra II, and Precalculus by the time they graduate? How many senior students are taking some mathematics course?

2.3 Attained Curriculum

As noted in Section 2.1, the math skills of U.S. students continue to improve. Nevertheless, they don't measure up to the performance of their peers in other industrialized countries at either eighth or twelfth grade. However, studies of adult mathematics literacy indicate U.S. citizens continue to grow in mathematical skill and by age 35 may be among the best in the world (OECD, 1995).

Mathematics knowledge can be classified in several ways. One frequently used scheme divides math knowledge into concepts, procedures, and problem-solving skills. We say that students are dealing with concepts when they discuss "what" something is. They exhibit *conceptual understanding* when they recognize, label, and generate examples of what concepts are and are not. They exhibit it when they use concepts and their representations to discuss or classify mathematical objects.

Conceptual understanding is used to compare and contrast objects, as well as to form interrelationships between concepts and principles. When students recognize symbolic representations or interpret words as signifying operations or concepts, they demonstrate understanding. Thus, students demonstrate conceptual knowledge when they know that "triangle" means a three-sided object. Young children even have rough ideas of a triangle being a three-sided object. As they continue through school, they learn to distinguish between right triangles, isosceles triangles, and scalene triangles. Along the way, students pick up the idea of segment sides, closed figure, convexity, angle sums, and other related triangle knowledge. When we talk of students' knowledge of concepts, we usually talk about the depth and breadth of their understanding of a particular concept.

Students demonstrate *procedural knowledge* when they select and apply procedures correctly—that is, when they recognize addition and find the correct sum, or when they use prime factorization to find the greatest common divisor of two positive integers. Students show procedural knowledge when they verify or justify the appropriateness of a procedure for carrying out a given task. They show it when they use concrete objects to model the steps in an algorithmic process. It is also present when they construct graphs to display a set of data, handle and read a measuring tool, carry out a construction with compass and straightedge or with geometric software. In each of these cases, students execute an algorithm to move from known materials to a desired final state. When we talk about procedural knowledge, we mean their level of skill in executing procedures correctly.

The third area of mathematical knowledge assessed, *problem solving*, is an amalgam of conceptual understanding and procedural knowledge. It requires students to recognize situations, abstract their core structure, model the relationships involved, manipulate those relationships, and communicate the results. In solving problems, students have to

- recognize and formulate the situation in mathematical terms
- determine which relationships are necessary and which are sufficient
- select relevant strategies, data, and models
- use reasoning (spatial, inductive, deductive, or statistical) in new settings
- judge the reasonableness and correctness of outcomes

Problem solving calls on students to confront new and challenging situations that involve mathematical concepts.

Student performance, the attained curriculum, is usually measured through large-scale assessments, such as standardized tests, state assessments, or NAEP. These assessments don't allow evaluators to ask follow-up questions like teachers do in classrooms. The interpretations of student performance make assumptions about what a student was thinking and doing when they responded to the item. Seeing student work is crucial in assessing achievement. However, attaining a representative sample of students' work is difficult to do in large-scale assessments.

Much of the information that we have about student problem-solving competence comes from large-scale assessments or interviews with individual students. Example 1 gives an item included on a recent NAEP assessment for twelfth grade students. Work the problem before continuing.

Example 1 ***NAEP Assessment Problem***

A certain machine produces 300 nails per minute. At this rate, how long will it take the machine to produce enough nails to fill 5 boxes of nails if each box will contain 250 nails?

A. 4 min

B. 4 min 6 sec

C. 4 min 10 sec

D. 4 min 50 sec

E. 5 min ■

Analysis of student performance on this item shows that overall 49% of twelfth graders answered the item correctly by selecting alternative C, 19% selected option B, and 12% selected option D. There is nothing tricky about this problem, but at least two steps are required to solve it. Research data indicates that problems with more than one step are considerably more difficult for students.

Two other problem formats used are called *regular student-constructed response* and *extended student-constructed response*. Both of these formats require students to construct their own answers and communicate them to the individuals evaluating, or grading, their work. Regular response problems are graded on a right-wrong basis with no partial credit. Extended response problems award partial credit depending on the answer. They require students to write a paragraph, draw or make a model, give an example, and then explain their significance. These problems usually give several levels of partial credit.

An example of a regular response problem and answer is shown in Figure 2.7. This problem, based on an incorrect graphic that appeared in a national newspaper, examines conceptual understanding. To successfully respond, the student must know that the volume of three-dimensional cans is related as the cube of their linear dimensions. In the incorrect graphic, the 1980 can is supposed to hold twice the amount of the 1960 can, but *every* dimension of the 1960 can has been doubled to construct the 1980 can, which gives a volume 8 times that of the 1960 can. Only 8% of eighth graders answered this item correctly.

An extended response problem is shown in Figure 2.8. Students are asked to determine patterns from the algebra and function domain for grade 12. Extended response items are assessed using a rubric, or scoring guide. The general format for the scoring rubric has six different rating levels.

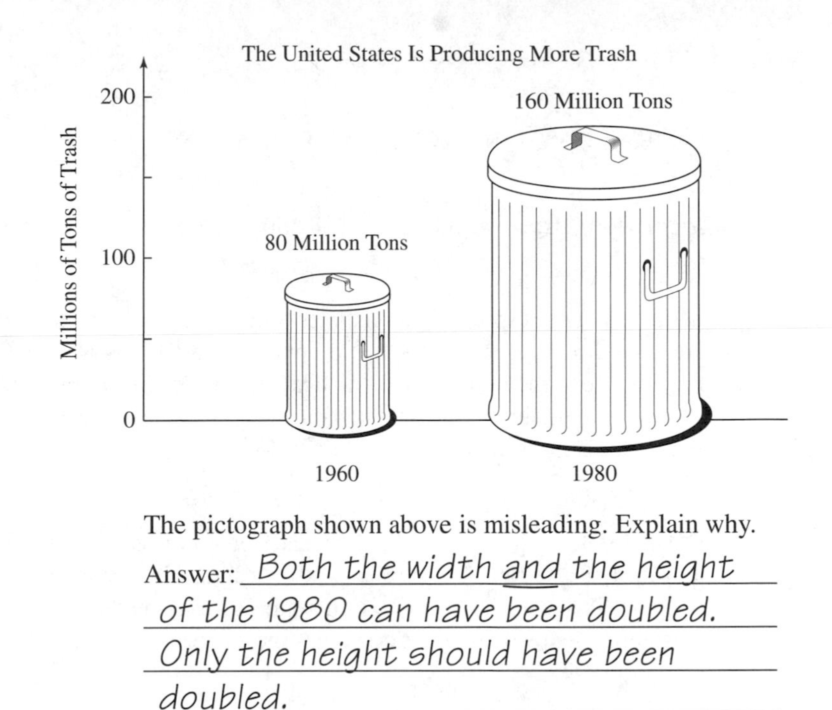

The United States Is Producing More Trash
200
100
0
Millions of Tons of Trash
160 Million Tons
80 Million Tons
1960
1980
The pictograph shown above is misleading. Explain why.
Answer: Both the width and the height
of the 1980 can have been doubled.
Only the height should have been
doubled.

Figure 2.7 Regular student-constructed response problem from a 1992 NAEP eighth grade assessment. *Source:* Dossey, J. A., Mullis, I. V. S., & Jones, C. O. (1993). *Can students do mathematical problem solving?* Washington, D.C.: Office of Educational Research and Improvement for the U.S. Department of Education.

The first 3 figures in a pattern of tiles are shown below. The pattern of tiles contains 50 figures.

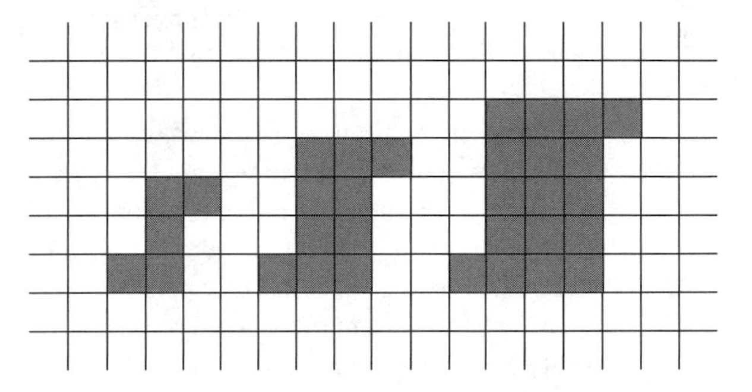

Describe the 20th figure in this pattern, including the total number of tiles it contains and how they are arranged. Then explain the reasoning that you used to determine this information. Write a description that could be used to define any figure in the pattern.

Figure 2.8 Extended student-constructed response problem from a 1996 NAEP twelfth grade assessment. *Source:* Mitchell et al. (1999). *Student work and teacher practices in mathematics.* Washington, D.C.: National Center for Education Statistics.

No Response This rating is given to blank papers.

Incorrect Response The student's work is incorrect or irrelevant.

Minimal The response demonstrates a minimal understanding of the problem posed, but does not suggest a reasonable approach to the problem. Although there may or may not be some correct mathematical statements in the response, the response is incomplete, contains major mathematical errors, or reveals serious flaws in reasoning. Examples are absent.

Partial The response contains evidence of a conceptual understanding of the problem in that a reasonable approach is indicated. However, on the whole, the response is not well developed. Although there are serious mathematical errors or flaws in the reasoning, the response does contain some correct mathematics. Examples provided are inappropriate.

Satisfactory The response demonstrates a clear understanding of the problem and provides an acceptable approach. The response is also generally well developed and coherent, but contains minor weaknesses in the development. Examples provided by the student are not fully developed, but are generally correct.

Extended The response demonstrates a complete understanding of the problem, is correct, and the methods of solution are appropriate and fully developed. These responses are logically sound, clearly written, and mathematically correct. Examples are well chosen and fully developed (Dossey, Mullis, & Jones, 1993).

The following examples of student work at the various levels are taken from the 1996 NAEP assessment.[1] Directions for the answer in Example 2 asked students to show all of their work and explain their reasoning. In addition, they were instructed that they could use drawings, words, or numbers in their explanations and that the answer needed to be clear enough that another person could read it and understand their thinking. Student responses were considered "extended" if the work included the following elements: (1) a correct count of 442 tiles for the 20th figure, and (2) a verbal or graphical explanation of their reasoning for this item. Of the students' submitted responses, 2% were rated extended.

---

[1] Examples 1–5 from Mitchell, J. et al. (1999). *Student work and teacher practices in mathematics*. Washington, D.C.: National Center for Education Statistics.

Example 2 ***Sample Extended Response***

* Each figure increases 1 layer in height and one middle layer in width for every succession, relative to the first.
For example, for the nth section the figure will be n+1 units across at the base, n units wide, n+1 units across at the top, and n+2 units high. This is the pattern
The 20th figure will be 21 units across on the bottom length, 20 units wide in the middle, 22 units high, and 21 units wide at the top. The increase is linear.
Total number of tiles it contains:
21 + (20×20) + 21 = 442
The inner square is always (n×n) units in area (n^2)

■

A response was considered "satisfactory" if the student described the 20th figure, gave the number of tiles, and provided some evidence of sound reasoning. However, satisfactory responses either included errors in computation or lacked clarity in the explanation. The sample "satisfactory" response in Example 3 contained most of the elements asked for, but lacked a clear explanation or generalization. In this test item, 2% of the students provided responses rated satisfactory.

Example 3 ***Sample Satisfactory Response***

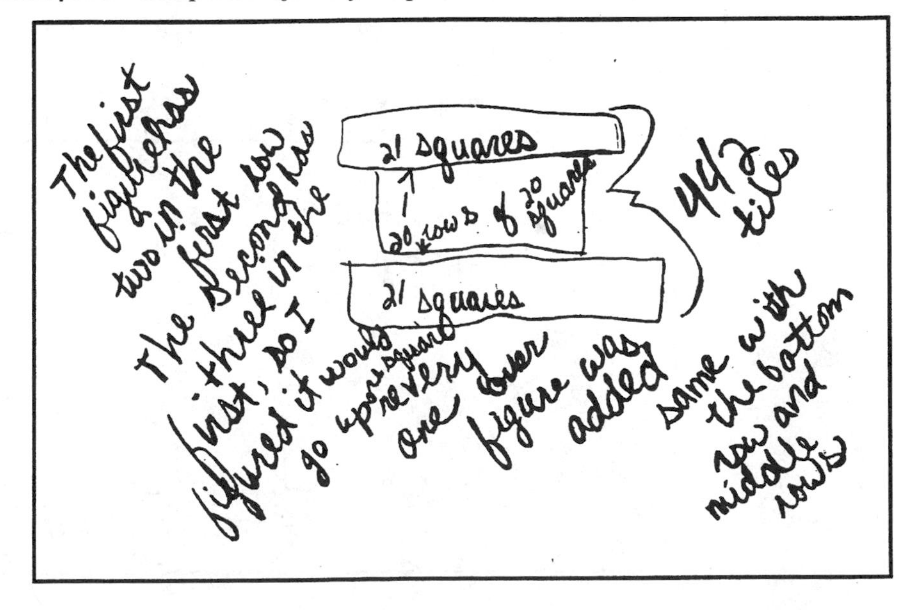

■

A response was considered "partial" if the student described at least one additional figure in the pattern correctly or stated that there are 442 tiles in the 20th figure but did no more. In the sample "partial" response in Example 4, the student correctly diagrammed the 20th figure, but did not state how many tiles were in the figure, explain his or her reasoning, or provide a generalization. Partial responses were given in 18% of the answers.

Example 4 ***Sample Partial Response***

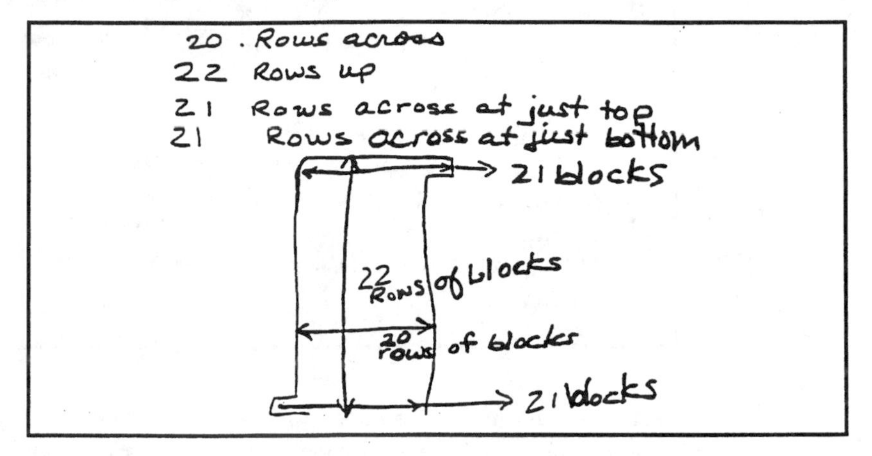

A response was considered "minimal" if the student attempted to draw or describe the pattern or an additional figure in the pattern or made at least some attempt to go beyond what was shown in the question. The "minimal" response in Example 5 shows an attempt to draw the 20th figure, but does not correctly describe it or state the number of tiles it contains. The student's reasoning was not clearly explained, and there was no description that could be generalized to any figure in the sequence; 58% of answers were in this classification.

Example 5 ***Sample Minimal Response***

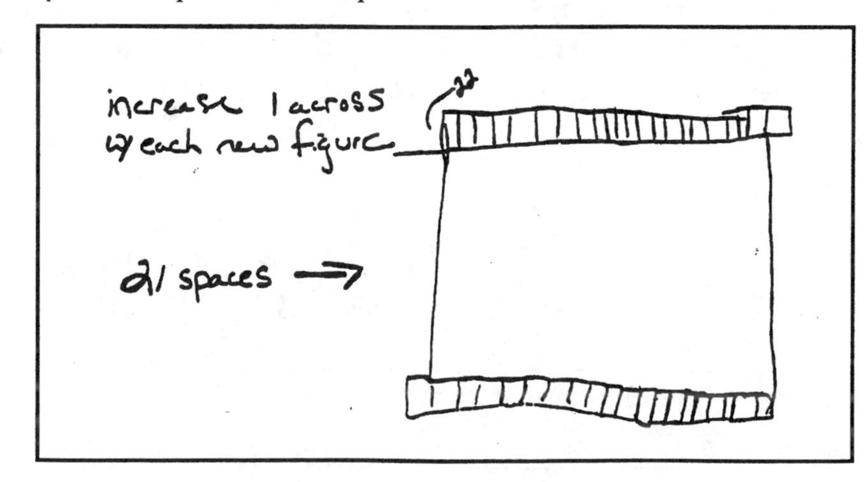

An additional 20% of the twelfth grade students left their papers blank, even though they had ample time to complete the item. Perhaps they had no idea of how to start working the problem, or perhaps this signals a lack of motivation. It may also be that these students had an idea, but were unable to express it or ran out of time.

Example 6 ***Sample Incorrect Response***

■

These responses give us a picture of the various ways assessments examine curriculum attainment. Student performance is higher when we look only at multiple-choice answers (Dossey, Mullis, & Jones, 1993), but student-constructed responses move us closer to finding out what they really know. In Chapter 14, we consider a variety of ways to examine student learning.

Exercises 2.3

1. Consider the following multiple-choice item and the percentage of students selecting each possible answer. What interpretations do you make of eighth and twelfth grade comprehension?

> Ken bought a used car for $5,375. He had to pay an additional 15% of the purchase price to cover both the sales and tax and extra fees. Of the following, which is closest to the total amount Ken paid?

		Student Responses	
		Grade 8	Grade 12
A.	$ 806	25%	20%
B.	$5,510	14%	4%
C.	$5,760	12%	3%
D.	$5,940	7%	3%
E.	$6,180	40%	69%

2. A test item asked the following:

   If $f(x) = 4x^2 - 7x + 5.7$, what is the value of $f(1/2)$?

   How would you establish a rubric for evaluating this item on an incorrect-correct basis? What issues surrounding student calculator use could occur in establishing the rubric?

3. You have a set of student responses to this extended response item:

   Treena won a 7-day scholarship worth $1000 to attend the Pro Shot Basketball Camp. Round-trip travel expenses to the camp are $335 by air or $125 by train. At the camp she must choose between a week of individual instruction at $60 a day or a week of group instruction at $40 per day. Treena's food and other expenses are fixed at $45 per day. If she does not plan to spend any money other than the scholarship, what are all choices of travel and instruction plans that she could afford to make? Explain your reasoning.

   What kind of grading scale would you establish for "extended," "satisfactory," "partial," "minimal," and "incorrect" responses for this item?

4. Design an extended student-constructed rubric for rating your papers. What would you look for in the student work for awarding each level of performance rating from incorrect to extended?

PROJECT Find Out for Yourself

Find information on your state's math assessment program.

1. How is student progress assessed?
2. How are the results reported?
3. Can you identify any trends in student performance?

2.4 Student Attitudes Toward Mathematics

The previous three sections consider the intended, implemented, and attained curriculum as a model for math education. Here we consider other aspects of the mathematics education: student attitudes toward mathematics. What do students believe about the nature of mathematics? Do they see mathematics as useful in their lives? Do they believe they can understand and use mathematics?

The beliefs students hold about math are important factors in the learning process. For several decades, teachers and researchers have noted that students' beliefs and attitudes affect their ability to learn mathematics (Lester, Garofalo, & Kroll, 1989;

McLeod, 1992). For instance, if students are often frustrated when they attempt to solve story problems, they are likely to believe that they can't solve story problems, and they will carry this belief through many years of schooling. Such students may not even attempt to solve story problems when they encounter them on class tests or standardized exams. Similarly, students who are not provided opportunities to experience math outside the classroom may believe that mathematics is unimportant and will be bored and disinterested in math class. These students are likely to pay less attention in class, which will severely hamper their learning.

The NCTM *Standards* stress two goals that directly address student beliefs and attitudes. One goal is to help students learn to value math. The other goal is to help them develop self-confidence in doing math (NCTM, 1989, 2000). If students are to attain the goals put forth by the NCTM, we must find ways to develop their curiosity about mathematics and confidence in their ability to do mathematics.

What do students think is the nature of mathematics? Many believe that math is about techniques for solving those mysterious equations or other obscure problems. They believe mathematics is about memorizing techniques and formulas. Beyond the grade they're given, they see little value in learning math. This all-too common view of mathematics arises because this is how the student experiences math in the classroom. Teachers present concepts and demonstrate techniques. Students practice these techniques repeatedly but never understand why they are important or useful.

Other students see mathematics as useful in routine tasks, such as doing simple calculations at the store, balancing a checkbook, or measuring a room to fit a carpet. They don't appreciate how math skills are important in other more sophisticated functions. However, students who experience math as real-world problem solving, like that presented in the NSF-funded curricula, tend to have a different view of mathematics. These students see how mathematics helps them make sense of the world around them.

Do students believe they can learn to understand and use mathematics? There will always be those who believe they cannot learn math, and those who believe that they can (even though it may be difficult at times). These beliefs develop over years of mathematical experiences in school (based on grades, frustrations, and boredom). And of course, teacher's beliefs in students' ability affect the student's beliefs. Research shows that teachers' conceptions about students influence the ways in which they interact with those students (Kenney & Silver, 1997; Reyes, 1980). In particular, Reyes (1980) found that teachers pay less attention to students who lack confidence in their ability to do mathematics and lack the determination to try.

Researchers have also addressed gender, racial, and ethnic issues that impact student attitudes toward math (Beaton et al., 1996; Kenney & Silver, 1997). In general, gender studies reveal that males are slightly more apt to believe they can do well in math whether it is due to natural ability, hard work, or memorization (Beaton et al., 1996). Females, on the other hand, are more likely to lack confidence in their math abilities. Our culture reflects this belief. Until recently, males were considered more mathematically talented than females. Parents thus had lower expectations for girls in math and science, and high school girls were less likely than boys to do well in math and less likely to pursue related careers (Campbell & Storo, 1994). Research on gender and mathematics achievement has led to major shifts in our attention to equal opportunity for participation for males and females (Fennema & Leder, 1990). Heightened awareness and encouragement from teachers has helped to address the issue; even so, gender inequities still exist and teachers need to continue to be vigilant.

Studies also indicate that students from various racial and ethnic groups appear to be self-confident about their ability to do well in mathematics, although this is not necessarily reflected in their scores (Kenney & Silver, 1997). Black students in the United States appear to be as confident as their White[2] counterparts in their ability to do well in mathematics. Hispanic students also appear to be confident in their mathematics ability, although less so. Even so, inequities associated with ethnicity are of continued concern. To help these students benefit from their positive attitudes toward math, teachers must create learning environments that support all students.[3]

As previously noted, beliefs about the nature of mathematics, its usefulness, and about one's ability to do math all influence student achievement and warrant careful consideration by all involved in mathematics education. When students' beliefs affect their performance in math classes, changes need to be made. Here are some changes that can help.

Make math meaningful. Help students see the value of studying mathematics. One way to do this is by providing opportunities for students to use math in real-life situations. Encourage and expect all students to make sense of mathematics.

Motivate learning. Show your love for math. Locate or create problem-solving tasks that catch student interest and inspire them to investigate the mathematics behind the problem.

Develop students' confidence in their ability to do math. Get to know your students. Don't ignore the student who has low self-confidence or who shows frustration. Instead, find ways to boost confidence through small successes.

Help dispel myths. Read the research and discover for yourself that many beliefs about math ability are myths. Share this information with the teachers and students in your school.

Exercises 2.4

1. What things might you do to interest your students in mathematics? Describe a project you might assign that would motivate them to learn math.
2. List some myths about learning math. Why do these myths exist? How can you dispel these myths?

---

[2] Subgroup designations (Black, White, Hispanic) follow those used in the National Assessment of Educational Progress.

[3] The topic of diverse learners is discussed in more detail in Chapter 13.

3. Suppose a student of yours has done poorly in past math classes, assumes she will also do poorly in your class, and as a result doesn't even try to make sense of the mathematics you present. What can you do to help her?

PROJECTS Find Out for Yourself

1. Test your beliefs about math by taking the questionnaire below. Compare your beliefs with those of a friend. Think about why you hold these beliefs. What experiences have you had that established or strengthened such beliefs?
   - **a.** Nothing new has been discovered in math for a long time.
   - **b.** Math helps us think logically.
   - **c.** Math will change rapidly in the near future.
   - **d.** There is always a rule to follow in solving a math problem.
   - **e.** Trial-and-error can often be used to solve a math problem.
   - **f.** Learning math mostly involves memorizing.
   - **g.** Math problems can be solved without using rules.
   - **h.** Mathematics is a good field for creative people.
   - **i.** Math helps us think according to strict rules.
   - **j.** Solving math problems doesn't require originality.
   - **k.** Mathematics is a set of rules.
   - **l.** Estimating is an important mathematical skill.
   - **m.** There are different ways to solve most math problems.
   - **n.** New discoveries in mathematics are constantly being made.
2. Interview a student to find out more about their beliefs and attitudes toward mathematics:
   - **a.** Before you interview the student, develop the set of questions you will ask to get at their beliefs and attitude. (You may want to use the questionnaire as a starting point.) For example, you might ask whether the student enjoys doing math. Why or why not? Does the student see math as useful?
   - **b.** After the interview, write up your discoveries and suggest ways to improve the student's attitudes or change their beliefs.

2.5 Planning for Change in the Teaching of Mathematics

The National Council of Teachers of Mathematics and others at the national and state levels have set ambitious goals for the teaching and learning of mathematics in the next few decades. To achieve these goals, teachers need to be aware of the changes that must take place. It is equally important that teachers realize that change is not a once-in-a-career event. The commitment to change the way we teach math must be firm, and the process of change must be continual because the needs of society are constantly evolving and shifting.

Many factors influence how we teach math at the middle school and high school levels. The workplace influences what we require in the way of math skills, whether it is a shift to new technology or the evolving needs of the workforce itself. Students' awareness of their future needs is another factor that drives how we teach mathematics. Parents and students may demand that schools focus on statistics, modeling, and calculus to meet new requirements for college. Curricular changes such as this influence how math is taught from kindergarten through high school. New ideas in science and mathematics as well as research on the learning of mathematics also spur changes. Teachers must be aware of these factors and their role in our changing needs.

Changing the way one teaches requires foresight and diligence. We must set goals, such as those in the NCTM *Principles and Standards for School Mathematics*, and we must periodically reflect on and update those goals. A first step for math teachers is to understand the NCTM goals (see the Intended Curriculum in Section 2.1) and how the goals impact process and content taught at the middle school and high school levels. In addition to their goals for a high-quality mathematics program, the NCTM has described the beliefs on which they based the *Principles and Standards*. The statement of beliefs, presented here, is a starting point for change in the classroom. As the last few lines of the Statement of Beliefs suggest, math teachers must work together for change to take place.

> As the primary professional organization for teachers of mathematics in grades K–12, the National Council of Teachers of Mathematics has the responsibility to provide broad national leadership in matters related to mathematics education. In meeting this responsibility, NCTM has developed a set of standards for school mathematics that address content, teaching, and assessment. These standards are guidelines for teachers, schools, districts, states, and provinces to use in planning, implementing, and evaluating high-quality mathematics programs for kindergarten through grade 12. The NCTM *Standards* are based on a set of core beliefs about students, teaching, learning, and mathematics.
>
> We believe the following:
>
> ■ Every student deserves an excellent program of instruction in mathematics that challenges each student to achieve at the high level required for productive citizenship and employment.
>
> ■ Every student must be taught by qualified teachers who have a sound knowledge of mathematics and how children learn mathematics and who also hold high expectations for themselves and their students.
>
> ■ Each school district must develop a complete and coherent mathematics curriculum that focuses, at every grade level, on the development of numerical, algebraic, geometric, and statistical concepts and skills that enable all students to formulate, analyze, and solve problems proficiently. Teachers at every grade level should understand how the mathematics they teach fits into the development of these strands.
>
> ■ Computational skills and number concepts are essential components of the mathematics curriculum, and a knowledge of estimation and mental computation are more important than ever. By the end of the middle grades, students should have a solid foundation in number, algebra, geometry, measurement, and statistics.

- Teachers guide the learning process in their classrooms and manage the classroom environment through a variety of instructional approaches directly tied to the mathematics content and to students' needs.

- Learning mathematics is maximized when teachers focus on mathematical thinking and reasoning. Progressively more formal reasoning and mathematical proof should be integrated into the mathematics program as a student continues in school.

- Learning mathematics is enhanced when content is placed in context and is connected to other subject areas and when students are given multiple opportunities to apply mathematics in meaningful ways as part of the learning process.

- The widespread impact of technology on nearly every aspect of our lives requires changes in the content and nature of school mathematics programs. In keeping with these changes, students should be able to use calculators and computers to investigate mathematical concepts and increase their mathematical understanding.

- Students use diverse strategies and different algorithms to solve problems, and teachers must recognize and take advantage of these alternative approaches to help students develop a better understanding of mathematics.

- The assessment of mathematical understanding must be aligned with the content taught and must incorporate multiple sources of information, including standardized tests, quizzes, observations, performance tasks, and mathematical investigations.

- The improvement of mathematics teaching and learning should be guided by ongoing research and by ongoing assessment of school mathematics programs.

Changing mathematics programs in ways that reflect these beliefs requires collaborative efforts and ongoing discussions among all the stakeholders in the process. NCTM stands ready to work with all those who care about improving mathematics education for all students. Through such dialogue and cooperative efforts, we can improve the mathematical competence of the students in mathematics classes across the continent (NCTM, 1998).

Mathematics teachers who model the doing of mathematics and who also encourage students to develop effective ways of doing mathematics have taken an important step toward quality teaching. Developing mathematical "habits of mind" is an important part of learning math (Cuoco, Goldenberg, & Mark, 1996). The math tasks that the teacher chooses, the teacher's attitude toward math, and the ways that the teacher introduces and explores the mathematics of the tasks determine, in part, the students' attitudes and interest in math, as well as their perception of what it means to do mathematics. Thus, it is important that the math teacher help students see how mathematics fits together. Students need to understand that the content and the processes learned in geometry class will be useful in precalculus. They also need to understand the underlying structure that guides all of mathematics, but that is often taken for granted. Weaving the math curriculum together in this way, by examining the connections between topics and revealing the overall structure of mathematics, is a second step toward quality mathematics teaching.

Another important step is for teachers to demonstrate math's usefulness. Students need to see how mathematics helps to explain our world. Using modeling as a basis for teaching and learning fits naturally with this demonstration. It also addresses the NCTM belief that "learning mathematics is enhanced when content is placed in context and is connected to other subject areas and when students are given multiple opportunities to apply mathematics in meaningful ways as part of the learning process" (NCTM, 1998).

We have thus suggested three major changes in teaching mathematics—modeling mathematical "habits of mind," revealing the structure of math by making connections, and modeling real-world phenomena. These changes reflect the vision of the NCTM *Principles and Standards*.

Developing a game plan for teaching requires that you build your own model for teaching math. It should parallel the models of your colleagues to some degree, as we need to bring the wide variety of approaches into a national focus and build a consensus on what is important for our students. However, that model should be one built on a solid and meaningful foundation, sound pedagogy, and a clear vision of the desired outcome. A first step might be to join the National Council of Teachers of Mathematics and your state or local council as well. Programs, meetings, and publications of these organizations will keep you informed about changes in mathematics education and give you access to the best materials for reaching your goals.

Exercises 2.5

1. Show the NCTM Statement of Beliefs to a person outside of mathematics and mathematics education. To what degree do they ascribe to the same set of beliefs for school mathematics? How do they interpret this statement?

2. How would you organize a parents' night discussion about mathematics education goals and contemporary programs for your classroom? What important issues would you highlight? What issues would you expect the parents to raise? How would you respond to their concerns?

3. What mathematical "habits of mind" have you developed? When you encounter a problem you have never seen before, what is the first thing you do? What is the second thing you do?

PROJECTS Find Out for Yourself

1. Locate a couple of recent issues of *Mathematics Teaching in the Middle School, The Mathematics Teacher*, and issues of the journal from your state mathematics teacher organization. What are the current topics of interest relative to curricular issues? Do any articles deal with new teaching methods or ideas? What information is provided for increasing student motivation?

2. Locate a recent program for your mathematics teacher organization's annual conference for your state. What topics were of greatest interest as indicated by the frequency of talks on particular topics? Did the areas of interest vary by grade level?

References

Beaton, A. E., Mullis, I. V. S., Martin, M. O., Gonzalez, E. J., Kelly, D. L., & Smith, T. A. (1996). *Mathematics achievement in the middle school years*. Chestnut Hill, MA: Boston College Center for the Study of Testing, Evaluation, and Educational Policy.

Blank, R., & Langesen, D. (1999). *State indicators of science and mathematics education:* 1999. Washington, D.C.: Council of Chief State School Officers.

Campbell, J. R., Hombo, C. M., & Mazzeo, J. (2000). *NAEP 1999 trends in academic progress: Three decades of student performance*. Washington, D.C.: National Center for Education Statistics.

Campbell, P. B., & Storo, J. N. (1994). *Girls are ... boys are ... : Myths, stereotypes and gender differences*. Washington, D.C.: Office of Educational Research and Improvement for the U.S. Department of Education.

Cuoco, A., Goldenberg, E. P., & Mark, J. (1996). Habits of mind: An organizing principle for a mathematics curriculum. *Journal of Mathematical Behavior* **15**(4):375–402.

Dossey, J. A. (1999). How should U.S. performance in school mathematics be interpreted? *In* Z. Usiskin (ed.). *Developments in school mathematics education around the world* (pp. 228–239). Reston, VA: National Council of Teachers of Mathematics.

Dossey, J. A., Mullis, I. V. S., Gorman, S., & Latham, A. (1994). *How school mathematics functions*. Washington, D.C.: Office of Educational Research and Improvement for the U.S. Department of Education.

Dossey, J. A., Mullis, I. V. S., & Jones, C. O. (1993). *Can students do mathematical problem solving?* Washington, D.C.: Office of Educational Research and Improvement for the U.S. Department of Education.

Fennema, E., & Leder, G. C. (eds.) (1990). *Mathematics and gender*. New York: Teachers College Press.

Flanders, J. (1987). How much of the content in mathematics textbooks is new? *Arithmetic Teacher* **35**(1):18–23.

Hawkins, E., Stancavage, F., & Dossey, J. (1998). *School policies and practices affecting instruction in mathematics*. Washington, D.C.: National Center for Education Statistics.

Kenney, P. A., & Silver, E. A. (eds.) (1997). *Results from the sixth mathematics assessment of the national assessment of educational progress*. Reston, VA: NCTM.

Kifer, E. (1992). Opportunities, talents, and participation. *In* L. Burstein (ed.), *The IEA study of mathematics III: Student growth and classroom processes*. Oxford, England: Pergamon Press (pp. 279–307).

Lee, V. E., & Smith, J. B. (1993). The effects of high school organization influences the equitable distribution of learning in mathematics and science. *Sociology of Education* **70**(2):164–187.

Lester, F. K., Garofalo, J., & Kroll, D. L. (1989). Self-confidence, interest, beliefs, and metacognition: Key influences on problem-solving behavior. *In* D. B. McLeod & V. M. Adams, *Affect and mathematical problem solving*. New York: Springer-Verlag (pp. 75–88).

McLeod, D. B. (1992). Research on affect in mathematics education: A reconceptualization. *In* D. A. Grouws (ed.), *Handbook of research on mathematics teaching and learning*. New York: Macmillian (pp. 575–596).

Mitchell, J., Hawkins, E., Jakwerth, P., Stancavage, F., & Dossey, J. (1999). *Student work and teacher practices in mathematics*. Washington, D.C.: National Center for Education Statistics.

National Council of Teachers of Mathematics (1989). *Curriculum and evaluation standards for school mathematics*. Reston, VA: NCTM.

——— (1998). NCTM statement of beliefs [online text]. Available: http//:www.nctm.org/about/nctm.beliefs.html.

——— (2000). *Principles and standards for school mathematics*. Reston, VA: NCTM.

Organization of Economic Cooperation and Development (1995). *Literacy, economy, and society: Results of the first international adult literacy survey*. Paris, France: OECD.

Reese, C. M., Miller, K. E., Mazzeo, J., & Dossey, J. (1997). *NAEP 1996 mathematics report card for the nation and the states*. Washington, D.C.: National Center for Education Statistics.

Reyes, L. H. (1980). Attitudes and mathematics. *In* M. M. Lindquist (ed.) *Selected issues in mathematics education*. Evanston, IL: National Society for the Study of Education (pp. 161–184).

Robitaille, D. F. (1997). *National contexts for mathematics and science education*. Vancouver, Canada: Pacific Educational Press.

Schmidt, W. H., McKnight, C. C., & Raizen, S. A. (1997). *A splintered vision: An investigation of U.S. science and mathematics education*. Dordrecht, The Netherlands: Kluwer Academic Publishers.

Weiss, I. R. (1994). *A profile of science and mathematics education in the United States: 1993*. Chapel Hill, NC: Horizon Research.

Wilson, L. D., & Blank, R. (1999). *Improving mathematics education using results from NAEP and TIMSS*. Washington, D.C.: Council of Chief State School Officers.

3 Doing Mathematics: Living the *Standards*

"Two of the major reasons why mathematics...has rapidly been losing student interest is that we have stripped the "fun" out of doing mathematics and we have furthermore neglected to convince students in a meaningful way that mathematics is a contemporary, dynamic and currently relevant subject."

William F. Lucas

Overview

What are the NCTM *Standards* and why should you study them? In the *Principles and Standards for School Mathematics* (NCTM, 2000), the NCTM discusses the state of mathematics education and paints a picture of what it can be if teachers and administrators strive to achieve quality programs for all students. The NCTM *Standards* spell out the many reasons for changing existing mathematics programs. As our society continues to change, the education community must also change in order to meet emerging educational needs. It is crucial that all members of society be given the opportunity to become proficient in mathematics. As modern life becomes ever more complex, citizens must become flexible thinkers to solve the problems that arise or to stay abreast of their profession. For thcsc rcasons and more, it is important to introduce current and future math teachers to the NCTM *Standards* and their implications in the classroom.

This chapter surveys the *Principles and Standards* from NCTM's most recent document (NCTM, 2000). Later chapters examine the *Standards* in detail and their relation to the math classroom, the current research on teaching and learning mathematics, the teacher's role in implementing the *Standards*, and the students' roles in doing and learning mathematics. Unit and lesson planning to incorporate the *Standards* is taken up in Chapters 4–8 and Chapter 12. At this point, we share the statement of the Principles and Standards, and a sense of the NCTM vision for mathematics education, and also make suggestions for putting the principles into practice. We also begin to involve you more deeply in the modeling approach to doing math. We emphasize this approach here and throughout the book because it incorporates the spirit of the *Standards* and gives you ways to teach core mathematics topics in realistic contexts.

Focus on the NCTM *Standards*

The NCTM *Standards* enable educators to establish, evaluate, and improve instruction at all levels. As such, they have become the basic framework of almost all state and local descriptions of what students should know and be able to do mathematically. At the national level, the *Standards* define the curriculum boundaries laid out by the National Assessment of Educational Progress and many of the state math assessments. The *Standards'* emphasis on authentic assessment has prompted the move to student-constructed response formats on examinations. In many states, the *Standards'* recommendations for subject matter–specific teacher knowledge has affected teacher certification regulations and the licensing examinations. Because of these, and other reasons, all teachers and teachers-to-be should be aware of what the NCTM *Standards* say and mean in terms of math education practices.

Focus on the Classroom

The NCTM *Principles and Standards* describe new goals for classroom teachers and students. The NCTM urges, for example, that teachers and students work together so that students learn to value mathematics, become confident in their math skills, become problem solvers, and learn to communicate and reason mathematically. Math teachers and math students should strive to create the classroom environment in which this learning can take place. The NCTM also provides a vision for this ideal environment. Although you must interpret the *Principles and Standards* for yourself, our goal in this chapter is to help you better understand the *Standards* and how they can be incorporated into daily activities in your classroom.

3.1 Principles That Guide the Teaching of Mathematics: More Than a Wish List

No two school districts or mathematics departments will have identical programs of mathematics instruction. Even so, educators can at least agree on the basic characteristics and principles on which quality math programs are built. As noted at the beginning of Chapter 1, the NCTM defines six principles to guide program developers and decision makers. These principles, namely, Equity, Mathematics, Teaching, Learning, Assessment, and Technology, support the core of quality math programs, especially those designed for classrooms of the 21st century.

Although these principles may seem obvious, do take the time to explore their recommendations, their underlying assumptions, and the perspectives they provide so that you can decide which ideas will work best in your classroom.

The Equity principle conveys the belief that "excellence in mathematics education requires equally high expectations and strong support for all students" (NCTM, 2000, p. 11). In a sense, this principle supports all the other principles. The Mathematics Curriculum principle focuses on mathematics content, and states that mathematics programs should teach key topics through meaningful yet comprehensive curricula. The Teaching and Learning principles address the idea that teachers are the most important component of a quality mathematics program. The Assessment principle describes how assessment accompanies quality math instruction. The last principle, Technology, supports the use of technology to enhance the learning of *all* students and to better prepare students for our high-tech world.

These principles are intended to guide decision makers at all levels of education, from teachers to district superintendents.

The Equity Principle

The Equity principle suggests that teachers should *expect* all students to be able to do mathematics. Such a belief is a major shift in thinking, a belief that has been central to the NCTM's *Standards* since the late 1980s (NCTM 1989, 1991), and is crucial to improving mathematics education. This expectation challenges *all* students to be successful in mathematics. If appropriately challenged, many more will be motivated to master mathematics.

The Equity principle implies much more than that teachers change their expectations. It suggests that teachers change how they do and teach mathematics. Diverse students have diverse needs. It is crucial that teachers be aware of these differences and that they address the needs of every student. Not always an easy task, several recent projects have focused on providing opportunities for all students, particularly those groups who haven't been successful in the past. The QUASAR project, for one, addressed the needs of mathematics students from economically disadvantaged areas (Silver, Smith, & Nelson, 1995). Through the QUASAR project, teachers from urban and rural areas collaborated with university professors and people from local businesses to find new ways to involve students.

The goal of the Algebra Project, started by Robert Moses in 1982, is to assist typically underserved students in acquiring the concepts and skills necessary to succeed in math-related high school courses. Moses's model includes rich concrete experiences that actively involve students in making sense of the world through mathematics. This model is now used in many rural and urban schools across the country. Teachers in places such as Jackson, Mississippi, Chicago, Illinois, and San Francisco, California, have restructured their middle school mathematics programs using the concepts and philosophies of the Algebra Project (Dossey & McCrone, 1999). The Algebra Project and QUASAR are just two examples of recent efforts to make mathematics learning equitable. On a smaller scale, we can better serve all students through how we structure and teach mathematics.

The Mathematics Curriculum Principle

To achieve equity, teachers and schools must assure that the mathematics curriculum is appropriate and meaningful; this is the Mathematics Curriculum principle. Teachers must carefully decide what content, processes, and related classroom experiences will help students grow in their ability to do math. Teachers must also have a deep understanding of the mathematics in order to assess the quality, coherence, and comprehensiveness of a chosen curriculum. For instance, teachers must know about students' prior experiences and knowledge in order to effectively design a curriculum that continues to build skill and understanding. Teachers must be aware of typical student difficulties and misunderstandings so they can present content appropriately and avoid misconceptions.

The NCTM highlights three aspects of a comprehensive mathematics curriculum that can guide teachers' decisions and the planning process. The first two aspects are the actual mathematics content and the mathematical processes that make up the curriculum. The content and process standards outlined in Sections 3.2 and 3.5 provide a fairly comprehensive core curriculum. Note that the five content areas in the *Principles and Standards* should not be taught in isolation. Students need to view mathematics from a broad perspective that demonstrates the interconnectedness of its many branches. The processes described in Section 3.2 can develop a more coherent view of the doing and learning of mathematics. A third aspect of a comprehensive curriculum is the balance of conceptual knowledge and skill development. Students must understand how to use the mathematics they encounter in school and in the world, but they will apply math better if they have a deeper, conceptual understanding. Current research suggests that a problem-solving approach to teaching and learning mathematics helps students develop both conceptual understanding and skill proficiency (Lester, 1994). A comprehensive curriculum provides opportunities for students to do mathematics and to understand what it means to do mathematics. That is, students need to experience mathematics by actively participating in the lesson, through not only doing but also through discussion. Applying math to real-life experiences enhances learning. The modeling focus of this book shows how a curriculum can provide such real-world connections.

The Teaching and Assessment Principles

The Teaching and Assessment principles lay out avenues for achieving equity in the mathematics classroom. The Teaching principle emphasizes the teacher's responsibility in making appropriate pedagogical choices that provide learning opportunities for all students. The principle emphasizes the fact that students' perceptions of math, their self-confidence in doing mathematics, and their dispositions toward mathematics are shaped by their teachers. Thus, teachers must continually monitor the effects their choices have on students. In order to make appropriate choices, math teachers at all levels must have a strong understanding of mathematics, of how students learn mathematics, and a commitment to continue to learn themselves.

An earlier document produced by the NCTM, the *Professional Standards for Teaching Mathematics*, provides specific details about the teacher's role in quality programs and their responsibility for making appropriate pedagogical choices (NCTM 1991). The *Professional Standards* outlines four essential components of teaching mathematics: (1) analysis and reflection, (2) worthwhile student tasks, (3) classroom discourse, and (4) the role of the teacher.

To make informed decisions in the classroom, teachers must continuously assess the classroom situation and the students' learning. Are you attaining your goals? Is your classroom environment conducive to learning? Is it a safe environment for all students? Do you know when your students are ready to move on to a new set of ideas? Ongoing assessment such as this should be a routine task. The Assessment principle emphasizes other important responsibilities of teachers to plan, develop, and effectively use assessment to promote student learning. Assessment should be viewed as a process through which evidence is gathered, decisions are made, plans are carried out, and more evidence is gathered. Evidence from assessments should come from various sources in order to get a more complete picture of student abilities. This information should then inform future planning. Assessment results can uncover student difficulties or guide timing for a particular activity. Assessment that compares students to each other is not useful and only exacerbates already existing inequities. Chapters 12–15 further elaborate on the teacher's role and the role of assessment in student learning and in the teaching of mathematics.

A second component of mathematics teaching is the ability to locate or construct worthwhile tasks for students. This process is discussed in more detail in Chapter 12. Classroom discourse is vital to promoting student learning, for it is through discourse that students learn to justify or validate statements, to see a concept from multiple perspectives, and to take responsibility for their own understanding (Yackel & Cobb, 1996; Corwin & Storeygard, 1994). The teacher's role in establishing the learning environment is the final component. Every teacher in every classroom establishes the classroom environment, but not all environments are conducive to learning mathematics. Teachers need to establish a relationship with their students in order to foster inquiry, to clarify student responsibilities, and to develop a community of learners.

The Learning Principle

Through the Learning principle, NCTM encourages providing opportunities for students to make sense of mathematics, to become actively engaged in doing mathematics, and to understand how and when to use mathematics. In helping students make sense of mathematics, current research suggests that teachers build from students' prior knowledge (Hiebert et al., 1997). With an understanding of students' prior knowledge, the teacher is better able to help students construct a solid conceptual understanding. For example, in learning algorithms, such as multiplying fractions or factoring polynomials, students need opportunities to develop proficiency in using algorithms and in making sense of them. Questions such as "Why does the algorithm work?" and "When should it be used?" should be encouraged and answered carefully. Then ask the students these same questions. Such discussions will help students choose appropriate algorithms and judge the reasonableness of their calculations.

Genuine problem solving (as opposed to mere exercises) actively engages students. Problems should require students to think about how to approach the problem and to solve it creatively. Research on problem solving indicates that when students do mathematics they're more likely to develop positive attitudes toward it, to develop creative solution strategies, and to build a deep understanding of the underlying concepts (Lester, 1994, Schoenfeld, 1992).

The Technology Principle

Using technology actively involves students in doing mathematics. Scientific calculators, graphing calculators, and a myriad of computer programs offer many opportunities for engaging with mathematics in useful ways. Always remember, however, that technological proficiency or lack of proficiency should not be a determining factor in learning mathematics.

The NCTM's Technology principle upholds that tools such as graphing calculators, dynamic computer geometry programs, or statistical programs should be used to enhance learning. For instance, a graphing calculator can help students interpret a graphical representation of a data set. Other software packages allow students to investigate ideas based on drawings they cannot represent or draw on their own.

Although technology can deepen understanding of certain concepts, it can also frustrate some students if they are not technologically literate. Nonetheless, it is important to keep in mind that not all students have access to such tools. This issue may require creative solutions such as business partnering. The NCTM vision of enhancing mathematics education through technology is a worthwhile goal and key to preparing students for their futures.

The six principles represent achievable goals for instituting a quality mathematics program. The principles are thus useful guides for decision makers at the classroom, school, and district levels. Fulfilling all of the six principles is a big job. But quality programs can only be achieved if classroom teachers make it happen.

Exercises 3.1

1. List three or four ways in which you can address the Equity principle in your classroom. Refer to the *Principles and Standards*, pages 12–14.
2. In your estimation, which principles do teachers most need to attend to at the classroom level? At the school level? Explain.
3. How might you reorder the six principles to reflect your responses to Exercise 2?

3.2 Doing Mathematics

What does it mean to "do mathematics"? What processes do we use as we do and learn mathematics? What abilities do we want our students to develop as they progress through middle school and high school? In response to these important questions, the NCTM elaborates five mathematical processes in its *Principles and Standards for School Mathematics*: Problem Solving, Reasoning and Proof, Communication, Connections, and Representations. These process standards are linked to the content standards described in Section 3.5. Problem solving, of course, is a key process in doing mathematics. Reasoning and proof represents another way we do and make sense of math. Being able to effectively communicate about mathematics—sharing ideas, listening carefully, and comprehending information—is essential to learning it. Student understanding is also greatly enhanced when they make connections within math problems and to the world around them. Mathematical representations communicate ideas and concepts and provide a framework for justification.

We have made a case, above, for the importance of each process described by the NCTM *Principles and Standards*. Next we elaborate on these processes and present ideas for encouraging and motivating high school math students.

Mathematics as Problem Solving

Mathematics as a problem-solving activity is a pretty obvious idea. When we think about doing mathematics, one of the first images that comes to mind is a person working to solve a problem, to find an answer to some question. Indeed, problem solving is at the very heart of "doing mathematics." Many kinds of mathematics problems exist, and the type of problem-solving strategy used depends on the skills of the person solving the problem.

For instance, if asked to find the least common multiple of 9 and 24, a sixth grader might list the first 10 multiples of each number and then look for the smallest match. A student in eleventh grade might find the factors of the two numbers; in fact, this might be a routine computation for the eleventh grader. A college student might divide the product of $9 \cdot 24$ by the greatest common divisor of the two numbers, 3, to get the quotient of 72.

The college student applied a theorem to find the solution. In symbols, this theorem is written as

$$\gcd(a, b) \cdot \text{lcm}\, ab = ab$$

To compute the least common multiple of 9 and 24 using this theorem, we get

$$\text{lcm}(a, b) = \frac{ab}{\gcd(a, b)}$$

Problem solving, then, is the process by which we answer the question or deal with the situation. A difficult and tedious problem for one person may be a routine and quick computation for another. The process of problem solving thus involves using prior knowledge in new or different ways, formulating a plan or strategy to reach the desired goal, and possibly acquiring new knowledge about the given situation. This process is complex and often requires creative connecting. As such, students need many opportunities to experience the process, and to develop various problem-solving strategies. The NCTM *Standards* describe four goals related to developing problem-solving abilities. We list them here.

Problem-Solving Standard

Instructional programs from prekindergarten through grade 12 should enable students to

- build new mathematical knowledge through problem solving;
- solve problems that arise in mathematics and in other contexts;
- apply and adapt a variety of appropriate strategies to solve problems;
- monitor and reflect on the process of mathematical problem solving. (NCTM, 2000)

To achieve these goals, problem solving must be a regular part of the curriculum at all grade levels. In his classic book *How to Solve It*, George Polya (1945) offers a model for problem solving (see Figure 3.1). It is a big-picture, flexible model and does not detail specific strategies. It is up to math teachers to guide students through the specifics of the problem-solving process (Schoenfeld, 1992).

Students must be encouraged to recall and use information and strategies previously taught. They must also learn (through instruction and practice) to construct and implement new strategies, and to think flexibly or creatively when initial attempts don't prove successful. Some commonly taught problem-solving strategies include working backwards, trying a simpler but similar problem, diagramming the situation, and looking for patterns. These and other strategies can help students develop a positive attitude toward problem solving.

As noted in the NCTM Problem Solving standard, problem solving builds new mathematical knowledge. Carefully chosen problems help students solidify previously learned concepts and procedures and challenge them to investigate unfamiliar ideas

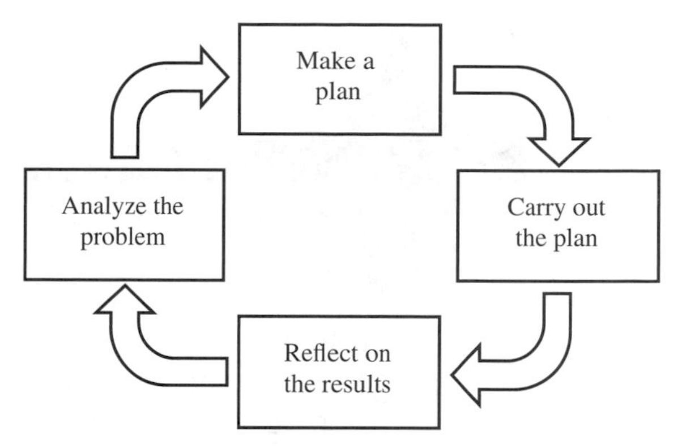

Figure 3.1 Polya's problem-solving framework.

and concepts. For example, procedures for solving systems of equations are seen as useful and less abstract when they are required to solve real-world problems. At the same time, students also learn new concepts within the problem's context. The problem situation shown in Figure 3.2 introduces students to calculating means, distances, and optimization.

As students develop multiple problem-solving strategies and find multiple ways to represent problems, they begin to make connections between the various representations and further enhance their repertoire for doing mathematics and their mathematical habits of mind as well.

All branches of mathematics involve problem solving, and hence students must encounter problem-solving situations throughout middle school and high school. This does not mean that students should be engaged in problem solving during every class hour of the school year. Rather, teachers should choose problem-solving situations that arise naturally from the study of key concepts. Often, the best and most realistic problem-solving situations will require time for students to investigate the situation and develop a solution plan or strategy. The modeling approach suggested in this book and in the textbook series *Mathematics: Modeling Our World* (COMAP, 1998) provides many wonderful examples of math problems that enhance learning by actively engaging students in doing mathematics and constructing meaning for the concepts involved.

Mathematics as Reasoning and Proof

Mathematical reasoning and proof are activities that should not be limited to high school geometry classes. Reasoning and justification should be a part of learning at all grade levels for all areas of mathematics. Making sense of the world around us is empowering, and mathematical reasoning is a key part of this process. This is also true in the classroom. Students want to make sense of the mathematics being taught. Hence, reasoning should be an explicit component of any mathematics curriculum.

During their middle school years, students need explicit demonstrations on how the mathematics that they are learning helps them make sense of the world around them.

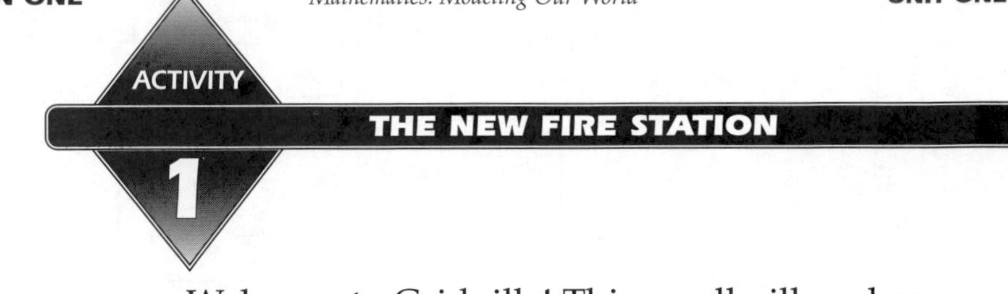

Welcome to Gridville! This small village has grown in the past year. The people of Gridville have agreed they now need to build a fire station. What is the best location for the fire station?

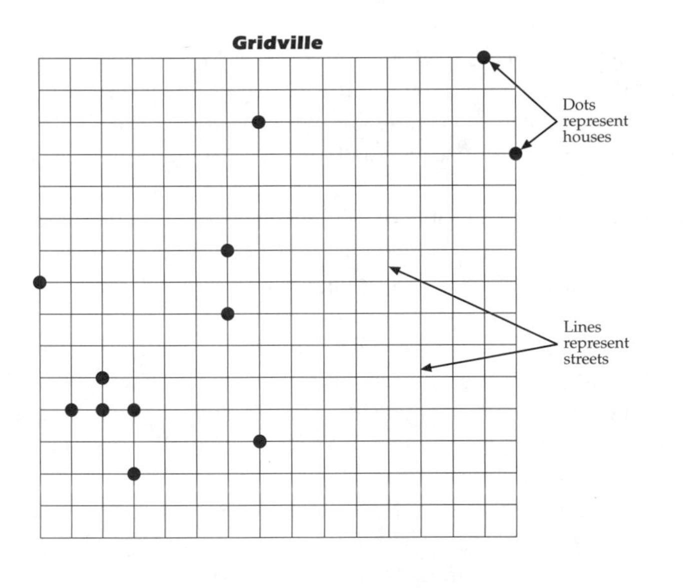

Figure 1.2. Map of Gridville.

All towns and cities are different. Gridville is a simplified version of a small village with streets running parallel and perpendicular to one another. It represents the start of the modeling process. What you learn in Gridville may be useful in other settings that require finding a best location. The streets of Gridville are two-way streets represented by the lines of the grid. Houses are represented by points and are located at the intersections of grid lines in order to identify their locations easily. Firetrucks may only travel along grid lines. Diagonal movement is not allowed. It is simpler than a real town in its size and in its layout (geometry), but it still has houses and roads, and it needs a fire station.

THE TASK:

The map of Gridville (**Figure 1.2**) shows the location of all houses. Determine the best location for Gridville's fire station, and write a persuasive argument defending your choice. The city leaders will follow the advice of the group that delivers the most convincing argument.

Figure 3.2 Sample real-world problem statement for high school students. (COMAP, 1998, p. 6–7)

ACTIVITY 1

THE NEW FIRE STATION

GUIDELINES:

1. Begin your argument by answering the questions, "What are important factors to consider in deciding the best location?" and "What does 'best' mean?"

2. You are encouraged to use charts, diagrams, tables, graphs, equations, calculations, and logical reasoning in making your decision.

3. Clearly state your choice of best location. Your written summary should include the arguments and mathematics that support your decision. The summary should also explain how your charts, diagrams, tables, graphs, equations, calculations, and logical reasoning relate to the factors you considered and led your group to your choice.

4. Your written presentation may be posted in a display area in the classroom.

5. In addition to the written presentation, your group will give an oral presentation of approximately two minutes. The oral presentation should summarize your arguments and explain the reasons for your decision.

Where do you build a new high school when the only high school in town becomes overcrowded? A class of students at Redlands High School in Redlands, California, answered this question and won $10,000. In 1993, the 25 students of teacher Donna St. George and teacher Judy Kanjo won second place in the American Express Geography Competition sponsored by American Express. The students conducted surveys of local residents and gathered information provided by local companies, local organizations, and Redlands Unified School District officials. They analyzed several sites proposed by the school district. Their 80-page report addressed major factors including traffic patterns, busing patterns, soil conditions, air traffic patterns, availability of public utilities, anticipated population growth, and potential for flood damage.

CONSIDER:

Answer the following questions based on the classroom presentations for Activity 1, *The New Fire Station*.

1. Which criteria were used most often to determine the location of the fire station?

2. Which factors invite mathematical investigation?

3. Even a model such as Gridville can be simplified further in order to study the essential elements of the location problem in detail. What are some ways you can simplify the Gridville model to investigate distance relationships?

Figure 3.2 *(Continued)*

This occurs when students are given opportunities to explore mathematics on their own, and are expected to verify the results of their explorations—a first step to developing a mathematical argument (Yackel, 1998). By the time students enter high school, they are able to justify conjectures by presenting arguments verbally or in writing. At the high school level, students should be able to justify a wide variety of mathematical statements and situations. Traditional two-column geometry proofs are a classic way to record or justify mathematical statements, but students should be encouraged to use many other strategies or methods as well.

The NCTM *Standards* emphasize four important components of mathematical reasoning and proof for all grade levels, as shown below.

Reasoning and Proof Standard

Instructional programs from prekindergarten through grade 12 should enable students to

- recognize reasoning and proof as fundamental aspects of mathematics;
- make and investigate mathematical conjectures;
- develop and evaluate mathematical arguments and proofs;
- select and use various types of reasoning and methods of proof. (NCTM, 2000)

The four components of reasoning and proof highlighted here are not merely four concepts to be taught or memorized. Rather, they emphasize four aspects of the reasoning *process* that are integrally linked. The connections suggested in Figure 3.3 show

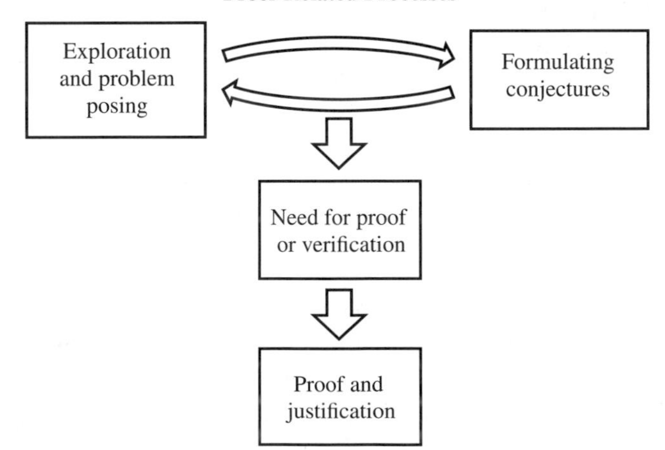

Figure 3.3 The cyclic process of developing a mathematical argument.

a somewhat cyclic process that can lead to the development of a valid mathematical argument (Martin & McCrone, 1999). One instructional strategy, then, is to recognize these various aspects of the reasoning process and choose or create activities that allow students to experience each aspect. The following Problem to Investigate encourages reasoning and justification in the exploration process. This problem is suitable for middle school or high school students.

Problem to Investigate: Overlapping Squares

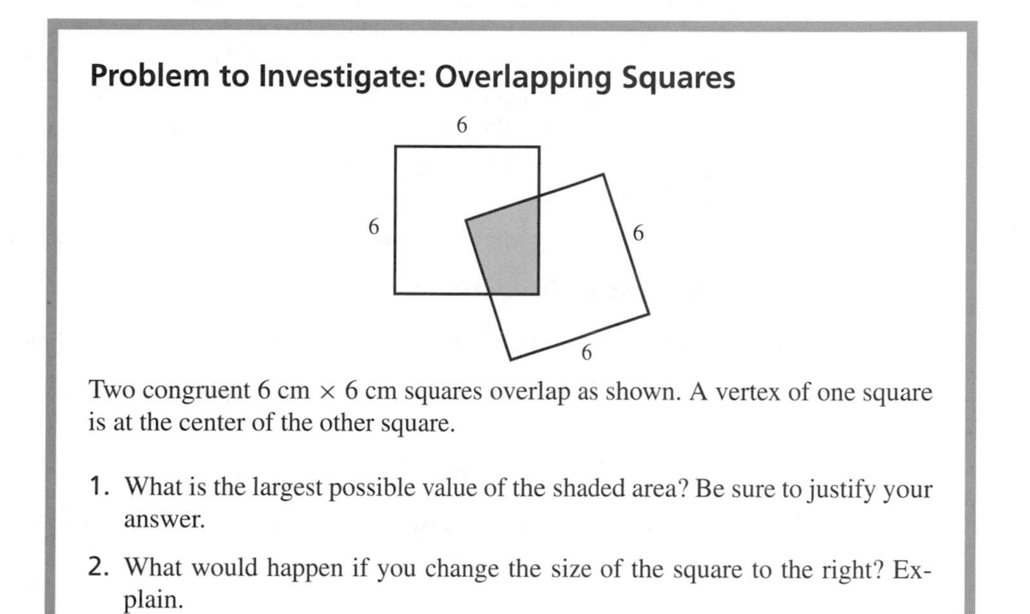

Two congruent 6 cm × 6 cm squares overlap as shown. A vertex of one square is at the center of the other square.

1. What is the largest possible value of the shaded area? Be sure to justify your answer.
2. What would happen if you change the size of the square to the right? Explain.

(Boyd et al., 1998, p. 533)

Solving the overlapping squares problem requires students to participate in a variety of activities that involve mathematical reasoning, such as investigating, conjecturing, problem solving, and justifying solutions (whether informally or through a more formal proof), as well as spatial reasoning, and inductive reasoning. Spatial reasoning, which entails manipulating objects mentally, is frequently overlooked in the high school math class. Yet, spatial reasoning is an important part of the mathematical tool kit.

Inductive reasoning, or generalizing from a series of examples, is often mistaken for proving a conjecture to be true. Rather, inductive reasoning relies on collecting examples or discerning patterns. This process does not verify or prove the validity of the statement. In the technology-driven classroom, teachers need to inform students about the strengths of inductive reasoning. In many geometry software packages, for example, users investigate classes of objects; this can lead to drawing conclusions that students might not see through other avenues. However, the ease with which the technology can lead students to conjectures often misleads them into thinking they have proved the conjecture. Thus, it is important for teachers to help students distinguish between inductive and deductive reasoning.

As with problem solving, strategies for justifying or proving statements abound. Some of these same strategies, such as working backwards, are helpful when first developing a formal argument. Techniques such as indirect proof or proof by contradiction should also be taught to students. In addition to developing strategies for justifying conjectures or solutions, students need experiences that give them a deeper sense of mathematical reasoning. For instance, students need to recognize what constitutes a valid argument (not just be able to write proofs). They also need to understand that reasoning and proof are methods of communicating mathematical ideas to others. Indeed, proofs are often useful for explaining mathematical concepts. After students have explored and developed conjectures, they should be given opportunities to explain or convince themselves and their peers of the statements' validity through communication and reasoning. Informal reasoning is usually a good starting place for developing more formal reasoning skills.

Problems to Investigate

1. How would you convince a classmate that $\sqrt{2}$ is an irrational number?
2. How would you develop an argument or a proof to help students in an advanced algebra class understand that $\sqrt{2}$ is irrational?

Investigating the irrationality of $\sqrt{2}$ is a good demonstration of mathematical reasoning. Let's show that $\sqrt{2}$ cannot be a rational number. This is an indirect proof.

Suppose $\sqrt{2}$ is a rational number. Then $2 = a^2/b^2$ where a and b are integers and $b \neq 0$. Hence,

$$2 = \frac{a^2}{b^2} \qquad \text{or} \qquad 2b^2 = a^2$$

Let $a_1a_2a_3 \ldots a_n$ be a prime factorization for a, and $b_1b_2b_3 \ldots b_m$ be a prime factorization for b. Our equation then becomes

$$2b_1^2b_2^2b_3^2 \ldots b_m^2 = a_1^2a_2^2a_3^2 \ldots a_n^2$$

For the left and right sides to be equal, we must have the same number of each prime factor on each side. Since each prime factor of a or b is squared, there will be an even number of each prime factor. Hence if any 2's exist in the prime factorization on the left side, there will be an odd number of them, but if any 2's exist on the right side, there will be an even number of them. This implies that the right and left sides of the equation are not equal, a contradiction that leads us back to our assumption that $\sqrt{2}$ is rational. That is, $\sqrt{2}$ cannot be rational because this assumption leads to a contradiction. Hence, $\sqrt{2}$ is irrational.

The proof above is rich in mathematical logic and reasoning. We must reason in general terms when we consider whether the values are even or odd. We also use symbolic manipulations in the argument to demonstrate the relationships between variables. And finally, we use the indirect proof to show the contradiction.

To foster this kind of mathematical reasoning, teachers need to cultivate a community of inquiry in their classroom. Students should strive to make sense of the mathematics they encounter, and seek to communicate this knowledge to others. When this is the case, students are more likely to acquire usable mathematical knowledge.

Mathematics as Communication

Mathematics is a language often spoken in symbols. Words also describe relationships between objects (two triangles are *similar*), classes of objects (even integers), and mathematical ideas (space, number). Mathematical terms are carefully defined, and symbols are used to interpret mathematical statements. Symbols are thus a powerful way of communicating mathematics. Symbols describe relationships or actions in mathematics (e.g., $3y - 4 < 5 + x$). A person who understands the language of mathematics has gone a long way toward understanding mathematics.

Students, however, often get bogged down in the symbols and the language of mathematics. Everyday terms take on new and subtle meanings in the math classroom. And the multitude of symbols, signs, and terms can be a lot for students to keep straight. Even so, the mathematical language that we take for granted, but that distracts many students, is crucial to student success.

Symbols and mathematical terms communicate ideas in shorthand. Nonetheless, there are many convenient methods for sharing mathematical ideas: graphs, charts, pictures, and technology. Technology offers a range of options, many of which are discussed in the following Representations section.

Graphs and charts represent data and other mathematical relationships visually. Newspapers, magazines, and many other publications use graphs and charts to communicate ideas or to describe trends in data. Charts and graphs make particular points. The mode of representations should be chosen carefully. The same data displayed differently communicates differently. For example, Figure 3.4a uses barrels to indicate bushels of barley harvested. Larger barrels show the increase in bushels harvested, but it is not easy

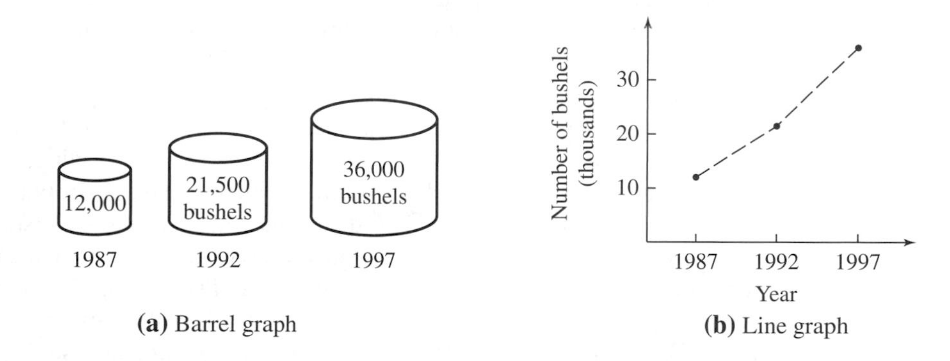

Figure 3.4 Two visual representations communicate different views of the increase of barley production in Cass County, Minnesota. (*Data source:* http://govinfo.library.orst.edu/cgi-bin/ag-state?Minnesota)

to see that the number of bushels has tripled between 1987 and 1997. The line graph in Figure 3.4b, on the other hand, shows the sharp increase clearly.

Students must thus understand how to interpret what they see in order to make appropriate judgments about data. This segues neatly into the next NCTM standard on Communication. This standard emphasizes the importance of learning to communicate mathematically, as well as the role of communication in building understanding.

Communication Standard

Instructional programs from prekindergarten through grade 12 should enable students to

- organize and consolidate their mathematical thinking through communication;
- communicate their mathematical thinking coherently and clearly to peers, teachers, and others;
- analyze and evaluate the mathematical thinking and strategies of others;
- use the language of mathematics to express mathematical ideas precisely. (NCTM, 2000)

When students work together to solve problems, and when they share their reasoning with classmates, they develop communication skills such as those outlined in the Communications standard. Teachers provide these opportunities and thus must continually look for ways to enhance and strengthen student outcomes. Through questioning, teachers help students learn to verbalize their actions and processes. Questioning encourages discussion; it lets students know they are responsible for sharing their thinking with others. With time, students grow more comfortable holding mathematical discussions, challenging each other's ideas, and working to interpret and incorporate another's suggestions. Developing communication proficiency takes time and guidance. Students need to work on communicating mathematical ideas throughout their schooling, beginning in the earliest grades.

Communicating in math class is not simply about learning the symbols and the terminology in order to respond appropriately to the teacher's questions. Communication skills in mathematics, and indeed in any discipline, involve reading, writing, speaking, listening, and modeling, as well as pictorial, symbolic, and possibly tabular representations. Students must learn to read written mathematics, and to understand what they are reading in the textbook, on the quiz, or in the newspaper. To communicate mathematical ideas to others, students must learn speaking and listening skills. Communication is a two-way process, so listening and comprehending is just as important as speaking. Finally, modeling helps students generalize or simplify situations so they can share them with others.

How do we teach students to be readers of mathematics, to listen to mathematical discussions, or to hone their speaking and presentation skills? What activities are appropriate for the classroom that will also help students learn content? A problem-solving or modeling environment is a great place to start for developing communication skills. Such an environment naturally lends itself to working together, sharing ideas, and negotiating. At times, the teacher may need to model sharing and negotiating, or to encourage such discussions. As with anything else, students need help as they learn to work cooperatively. A classroom in which justification and the sharing of reasoning are expected fosters good communication skills. Again, teachers hone student listening skills and help them learn when and how to share ideas.

Taken together, the first three process standards, Problem Solving, Reasoning and Proof, and Communication, describe a classroom environment that promotes the learning of mathematics for all students. As students find ways to express their ideas, share them with others, and make sense of the ideas of others, their understanding of mathematical ideas will grow and strengthen. Mathematicians continually look for new ways to express ideas that were inaccessible in the past. Making sense of others' ideas is one way that enables the discipline of mathematics to continue to explore new frontiers.

Mathematics as Connections

Education researchers claim that mathematical understanding builds from current understanding or prior knowledge. In other words, we learn mathematics by making connections between new ideas and those ideas that are already a part of what we know. From this perspective, the Connections standard is a key element in the teaching and learning of mathematics. The NCTM outlines three types of connections that build mathematical knowledge: (1) connections among various mathematical concepts and ideas; (2) connections within and among various areas that demonstrate the overall coherence of the subject; and (3) connections between mathematics and other contexts, be they academic subjects or real-world situations. Each type of connection builds student understanding, and should be included in the teaching of mathematics at all levels.

In the modeling sections that follow, we look at connections and explore ways of introducing students to the depth and breadth of these.

Connections Standard

Instructional programs from prekindergarten through grade 12 should enable students to

- recognize and use connections among mathematical ideas;
- understand how mathematical ideas interconnect and build on one another to produce a coherent whole;
- recognize and apply mathematics in contexts outside of mathematics. (NCTM, 2000)

A student who can connect various mathematical concepts is more likely to develop a deeper mathematical understanding. Such connections give students greater flexibility in problem solving and in constructing mathematical arguments. Besides being a tool for the problem-solving kit, connecting key concepts or areas of mathematics helps us view mathematics as an integrated whole. We often divide mathematics into numerous subcategories. Students take Pre-algebra, then Algebra, then Geometry, and so on, and frequently view them as unrelated subjects. Although earlier courses lay the foundation for understanding advanced mathematics, the separation of topics is somewhat artificial. This separation can discourage students from making connections that promote big-picture understanding. Emphasizing connections helps students make sense of their experiences and builds a stronger foundation for continued learning.

The distributive property is a perfect example of a mathematical concept that connects many areas of mathematics. The distributive property plays a vital role in such basic arithmetic as whole-number multiplication. Often multiplication with two- and three-digit numbers such as 8×172 is simplified when we recognize that the distributive property allows us to compute $(8 \times 100) + (8 \times 70) + (8 \times 2)$. One version of the distributive property helps students find the area of a figure by dividing it into simpler shapes and adding the area of each piece (see Figure 3.5). The distributive property not only enables us to factor and expand algebraic expressions, it also allows us to work with scalar multiples of vectors and matrices in more advanced mathematics classes. Teachers who help students connect these situations are helping them develop big-picture understanding.

Students need to see and begin to make connections early in the elementary grades. At times, it is helpful to point out connections explicitly, like the teacher who reminds her students about coordinate geometry as they struggle to justify that the diagonals of a parallelogram bisect each other. Students who are having trouble with a synthetic approach might quickly recognize that finding the midpoints of the two diagonals (see Figure 3.6) helps verify the claim.

Teachers can guide students in the connecting process through the tasks they choose. Carefully designed tasks demonstrate connections with previously learned con-

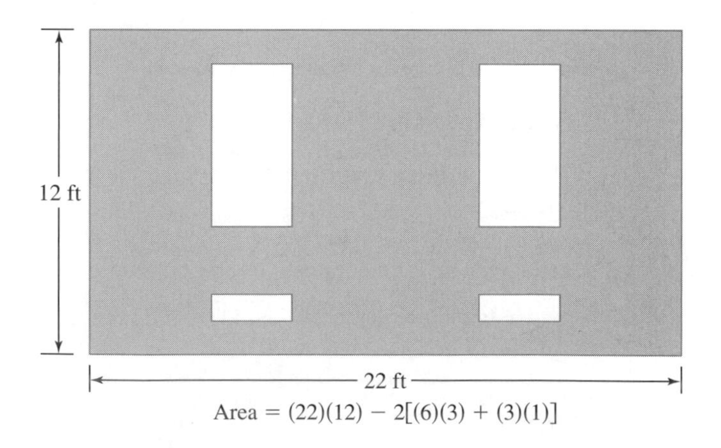

Figure 3.5 Use of distributive property to find area of shaded region.

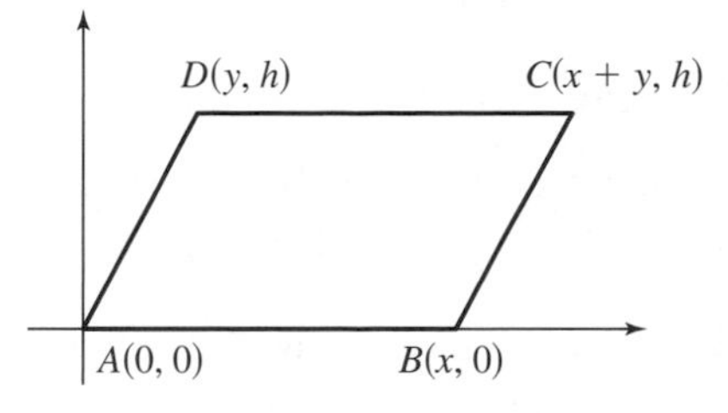

Figure 3.6 Analytic approach to proving that the diagonals of a parallelogram bisect each other.

cepts. Explorations of the Fibonacci sequence (1, 1, 2, 3, 5, 8, 13, . . .) can be used to connect pattern finding and naming recursive functions. This exploration might lead to work with the golden ratio, the golden rectangle, and to its implications for art and the world of design. Connections to other disciplines and to the world around us are rich with possibilities for enhancing the learning of mathematics.

As noted earlier, the Curriculum standard emphasizes that instructional programs should help students build mathematical ideas across grades. As students work to make sense of the mathematics they experience in school and elsewhere, they connect new ideas to concepts and ideas that are already a part of their mathematical and nonmathematical repertoire. If students are unable to make these connections, their mathematical understanding will be severely jeopardized. Thus, teachers must build a curriculum and maintain an environment that provides opportunities for making connections.

Mathematics as Representations

We use symbols to represent and understand mathematical ideas. Hence, mathematical literacy is crucial to student success. Many representational forms describe mathematical situations and ideas. The importance of various forms were discussed in the Communication and Connections sections above. Here we note that representations are useful tools for learning and doing mathematics, as well as for communicating and making connections. Representations extend a student's understanding of a concept, and shed light on an idea not fully understood in another form. Thus, students' ability to develop and interpret various representations increases their ability to do and understand mathematics.

The NCTM standard on Representation emphasizes that students should routinely create, interpret, and use representations in a variety of mathematical settings. By the time students are in high school, they should have a wide variety of representations available to them, and they should be flexible in using the various modes. Note that a representation is an object that describes or models a situation. The process of representing, as described in the Representation standard, is just as important as the product or object. This process is the act of capturing a mathematical concept or relationship in some form that conveys an idea, a picture, or a mathematical connection to the viewer.

Representation Standard

Instructional programs from prekindergarten through grade 12 should enable students to

- create and use representations to organize, record, and communicate mathematical ideas;
- select, apply, and translate among mathematical representations to solve problems;
- use representations to model and interpret physical, social, and mathematical phenomena. (NCTM, 2000)

Representations develop student understanding. As Figure 3.7 implies, different representations emphasize various facets of the mathematics. The edges of the tetrahedron symbolize the transformation from one representation to another. Students need to learn to shift from one representation to another (one vertex to another) in order to view the different facets of mathematics, and to make the appropriate connections. As students work with and create different representations of a concept or situation, they move to the different vertices and, along the way, discover new aspects of the concept. And when students analyze representations (see how the various modes enhance each other) they can better decide which representations provide useful information, and which do not. For example, after using graphical and symbolic representations to find the solution set of a system of equations, students may realize that a symbolic representation is more accurate. You may find the tetrahedron model is useful as you develop lessons and assessments.

Students often need to see multiple representations of a concept before their sense of the concept begins to take shape. Some students develop stronger understandings when they see drawings or graphs. Others prefer algebraic or symbolic representations. Still others need both representations.

Viewing a variety of representations is also useful after students have developed a sense of the concept because representations strengthen student understanding. Representations also help organize and record data that is cumbersome or noninformative.

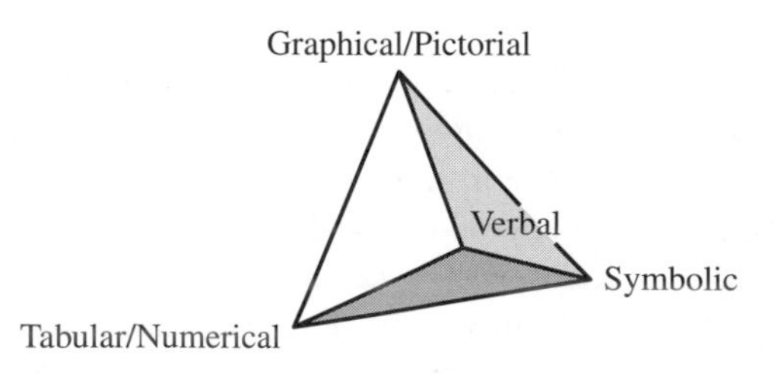

Figure 3.7 Representations in mathematics.

Box plots or stem-and-leaf graphs reveal aspects of the data not immediately obvious in the raw data. To take advantage of multiple representations, students need to see and explore numerous representations. Thus, mathematics teachers must work with students to develop their ability to create, interpret, and use a variety of representations.

Technology opens many doors when it comes to creating representations. Students with even basic graphing calculators can easily jump from symbolic to graphical representations for a wide variety of functions. More advanced graphing calculators and many software packages provide three-dimensional representations that are quite realistic. Dynamic geometry software allows users to "drag" an object in a drawing to discover more general properties of the figure. Users can also display multiple representations on a single screen. With this sometimes bewildering array of possibilities, it is crucial that teachers help students recognize important features so they can make better decisions about when one representation is more appropriate than another.

With or without technology, the process of representing real-world situations through mathematics is called modeling. Modeling allows students to characterize real-world phenomena via symbols and diagrams. Students then use mathematics to analyze the situation and make decisions (problem-solve). Often, students must oversimplify the actual situation in order to create a model. Therefore, the limitations of modeling must also be examined. For instance, a predator-prey model of mice and hawks may consider a restricted habitat, but not take into account predators of the hawk or other predators of the mice. Even so, models give students a sense of the strength of modeling and its usefulness in mathematics. We discuss the process of creating and interpreting models in more detail later in this chapter.

Exercises 3.2

1. Being able to communicate mathematically is a crucial part of learning mathematics. Communication involves sharing ideas as well as listening to and incorporating the ideas of others.

a. Name a couple of common obstacles to mathematical communication in a typical high school classroom. How would you address these to aid students' ability to communicate?

b. How can you encourage your students to listen actively when others are sharing ideas?

c. How can you help students develop effective communication skills?

2. What kinds of technology do you want to use to promote problem solving? Give an example and discuss how you might use it.

3. What do you need to be aware of in terms of the strengths and limitations of technology so that you can use it effectively to promote mathematical reasoning?

4. The discussion of the Problem-Solving standard lists problem-solving strategies that include working backwards, trying a simpler but similar problem, diagramming the situation, and looking for patterns.

a. What problem-solving strategies have you found helpful as a mathematics student? What aspects of these strategies are particularly appealing?
b. Add at least three more strategies to this list that you might teach to high school students.

5. Take a 5×13 rectangle and draw the diagonal. Then on the base find the point 8 units out from the most acute angle of the diagonal and the base. Construct a perpendicular from the base to the diagonal at this point. Repeat this procedure working from the acute angle on the upper base as well. Cut along these lines (that is, down the diagonal and then from the bases up to the diagonal.) Rearranging the pieces, you can form an 8×8 square. The 5×13 had an area of 65, whereas the resulting square has an area of 64. What happened to the missing unit?

6. Refer to Exercise 5. What mathematical connections can you make in this problem? What other kinds of connections could you add to a classroom discussion?

PROJECTS Find Out for Yourself

1. Visit a middle school classroom and watch how the teacher helps students develop reasoning skills.
   a. Does the teacher ask students to justify their responses?
   b. Does the teacher model the process of investigating, conjecturing, and justifying results?
   c. Do students follow such a model?

2. What kind of a mathematics learner are you? Do you prefer to see math as graphs, drawings, and other diagrams? Or is it easier to manipulate it symbolically?

3.3 The Mathematical Behavior of Dynamical Systems

As noted in Section 3.2, the modeling process incorporates the five process standards of Problem Solving, Reasoning and Proof, Communication, Connections, and Representation. In this section and Section 3.4, we explore more modeling techniques appropriate for high school students. The modeling process itself is discussed in Section 3.6.

In this section we build numerical solutions to dynamical systems by starting with an initial value and iterating a sufficient number of values in order to discern patterns. You will see that the behavior predicted by dynamical systems can often be characterized by the mathematical structure of the system. In other cases, you will see wild variations in the behavior caused by only small changes in the initial values. We also examine dynamical systems for which small changes in the proportionality constants cause widely different predictions.

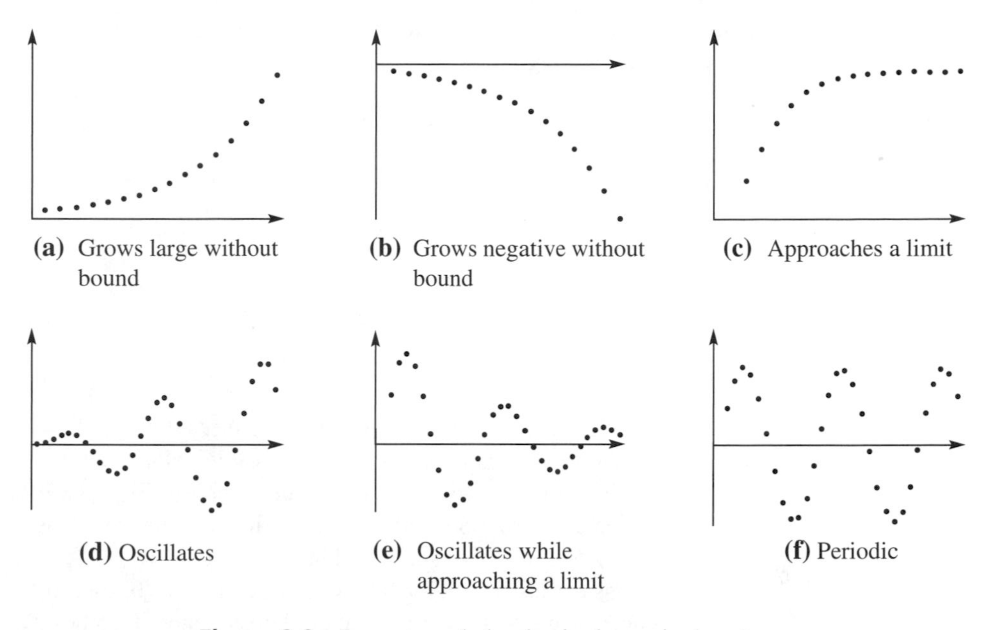

Figure 3.8 Long-term behavior in dynamical systems.

Long-Term Behavior

We use dynamical systems to represent real-world behavior that we are trying to understand. Often we want to predict future behavior and gain insight into how to influence the observed behavior. Consequently, we have great interest in the predictions of a model. Does the sequence in the dynamical system grow without bound (Figure 3.8a)? (A yes answer here would be of great interest if the model were of global warming.) Does the sequence become increasingly negative without bound (that is, negative with an absolute value that grows large as illustrated in Figure 3.8b)? Does the behavior approach a limiting value (Figure 3.8c)? If we are modeling the motion of an automobile suspension system damped by springs and shock absorbers, we would want a design that damps external disturbances. Does the motion oscillate (Figure 3.8d)? Oscillations, if too large or frequent, can cause great damage to structures such as bridges, buildings, or automotive systems. Does the sequence oscillate as it approaches a limit (Figure 3.8e)? We might expect this to happen to the vertical displacements in a properly designed automobile after hitting a pothole. If the model is of global warming, it would be reassuring to know that the temperature will return to an equilibrium value after soaring past it. Is the motion periodic, such as in a swinging undamped pendulum, the seasonal variations in weather, or locations of heavenly bodies (Figure 3.8f)? These are but a few of the interesting possibilities summarized in Figure 3.8.

Linear Dynamical Systems $a_{n+1} = ra_n$, for r Constant

We begin our exploration by looking at equations having the form $a_{n+1} = ra_n$, where r can be any positive or negative constant. We illustrate here an exploration process you

Table 3.1 Long-term behavior of $a_{n+1} = ra_n$ for key values of r.

Values of r	Long-term behavior
$r = 0$	?
$r = 1$	?
$\|r\| > 1$	?
$\|r\| < 1$	?
$r < 0$	?

can use with other dynamical systems in the exercises. Our goal is to gain insight into the nature of the long-term behavior for all possible values of r and any initial value. That is, what is the long-term behavior (as described in Figure 3.9), and how does it vary with the parameter r? Second, for what critical values of r does the nature of the long-term behavior change, as suggested in Table 3.1?

Let's consider some significant values of r. If $r = 0$, then the values of the sequence beyond a_0 are zero. If $r = 1$, then the sequence becomes $a_{n+1} = a_n$. This is a rather interesting case because, no matter where the sequence starts, it stays there forever (as illustrated in Figure 3.9 where the sequence is enumerated and graphed for $a_0 = 50$, and for several other starting values as well). Values for which a dynamical system remains constantly at that value, once reached, are called **equilibrium values** of the system. We'll define that term more precisely later, but for now note that in Figure 3.9, any starting value is an equilibrium value.

We encountered the form $a_{n+1} = ra_n$, r some constant, in some of our previous modeling applications (Section 1.2, Example 1). If a_n represents the amount in a savings account n months after investing \$1000 at 1% monthly interest, then the sequence can be iterated and graphed, as shown in Figure 3.10. From the graph and an analysis of the model, we expect that if $r > 1$, then the sequence grows without bound.

But what happens if r is negative? If we replace 1.01 with -1.01, we obtain the graph in Figure 3.11. Note the oscillation between both positive and negative values.

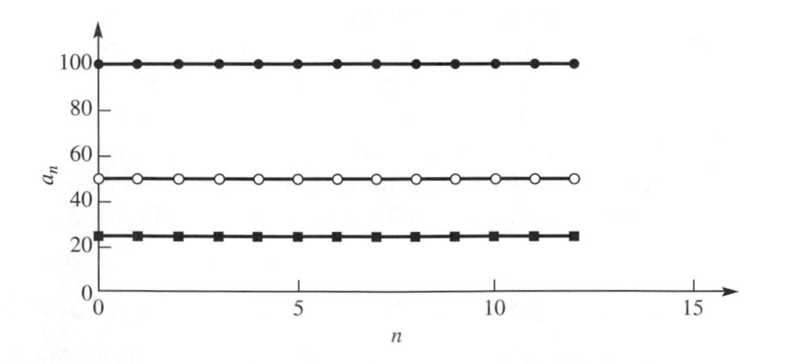

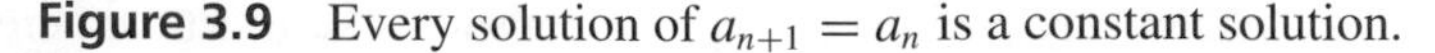

Figure 3.9 Every solution of $a_{n+1} = a_n$ is a constant solution.

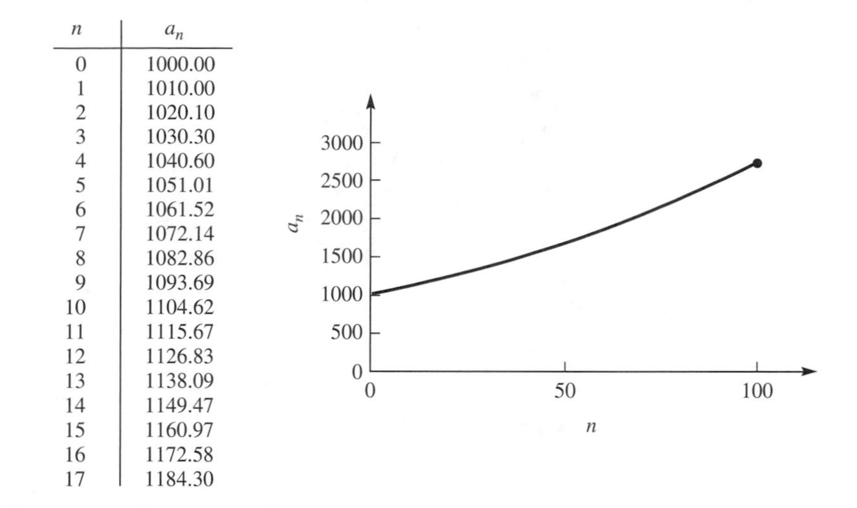

n	a_n
0	1000.00
1	1010.00
2	1020.10
3	1030.30
4	1040.60
5	1051.01
6	1061.52
7	1072.14
8	1082.86
9	1093.69
10	1104.62
11	1115.67
12	1126.83
13	1138.09
14	1149.47
15	1160.97
16	1172.58
17	1184.30

Figure 3.10 For $r > 1$ and constant, the solutions to $a_{n+1} = ra_n$ grow without bound.

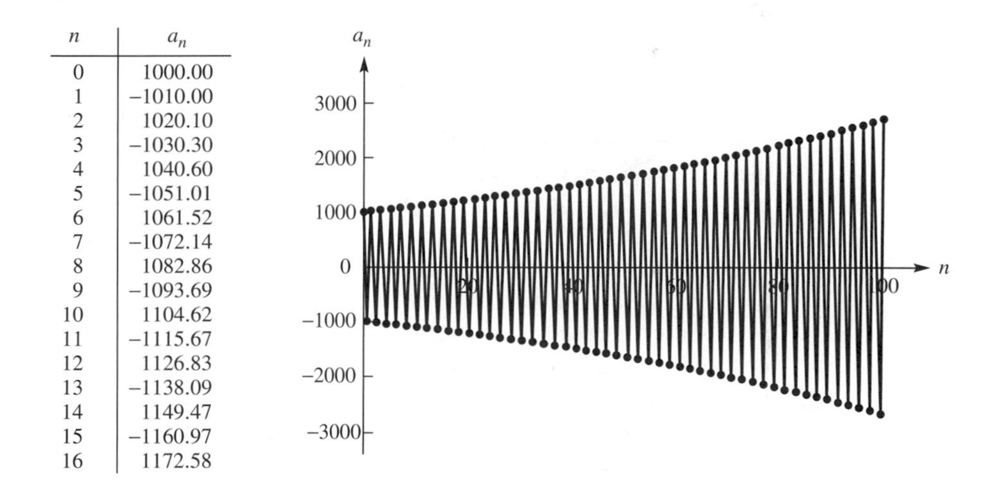

n	a_n
0	1000.00
1	−1010.00
2	1020.10
3	−1030.30
4	1040.60
5	−1051.01
6	1061.52
7	−1072.14
8	1082.86
9	−1093.69
10	1104.62
11	−1115.67
12	1126.83
13	−1138.09
14	1149.47
15	−1160.97
16	1172.58

Figure 3.11 A negative value of r causes oscillation.

Since the negative sign causes the next term in the sequence to be of opposite sign from the previous term, we conclude that, in general, negative values of r cause oscillations in the linear sequence $a_{n+1} = ra_n$.

What happens if $|r| < 1$? We know what happens if $r = 0$, if $r > 1$, if $r < -1$, and if r is negative in general, so let's next consider positive fractions. Suppose digoxin decays in the bloodstream such that each day one-half of the concentration of digoxin remains from the previous day. If we start with 0.6 mg, and a_n represents the amount after n days, we can represent the behavior with the following model (a numerical solution appears in Figure 3.12):

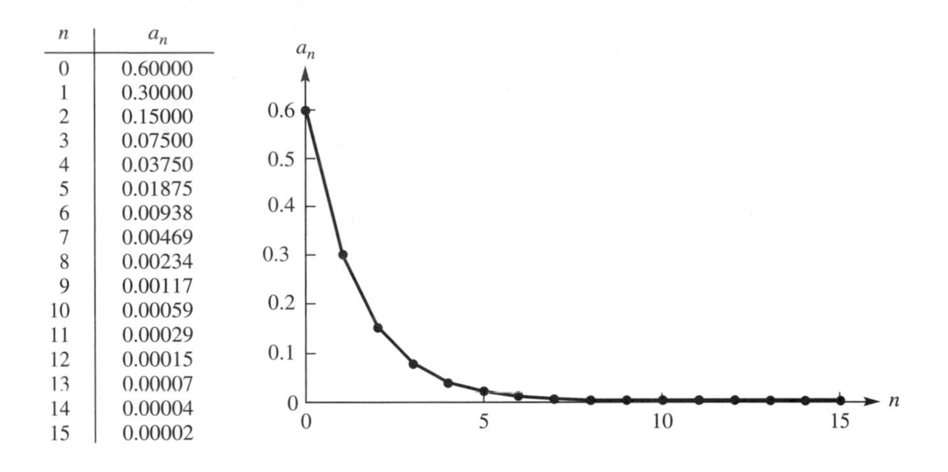

n	a_n
0	0.60000
1	0.30000
2	0.15000
3	0.07500
4	0.03750
5	0.01875
6	0.00938
7	0.00469
8	0.00234
9	0.00117
10	0.00059
11	0.00029
12	0.00015
13	0.00007
14	0.00004
15	0.00002

Figure 3.12 A positive (proper) fractional value of r causes decay.

$$a_{n+1} = 0.5a_n \quad n = 0, 1, 2, 3, \ldots$$
$$a_0 = 0.6$$

Note that the behavior approaches the value 0 (which is an *equilibrium value* for this system since if we start with no digoxin in the bloodstream, we will never have any). The value 0 is a *limit* for the system, meaning that the difference between the value 0 and the value for a_n (starting with $a_0 = 0.6$) can be made as small as we please once n is large enough. A knowledge of equilibrium values and limits immensely aids our understanding of the behavior of dynamical systems.

Growth and decay problems of the form investigated in this section are very important in modeling. For example, carbon 14 is contained in all living organisms and begins to decay after death at such a predictable rate that the amount remaining is used to date archaeological findings, rare paintings, documents, and the like. Thus far, we have made observations for a few values of r and only a few starting values. But from those examples, can you guess the nature of the behavior for all values of r and all possible starting values? Can you complete Table 3.1 now? Try it and see, based on our examples. Our observations are summarized in Table 3.2.

Table 3.2 Behavior for all values of r.

Values of r	Long-term behavior
$r = 0$	Constant solution and equilibrium value at 0
$r = 1$	All initial values are constant solutions
$r < 0$	Oscillation
$\lvert r \rvert < 1$	Decay to limiting value of 0
$\lvert r \rvert > 1$	Growth without bound

Finding Equilibrium Values

Consider a dynamical system of the form $a_{n+1} = ra_n$. For an equilibrium value to occur, eventually $a_{n+1} = a_n$. Let's call such an equilibrium value a. Substituting $a_{n+1} = a_n = a$ and solving gives

$$a = \frac{0}{1-r}, \qquad r \neq 1$$

Thus if $r \neq 1$, the number 0 is an equilibrium value (as we saw in Figure 3.12). If $r = 1$, then every initial value results in a constant solution (as we saw in Figure 3.9). Hence, every value is an equilibrium value. Summarizing our observations, we have the following theorem.

THEOREM 3.1 For the dynamical system $a_{n+1} = ra_n$:

If $r \neq 1$, an equilibrium value exists at $a = 0$.
If $r = 1$, every number is an equilibrium value.

Dynamical Systems of the Form $a_{n+1} = ra_n + b$, Where r and b Are Constants

Now let's add a constant b to the dynamical system we just studied. Again we'd like to classify the nature of the long-term behavior for all possible cases. We use three examples to gain some insight into the behavior.

Example 1 *Prescription for Digoxin*

Consider again the digoxin problem. Digoxin is used to treat heart patients. The objective of the problem is to consider the decay of digoxin in the bloodstream in order to prescribe a dosage that keeps the concentration between acceptable (safe and effective) levels. Suppose we prescribe a daily drug dosage of 0.1 mg. This results in the dynamical system

$$a_{n+1} = 0.5a_n + 0.1$$

Now let's consider three starting values, or initial doses:

$$\text{A:} \quad a_0 = 0.1$$

$$\text{B:} \quad a_0 = 0.2$$

$$\text{C:} \quad a_0 = 0.3$$

In Figure 3.13, we compute the numerical solutions for each case. ■

Note that the value 0.2 is an equilibrium value, since once reached the system remains at 0.2 forever. Further, if we start below the equilibrium (as in case A) or above the equilibrium (as in case C), we apparently approach the equilibrium value as a *limit*.

	A	B	C
n	a_n	a_n	a_n
0	0.10000	0.20000	0.30000
1	0.15000	0.20000	0.25000
2	0.07500	0.20000	0.22500
3	0.03750	0.20000	0.21250
4	0.01875	0.20000	0.20625
5	0.00938	0.20000	0.20313
6	0.00469	0.20000	0.20156
7	0.00234	0.20000	0.20078
8	0.00117	0.20000	0.20039
9	0.00059	0.20000	0.20020
10	0.00029	0.20000	0.20010
11	0.00015	0.20000	0.20005
12	0.00007	0.20000	0.20002
13	0.00004	0.20000	0.20001
14	0.00002	0.20000	0.20001
15	0.00001	0.20000	0.20000

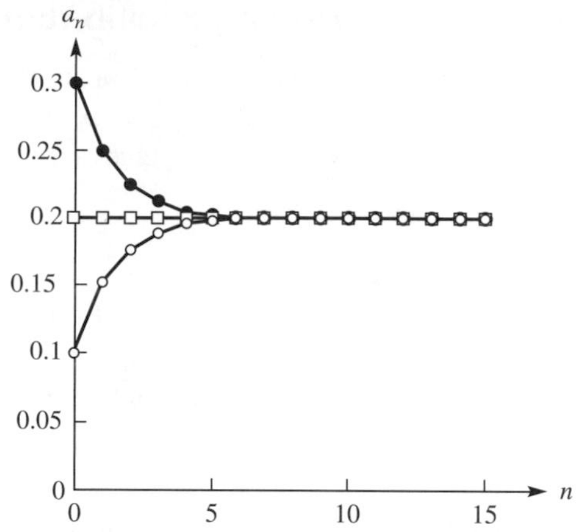

Figure 3.13 Three initial digoxin doses.

In the exercises you are asked to compute solutions for starting values even closer to 0.2, lending evidence that 0.2 is a *stable* equilibrium value. When prescribing digoxin, the concentration level must stay above an "effective level" for a period of time without exceeding a "safe" level. You will also find initial and subsequent doses that are both *safe* and *effective.*

Example 2 *An Investment Annuity*

Let's return to the bank account problem (Example 1, Section 1.2). Consider what is known as an *annuity*, which is often used for retirement purposes. Annuities are basically savings accounts that pay interest on the amount present, but allow the investor to withdraw a fixed amount each month until the account is depleted. An interesting issue (posed in the exercises) is to know how much to save each month in order to build an annuity that allows for withdrawals, beginning at a certain age with a specified amount for a desired number of years, before the account is depleted. For now, consider 1% as the monthly interest rate, and a monthly withdrawal of $1000. This gives the dynamical system

$$a_{n+1} = 1.01a_n - 1000$$

Suppose you make the following initial investments:

A: $a_0 = 90{,}000$

B: $a_0 = 100{,}000$

C: $a_0 = 110{,}000$

The numerical solutions for each case are graphed in Figure 3.14. ■

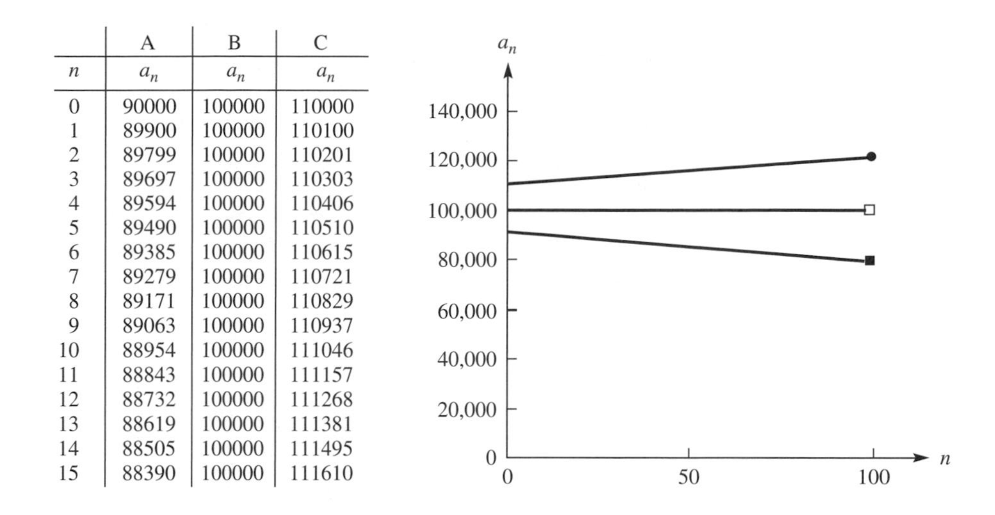

	A	B	C
n	a_n	a_n	a_n
0	90000	100000	110000
1	89900	100000	110100
2	89799	100000	110201
3	89697	100000	110303
4	89594	100000	110406
5	89490	100000	110510
6	89385	100000	110615
7	89279	100000	110721
8	89171	100000	110829
9	89063	100000	110937
10	88954	100000	111046
11	88843	100000	111157
12	88732	100000	111268
13	88619	100000	111381
14	88505	100000	111495
15	88390	100000	111610

Figure 3.14 An annuity with three initial investments.

Note that the value 100,000 is an equilibrium value since once reached, the system remains there for all subsequent values. However, if we start above that equilibrium, there is growth without bound. (Try plotting with $a_0 = \$100{,}000.01$.) On the other hand, if we start with an account below \$100,000, the savings are used up at an increasing rate! (Try \$99,999.99.) Note how drastically the long-term behaviors differ even though the starting values differ by only \$0.02! In this situation we say that the equilibrium value 100,000 is *unstable*: If you start close to the value (even within a penny!), you don't remain close. Look at the different results in Figures 3.13 and 3.14. Both systems show equilibrium values, but the first is stable and the second is unstable.

We've considered here the cases where $|r| < 1$ and $|r| > 1$. Let's see what happens when $r = 1$.

Example 3 ***A Checking Account***

Most students can't keep enough cash in their checking account to earn any interest. Suppose you have an account that pays no interest, and that each month you pay only your dorm rent of \$300, giving the dynamical system

$$a_{n+1} = a_n - 300$$ ■

The final result is no doubt obvious, but now compare its graph (Figure 3.15) with those in Figures 3.13 and 3.14. Do you see how drastically the graph of this solution differs from the previous examples? Can you find an equilibrium, as in Examples 1 and 2?

Our observations are summarized in Table 3.3, where we have classified the three examples and their long-term behaviors according to values of r.

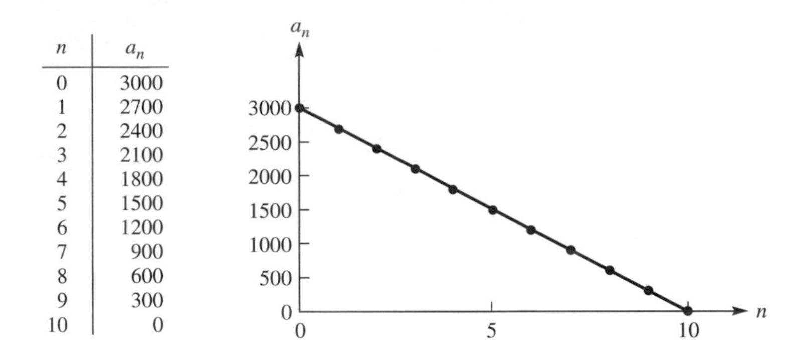

n	a_n
0	3000
1	2700
2	2400
3	2100
4	1800
5	1500
6	1200
7	900
8	600
9	300
10	0

Figure 3.15 A checking account for paying dorm rent.

Table 3.3 Dynamical system $a_{n+1} = ra_n + b$, $b \neq 0$.

Value of r	Long-term behavior observed
$\lvert r \rvert < 1$	Stable equilibrium
$\lvert r \rvert > 1$	Unstable equilibrium
$r = 1$	Graph is a line with no equilibrium
$r = -1$	Oscillation between two values

Finding Equilibrium Values

Determining whether equilibrium values exist, and classifying them as stable or unstable, immensely assists in analyzing the long-term behavior of a dynamical system. Consider again Examples 1 and 2. In Example 1, how did we know that a starting value of 0.2 would result in a constant solution or equilibrium value? Similarly, how do we know the answer to the same question for an investment of \$100,000 in Example 2? For a dynamical system of the form

$$a_{n+1} = ra_n + b \tag{1}$$

let's use a to denote the equilibrium value, if one exists. From the definition of equilibrium value, if we start at a, we must remain there for all n; that is, $a_{n+1} = a_n = a$ for all n. Substituting a for a_{n+1} and a_n in equation (1) yields

$$a = ra + b$$

and solving for a we find,

$$a = \frac{b}{1-r}, \qquad \text{if } r \neq 1$$

Thus for Example 1, the equilibrium value is

$$a = \frac{0.1}{1 - 0.5} = 0.2$$

For Example 2, the equilibrium value is

$$a = \frac{-1000}{1 - 1.01} = 100{,}000$$

In Example 3, $r = 1$ and no equilibrium exists. Theorem 3.2 summarizes our observations.

THEOREM 3.2

For the dynamical system $a_{n+1} = ra_n + b$:

If $r \neq 1$, an equilibrium exists at $a = \dfrac{b}{1 - r}$.

If $r = 1$, no equilibrium value exists.

Nonlinear Systems

An important advantage of discrete systems is that numerical solutions can be constructed for any dynamical system when given an initial value. We have seen thus far that long-term behavior can be very sensitive to the starting value and to the values of the parameter r. Recall the model for the yeast biomass from Section 1.3:

$$p_{n+1} = p_n + 0.00082(665 - p_n)p_n$$

This model transforms algebraically into the dynamical system

$$a_{n+1} = r(1 - a_n)a_n$$

In Figure 3.16 we plot numerical solutions for the four values of r beginning with $a_0 = 0.2$.

Note the remarkably different behaviors of each case. In Figure 3.16a, for $r = 2$ the behavior approaches a limit. In part b, for $r = 3.25$ the behavior oscillates between about 0.49 and 0.81. We call such behavior periodic with a 2-cycle. In part c, the motion repeats itself every four iterations, or is periodic with a 4-cycle. Although it is not evident in part d, the value of r is sufficiently large that no periodicity or pattern exists. Thus, it is impossible to predict long-term behavior from the model. We call such behavior *chaotic*. Chaotic systems demonstrate sensitivity to the constant parameters of the system. A chaotic system can be quite sensitive to initial conditions as well.

Exercises 3.3

1. For the following problems, find an equilibrium value if one exists. Classify the equilibrium value as stable or unstable.

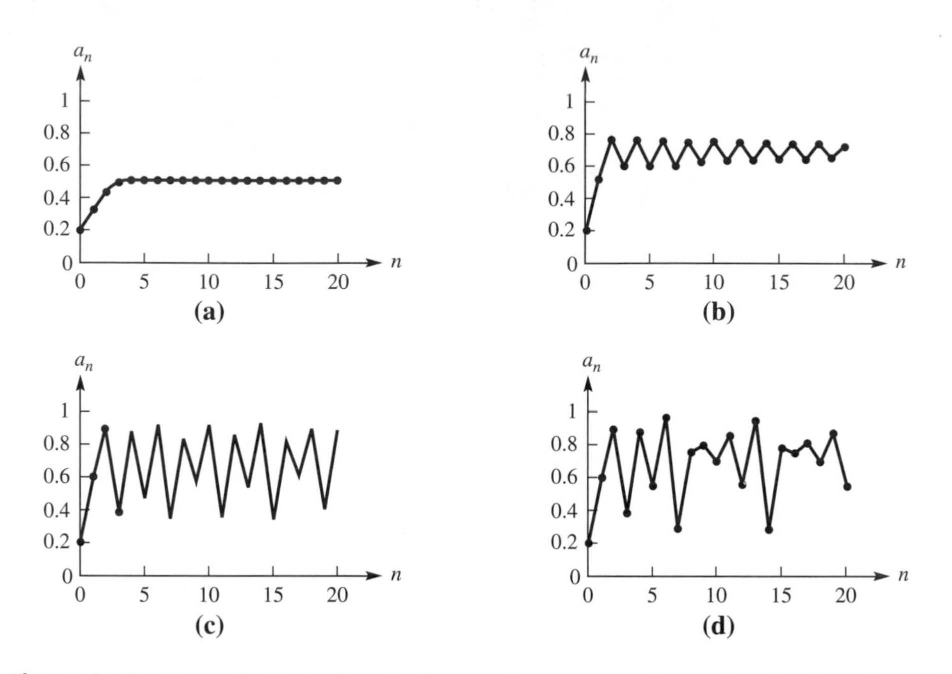

Figure 3.16 Numerical solutions to the equation $a_{n+1} = r(1 - a_n)a_n$ for the values of r: (a) $r = 2$, (b) $r = 3.25$, (c) $r = 3.5$, (d) $r = 3.75$.

a. $a_{n+1} = 1.1a$
b. $a_{n+1} = 0.9a_n$
c. $a_{n+1} = -0.9a_n$
d. $a_{n+1} = a_n$
e. $a_{n+1} = -1.2a + 50$
f. $a_{n+1} = 1.2a_n - 50$
g. $a_{n+1} = 0.8a_n + 100$
h. $a_{n+1} = 0.8a_n - 100$
i. $a_{n+1} = -0.8a_n + 100$
j. $a_{n+1} = a_n - 100$
k. $a_{n+1} = a_n + 100$

2. Build a numerical solution for the following initial-value problems. Plot your data to observe patterns in the solution. Is there an equilibrium solution? Is it stable or unstable?

a. $a_{n+1} = -1.2a + 50,\ a_0 = 1000$
b. $a_{n+1} = 0.8a_n - 100,\ a_0 = 500$
c. $a_{n+1} = 0.8a_n - 100,\ a_0 = -500$
d. $a_{n+1} = -0.8a_n + 100,\ a_0 = 1000$
e. $a_{n+1} = a_n - 100,\ a_0 = 1000$

3. You currently have \$5000 in a savings account that pays 0.5% interest each month. You add another \$200 each month. Build a numerical solution to determine when the account reaches \$20,000.

4. You owe \$500 on a credit card that charges 1.5% interest each month. You can pay \$50 each month and you make no new charges. What is the equilibrium value? What does the equilibrium value mean in terms of the credit card? Build a numerical solution. When will the account be paid off? How much is the last payment?

5. Your parents are considering a 30-year $100,000 mortgage that charges 0.5% interest each month. Formulate a model in terms of a monthly payment p that allows the mortgage (loan) to be paid off after 360 payments. [*Hint:* If a_n represents the amount owed after n months, what are a_0 and a_{360}?] Experiment by building numerical solutions to find a value of p that "works."

6. Your parents are considering a 30-year mortgage that charges 0.5% interest each month. Formulate a model in terms of a monthly payment p that allows the mortgage (loan) to be paid off after 360 payments. Your parents can afford a monthly payment of $1500. Experiment to determine the maximum amount of money they can borrow. [*Hint:* If a_n represents the amount owed after n months, what are a_0 and a_{360}?]

7. Your grandparents have an annuity. The value of the annuity increases each month as 1% interest on the previous month's balance is deposited. Your grandparents withdraw $1000 each month for living expenses. Presently, they have $50,000 in the annuity. Model the annuity with a dynamical system. Find the equilibrium value. What does the equilibrium value represent for this problem? Build a numerical solution to determine when the annuity is depleted.

8. *Continuation of Example 1, Section 3.3.*
   a. Find the equilibrium value of the digoxin model. What is the significance of the equilibrium value?
   b. Experiment with different initial and maintenance doses. Find a combination with time between doses and amount to be taken considered as measures of convenience. (Rounded amounts are considered to be more convenient.)

9. *Continuation of Example 3, Section 1.3.* Determine the age of the humanoid skull found near the remains of the ancient campfire.

PROJECTS Modeling Explorations

1. You plan to invest part of your paycheck to finance your children's education. You want enough money in the account to be able to draw $1000 a month, every month for 8 years beginning 20 years from now. The account pays 0.5% interest each month.
   a. How much money will you need 20 years from now to finance one child's education? Assume you stop investing when your first child begins college (a safe assumption!).
   b. How much must you deposit each month over the next 20 years?

2. Consider the following dynamical system

$$a_{n+1} = r(a_n)(1 - a_n)$$

$$a_0 = 100$$

For the following values of r, build a numerical solution and plot the results: $r = 0.75, 1.5, 2.75, 3, 3.25, 3.45, 3.55, 3.57, 3.59, 3.65$. Observe the pattern of your numerical solution for each value of r. Does there appear to be an equilibrium value? Would you be able to make predictions from the model? Experiment with different values for a_0. Are the results sensitive to the starting value?

3. For an ecology project, you consider the survival of whales and assume that if the population falls below a minimum survival level m, the species will become extinct. You also assume that the population is limited by the carrying capacity M of the environment. That is, if the whale population is above M, then it will experience a decline because the environment cannot sustain that large a population. In your model, a_n represents the whale population after n years. Discuss your model:

$$a_{n+1} = k(M - a_n)(a_n - m)$$

Now experiment with different values for m, M, and k. Try several starting values. What does your model predict?

Further Reading

Tuchinsky, P. M. (1981). *Man in competition with the spruce budworm*, UMAP Expository Monograph. A population of tiny caterpillars periodically explodes in the evergreen forests of Eastern Canada and Maine. They devour tree needles and cause great damage to forests that are central to the region's economy. The province of New Brunswick is using mathematical models of budworm/forest interaction in an effort to plan for and control the damage. The monograph surveys the ecological situation and examines the computer simulation and models that are currently in use.

3.4 Systems of Difference Equations

In this section we consider systems of difference equations. For selected starting values, we build numerical solutions to get an indication of the system's long-term behavior. As noted in the last section, equilibrium values are values of the dependent variable(s) for which there is no change in the system once the equilibrium values are obtained. For the systems we consider here, we first find the equilibrium values, then we "explore" starting values in the vicinity of the equilibrium values. If we start close to an equilibrium value, we want to know whether the system will

1. remain close
2. approach the equilibrium value
3. not remain close

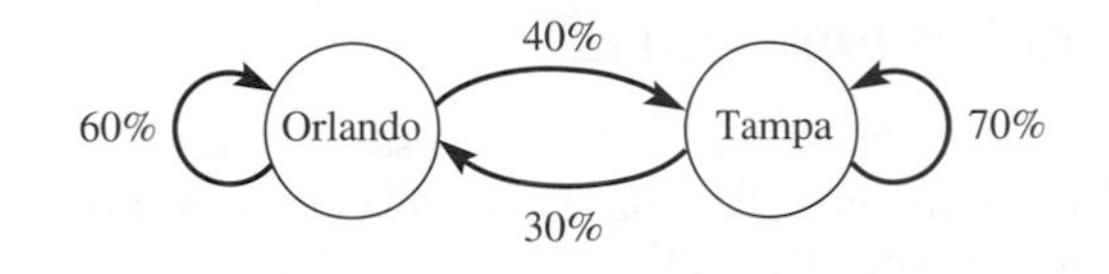

Figure 3.17 Car rental offices in Orlando and Tampa.

What happens near these values gives us great insight into a system's long-term behavior. Does the system demonstrate periodic behavior? Are there oscillations? Does the long-term behavior appear to be sensitive to

- initial conditions?
- small changes in the proportionality constants?

Our goal is to model the behavior with difference equations and then explore the predicted behavior.

Example 1 ***A Car Rental Company***

A car rental company with distributorships in Orlando and Tampa caters to travel agents who arrange tourist activities in both cities. Consequently, a tourist may choose to rent a car in one city and drop it off in the second city. Tourists can begin their itinerary in either city. The company wants to determine how much they should charge for the "drop-off" convenience. Since cars are dropped off in both cities, will a sufficient number of cars end up in each city to satisfy the demand for cars in that city? If not, how many cars must the company transport from Orlando to Tampa, and from Tampa to Orlando? The answers to these questions will help the company predict its expected costs.

As the company's consultant, you have analyzed their records and determined that 60% of the cars rented in Orlando are returned to Orlando, while the remaining 40% end up in Tampa. Of the cars rented in Tampa, 70% are returned to Tampa while 30% end up in Orlando. When analyzing problems such as this, you may find a diagram like the one in Figure 3.17 helpful.

Dynamical Systems Model

To develop a model of the system, we let n represent the number of business days, and define

O_n: the number of cars in Orlando at the end of day n

T_n: the number of cars in Tampa at the end of day n

Thus, the historical records reveal the system

$$O_{n+1} = 0.6O_n + 0.3Tn$$
$$T_{n+1} = 0.4O_n + 0.7T_n$$

Equilibrium Values

The equilibrium values for the system are those values of O_n and T_n for which no change in the system takes place. We'll call the equilibrium values (if they exist) O and T, respectively. Then $O = O_{n+1} = O_n$ and $T = T_{n+1} = T_n$ simultaneously. Substitution in our model yields the following requirements for the equilibrium values:

$$O = 0.6O + 0.3T$$
$$T = 0.4O + 0.7T$$

This system is satisfied whenever $O = \frac{3}{4}T$. For example, if the company owns 7000 cars, and starts with 3000 in Orlando and 4000 in Tampa, then our model predicts that

$$O_1 = 0.6(3000) + 0.3(4000) = 3000$$
$$T_1 = 0.4(3000) + 0.7(4000) = 4000$$

Thus, this system remains at $(O, T) = (3000, 4000)$ if we start there.

Next let's explore what happens if we start at values other than the equilibrium values. We'll iterate the system for the initial four cases shown in Table 3.4. A numerical solution, or table of values, for each starting value is graphed in Figure 3.18.

Sensitivity to Initial Conditions and Long-Term Behavior

Note that within a week, the system is very close to the equilibrium value (3000, 4000) even if initially no cars were at one site! Our results suggest that the equilibrium value is stable and insensitive to the starting values. Based on these explorations, we would be inclined to predict that the system will approach the equilibrium value where $\frac{3}{7}$ of the fleet ends up in Orlando and the remaining $\frac{4}{7}$ in Tampa. This information is quite useful. Because the company now knows the demand patterns in each city, they can estimate how many cars they need to ship. In the exercises, we ask you to explore the system to determine whether it is sensitive to the coefficients in the equations. ■

In Section 1.3 we developed several models for a spotted owl population. One model assumed that a second species was competing for habitat resources (such as

Table 3.4 Starting values for the car rental problem.

Case	Orlando	Tampa
1	7000	0
2	5000	2000
3	2000	5000
4	0	7000

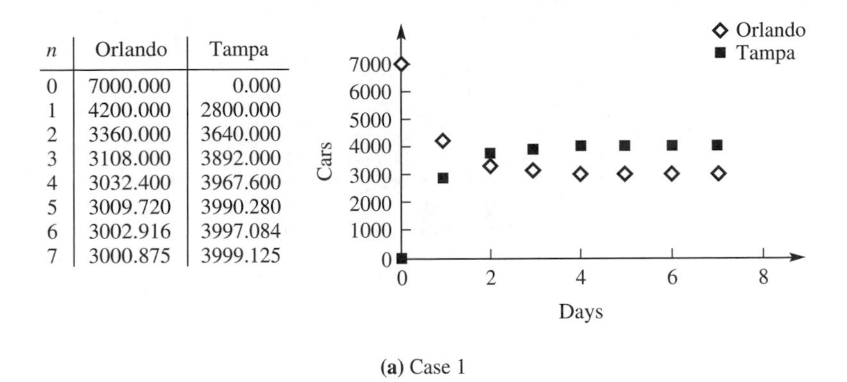

n	Orlando	Tampa
0	7000.000	0.000
1	4200.000	2800.000
2	3360.000	3640.000
3	3108.000	3892.000
4	3032.400	3967.600
5	3009.720	3990.280
6	3002.916	3997.084
7	3000.875	3999.125

(a) Case 1

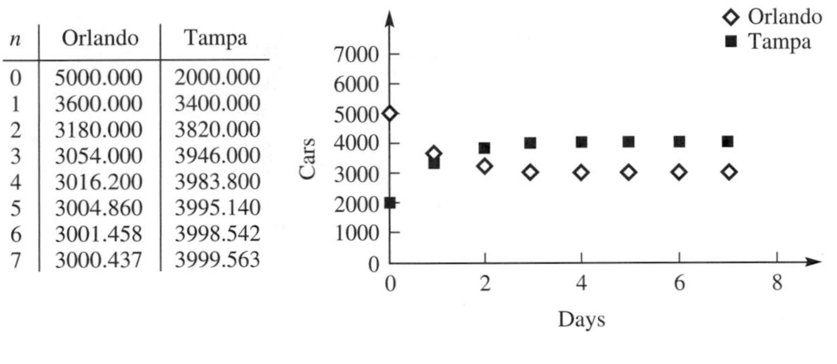

n	Orlando	Tampa
0	5000.000	2000.000
1	3600.000	3400.000
2	3180.000	3820.000
3	3054.000	3946.000
4	3016.200	3983.800
5	3004.860	3995.140
6	3001.458	3998.542
7	3000.437	3999.563

(b) Case 2

Figure 3.18 The rental car problem.

food). The presence of the second species limits the growth rate of the spotted owl population. Let's look at an example.

Example 2 *Competitive Hunter Model—Spotted Owls and Hawks*

Let O_n be the value of the spotted owl population at the end of day n and let H_n denote the competing hawk population. Based on the competing species model in Example 2 of Section 1.3, we have

$$O_{n+1} = (1 + k_1)O_n - k_3 O_n H_n$$
$$H_{n+1} = (1 + k_2)H_n - k_4 O_n H_n$$

where k_1, k_2, k_3, and k_4 are positive constants. Let's choose specific values for the proportionality constants and consider the system

$$\begin{aligned} O_{n+1} &= 1.2O_n - 0.001 O_n H_n \\ H_{n+1} &= 1.3H_n - 0.002 O_n H_n \end{aligned} \tag{1}$$

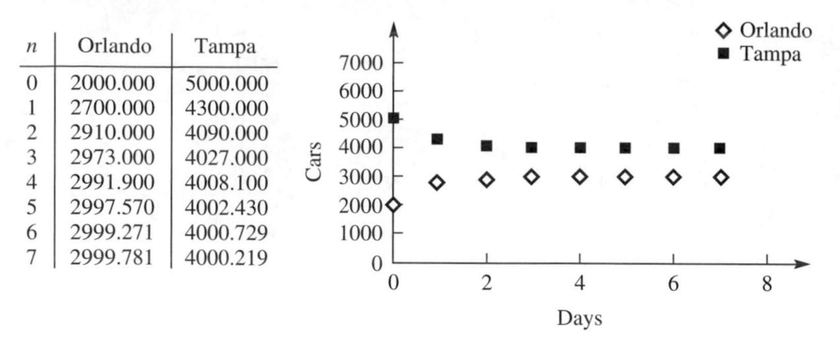

n	Orlando	Tampa
0	2000.000	5000.000
1	2700.000	4300.000
2	2910.000	4090.000
3	2973.000	4027.000
4	2991.900	4008.100
5	2997.570	4002.430
6	2999.271	4000.729
7	2999.781	4000.219

(c) Case 3

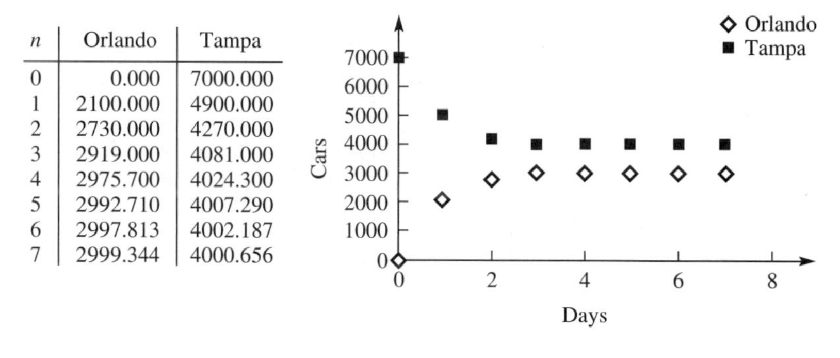

n	Orlando	Tampa
0	0.000	7000.000
1	2100.000	4900.000
2	2730.000	4270.000
3	2919.000	4081.000
4	2975.700	4024.300
5	2992.710	4007.290
6	2997.813	4002.187
7	2999.344	4000.656

(d) Case 4

Figure 3.18 *(Continued)*

Note that in the absence of other species, the owls and hawks exhibit a positive growth rate, and the presence of each species curtails the other species' growth rate.

Equilibrium Values

If we call the equilibrium values (O, H), then $O = O_{n+1} = O_n$ and $H = H_{n+1} = H_n$ simultaneously. Substituting into system (1) yields

$$\begin{aligned} O &= 1.2O - 0.001OH \\ H &= 1.3H - 0.002OH \end{aligned} \tag{2}$$

or

$$\begin{aligned} 0 &= 0.2O - 0.001OH = O(0.2 - 0.001H) \\ 0 &= 0.3H - 0.002OH = H(0.3 - 0.002O) \end{aligned} \tag{3}$$

System (2) indicates that the owl population doesn't change if $O = 0$ or $H = 0.2/0.001 = 200$. System (3) indicates the hawk population doesn't change if $H = 0$

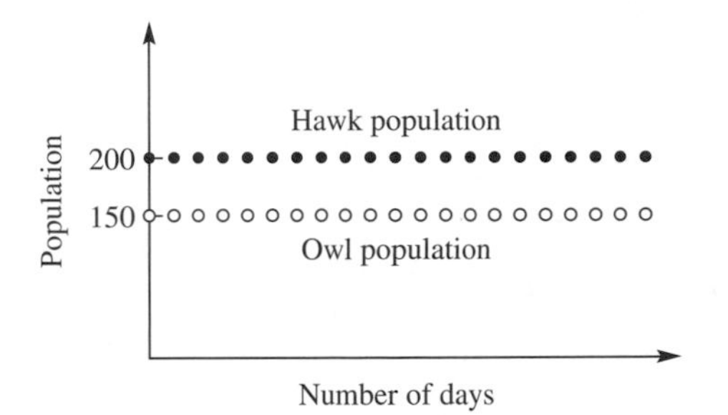

Figure 3.19 If initial owl population is 150 and the initial hawk population is 200, the two populations remain at these levels.

or $O = 0.3/0.002 = 150$, as depicted in Figure 3.19. Equilibrium values exist at $(O, H) = (0, 0)$ and $(O, H) = (150, 200)$ because *neither* population changes at those points. [Substitute the equilibrium values into system (1) to check that the system indeed remains at (0, 0) and (150, 200) if these points are the initial values.]

Let's see what happens in the vicinity of the equilibrium values. To do this, we build numerical solutions for the three initial populations given in Table 3.5. Note that the first two cases are close to the equilibrium value (150, 200), while the third is near the equilibrium (0, 0).

Iterating system (1) beginning with the given values results in the numerical solutions graphed in Figure 3.20. Note that, in each case, eventually one species drives the other to extinction.

Sensitivity to Initial Conditions and Long-term Behavior

Suppose 350 owls and hawks are to be placed in a habitat modeled by system (1). If 150 of the birds are owls, our model predicts the owls will remain at 150 forever. If one owl is removed from the habitat (149 remain), then the model predicts that the owl population will die out. However, if 151 owls are placed in the habitat, the model predicts that the owls will grow without bound and the hawks will disappear! This model is extremely sensitive to initial conditions. The equilibrium values are unstable in the sense that if we start close to either equilibrium value, we don't remain close. Note how the model predicts that coexistence of the two species in a single habitat is highly

Table 3.5 Initial values for the owl and hawk populations.

Case	Owls	Hawks
1	151	199
2	149	201
3	10	10

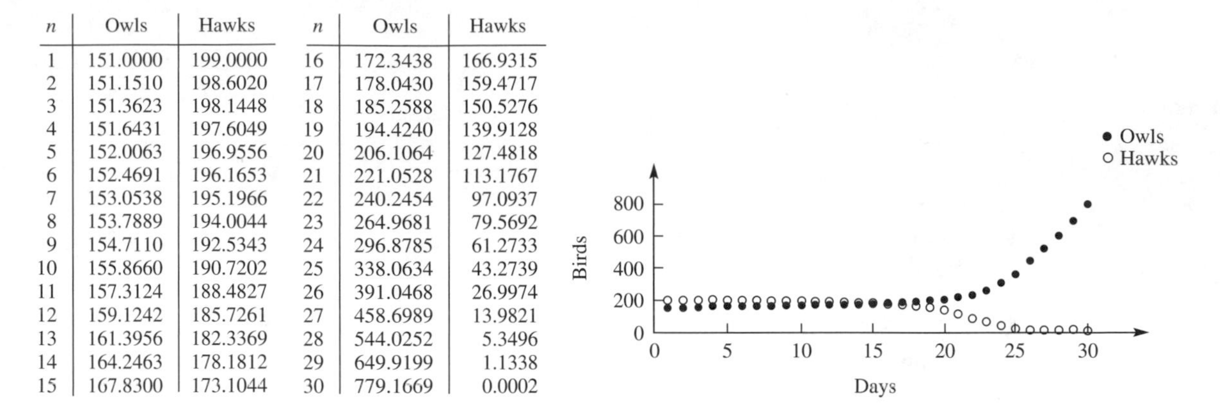

n	Owls	Hawks	n	Owls	Hawks
1	151.0000	199.0000	16	172.3438	166.9315
2	151.1510	198.6020	17	178.0430	159.4717
3	151.3623	198.1448	18	185.2588	150.5276
4	151.6431	197.6049	19	194.4240	139.9128
5	152.0063	196.9556	20	206.1064	127.4818
6	152.4691	196.1653	21	221.0528	113.1767
7	153.0538	195.1966	22	240.2454	97.0937
8	153.7889	194.0044	23	264.9681	79.5692
9	154.7110	192.5343	24	296.8785	61.2733
10	155.8660	190.7202	25	338.0634	43.2739
11	157.3124	188.4827	26	391.0468	26.9974
12	159.1242	185.7261	27	458.6989	13.9821
13	161.3956	182.3369	28	544.0252	5.3496
14	164.2463	178.1812	29	649.9199	1.1338
15	167.8300	173.1044	30	779.1669	0.0002

(a) Case 1

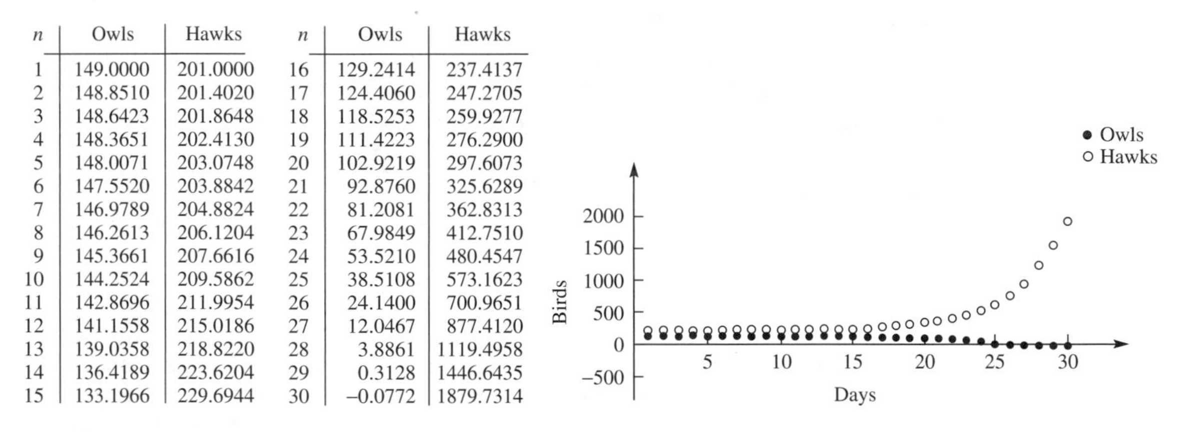

n	Owls	Hawks	n	Owls	Hawks
1	149.0000	201.0000	16	129.2414	237.4137
2	148.8510	201.4020	17	124.4060	247.2705
3	148.6423	201.8648	18	118.5253	259.9277
4	148.3651	202.4130	19	111.4223	276.2900
5	148.0071	203.0748	20	102.9219	297.6073
6	147.5520	203.8842	21	92.8760	325.6289
7	146.9789	204.8824	22	81.2081	362.8313
8	146.2613	206.1204	23	67.9849	412.7510
9	145.3661	207.6616	24	53.5210	480.4547
10	144.2524	209.5862	25	38.5108	573.1623
11	142.8696	211.9954	26	24.1400	700.9651
12	141.1558	215.0186	27	12.0467	877.4120
13	139.0358	218.8220	28	3.8861	1119.4958
14	136.4189	223.6204	29	0.3128	1446.6435
15	133.1966	229.6944	30	–0.0772	1879.7314

(b) Case 2

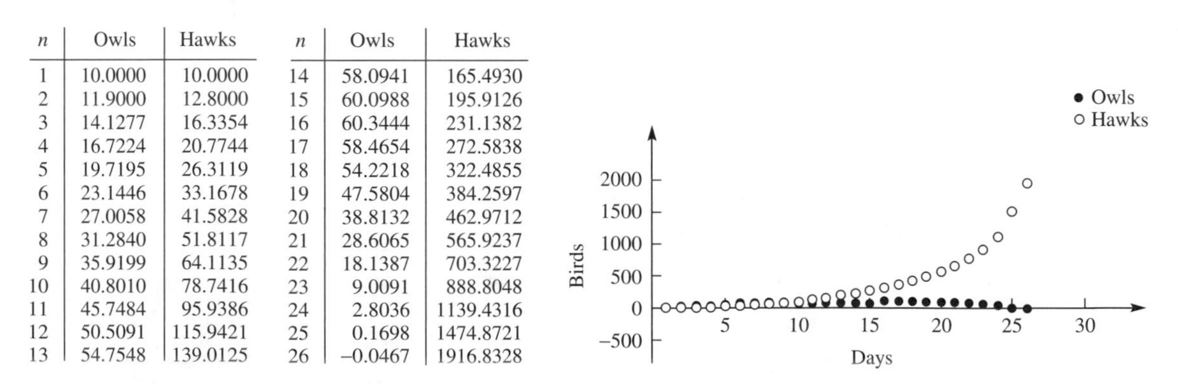

n	Owls	Hawks	n	Owls	Hawks
1	10.0000	10.0000	14	58.0941	165.4930
2	11.9000	12.8000	15	60.0988	195.9126
3	14.1277	16.3354	16	60.3444	231.1382
4	16.7224	20.7744	17	58.4654	272.5838
5	19.7195	26.3119	18	54.2218	322.4855
6	23.1446	33.1678	19	47.5804	384.2597
7	27.0058	41.5828	20	38.8132	462.9712
8	31.2840	51.8117	21	28.6065	565.9237
9	35.9199	64.1135	22	18.1387	703.3227
10	40.8010	78.7416	23	9.0091	888.8048
11	45.7484	95.9386	24	2.8036	1139.4316
12	50.5091	115.9421	25	0.1698	1474.8721
13	54.7548	139.0125	26	–0.0467	1916.8328

(c) Case 3

Figure 3.20 Either the owls or the hawks dominate.

unlikely because one species will eventually dominate. In the exercises, you are asked to explore this system further by examining other initial-value combinations and by changing the coefficients of the model. ■

In the model we developed in Section 1.3, we assumed mice were the owl's sole source of food. We now consider a specific example.

Example 3 ***Predator-Prey Model: Owls and Mice***

Let O_n represent the size of the spotted owl population at the end of day n, and M_n denote the size of the mouse population. Then from Section 1.3,

$$M_{n+1} = (1 + k_1)M_n - k_3 O_n M_n$$
$$O_{n+1} = (1 - k_2)O_n + k_4 O_n M_n$$

where k_1, k_2, k_3, and k_4 are positive constants. Choosing specific values for the proportionality constants, we obtain the system

$$\begin{aligned} M_{n+1} &= 1.2M_n - 0.001 O_n M_n \\ O_{n+1} &= -0.7 O_n + 0.002 O_n M_n \end{aligned} \tag{4}$$

Note that in the absence of owls, the mice grow without bound. Similarly, in the absence of mice, the owls die out. Notice also that this system resembles our model for competing species (the owl/hawk competition) except that the signs differ for certain coefficients. How do you think this will affect the model's predictions? To find out, we analyze the equilibrium values.

Equilibrium Values

If we call the equilibrium values (M, O), then $M = M_{n+1} = M_n$ and $O = O_{n+1} = O_n$ simultaneously. Substituting into system (4) gives

$$\begin{aligned} M &= 1.2M - 0.001MO \\ O &= -0.7O + 0.002MO \end{aligned} \tag{5}$$

or

$$\begin{aligned} 0 &= 0.2M - 0.001MO = M(0.2 - 0.001O) \\ 0 &= -1.7O + 0.002MO = O(-1.7 + 0.002M) \end{aligned} \tag{6}$$

System (5) indicates that the mouse population doesn't change if $M = 0$ or $O = 0.2/0.001 = 200$; system (6) indicates that the owl population doesn't change if $O = 0$ or $M = 1.7/0.002 = 850$. Thus, equilibrium values exist at $(M, O) = (0, 0)$ and $(M, O) = (850, 200)$ since *neither* population changes at those points. [Substitute the equilibrium values into system (4) to check that the system indeed remains at (0, 0) and (850, 200) if either of those points represents the starting values.]

Table 3.6 Initial values for mice and owls.

Case	Mice	Owls
1	850	200
2	849	201
3	851	199
4	860	190

Now let's see what happens in the vicinity of the equilibrium values. We'll build numerical solutions of the starting populations in Table 3.6. Note that three cases are near the equilibrium point (850, 200).

Iterating system (4), beginning with the values given in Table 3.6, results in the numerical solutions graphed in Figure 3.21. In cases 2, 3, and 4, the sizes of the two populations oscillate, and the amplitudes of the oscillations seem to increase with time. Does one species eventually become extinct? ■

In the exercises you are asked to continue this exploration and determine the long-term behavior and sensitivity of the model to initial conditions (as we did for the competing species model).

Example 4 ***Voting Tendencies***

Consider a three-party system with Republicans, Democrats, and Independents. Suppose that in the next election, 75% of those who previously voted Republican do so

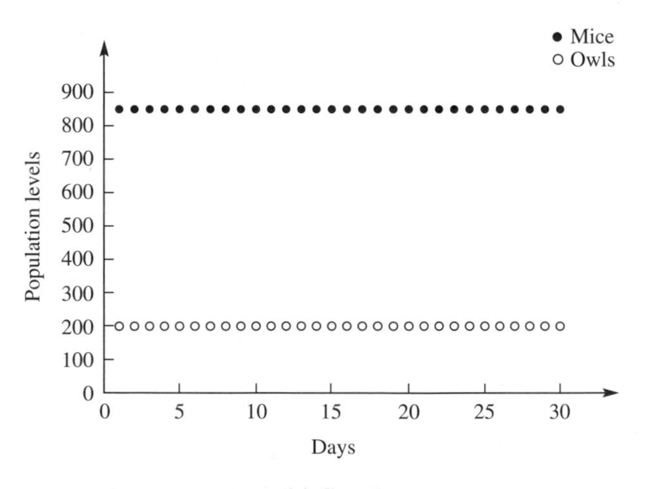

(a) Case 1

Figure 3.21 Three cases for the owl/mice predator-prey model.

n	Mice	Owls	n	Mice	Owls
1	849.0000	201.0000	16	837.6558	193.1959
2	848.1510	200.5980	17	843.3552	188.4262
3	847.6438	199.8562	18	853.1160	185.9221
4	847.7657	198.9144	19	865.1261	187.0808
5	848.6861	198.0255	20	876.3028	192.7404
6	850.3618	197.5051	21	882.6644	202.8797
7	852.4833	197.6480	22	880.1227	216.1336
8	854.4883	198.6297	23	865.9232	229.1546
9	855.6592	200.4127	24	840.6775	236.4523
10	855.3061	202.6811	25	810.0329	232.0437
11	853.0130	204.8320	26	784.0765	213.4954
12	848.8913	206.0663	27	773.4950	185.3467
13	843.7416	205.6093	28	784.8293	156.9868
14	839.0088	203.0358	29	818.5873	136.5249
15	836.4618	198.5726	30	870.5472	127.9477

● 849 Mice
○ 201 Owls

Population levels

Days

(b) Case 2

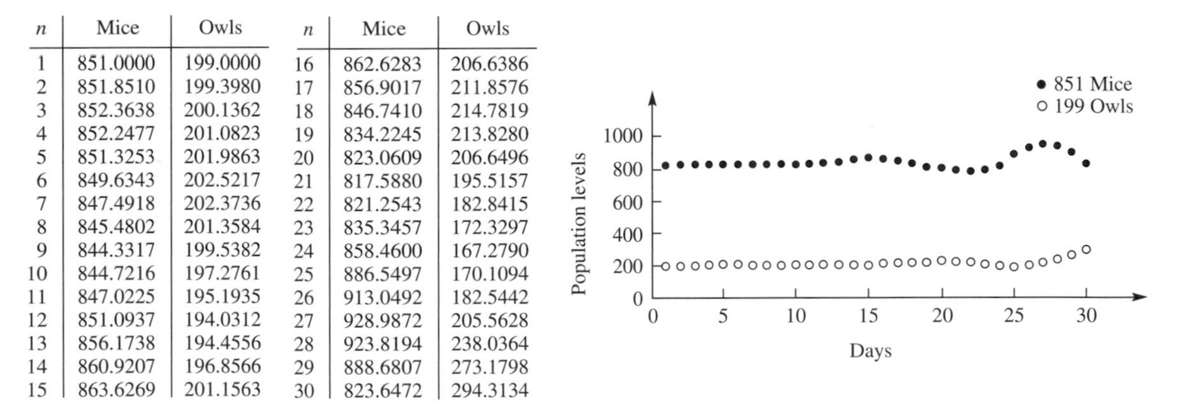

n	Mice	Owls	n	Mice	Owls
1	851.0000	199.0000	16	862.6283	206.6386
2	851.8510	199.3980	17	856.9017	211.8576
3	852.3638	200.1362	18	846.7410	214.7819
4	852.2477	201.0823	19	834.2245	213.8280
5	851.3253	201.9863	20	823.0609	206.6496
6	849.6343	202.5217	21	817.5880	195.5157
7	847.4918	202.3736	22	821.2543	182.8415
8	845.4802	201.3584	23	835.3457	172.3297
9	844.3317	199.5382	24	858.4600	167.2790
10	844.7216	197.2761	25	886.5497	170.1094
11	847.0225	195.1935	26	913.0492	182.5442
12	851.0937	194.0312	27	928.9872	205.5628
13	856.1738	194.4556	28	923.8194	238.0364
14	860.9207	196.8566	29	888.6807	273.1798
15	863.6269	201.1563	30	823.6472	294.3134

(c) Case 3

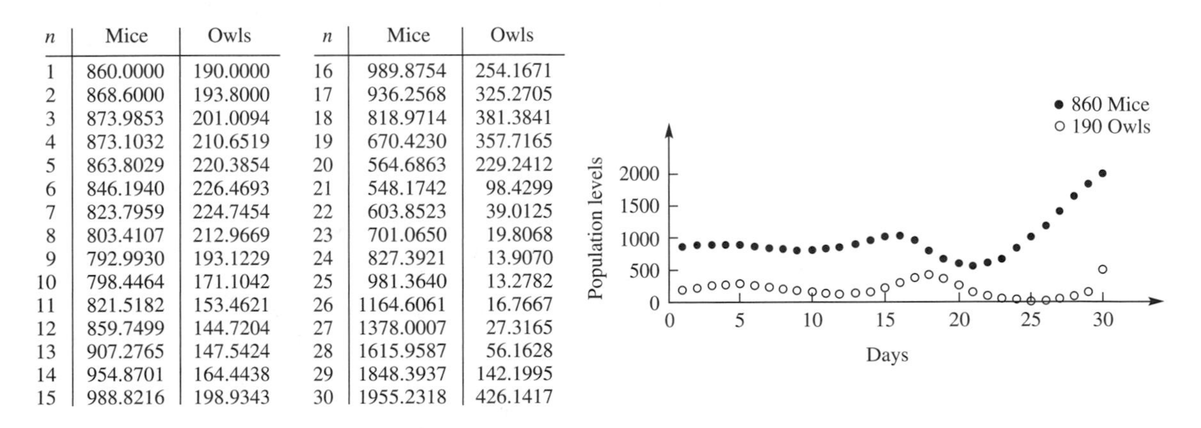

n	Mice	Owls	n	Mice	Owls
1	860.0000	190.0000	16	989.8754	254.1671
2	868.6000	193.8000	17	936.2568	325.2705
3	873.9853	201.0094	18	818.9714	381.3841
4	873.1032	210.6519	19	670.4230	357.7165
5	863.8029	220.3854	20	564.6863	229.2412
6	846.1940	226.4693	21	548.1742	98.4299
7	823.7959	224.7454	22	603.8523	39.0125
8	803.4107	212.9669	23	701.0650	19.8068
9	792.9930	193.1229	24	827.3921	13.9070
10	798.4464	171.1042	25	981.3640	13.2782
11	821.5182	153.4621	26	1164.6061	16.7667
12	859.7499	144.7204	27	1378.0007	27.3165
13	907.2765	147.5424	28	1615.9587	56.1628
14	954.8701	164.4438	29	1848.3937	142.1995
15	988.8216	198.9343	30	1955.2318	426.1417

(d) Case 4

Figure 3.21 *(Continued)*

again, but 5% vote Democrat, and 20% vote Independent. Of those who previously voted Democrat, 20% now vote Republican, 60% again vote Democrat, and 20% vote Independent. Similarly, previous Independents vote 40% Republican, 20% Democrat, and 40% Independent. Assume these tendencies continue from election to election and no additional voters enter or leave the system. These tendencies are depicted in Figure 3.22.

To formulate a system of difference equations, we let n represent the nth election and define

$$R_n = \text{number of Republican voters in the } n\text{th election}$$
$$D_n = \text{number of Democrat voters in the } n\text{th election}$$
$$I_n = \text{number of Independent voters in the } n\text{th election}$$

Formulating the system of difference equations gives the dynamical system

$$R_{n+1} = 0.75R_n + 0.20D_n + 0.40I_n$$
$$D_{n+1} = 0.05R_n + 0.60D_n + 0.20I_n$$
$$I_{n+1} = 0.20R_n + 0.20D_n + 0.40I_n$$

Equilibrium Values If we call the equilibrium values (R, D, I), then $R = R_{n+1} = R_n$, $D = D_{n+1} = D_n$, and $I = I_{n+1} = I_n$ simultaneously. Substituting into the dynamical system yields

$$-0.25R + 0.20D + 0.40I = 0$$
$$0.05R - 0.40D + 0.20I = 0$$
$$0.20R + 0.20D - 0.60I = 0$$

This system of equations has an infinite number of solutions. Letting $I = 1$, the system is satisfied if $R = 2.2221$ and $D = 0.7777694$ (approximately). Let's say the system has 399,998 voters. Then $R = 222{,}221$, $D = 77{,}777$, and $I = 100{,}000$ voters

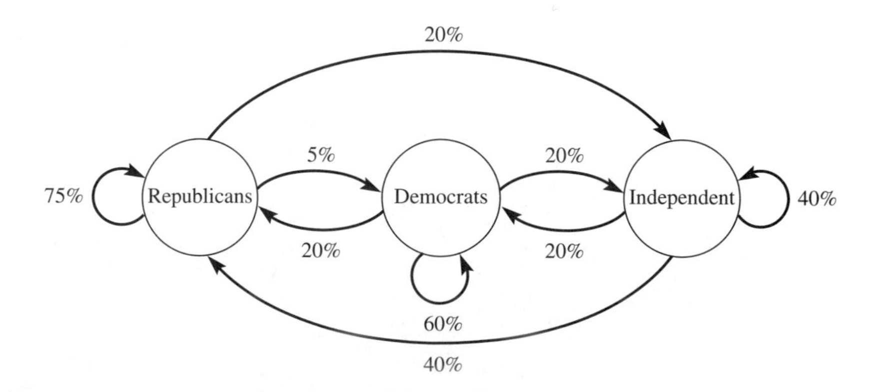

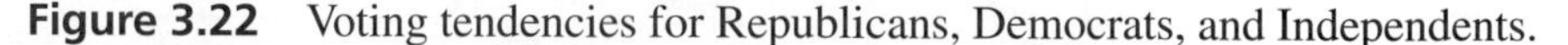

Figure 3.22 Voting tendencies for Republicans, Democrats, and Independents.

Table 3.7 Starting values for voting.

Case	Republicans	Democrats	Independents
1	222,221	77,777	100,000
2	227,221	82,777	90,000
3	100,000	100,000	199,998
4	0	0	399,998

should approximate the equilibrium values. A spreadsheet could be used to check the equilibrium values and several other values as well. The total voters in the system is 399,998 with the initial voting as shown in Table 3.7. The numerical solutions for the starting values are graphed in Figure 3.23.

Sensitivity to Initial Conditions and Long-Term Behavior

Suppose there are 399,998 voters in the system initially and all remain in the system. At least for the starting values we investigated, the system approaches the same result, even if there are initially no Republicans or Democrats in the system! This particular equilibrium appears to be stable, and starting values in its vicinity appear to approach it. What about the origin: Is it stable? ■

In the exercises you will explore this system further by examining other starting points and changing the model coefficients. Additionally, you'll investigate systems where voters enter and leave the system as well.

Summary

In this section we explored several dynamical systems and observed some interesting, perhaps surprising, results. We analyze these systems further in Chapter 11 where we

n	Republicans	Democrats	Independents
0	222221.00	77777.00	100000.00
1	222221.15	77777.25	99999.60
2	222221.15	77777.33	99999.52
3	222221.14	77777.36	99999.50
4	222221.13	77777.37	99999.50
5	222221.12	77777.38	99999.50
6	222221.12	77777.38	99999.50
7	222221.11	77777.39	99999.50
8	222221.11	77777.39	99999.50
9	222221.11	77777.39	99999.50
10	222221.11	77777.39	99999.50

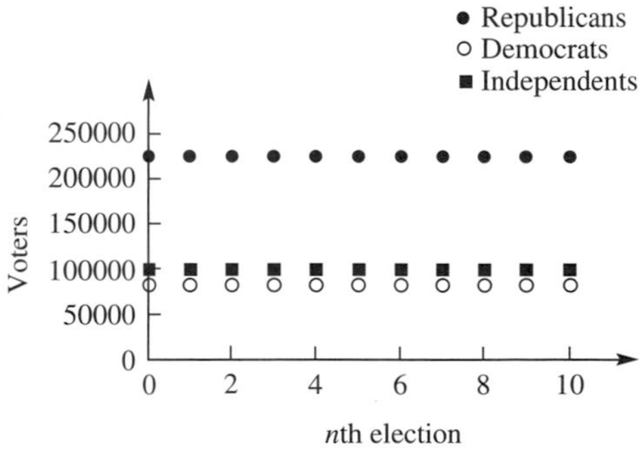

(a) Case 1

Figure 3.23 Four cases for the voting tendencies model.

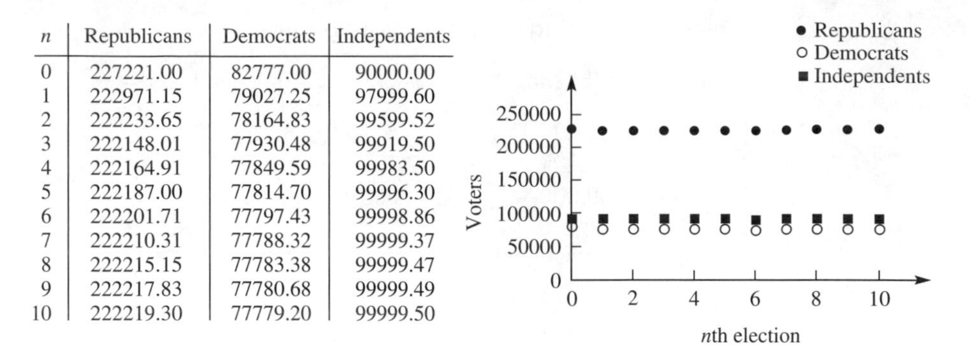

n	Republicans	Democrats	Independents
0	227221.00	82777.00	90000.00
1	222971.15	79027.25	97999.60
2	222233.65	78164.83	99599.52
3	222148.01	77930.48	99919.50
4	222164.91	77849.59	99983.50
5	222187.00	77814.70	99996.30
6	222201.71	77797.43	99998.86
7	222210.31	77788.32	99999.37
8	222215.15	77783.38	99999.47
9	222217.83	77780.68	99999.49
10	222219.30	77779.20	99999.50

(b) Case 2

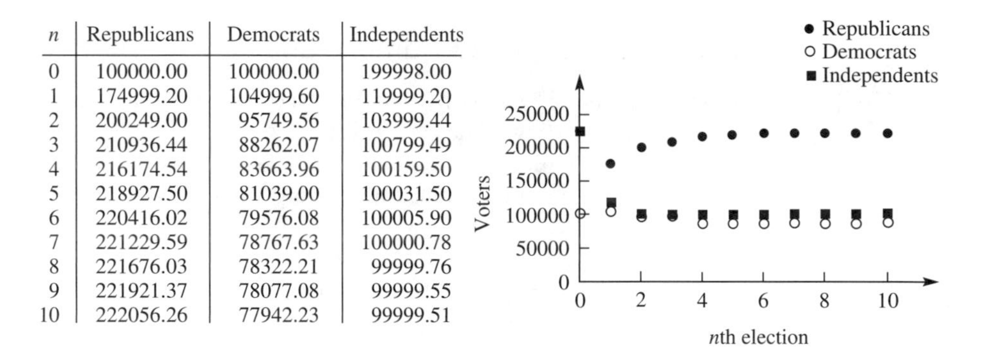

n	Republicans	Democrats	Independents
0	100000.00	100000.00	199998.00
1	174999.20	104999.60	119999.20
2	200249.00	95749.56	103999.44
3	210936.44	88262.07	100799.49
4	216174.54	83663.96	100159.50
5	218927.50	81039.00	100031.50
6	220416.02	79576.08	100005.90
7	221229.59	78767.63	100000.78
8	221676.03	78322.21	99999.76
9	221921.37	78077.08	99999.55
10	222056.26	77942.23	99999.51

(c) Case 3

n	Republicans	Democrats	Independents
0	0.00	0.00	399998.00
1	159999.20	79999.60	159999.20
2	199999.00	87999.56	111999.44
3	212398.94	85199.57	102399.49
4	217298.91	82219.59	100479.50
5	219609.90	80292.60	100095.50
6	220804.15	79175.15	100018.70
7	221445.62	78549.04	100003.34
8	221795.36	78202.37	100000.27
9	221987.10	78011.25	99999.65
10	222092.44	77906.03	99999.53

(d) Case 4

Figure 3.23 *(Continued)*

assume the behavior under study is taking place continuously. In many cases, equilibriums for continuous systems can be classified. The nature of equilibriums for **nonlinear discrete** systems (such as the competitive hunter and predator-prey models) can be highly dependent on the model's coefficients. You are asked to explore such systems in the exercises.

Exercises 3.4

1. Consider Example 1, the car rental model. Experiment with different values for the coefficients. Iterate the resulting dynamical system for the starting values given on page 100. Then experiment with different starting values. Do your experimental results indicate that the model is sensitive
a. to the coefficients?
b. to the starting values?

2. Consider Example 2, the spotted owl/hawk competitor model. Experiment with different values for the coefficients using the starting values given on page 103. Then try different starting values. Describe the model's long-term behavior. Do your experimental results indicate that the model is sensitive
a. to the coefficients?
b. to the starting values?

3. Consider Example 3, the owl/mice predator-prey model. For the starting values given on page 106, continue to iterate the system until you are confident that one species becomes extinct or that an equilibrium is achieved. Now experiment with different values for the coefficients, again using the starting values given. Try different starting values. What is the model's long-term behavior? Do your experimental results indicate that the model is sensitive
a. to the coefficients?
b. to the starting values?

4. Consider Example 4, the voting tendencies model. Experiment with starting values near the origin. Does the origin appear to be a stable equilibrium? Explain. Now experiment with different values for the coefficients using the starting values given on page 109. Try different starting values. Describe the model's long-term behavior. Do your experimental results indicate that the model is sensitive
a. to the coefficients?
b. to the starting values?

Suppose each party recruits new party members and that the total number of voters increases initially as each party recruits unregistered citizens. Experiment with different values for new party members. What is the model's long-term behavior? Does it seem to be sensitive to recruiting rates? How would you adjust your model to reflect that the total number of citizens in the voting district is constant? Adjust the model to reflect what you think is happening in your voting district. What do you think will happen in your district over the long run?

5. *Continuation of Exercise 11, Section 1.3.* Build a numerical solution to the resulting model for different starting values. Does your model indicate that coexistence of the two species is possible? Likely? Vary the k_i. Interpret the results in terms of the given situation.

6. *Continuation of Exercise 10, Section 1.3.* Pick values for your coefficients and try several starting values. What long-term behavior does your model predict? Vary the coefficients. Do your experimental results indicate that the model is sensitive
   a. to the coefficients?
   b. to the starting values?

PROJECTS Modeling Explorations

1. Complete the requirements of the UMAP module, "Graphical Analysis of Some Difference Equations in Biology," by Martin Eisen, UMAP 553. The growth of many biological populations can be modeled by difference equations. This module shows how the behavior of the solutions to certain equations can be predicted by graphical techniques.

2. Prepare a summary of a paper by May et al. listed in the references for this section.

Further Reading

Clark, C. W. (1968). *Mathematical bioeconomics: The optimal management of renewable resources*. New York: Wiley.

May, R. M. (1973). *Stability and complexity in model ecosystems*. Monographs in Population Biology VI, Princeton, NJ: Princeton University Press.

May, R. M., Beddington, J. R., Clark, C. W., Holt, S. J., Lewis, R. M. Management of multispecies fisheries. *Science* **205**(July 1979):267–277.

May, R. M., ed. (1976). *Theoretical ecology: Principles and applications*. Philadelphia: Saunders.

3.5 Curriculum Content for Grades 6–12

Mathematical ideas and discoveries have proliferated over the past century. Using supercomputers and the discoveries of new connections between mathematics and physical phenomena, mathematicians continue to broaden their field. Also, as a result of technology and our enhanced understanding of student learning abilities, mathematics has never been more accessible to students. So what content is appropriate and/or most important for high school students? Should today's students study the basic concepts taught at the beginning of the 20th century? If not, what content should be retained? What content is no longer relevant? What new ideas should they be taught?

These questions are not always easy to answer. With the field of mathematics continuously expanding, teachers must make careful decisions about what is most important and what will benefit students in the future. The NCTM *Principles and Standards* (2000) spell out five major topics of mathematics that are important for students from prekindergarten through high school: Number and Operations, Algebra, Geometry, Measurement, and Data Analysis and Probability. These topics and the ideas presented in the document are not exhaustive, but the document does provide a strong basis on which to build a quality mathematics curriculum.

Note that dividing mathematics into five content areas is somewhat artificial. In practice, mathematical ideas don't divide so neatly. However, the five areas outlined by the NCTM provide a way to organize curricula. Nonetheless, teachers must continue to make intelligent choices about the content taught at the high school level. It is important that teachers stay abreast of new concepts, as well as new applications of the mathematics that are already part of the curriculum.

The world outside the classroom should drive the curriculum. Business and industry play a vital role in keeping school districts aware of the mathematics needed in the workplace. Researchers in science and mathematics also contribute continually with recent discoveries. Many other aspects of the world also help enrich a quality mathematics curriculum. A modeling approach to math instruction helps emphasize the connections between the high school math curriculum and real-world phenomena.

Chapters 4–7 examine the NCTM Content standards in more depth and make recommendations for middle school and high school curricula. These chapters also investigate teaching the content, and review the implications of research on student learning. Chapter 8 addresses advanced mathematics for the high school, while Chapters 9–11 continue to introduce techniques for mathematical modeling. Throughout Chapters 4–8, a modeling approach to mathematics instruction is explored.

3.6 The Modeling Process

In previous sections we used models to help us understand real-world behavior. We performed the following steps:

1. Through observation, identify the primary factors involved in the real-world behavior, possibly making simplifications.
2. Conjecture tentative relationships among the factors.
3. Apply mathematical analysis to the resultant model.
4. Interpret mathematical conclusions in terms of the real-world problem.

We now examine the modeling process more closely.

Figure 3.24 portrays the entire modeling process as a closed system. Given some real-world system, we gather sufficient data to formulate a model. Next we analyze the model and reach conclusions. Then we interpret the model and make predictions or offer explanations. Finally, we test our conclusions about the real-world system against new

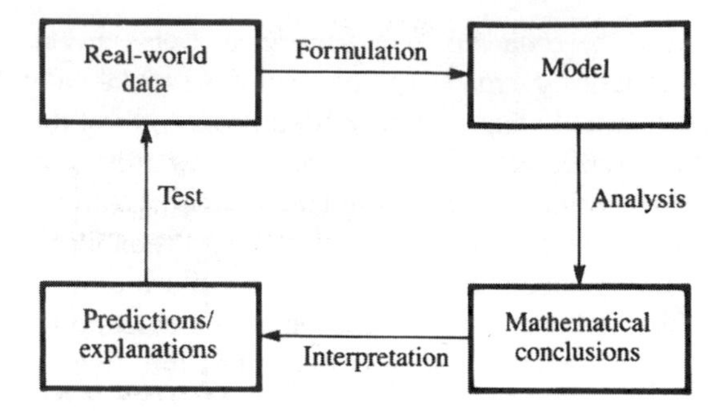

Figure 3.24 The modeling process as a closed system.

observations and data. We may then find we need to refine the model in order to improve its predictive or descriptive capabilities. Or perhaps we'll discover that the model really doesn't "fit" the real world at all, so we must formulate a new model. You will study the various components of this modeling process in detail throughout the book.

Mathematical Models

For our purposes, we define a **mathematical model** as a mathematical construct designed to study a particular real-world system or phenomenon. We include graphical, symbolic, simulation, and experimental constructs. Mathematical models can be differentiated further. Mathematical models already exist that can be identified with some particular real-world phenomenon and can be selected to study it. Then there are mathematical models that we construct specifically to study a phenomenon. Starting with some real-world phenomenon, we can represent it mathematically by constructing a new model or selecting an existing model as shown in Figure 3.25. On the other hand, we can also replicate the phenomenon experimentally or with some kind of simulation.

When it comes to constructing a mathematical model, a variety of conditions can cause us to abandon hope of achieving any success. The mathematics involved may be so complex and intractable that there is little hope of analyzing or solving the model, thereby defeating its utility.

This complexity occurs when models given by a system of partial differential equations or a system of nonlinear algebraic equations are used. Or the problem may be so large (in terms of the number of factors involved) that it is impossible to capture all

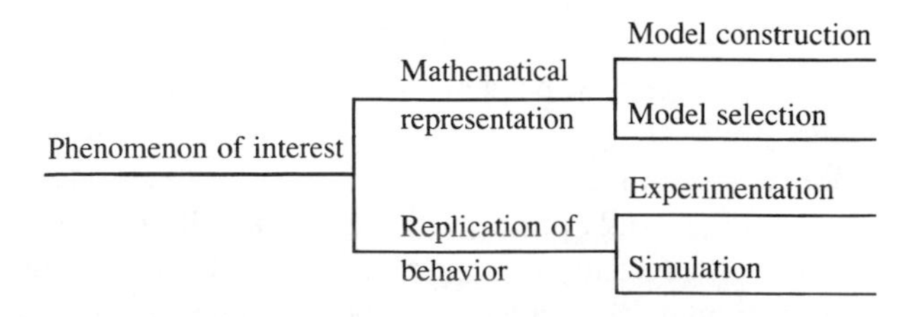

Figure 3.25 The nature of the model.

the necessary information in a single mathematical model. Predicting the global effects of the interactions of a population, the use of resources, and pollution exemplifies such a situation. In such cases, we may attempt to replicate the behavior *directly* by conducting various experimental trials. Then we analyze our data from these trials, possibly with statistical techniques or curve-fitting procedures. This limited replication allows us to make certain limited conclusions.

In other cases, we may replicate the behavior *indirectly*. We might use an analogue device such as an electrical current to model a mechanical system. We might use a scaled representation of a jet aircraft in a wind tunnel. Or we might replicate a behavior on a digital computer—for instance, simulating the global effects of the interactions of population, use of resources, and pollution, or simulating the operation of an elevator system during morning rush hour.

The Construction of Models

In the preceding discussion we viewed modeling as a process and briefly considered the form of the model itself. Now let's focus on constructing mathematical models. We begin by outlining a helpful procedure.

Step 1. Identify the Problem

What do you want to find out? Typically, this is a troublesome step because people often have great difficulty sorting through parameters and deciding on criteria. In real life, no one simply hands us a mathematical problem to solve. Usually we sort through large amounts of data and identify some particular aspect of the situation we wish to study. Moreover, we must be sufficiently precise (ultimately) in our problem formulation so that we can translate the problem into mathematical statements. This translation is accomplished through the next several steps. It is important to realize that answers to the posed questions don't lead directly to a usable problem formulation.

Step 2. Make Assumptions

Generally you cannot hope to capture in a usable mathematical model all of the problem factors that have been identified. You can, however, simplify the task by reducing the number of factors under consideration. Then relationships between the remaining variables must be determined. Again, the complexity of the problem can be reduced by assuming relatively simple relationships. Thus, the assumptions fall into two main activities:

a. Classify the Variables. What things influence the problem situation you identified in step 1? List them as variables. The variables the model seeks to explain are the *dependent variables*, and there may be several of these. The remaining variables are the *independent variables*.

You may choose to ignore some of the independent variables for one of two reasons. First, the effect of the variable may be relatively small compared to other factors involved. You may also ignore variables that affect all the alternatives in about the same way, even though they may have important effects on the problem situation. For example, consider the problem of determining the optimal shape for a lecture hall where readability of a chalkboard or overhead projection is a dominant criterion. Lighting is certainly a crucial factor, but perhaps you know that it affects all possible shapes in about the same way. You can simplify the analysis considerably by initially ignoring this variable, possibly incorporating it later in a separate, more refined model.

b. Determine Interrelationships among the Variables. Before you can hypothesize relationships between the variables, you generally must make some additional simplifications. Also, the problem may be sufficiently complex so that you don't initially see a relationship among the variables. In such cases, it may be possible to study **submodels**. This enables you to study one or more independent variables separately. Eventually you will connect the submodels.

Step 3. Solve or Interpret the Model

Now put together all the submodels to see what the model is telling you. In some cases, the model may consist of mathematical equations or inequalities that must be solved in order to find the information. Often a problem statement requires a "best" or *optimal solution* to the model.

You frequently will find that you need to do more work on your submodels before you can complete this step. Or you may end up with a model so unwieldy you cannot solve or interpret it. In such situations, return to step 2 and make additional simplifying assumptions. You may even want to return to step 1 to redefine the problem. This point will be amplified in the following discussion.

Step 4. Verify the Model

Before you use the model, you must test it. There are several questions you should ask before you design the tests and collect data—a process that can be expensive and time-consuming. First, does the model answer the problem you identified in step 1? Think carefully about your answer here. It's easy to stray from the key issue during model construction. Second, is the model usable in a practical sense; that is, can you really gather the data necessary to operate the model? Third, does the model make sense?

Once the commonsense test is passed, you will want to test your model using data obtained from empirical observations. Design the test carefully; make sure you include observations over the *same range* of values of the various independent variables you expect to encounter when you actually use the model. The assumptions you made in step 2 may be reasonable over a restricted range of the independent variables, but very poor outside those values. For instance, a frequently used interpretation of Newton's second law states that the net force acting on a body is equal to the mass of the body times its acceleration. This interpretation is a reasonable model until the speed of the object approaches the speed of light.

Be very careful about the conclusions you draw from tests. Just as you can't prove a theorem simply by demonstrating many cases where it holds, likewise, you can't extrapolate broad generalizations from the particular evidence you gather about your model. A model doesn't become a law just because it is verified repeatedly in some specific instances. Rather, you *corroborate the reasonableness* of your model through the data you collect.

Step 5. Implement the Model

Of course, your model is of no use sitting in a filing cabinet. You will want to explain your model in terms that decision makers and users can understand if it is ever to be of use to anyone. Further, unless the model is user-friendly, it will quickly fall into disuse. Expensive computer programs sometimes suffer such a demise. Often the inclusion of an additional step to facilitate the collection and input of the data necessary to operate the model determines its success or failure.

Step 6. Maintain the Model

Note that your model is derived from the problem you identified in step 1 and from the assumptions you made in step 2. Has the original problem changed in any way, or have previously ignored factors become important? Does a submodel need to be adjusted?

We summarize the steps for constructing mathematical models in the box. We should not be too enamored with our work. Like any model, our procedure is an approximation process and therefore has its limitations. For example, the procedure seems to consist of discrete steps leading nicely to a usable result, but that's rarely the case in practice. Before offering an alternative procedure that emphasizes the iterative nature of the modeling process, let's discuss the advantages of this methodology.

The Model Construction Process

Step 1. Identify the problem.

Step 2. Make assumptions.

- a. Identify and classify the variables.
- b. Determine interrelationships between the variables and submodels.

Step 3. Solve the model.

Step 4. Verify the model.

- a. Does it address the problem?
- b. Does it make sense?
- c. Test it with real-world data.

Step 5. Implement the model.

Step 6. Maintain the model.

The process outlined in the box enables you to progressively focus on those aspects of the problem you wish to study. Furthermore, it demonstrates the rather curious blend of creativity with the scientific method used in the modeling process. The first two steps are a bit of an art. They involve abstracting the essential features of the problem, ignoring factors judged to be unimportant, and postulating relationships that are precise enough to answer the questions posed yet simple enough to permit the completion of the remaining steps. While these steps admittedly involve a degree of craft, certain scientific techniques will help you assess the importance of a particular variable and the preciseness of an assumed relationship. Nevertheless, when generating numbers in steps 3 and 4, remember that the preceding process has largely been inexact and intuitive.

Let's contrast the modeling process with the scientific method. One version of the **scientific method** is as follows:

Step 1. Make some general observations of a phenomenon.

Step 2. Formulate a hypothesis about the phenomenon.

Step 3. Develop a method to test that hypothesis.

Step 4. Gather data to use in the test.

Step 5. Test the hypothesis using the data.

Step 6. Confirm or deny the hypothesis.

By design the modeling process and scientific method have some obvious similarities. For instance, both processes make assumptions or hypotheses, gather real-world data, and test or verify hypotheses using that data. These similarities should not surprise you; while recognizing that the modeling process is an art, we do attempt to be scientific and objective whenever possible. Nevertheless, some subtle differences exist between the two procedures. One is in their primary goals. In the modeling process, we make assumptions that may or may not be valid when we select which variables to include or ignore and postulate the relationships between the remaining variables. The goal in modeling is to *hypothesize a model*, and like the scientific method, evidence is gathered to corroborate that model. However, unlike the scientific method, the objective is not to *confirm* or *deny* the model (we already know it is not precisely correct because of our simplifying assumptions), but rather to test its *reasonableness*. We may decide that the model is quite satisfactory and useful, and elect to accept it. Or we may decide that the model needs to be refined or simplified. In extreme cases, we may even redefine the problem, in a sense rejecting our original model altogether. This decision process really constitutes the heart of mathematical modeling.

Iterative Nature of Model Construction

Model construction is an iterative process. We begin by examining some system and identifying the particular behavior we wish to predict or explain. Next we identify the variables and simplifying assumptions, and then we generate a model. We generally start with a rather simple model, progress through the modeling process, and then refine the model as our results dictate.

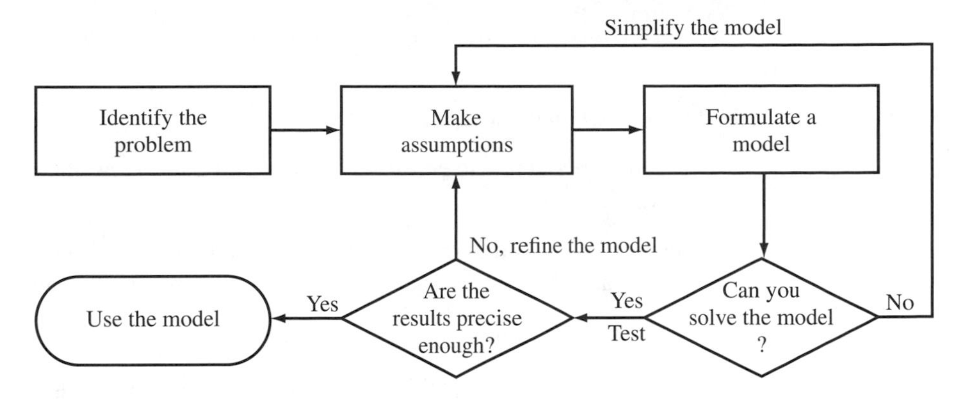

Figure 3.26 The iterative nature of model construction.

Table 3.8 The art of mathematical modeling: simplying or refining the model as required.

Model Simplification	Model Refinement
1. Restrict problem identification.	**1.** Expand the problem.
2. Neglect variables.	**2.** Consider additional variables.
3. Conglomerate effects of several variables.	**3.** Consider each variable in detail.
4. Set some variables to be constant.	**4.** Allow variation in the variables.
5. Assume simple (linear) relationships.	**5.** Consider nonlinear relationships.
6. Incorporate more assumptions.	**6.** Reduce the number of assumptions.

It is therefore important to keep in mind that if you can't come up with a model or solve the one you have, you must *simplify* it (see Figure 3.26). This is done by treating some variables as constants, by ignoring or aggregating some variables, by assuming simple relationships (such as linearity) in submodels, or by further restricting the problem. On the other hand, if your results aren't precise enough, you must *refine* the model (see Figure 3.26). Refinement is generally achieved in the opposite direction: You introduce additional variables, assume more sophisticated relationships among the variables, or expand the scope of the problem. By simplification and refinement, you determine the generality, realism, and precision of your model. This process cannot be overemphasized; this is the art of modeling and key to successful models. These ideas are summarized in Table 3.8.

Exercises 3.6

In Exercises 1–8, the scenarios are vaguely stated. From these, identify a problem you would like to study. What variables affect the behavior you have identified in the problem statement? Which variables are the most important? Remember, there are no "right" answers.

1. The population growth of a single species.
2. A retail store intends to construct a new parking lot. How should the lot be illuminated?
3. A farmer wants to maximize the yield of a certain crop of food grown on his land. Has the farmer identified the correct problem? Discuss alternative objectives.
4. How would you design a lecture hall for a large class?
5. An object is to be dropped from a great height. When and how hard will it hit the ground?
6. How should a manufacturer of some product decide how many items of that product should be manufactured each year and how much to charge for each item?
7. The U.S. Food and Drug Administration is interested in knowing whether a new drug is effective in controlling a certain disease in the population.
8. How fast can a skier ski down a mountain slope?

For the scenarios presented in Exercises 9–17, identify a problem worth studying and list the variables that affect the problem behavior. Which variables would be ignored completely? Which might be initially considered as constants? Can you identify any submodels you would want to study in detail? What data you would want collected?

9. A botanist is studying the shapes of leaves and the forces that mold them. He clips some leaves from the bottom of a white oak tree and finds the leaves to be rather broad, not very deeply indented. When he goes to the top of the tree, he gets very deeply indented leaves with hardly any broad expanse of blade.
10. Animals of different size work differently. The small ones talk in squeaky voices, their hearts beat faster, and they breathe more often than large animals. On the other hand, the skeleton of a larger animal is more robustly built than that of a small animal. The ratio of the diameter to the length in a larger animal is greater than it is in a smaller one. Thus, there are regular distortions in the proportions of animals as the size increases from small to large.
11. A physicist is studying properties of light. She wants to understand the path of a ray of light as it travels through the air into a smooth lake, particularly at the interface of the two different media.
12. A company with a fleet of trucks faces rising maintenance costs as the age and mileage of the trucks increase.
13. Man is fixated by speed. Which computer systems offer the most speed?
14. How can we improve our ability to sign up for the "right classes" each term?
15. How should we save a portion of our earnings?

16. Consider a start-up company that produces a single product in a competitive market situation. Discuss some of the short- and long-term goals the company might identify. How do these goals affect employee job assignments? Would the company necessarily decide to maximize profits in the short run?

17. Discuss the differences between using a model to predict, versus to explain, a real-world system. Think of some situations in which you would like to explain a system; likewise, imagine others in which you would want to predict a system.

PROJECTS Modeling Explorations

1. Consider the taste of brewed coffee. What variables affect its taste? Which variables might be ignored initially? Suppose you hold all variables fixed except water temperature. Most coffeepots use boiled water in some manner to extract flavor from the ground coffee. Is boiled water optimal for producing the best flavor? How would you test this submodel? What data would you collect and how would you gather it?

2. A company is considering transporting people between skyscrapers in New York City via helicopter. You are hired as a consultant. Can such an idea possibly make sense? Could the company make a profit? What about safety problems, noise restrictions, and so forth? Identify an appropriate problem precisely. Use the model construction process to identify the data you would like to have in order to determine the relationships between the variables you select. You may want to redefine your problem as you proceed.

3. Consider wine making. Suggest some objectives a commercial producer might have. Consider taste as a submodel. What variables affect taste? Which variables might be ignored initially? How would you relate the remaining variables? What data would be useful to determine the relationships?

4. Should a couple buy or rent a home? If mortgage costs rise, intuitively it would seem that there is a point where it no longer "pays" to buy a house. What variables determine the total cost of a mortgage?

5. Consider the operation of a medical office. Detailed records are kept on individual patients, and accounting procedures are a daily task. Should the office buy or lease a small computer system? Suggest factors that might be considered. What variables would you consider? How would you relate the variables? What data would you like to have in order to determine the relationships between the selected variables? Why might solutions to this problem differ from office to office?

6. When should people replace their vehicles? What factors should affect the decision? Which variables might be neglected initially? Identify the data you need to determine the relationships among the variables.

7. How far can a person long-jump? In the 1968 Olympic Games in Mexico City, Bob Beamon of the United States increased the record by a remarkable 10%. List the

variables that affect the length of the jump. Do you think the low air density of Mexico City accounts for the 10% difference?

8. Is college a financially sound investment? Income is forfeited for four years, and college costs are high. What factors determine the total cost of a college education? How would you determine the circumstances necessary for the investment to be profitable?

References

Boyd, C. J., Burrill, G. F., Cummins, J. J., Kanold, T. D., Malloy, C. (1998). *Geometry: Integration applications, and connections*. New York: Glencoe/McGraw-Hill.

COMAP (1998). *Mathematics: Modeling our world course 2*. Cincinnati, OH: South-Western Educational Publishing.

Corwin, R. B., & Storeygard, J. (1995). Talking mathematics: Supporting discourse in elementary school classrooms. *Hands On!* **8**(1):6.

Dossey, J. A., & McCrone, S. M. (1999). *Mathematics: A chapter of the ASCD curriculum handbook*. Alexandria, VA: Association for Supervision and Curriculum Development.

Hiebert, J., Carpenter, T. P., Fennema, E., Fuson, K., Wearne, D., Murray, H., Olivier, A., & Human, P. (1997). *Making sense: Teaching and learning mathematics with understanding*. Portsmouth, NH: Heinemann.

Lester, F. K. (1994). Musings about mathematical problem-solving research: 1970–1994. *Journal for Research in Mathematics Education*, **25**(6), 660–675.

Martin, T. S., & McCrone, S. (April 1999). *Preludes to proof*. Presentation made at the Annual Meeting of the National Council of Teachers of Mathematics. San Francisco, CA.

Mathematics Association of America. (1997). Second report from the task force [On-line text available: http://www.maa.org/past/maanctm3.html.]

National Council of Teachers of Mathematics (1989). *Curriculum and evaluation standards for school mathematics*. Reston, VA: NCTM.

———. (2000). *Principles and standards for school mathematics*. Reston, VA: NCTM.

———. (1991). *Professional standards for teaching mathematics*. Reston, VA: NCTM.

Polya, G. (1945). *How to solve it*. Princeton, NJ: Princeton University Press.

Schoenfeld, A. H. (1985). *Mathematical problem solving*. Orlando, FL: Academic Press.

Silver, E., Smith, M., & Nelson, B. S. (1995). The QUASAR project: Equity concerns meet mathematics education reform in the middle school. *In* W. G. Secada, E. Fennema, and L. B. Adajian (eds.), *New directions for equity in mathematics education* (pp. 9–56). Cambridge, England: Cambridge University Press.

Yackel, E. (1998). *Reasoning and proof*. White paper prepared for the National Council of Teachers of Mathematics, Reston, VA.

Yackel, E., & Cobb, P. (1996). Sociomathematical norms, argumentation, and autonomy in mathematics. *Journal for Research in Mathematics Education* **27**(4):458–477.

Part II

Middle and High School Mathematics Curricula: Fitting Mathematics to Reality

4
Number and Operations

Without question the most important goal of school mathematics is to develop students' ability to reason intelligently with quantitative information. The mathematical concepts, techniques, and principles that model quantitative aspects of experience are provided by structures of number systems, algebra, and measurement that have long been the heart of school curricula. However, the emergence of electronic calculators and computers as powerful tools for representing and manipulating quantitative information has challenged traditional priorities for instruction in those subjects.

James T. Fey

Overview

Number and operations form a basis for doing mathematics at all levels of middle school and high school. Students encounter many situations both in and out of school where fluency with numbers and calculations is helpful. As noted in Chapter 1, today's society requires us to be mathematically literate. One aspect of mathematical literacy deals with an individual's ability to understand and work with numerical relationships. For this reason, continued attention to number and operations is crucial throughout middle school and high school.

In this chapter we consider the role of number and operations in the middle school and secondary curriculum, including how topics arise in the traditional curriculum, and where new ideas and topics might be investigated. In addition, we discuss teaching this content area, including a look at the relevant research on learning algorithms and techniques for estimating. This chapter and the next four also contain sample lesson plans that exemplify teaching as described by the NCTM content standards. These lesson plans, found at the end of each chapter, provide ideas for shaping instructional programs at the middle school and high school levels. Lesson planning and related issues are discussed in more detail in Chapters 12–15.

Focus on the NCTM *Standards*

It is fitting that Number and Operations is the first of the five content standards in the NCTM *Principles and Standards for School Mathematics* since so much of school mathematics depends on student abilities to understand numbers and number systems and to operate within these systems. The Number and Operations standard shown here outlines goals for students at all levels.

Number and Operations Standard

Instructional programs from prekindergarten through grade 12 should enable all students to

- understand numbers, ways of representing numbers, relationships among numbers, and number systems;
- understand meanings of operations and how they relate to one another;
- compute fluently and make reasonable estimates. (NCTM, 2000)

The study of number and operations has long been a prominent part of the school mathematics tradition in the United States. Their importance in all areas of schooling and life outside the classroom are emphasized in the NCTM *Standards*. The NCTM also acknowledges that the study of number and operation is not only an appropriate topic for the elementary grades. Although teachers at the high school level may expect student fluency in whole numbers, integers, and basic operations, they don't always get it. So some review may be necessary even as teachers move students to other areas of mathematics and more complex aspects of number and operations.

Focus on the Classroom

The importance of number and operations in the middle and high school classroom is obvious. We encounter numbers in all areas of mathematics. Computational fluency and an ability to estimate are crucial as students encounter situations that require critical thinking coupled with basic number sense. Computational fluency and estimation skills are also needed in other subjects, such as the natural and physical sciences. Hence, students need continued practice working with numbers and operations. Focusing on number systems and operations through complex numbers and matrices helps students build on their knowledge of number and operation systems and deepens their mathematical understanding. Number theory and counting strategies should also be part of the curriculum.

4.1 Number and Operations in the Curriculum

By middle school, students should be very familiar with whole numbers and integers. Continued work in these number systems, whether listing factors of composite numbers or finding primes or squares, strengthens student understanding. During middle school, students begin to encounter rates, ratios, and proportion. This leads students from the more familiar whole numbers and integers to the rational numbers and their various forms such as fractions, percents, and decimals. Consequently, middle school students have a new number system to make sense of. Just as they learned about negative numbers, students must learn about the basic structure of rational numbers and how they behave. One way to help students make sense of the concepts and properties of rational numbers is to make comparisons across number systems. For instance, students might investigate which properties of integers carry over to their work with rational numbers and which properties don't.

At the high school level, students should be encouraged to use their knowledge of integers and rational numbers to make sense of other number systems, such as the real numbers and the complex numbers. In addition, new ways of representing numbers, such as set notation and matrices, help students develop flexibility in working with and operating on numbers. For example, students can compare the relative size of rational and irrational numbers based on various representations. Students can also investigate the commutative and associative properties for matrices, and consider which familiar properties hold for matrices, which don't, and why.

Additional new topics such as combinatorics and sequences extend familiar counting ideas. Counting techniques of combinatorics introduce students to new strategies for collecting and working with data. These techniques offer ways of organizing the counting process. Sequences and series also describe ways to organize a process through pattern finding. These ideas may be new to students at the middle school or high school level, but can be approached from a very basic level using familiar number systems.

Students in the middle school and high school have widely varying abilities to use and make sense of numbers. Some students rely on calculators for arithmetic computations while others quickly compose or decompose numbers mentally. With practice, students develop "good intuition about numbers and their relationships" (Howden, 1989), and maintain their fluency and proficiency in computation.

4.2 Number Systems, Their Properties, and Operations

Number Systems

Through work with numbers and number systems, students gain a sense of the relationships and differences between numbers and number systems. In particular, middle school students need to continue to investigate the familiar operations of multiplication and division and to compare integers and rational numbers. They are often surprised when they discover that multiplication doesn't always result in a number larger than the two factors, and division doesn't always give a smaller result. How can we explain this phenomenon to our students?

One way to help students understand the changes that take place in rational-number multiplication and division, as opposed to integers, is to make connections between these number systems. For instance, we refer to multiplication as repeated addition. Students often wonder if this definition carries over to multiplying fractions. You might ask the student to interpret the definition in terms of a specific problem involving a combination of integers and rational numbers. When finding the product of an integer and a fraction, we certainly think about repeated addition of the fraction.

Suppose a student is asked to find the product of 6 and $\frac{1}{4}$. The student can think of this as adding the quantity $\frac{1}{4}$ six times. Realizing that 4 quarters make a whole, the student gets 1 whole, plus 2 more quarters or $\frac{1}{2}$. So, $6 \cdot \frac{1}{4} = 1\frac{1}{2}$.

Does the idea of multiplication as repeated addition help explain the product of two rational numbers such as $\frac{7}{8}$ and $\frac{3}{5}$? Not really. But repeated addition can be used to introduce middle school students to new terminology and a new way of viewing multiplication: $7 \cdot \frac{3}{5}$ means we want 7 of the quantity $\frac{3}{5}$ (or seven $\frac{3}{5}$-cup servings of rice), and we can use repeated addition, as above, to find the result. So, $\frac{7}{8} \cdot \frac{3}{5}$ means we want $\frac{7}{8}$ of the quantity $\frac{3}{5}$. We usually shorten this to read "$\frac{7}{8}$ *of* $\frac{3}{5}$." Students should be encouraged to find their own ways for determining the value of this product. They may choose to use a pictorial representation such as the one in Figure 4.1, or they may develop an algorithm such as finding the value $7 \cdot \frac{3}{5}$ using repeated addition, and then taking $\frac{1}{8}$ of that value to arrive at a final answer. However students make sense of multiplying fractions, they should also be shown how and why their algorithms connect to the standard algorithm: $\frac{7}{8} \cdot \frac{3}{5} = \frac{(7 \cdot 3)}{(8 \cdot 5)}$. This connection is shown in Figure 4.1.

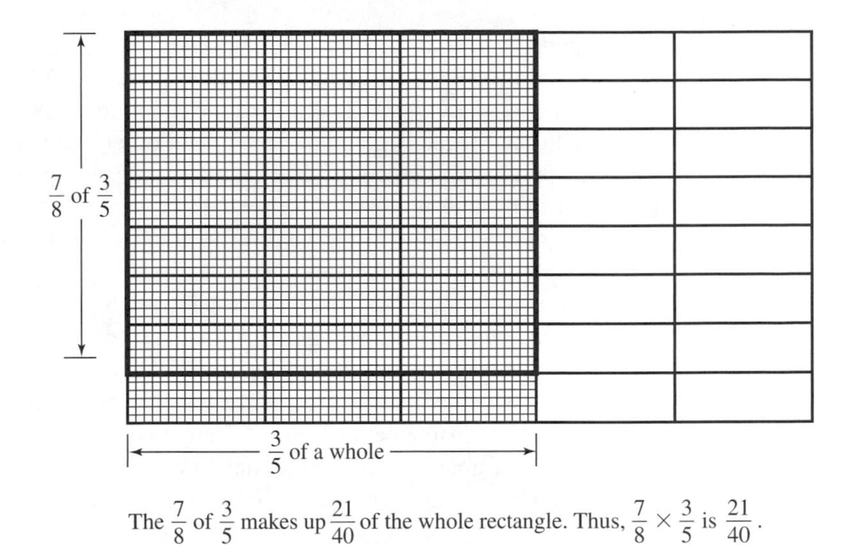

The $\frac{7}{8}$ of $\frac{3}{5}$ makes up $\frac{21}{40}$ of the whole rectangle. Thus, $\frac{7}{8} \times \frac{3}{5}$ is $\frac{21}{40}$.

Figure 4.1 A pictorial algorithm for multiplying fractions.

Problem to Investigate

The standard division algorithm for fractions (invert and multiply) is often taught to students with no attempt to foster conceptual understanding. Why does this procedure work? How does it connect to division of integers? How would you share these connections with middle school or even high school students? How might you help students explain the basis of the algorithm?

At the high school level, students need to be familiar with rational numbers, including their properties and their operations, before they encounter irrational numbers, complex numbers, and different representations of numbers such as vectors and matrices. Working with these new number systems and representations will be easier if students have a solid understanding of proportions, the distributive property, and other similar properties, as well as the familiar operations of addition, subtraction, multiplication, and division. Typically, these experiences occur in middle school or in pre-algebra as well as in algebra. But students should receive plenty of practice working within the real-number system in any mathematics course.

Properties of Number Systems

All grade levels of students should be provided opportunities to explore properties of numbers. In the middle school, students encounter prime numbers, square numbers, and the factorability of whole numbers. In high school, students continue to investigate these properties, but they extend their investigations to a wider range of properties, including equivalent fractions, relative primes, finding the square and cube roots of numbers, and representing complex conjugates. With the introduction of each new number system, students should investigate which properties from other number systems still hold and what new properties can be defined. For example, students might wonder if primes are a property of nonwhole numbers. Can we talk about the prime factors of a fraction or a complex number? Why don't we refer to fractions as being prime or composite numbers?

Properties of number systems make for interesting investigations. For instance, you can ask students what important properties are preserved when they move from whole numbers to integers, from integers to rational numbers, or from rational numbers to real numbers. Which properties are not preserved? A property of real numbers important in many areas of mathematics is that, for any two rational numbers, no matter how close they are, we can always find a rational number between them. More precisely, if a and b are rational numbers with $a < b$, then there exists a rational number x such that $a < x < b$. After working with complex numbers, students should come to realize that this property does not hold for the complex numbers. In fact, there is no order relation for complex numbers. For instance, we cannot determine if $3 - 5i$ is greater than, less than, or equal to $2 - 3i$. We can represent these numbers in Cartesian coordinates (Figure 4.2), but how does this help us determine their relative sizes?

Operations

Often, number and operations are not a significant part of the high school curriculum except for students in low-level courses who are not fluent in basic computations or who haven't mastered basic algorithms. Hence, not much time is spent on strengthening student understanding of number systems, their operations, and their properties. Consequently, gaps in student skill and understanding show up in related work. When students work in new systems and have difficulty making sense of the fundamentals, such as simplifying expressions through arithmetic, factorization, and combining "like" terms, this indicates a flawed understanding of similar systems taught in earlier grades. Since these concepts are fundamental, teachers should continue to stress reasonable

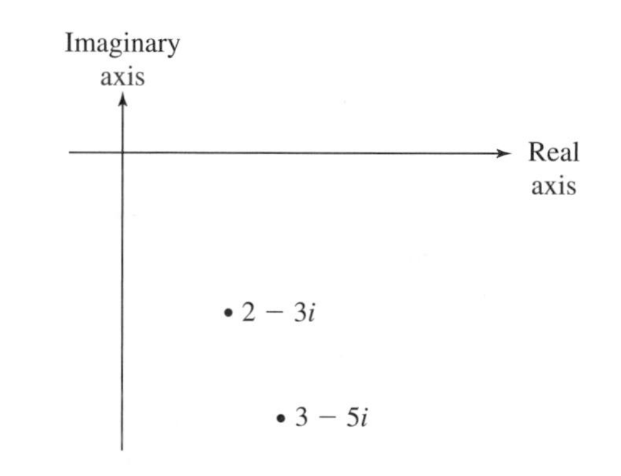

Figure 4.2 Coordinate representation of complex numbers.

levels of fluency with basic arithmetic as well as facility with number systems. Students who obtain a deep conceptual understanding of the operations are more likely to carry this knowledge to new number systems. Such students adapt their understanding to new situations, and develop their own algorithms when traditional algorithms don't make sense.

To develop students' conceptual understanding of number and operations, students should be given opportunities to compare the relative size of numbers, and to work with operations and number in multiple situations using multiple representations. The models and techniques presented in Chapters 2 and 3, for instance, require some facility with number and operations. Such problems help students sharpen their skills and apply this knowledge to context-based problems (not just computation for computation's sake). Modeling frequently uses various representations of number and operations, such as pictures, graphs, symbols, and tables. Many rich mathematical situations exist that will strengthen student understanding of number and operations. Some of these are discussed below and are shared in the lesson plans at the conclusion of this chapter.

Exercises 4.2

1. How would you explain this statement to a class of pre-algebra students?

$$\frac{1}{a+b} \neq \frac{1}{a} + \frac{1}{b}$$

2. Find a way to explain why multiplying fractions (with values less than 1) results in a number smaller than the two factors being multiplied.

3. Integer and rational values can be presented pictorially using squares or other objects. How might you present the value $\sqrt{3}$?

4. If a number system is commutative, must it also be associative? How could you find out?

5. Does every nonempty set of whole numbers have a least element? What about each nonempty set of integers? Or each nonempty set of positive rational numbers?

PROJECT Find Out for Yourself

Visit a junior high school or middle school classroom.

1. In what ways are number and operations an explicit part of the curriculum? In other words, does the teacher work with students on particular skills or concepts related to number systems? Do students practice operating within a given system?

2. In what ways are number and operations an implicit part of the curriculum? Are students expected to use rational numbers in problem solving? Do students employ algorithms or familiar operations without explicit mention of the algorithm or method? That is, they're effectively using them but don't know it.

3. What suggestions would you give for making number and operations a more central component of the middle school or high school curriculum?

4.3 Number Theory in Problem Solving

In the United States we value knowledge of number and operations for its own sake. Arithmetic (one of the three R's of education) has long been a school tradition. Even so, we should continue to question the value of teaching arithmetic in our changing society. With so much reliance on technology to do our computing for us, why must we continue to teach standard arithmetic in school? The value of arithmetic lies in its usefulness in problem solving. Whether we are balancing a checkbook, calculating our semester average in a class, or working on an intriguing number puzzle, we use our knowledge of number and operations. Working on number puzzles also helps us develop problem-solving techniques while deepening our understanding of number properties. Elementary number theory, including various number puzzles, is an area of mathematics that fits well with a middle school or early high school curriculum. In this section we discuss several aspects of number theory that are essential for developing number sense preparatory to algebra learning.

One area of number theory that helps to build a sense of number and operations is the basic structure of consecutive integers. A sense of the properties of consecutive integers or whole numbers can be helpful in making sense of how numbers work when we put them together with operations. Consider the following problem and how it supports student understanding of compositions of consecutive numbers. Since consecutive integers are alternately even and odd, this problem also highlights characteristics of evens and odds.

Problem to Investigate: Partitioning Integers

In the set of all integers from 1 to 15, how can you partition the set into five sets of three elements such that the sum of the integers in each subset is equal? Is it possible to find more than one solution?[1]

When students begin working on a problem such as this, they instinctively fall back on prior knowledge. One student may list the integers between 1 and 15, and then choose groups of three numbers whose sums are equal. He may form two or three groups before he realizes that not all five groups have equal sums. A pair of students may brainstorm ways to arrange the numbers such as grouping the lowest and the highest

[1] This problem adapted from Butts, T. (1973), *Problem solving in mathematics*, Glenview, IL: Scott-Foresman.

with a number in the middle of the pack. Another pair of students considers whether they may group the numbers so that each grouping of three will be either odd or even. If they find it impossible to group numbers so that all sums are even (two odds and an even or three evens), then as far as they are concerned, the problem is not solvable. These students are vastly different in their prior knowledge.

It is important to remember that each student brings techniques to your classroom that may or may not be useful. But in all cases, as they begin to work on the problem, their knowledge of number and operations comes into play. The first student understands addition and how to compare the resulting values. The first student pair have a problem-solving technique in mind, to create groups that are just about equal. They understand the set of integers as a well-ordered set $\{n, n + 1, (n + 1) + 1, \ldots\}$. The second pair used a more sophisticated property of the integers to first ascertain the possibility of successfully completing the problem. They at least knew of the number properties that an even plus an even results in an even number, and so on.

This number puzzle demonstrates how an understanding of number and operations can be useful in problem solving. The problem strengthens and extends student understanding of number and operations when ideas, concepts, and skills are put to use in novel situations. As an extension, you might ask your students to use the idea of evens and odds to find a solution for the partitioning problem. What other strategies might arise as students discuss their results?

Here is another number puzzle that requires some thought, and an understanding of whole numbers. Try it first before you read the discussion that follows.

Problem to Investigate: Running Around the Track

You and your father go to the track at the neighborhood school for a run. You both start running from the same place on the track at the same moment. Your father runs around the track in 120 seconds while it takes you only 90 seconds to run around the track once. After how many seconds of running will you both cross your starting point at the same moment? How many times will you and your father go around the track before you both cross the starting point at the same moment?[2]

Although there are many ways to solve this problem, students will begin to recognize the usefulness of multiples when they work on this one. The least common multiple of 120 and 90, namely 360, gives the time in seconds before the father and child pass the starting point together. Thus, the father runs around the track $\frac{360}{120}$ or 3 times while the child runs around $\frac{360}{90}$ or 4 times. It is helpful to note that the factors of 90 are 9 and 10 while the factors of 120 are 12 and 10. What knowledge of number and operations do students need in order to realize that they can simply find the least common multiple of 9 and 12 (that's 36) and multiply this by 10 to get the correct answer to the prob-

[2] This problem is adapted from Lappan et al. (1996). *Connected mathematics: Prime time*, Palo Alto, CA: Dale Seymour.

lem? How would you justify this method? As with the previous integer problem, this one requires students to have an understanding of factors, multiples, and the distributive property.

We next look at a number theory problem that deals with lockers. On a small scale, say 10 lockers, this problem can be solved by creating some sort of model—using drawings, charts, or manipulatives such as blocks painted with different colors on alternate sides. To extend the problem beyond a small number, however, the student must investigate the results and any relationships they found as they worked the problem.

Problem to Investigate: A Locker Puzzle

On graduation day, 1000 seniors line up outside the school. As they enter the school, they pass the school lockers, aptly numbered 1 to 1000. The first student opens all of the lockers. The second student closes every other locker beginning with the second locker. The third student changes the status of every third locker beginning with every third one (if opened, the student closes it; if closed, the student opens it). The fourth student changes the status of every fourth locker, and so on. Which lockers remain open after all 1000 students entered the school?

The results of this problem may surprise students—all the lockers that are perfect squares remain open. This pattern may be quickly recognized by students, but the pattern itself does not answer the question "Why is this the case?" To discover why, students must investigate not only lockers that remain open, but the process of opening and closing lockers. It is the relationship between the number of times lockers are opened or closed and the number of prime factors of locker numbers that is key to the solution. Using the many variations of this problem helps students investigate other aspects of integers and their properties. Some variations and extensions are included in the exercises that follow this section.

Novel problems such as these develop critical thinking skills, particularly when students brainstorm ideas. Students learn about problem-solving strategies, such as simplifying the problem or using a physical model. The exercises illustrate various aspects of number theory that facilitate problem solving. As you work them, consider the number and operations knowledge that you are using. How can you help develop students' problem-solving skills for these types of number puzzles? How can you integrate these problems into your lesson plan? How do these problems connect to aspects of algebra, geometry, or other areas of the curriculum?

Exercises 4.3

1. Given the set of positive integers $\{1, 2, 3, 4, , n\}$, for what values of n can the set be partitioned into subsets of three numbers such that the sum of each subset is equal?
2. Using your response to Exercise 1, choose two values of n that work. For these values of n, how many different partitions can you make?

3. If a divides evenly into b and b divides evenly into c, is it true that a divides evenly into c? Explain.

4. Jenna is planning a sleepover for her daughter Kate and some of her friends. To get ready for the sleepover, Jenna ordered 14 mini pizzas from the local grocery store and 2 packages of popsicles (each package contains 12 popsicles). Kate wants to make sure that she and her friends each receive one mini pizza and the same number of popsicles. What is the largest number of friends that can be invited? How many popsicles will each friend get? If Kate wants each friend to have at least two popsicles each, what possible numbers of friends can she invite? How many mini pizzas will each friend receive?

5. You build a model of a small solar system for the school science fair and attach a small motor so that the planets will actually orbit the sun. The first planet orbits every 6 seconds, the second one every 10 seconds, the third every 15 seconds, and the fourth every 24 seconds. At the start of the science show, you place the planets in alignment (called conjunction) and start the motor. After how many seconds will all four planets be in conjunction again?

4.4 Counting Theory and Techniques

The problem-solving techniques discussed in Section 4.3 rely on a solid understanding of numbers and their properties as well as basic operations and pattern finding. Many mathematics problems require calculations that go beyond simple arithmetic. For example, calculate the number of options you have if you can choose 12 of 18 equally talented players for a baseball team. Obviously this situation is simplified because other factors frequently determine who gets picked. But similar situations do arise when it is helpful to have an understanding of counting theory.

Addition and Multiplication

Some basic counting problems are easily solved with a little critical thinking about the situation. Nonetheless, it is usually more efficient to develop techniques for solving counting problems. For instance, it would be extremely tedious to count the number of different 2-topping pizzas that are possible if you can choose from 8 different toppings and 4 different crust choices as well as 4 different sizes of pizza. It would take a while to list all 128 possibilities.

Addition is one of the basic counting principles. It is appropriate for middle school students and helps them think about the ways we count objects. Once students understand the principle, they will be well on their way to solving problems that require more sophisticated counting principles.

Example 1 ***Selecting from Disjoint Sets***

Suppose you have two disjoint sets of objects, that is, two sets with no elements in common. Set $A = \{a, b, c\}$ and set $B = \{w, x, y, z\}$. In how many ways can one letter be chosen from the sets A and B? ■

Students may quickly reason that since you essentially have seven letters to choose from, the answer is seven. This is often referred to as the basic addition principle of counting. It is stated more formally below.

Addition Principle

If a choice from set A can be made in n ways and a choice from set B can be made in m ways, then the number of ways one element can be chosen from set A or B is $n + m$, provided no element of set A is also an element of set B.

If, however, sets A and B have items in common, the answer is not as obvious. The basic counting principle must be modified, which we do in Example 2.

Example 2 ***Selecting from Overlapping Sets***

Suppose set $A = \{1, 2, 3, 4, 5, 6\}$ and set $B = \{1, 3, 5, 7, 9\}$. How many ways can an integer be chosen from sets A and B? ■

As they compare Examples 1 and 2, students will notice that sets A and B are not disjoint. Students may reason that since choosing a 1 from set A is not really different from choosing a 1 from set B, the number of ways to choose elements from sets A or B is no longer the sum of the number of elements in the two sets. What adjustments have to be made to this algorithm to account for the overlap in the two sets? The first step in solving this problem is to determine the overlap. In this case, sets A and B have three elements in common. Hence, when we add the number of elements in set A to the number of elements in set B, we account for these three elements twice. Thus students can subtract 3 from the sum of the elements to arrive at an appropriate answer to the example. How might a middle school student write a generalized algorithm that modifies the basic addition principle?

Multiplication is another counting principle. The problem in Example 3 uses the basic multiplication principle.

Example 3 ***Counting Meal Choices***

Suppose a local sandwich shop offers a selection of 3 soups, 3 sandwiches, and 4 desserts (see the menu in Figure 4.3). How many different meals consisting of soup, sandwich, and dessert can be created? ■

To begin, students should be given plenty of time to work the problem so they can come up with various ways to count the possibilities. Some will approach this problem in an unorganized manner, simply combining various choices from each list without tracking which combinations have been made. Others will find more organized ways to count. Still others will notice that if they find all possible combinations with one soup, they need only multiply this by 3 to find the total number of possible combinations.

These solution strategies all have one thing in common. They employ listing or counting techniques that are not very sophisticated; listing every choice, for example,

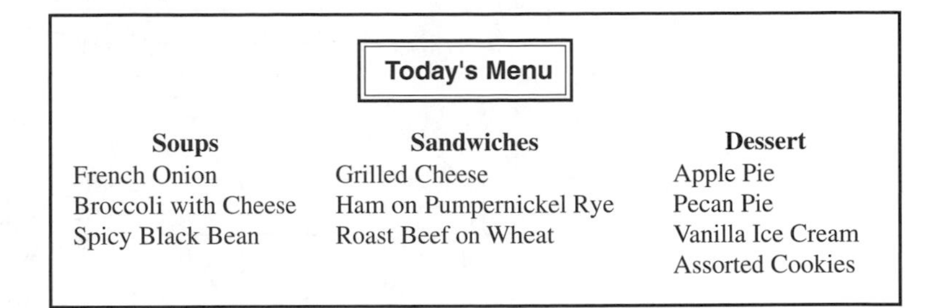

Today's Menu

Soups	Sandwiches	Dessert
French Onion	Grilled Cheese	Apple Pie
Broccoli with Cheese	Ham on Pumpernickel Rye	Pecan Pie
Spicy Black Bean	Roast Beef on Wheat	Vanilla Ice Cream
		Assorted Cookies

Figure 4.3 Sample menu for Example 3.

is pretty tedious! A more sophisticated way to approach the problem is to create a tree diagram of the possibilities (see Figure 4.4). Tree diagrams organize data quickly and clearly, and eliminate the tedium of listing *all* choices. Notice that the tree diagram shows that for any given choice of soup, there are 12 possible meals (3 sandwiches and 4 dessert choices, or $3 \cdot 4 = 12$). Since there are three soups to choose from, there are $3 \cdot 12$ or 36 total possible meals.

Once students have a sense of how to count the possibilities, they should be encouraged to find an algorithm that describes the counting process. They should then try out their algorithm on problems with more choices. The algorithm for Example 3 is find the product of the numbers of choices for each of the three items (soup, sandwich, dessert). This is summarized in the multiplication principle.

Multiplication Principle

If a task involves two steps and the first step can be completed in n ways and the second step can be completed in m ways, then there are $n \cdot m$ ways to complete the task, provided that each step can be completed independent of the other.

If a task involves several steps such that each step can be completed in $n_1, n_2, n_3, \ldots$ ways, then there are $n_1 \cdot n_2 \cdot n_3 \cdot \cdots$ ways to complete the task, provided each step can be completed independent of the other.

Permutations and Combinations

Two other well-known counting techniques are called *permutations* and *combinations*. In a sense, they are both modifications of the multiplication principle. Mathematical reasoning and an understanding of the multiplication operation are helpful in understanding these techniques. Examples 4 and 5 illustrate the usefulness of these counting techniques.

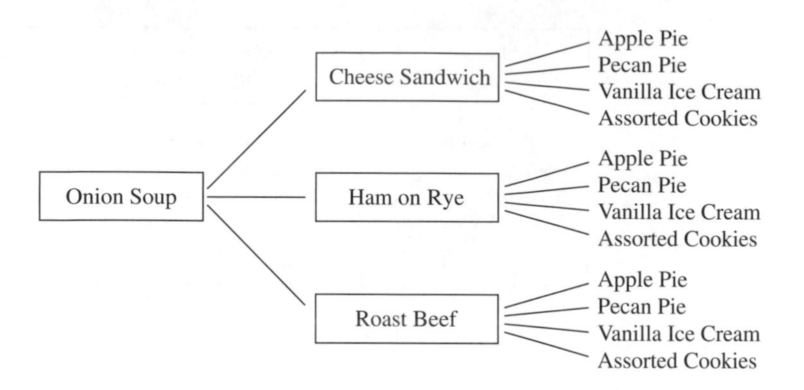

Figure 4.4 A tree diagram of all possible meal combinations for one soup.

Example 4 ***Arranging Paintings***

You inherit three paintings from your grandmother, a painting of a cow jumping over the moon, a painting of dancing bears, and a portrait of your great-great Aunt Sophie. In how many ways can you arrange the three paintings from left to right on your dining room wall? ■

Example 5 ***More Paintings***

Oops, you really inherited seven paintings but you only have space for three on your wall. In how many different ways can you choose three paintings and arrange the three on the wall? Keep in mind that the arrangement {Dancing Bears, Aunt Sophie, Cow} is a different arrangement from {Dancing Bears, Cow, Aunt Sophie}. ■

Middle school and high school students may quickly solve Example 4 by listing all possible ways to arrange the letters D (dancing bears), A (Aunt Sophie) and C (cow jumping over moon). Students can then be asked to relate this solution to the multiplication principle. They may need help to come up with the following reasoning: (1) Since there are three paintings, I have 3 choices for the first spot on the wall; (2) as soon as a painting is selected for the first spot, I now have only 2 paintings to choose from for the second spot; (3) filling the first and second spots on the wall leaves me with only 1 choice for the third spot. Using the multiplication principle, I have $3 \cdot 2 \cdot 1 = 6$ different arrangements.

This reasoning can then be applied to Example 5. There are 3 spaces to fill and 7 choices for the first space, only 6 choices for the second space (one painting is already chosen), and 5 choices for the third space. This gives $7 \cdot 6 \cdot 5 = 210$ possible choices. This modification of the multiplication principle is known as a permutation. How can these techniques be generalized?

When there are n spaces to fill with n objects, students may conclude that there are $n \cdot (n-1) \cdot (n-2) \cdot \cdots \cdot 2 \cdot 1 = n!$ possible arrangements. It is not as obvious that with 7 items to fill only 3 spaces, we get $\frac{7!}{4!}$ possible arrangements. In general, we use the notation $P(n, r)$ to represent the number of ways to arrange r objects from a set of n objects when the order matters. Notice from the solution to Example 5 that

$$P(n,r) = n \cdot (n-1) \cdot (n-2) \cdot \cdots \cdot (n-r+1) = \frac{n!}{(n-r)!}$$

Permutation

A linear arrangement of elements for which the order of the elements must be taken into account. We write $P(n,r) = n!/(n-r)!$

Another counting technique similar to permutations is known as combinations. This technique provides a quick and efficient way to calculate the number of ways we can choose elements from a set when the order doesn't matter. Suppose, for instance, there are 3 topping choices for a pizza and you wish to know how many different 2-topping pizzas you can make with the 3 toppings. In this case, pepperoni and onion is the same choice as onion and pepperoni.

In more general terms, we may want to know how many two-object combinations we can make from three objects. This type of counting problem is appropriate for many grade levels. Students with minimal knowledge of counting techniques can find ways to pair the three choices and count all possible pairings.

Example 6 ***Two-Topping Pizzas***

You have a choice of 3 pizza toppings: pepperoni, mushroom, and onion. How many different 2-topping pizzas are possible? ■

To solve Example 6, students might count possible pairs (different 2-topping pizzas) in the following way:

Pepperoni and Mushroom
~~*Mushroom and Pepperoni*~~
~~*Onion and Pepperoni*~~
Pepperoni and Onion
Mushroom and Onion
~~*Onion and Mushroom*~~

Realizing that a pizza with onion and pepperoni is the same as a pizza with pepperoni and onion, students will cross out the duplicates and arrive at the answer of 3 possible different pairings. Unlike Examples 4 and 5, Example 6 does not require us to take into account the order of our arrangements. Hence there are fewer than $\frac{3!}{1!} = 3!$ arrangements. In this case, there are half as many possibilities.

When this problem is extended to 4 or 5 toppings, students can look for patterns in the data or in the results. These patterns can be used to formulate an algorithm. Students should expect to explain and justify their results using logical reasoning. Example 7 and the following discussion illustrate how teachers help students realize and understand the combinations counting technique.

Table 4.1 Combinations for Example 7.

Number of objects	Number of possible distinct pairs
3	$\frac{3 \cdot 2}{2} = 3$
4	$\frac{4 \cdot 3}{2} = 6$
5	$\frac{5 \cdot 4}{2} = 10$
n	$\frac{n \cdot (n-1)}{2} = \frac{n^2 - n}{2}$

Example 7 ***Combinations Algorithm***

State an algorithm so that if given a specific number of objects, you can find exactly how many different pairings of the objects are possible. ■

In response to this problem, students might generate Table 4.1 and use it to formulate the algorithm. Students might reason as follows: "The formula

$$\frac{n \cdot (n-1)}{2}$$

makes sense since we take each of n items and match them up with the $(n-1)$ items that are left, creating $n \cdot (n-1)$ pairs. But half of these pairs are repeats since the ith item paired with the jth item is the same as the jth item paired with the ith item. So, we divide all possible pairs by 2, to get the number of distinct pairs."

What is the relationship between combinations and permutations? Can we find a more general algorithm for computing combinations that derives from our work with permutations? To relate combinations to permutations, let's choose 3 pizza toppings from a set of 5 possible toppings. We'll consider all possible permutations for this problem. This gives us $\frac{5!}{2!} = 60$ possibilities. However, since order doesn't matter here, many of the 60 possibilities will be duplicates. So we consider the fact that there are 6 possible ways to arrange any set of three toppings. Thus the choice {P, M, O} is the same as {P, O, M}, {O, M, P}, {O, P, M}, {M, O, P}, and {M, P, O}. Hence, for every set of three toppings, we would have to cross out five duplicates. That is,

1. There are $P(5, 3) = 60$ ordered arrangements of 3-element subsets, and
2. Within the 60 ordered arrangements, 10 groups of 6 arrangements use the same 3-letter subset: $\frac{60}{6} = 10$ unique 3-letter subsets. Hence we write

$$C(5, 3) = \frac{P(5, 3)}{P(3, 3)} = \frac{5!}{2!\,3!}$$

Combinations

If r elements are collected or arranged from a set of n elements, then the number of combinations of n elements taken r at a time, $C(n, r)$, is related to the number of permutations of n elements taken r at a time, $P(n, r)$, according to the equation

$$C(n, r) = \frac{P(n, r)}{P(r, r)} = \frac{n!}{(n-r)!r!}$$

As illustrated in our examples, it is crucial that students find ways to ascertain that all possible pairings or groupings have been found and none counted more than required. To do so, students must organize their counting methods. The principles and techniques just described are but a few of the counting algorithms within the grasp of middle school and high school mathematics students. These techniques make counting efficient and accurate. It is important that students understand these techniques so that they can choose the appropriate method, or modify a method, when solving counting problems.

Other Counting Techniques

Counting techniques are basically algorithms that help us count more easily all possible solutions for a situation. There are many situations where we need to list and count all possibilities but where no particular counting technique seems appropriate. However, even these situations can be approached systematically with the aid of diagrams, lists, and models.

Consider the yellowbellied gnat problem posed next. What techniques can you use to keep track of the number of gnats during each successive month of the breeding season? Would a diagram or a list be more helpful? Which techniques are more efficient? Why? Which are less efficient? Why?

Problem to Investigate: Yellowbellied Gnat

Each year the yellowbellied gnat harasses residents in Bloomville. Cats and dogs suffer voracious attacks as well. In early March, the yellowbellied gnat population is estimated to be at 100,000. The March gnats are newborns. Gnats reproduce beginning at 2 months and reproduce monthly until they die. The 2-month-old gnats reproduce at a rate of 200%; 3-month-old gnats reproduce at 250%; and 4-month-old gnats at 80%. At 5 months, all gnats die and (thankfully) the entire population is gone by the first hard freeze (no earlier than mid-October). How many gnats are in the Bloomville area in early October?[3]

[3] Thanks to John Lannin for suggesting this problem.

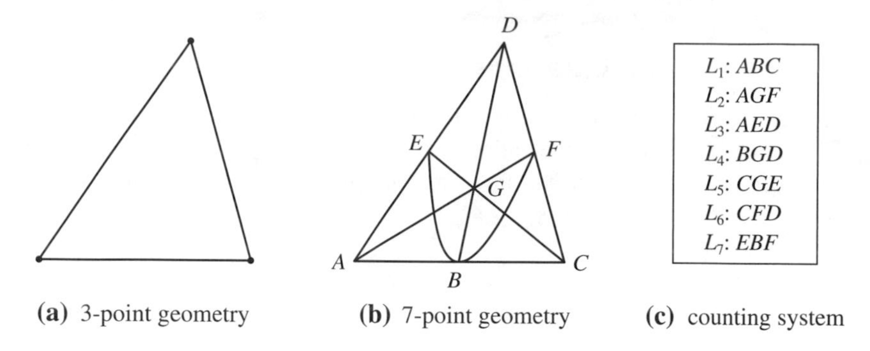

Figure 4.5 Finite geometries as counting problems.

When students first attempt to solve this problem, they typically set up some sort of chart for tracking the number of gnats who are newborns, 1 month old, and so on. This is a tedious process, but with good record keeping, students reliably arrive at the correct solution. You may wish to encourage students to use a spreadsheet program, which organizes data well, and this is the crux of this problem. Unlike many counting problems, this one requires counting data that change in size and type with each iteration. Exercise 2 in the exercise set gives an efficient counting technique for this problem.

Finite geometries are another place where counting techniques can be quite useful. These are geometries that consist of a finite number of objects (such as points and lines). An example is a three-point geometry that can be modeled using a triangle. Suppose only three noncollinear points exist, and lines join the points. One possible model of the system is shown in Figure 4.5a. A more complex finite geometry is shown in Figure 4.5b, which shows a 7-point geometry. Here the lines contain exactly 3 points, any set of 2 points is contained on exactly 1 line, and each pair of lines share exactly 1 point. In this case, we can think of all possible combinations of 3 objects chosen from a set of 7, but the other conditions further restrict the number of possible combinations. Taken as a combinations problem, we find out how many lines we should have, but this doesn't tell us which points determine the lines. To do this, we can either diagram the system, as in Figure 4.5b, or develop a system for listing the combinations, as in Figure 4.5c.

Counting techniques such as those presented in this section simplify number problems that require counting. However, students must solve a variety of problems that don't rely on permutations and combinations. Students need to be able to develop efficient counting methods in a variety of situations. Test your abilities with the exercises.

Exercises 4.4

1. Tess, Ryan, Michele, and Jake are the four finalists in an art contest at Prairieside Middle School. How many possible ways can the first- and second-place ribbons be distributed? Find an algorithm for determining all possible pairings of n objects if order matters.

2. How can you use matrices to represent and solve the yellowbellied gnat problem?

3. Suppose some yellowbellied gnats die from disease or from carefully placed swats by pets and people; that is, 10% die at 1 month, 10% at 2 months, 20% at 3 months, 30% at 4 months, and the remainder die at 5 months. Assuming the reproduction rates given on page 141, how many gnats will there be in Bloomville in early October?

4. Ten students at your school are trying to decide who among them will represent all students at school board meetings. The school board has agreed to a committee of 6 as long as at least 2 members are parents. Ten parents agree to serve if they are chosen. How many ways can the 6-person committee be selected from 10 students and 10 parents so that it includes at least 2 parents?

5. You are given 25 points in a plane, no 3 collinear.
   a. How many straight lines are determined?
   b. How many triangles are determined?

6. A classroom has 2 rows, each with 8 seats. Of 14 students, 5 always sit in the front row, 4 always sit in the back row, and the rest sit in either row. How many ways can the students be seated?

7. A finite geometry consists of exactly 9 points. As with the 7-point geometry illustrated in Figure 4.5, every set of 3 points is contained in a line, and every line contains exactly 3 points. List or draw the lines in this geometry.

4.5 Estimation and Algorithms

Estimation and algorithms have long been a part of the mathematics curricula. Even so, issues surrounding the teaching of these two topics still give rise to debate among educators. Three yearbooks produced by the National Council of Teachers of Mathematics document this debate (Morrow, 1998; Schoen, 1986; Suydam, 1978). Beginning in the 1980s, some educators advocated that algorithms should not be given to students, but rather, students should derive their own algorithms through a discovery process (Kamii, 1985; Madell, 1985). These educators believed that students develop better understanding if they use algorithms that they develop themselves. The educators reasoned that such algorithms would make more sense to the students and they would be less reliant on memorizing formulas because they would know how to think through problems. Some student-derived algorithms would be more efficient than others, but a less efficient way of arriving at answers would be preferable to memorizing without understanding. Yet, even when students derived their own algorithms, they still used the algorithms incorrectly.

One branch of this research focused on errors students typically make when using algorithms for computation (Ashlock, 1982). Ashlock developed a teaching protocol in which the teacher (a) identifies the error, (b) analyzes the error, and (c) helps the student correct the error. Today error analysis is a routine student task for all types of problem solving, not only basic computation. An expanded version of this protocol places the teacher in the role of helping students identify, analyze, and correct errors. This student-driven error analysis encourages students to reflect on and analyze computations and the validity of their results. One tool that is particularly effective for student-driven error analysis is estimation.

Estimation

Estimation has long been thought to be appropriate subject matter for elementary school children. But estimation is invaluable in all areas of calculation. At the secondary level, estimation aids students in familiar computations, in gaining insights into number and operations, and in attaining better problem-solving skills (Trafton, 1978). Facility with estimation demonstrates facility in computation, but also indicates a student's ability to think quantitatively and reason numerically.

The teaching of estimation and algorithms have come to the attention of educators in both good and bad lights over the years. It is clear, however, that students who develop estimation skills and who understand and use algorithms are more effective problem solvers.

Estimating has been a part of mathematics since ancient times. Archimedes estimated the value of π as $\frac{223}{71} < \pi < \frac{22}{7}$. In his development of calculus, Newton used successive estimates to approximate the area under a curve. And in modern times, estimation plays a key role in our statistics-driven society where statisticians use small samples to estimate the responses of larger populations. All of us use estimation on a daily basis, when we decide if we have enough money to buy one or two burgers, or when we shop sales without a calculator in hand.

What place should estimation occupy in school mathematics programs? Students typically (and perhaps all too readily!) use calculators to do their computations. Besides, isn't mathematics about exactness and precision? The short answer here is, not always. Many contexts require estimates over exact calculations, and students need to know when approximate answers are good and when precision is needed. Estimation is helpful for

- arriving at a "ball park" figure when information is needed in a short amount of time and a calculator is not available;
- getting a sense of the magnitude of an answer so its "reasonableness" can be checked;
- checking an intermediate value when a long string of calculations depend on the accuracy of earlier calculations;
- pencil-and-paper calculations when an exact value is known but not useful ($\pi \approx 3.14$);
- approximating a value when an exact value is possible but unknown (the age of a tree);
- deriving a value when an exact value is impossible (life of a light bulb) (Usiskin, 1986).

Problem to Investigate

Chen is asked to find the area of a rectangle that has length 12.5 units and width 19.5 units. He decides he can make the problem easier if he subtracts 0.5 from 12.5 and adds this amount to the width. He then calculates 12×20 and arrives at an area of 240 square units. Maria looks at what was done and comments that this method only gives an estimate of the original rectangle's area. She wonders "How can I find out if this is an overestimate or an underestimate?"

How would you help Chen recognize that his method only estimates the rectangle's area? How would you answer Maria's question?

Often, the terms *estimation* and *approximation* are used interchangeably. Indeed, most dictionaries list these words as synonyms. Yet, we usually refer to approximations as rounded values of exact calculations. That is, approximation happens *after* calculations. On the other hand, we frequently associate estimation with rounding values *before* the calculations are performed. To estimate an answer to a complex calculation, we reduce the quantities to ones that are easy to work with. For example, to estimate $3 \cdot \$0.88$, we can round \$0.88 to \$1.00 and multiply by 3 to get a quick result of \$3.00.

To understand how the terms *estimation* and *approximation* are typically used in math class, consider the following: Instead of taking the square root of 72, recall that 8 is the principal[4] square root of 64 and 9 is the principal square root of 81. Thus, $\sqrt{72}$ will be between 8 and 9, and a reasonable estimate for $\sqrt{72}$ is 8.5. On the other hand, we approximate $\sqrt{72}$ by performing the calculation and rounding the result. In this case, $\sqrt{72}$ on the calculator gives 8.485281374. Thus 8.5 is an approximated value. The answers here happen to be the same (8.5), the processes for deriving them are different. Students need to know why and when they should estimate or approximate values, and they should know how to indicate that a result is not an exact answer.

Strategies for Estimating

People use many strategies when they estimate. Some of these can be taught by brainstorming examples. Students also develop viable strategies that can be shared with others. When teaching strategies for estimating, be sure to stress when one strategy is more appropriate than another.

The most common estimation strategy is rounding. Mentally calculating the sum of a string of numbers can be quite difficult, but students can quickly estimate the sum by rounding to the nearest whole number, or to the nearest hundred or thousand, depending on the problem. Thus $447 + 109 + 593$ might cause students to pause if they are asked to decide quickly whether the sum is closer to 1000 or 1100. By rounding to the nearest hundred they get $400 + 100 + 600 = 1100$. It's also clear that this is an underestimate because 447 is almost as close to 500 as it is to 400. Hence, the sum of 447, 109, and 593 is closer to 1100.

[4] The principal square root of a number r is the positive value b such that b^2 is r.

A second common strategy, called "front-end" estimating (Trafton, 1978), derives from an addition algorithm. To add the numbers 7.49, 2.15, and 3.78, start with the front end, or the digits 7, 2, and 3. This sum is 12. Now move to the second digit in each number, $4 + 1 + 7 = 12$, or notice simply that the sum of second digits is greater than 10 (10 tenths, which is equal to 1 in this case) so $7.49 + 2.15 + 3.78 > 12 + 1 = 13$. The sum of the third digits $9 + 5 + 8$ gives 22. This does not add anything significant to the original sum since 22 hundredths is less than 1. In determining that the sum of the second digits $4 + 1 + 7$ is greater than 10, we considered a range for the sum, or the magnitude of the sum.

Determining a range for a calculation is often helpful when checking the reasonableness of a result. What is the product of 40 and 80? Is it 320 or 3200? Both seem reasonable since 4×8 is 32. But what is the correct magnitude of the answer? An understanding of the associative and commutative properties of multiplication resolves this issue: $40 \times 80 = (4 \times 10)(8 \times 10) = (4 \times 8)(10 \times 10) = (32)(100) = 3200$. More generally, estimating that focuses on the possible magnitude of a computation is helpful for checking both paper-and-pencil calculations and calculator results. If the computation is lengthy, estimating answers quickly identifies when something is out of kilter in a computation. For example, if a student intends to enter 78×0.5 and the calculator reads 390, estimation helps the student recognize immediately that 390 is incorrect since $\frac{1}{2}$ of 78 has to be a quantity smaller than 78.

Algorithms

Like estimation, algorithms provide a quick and efficient means for solving problems. In general, an algorithm is a strategy for calculating a particular value. A general algorithm is one that can be used over and over in similar situations. For example, the quadratic formula is a general algorithm that can be used whenever we want to find all real and complex roots of a polynomial of degree 2. The recursive formula $a_n = a_{n-1} \cdot 2$, for $a_0 = 1$, is an algorithm for calculating successive terms of the sequence $1, 2, 4, 8, 16, \ldots$. Algorithms also describe processes for generating objects, like a fractal. The fractal is a result of continually repeating a recursively defined algorithm. Since we can't repeat the algorithm forever, we only see stages of the fractal. One algorithm for creating fractals begins with a line segment. The middle third of the segment is removed and replaced with two segments equal in length to the piece removed (see stage 2 of Figure 4.6). At each successive stage, the middle third of all segments are removed and replaced with two segments of length equal to the piece removed. Figure 4.6 shows the first three stages produced by this algorithm.

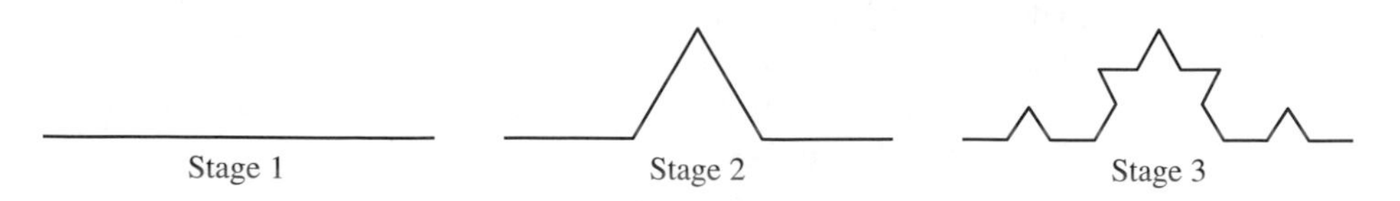

Figure 4.6 An algorithm for generating a basic fractal.

Students should experience both using and creating algorithms. Mathematics is full of algorithms; students begin to learn them in the earliest grades. And children frequently create algorithms, although they don't recognize them as such. When students approach problem situations consistently, or modify a formula to fit new situations, they are developing algorithms. Creating algorithms helps students "own" the process of problem solving. Using algorithms developed by others is by no means a failure. Students can't redevelop every existing algorithm, but they better understand the power of algorithms if they "invent" a few of their own.

Nonroutine mathematical problems are a rich domain in which students can "invent" algorithms. These algorithms may be modifications of others they have used before, or they may solve only the problem at hand. Suppose you are planning a dinner for 5 friends and you want to know how many different arrangements of the 6 of you are possible. You will all be seated at a round table. This resembles a permutation problem, but the round table requires that we modify the algorithm since ABCDEF in a circular arrangement is the same as BCDEFA arranged in a circle. Now consider the crossing-the-river problem. How can an algorithm determine a best solution?

Problem to Investigate: Crossing the River

Eight adults and 2 children need to cross a river. A small boat is available, but it only holds at most 1 adult, or 2 children. Everyone can row the boat. What minimum number of trips will get all the adults and children across the river? From your solution, develop a strategy so that you can quickly decide how many trips are needed for 15 adults and 2 children.

Exercises 4.5

1. Given real numbers x, y, and a. When will $(x + a)(y - a)$ have a value greater than xy, and when will it have a value less than xy? Explain.

2. How would you assess Jonathon and Shatara's work if Jonathon's answer is the result of successive approximations and Shatara uses exact values?

3. You are painting all the lockers on your high school's first floor (use the floor plan shown on page 148). The lockers are located along the walls of the halls. Five-gallon buckets of paint and your other equipment are located in the equipment room. You have to move this bulky equipment with you as you paint and return it when you're done. If the lockers are on both sides of a hallway, they must be painted one row at a time so that the hall isn't blocked by your painting equipment. You are being paid by the job, not by the hour, so you want to finish this job as quickly and efficiently as possible. How can you do this (Hart, 1998)?

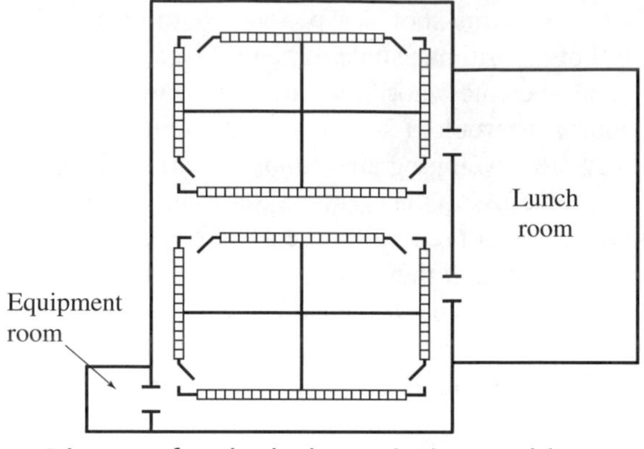

Diagram for the locker painting problem.

4. Find the sum of the first 500 whole numbers. Describe the method you used to calculate this sum. Can you extend this method to find the sum of the multiples of 4 between 28 and 2,800? [*Hint:* Develop a reliable algorithm, a step-by-step process, or derive a formula.]

4.6 Sequences and Series

The concepts of sequences and series appear often throughout middle and secondary school, although students don't always make the connection. When students are asked to create a table of values or find a pattern, they are working with sequences. When they find the pattern, what they've really found is an algorithm for describing the sequence.

In earlier grades, students typically have plenty of practice finding patterns and writing simple formulas that describe them. In high school, work with more complex patterns should be tied to meaningful problem situations. It's often helpful for students to graph patterns so they can "see" what algebraic statement describes the sequence. Students can use the graph to describe the relationship between successive terms of the sequence, or to develop a formula for the sequence.

As students begin to identify patterns in number sequences, they are also relating the sequence of numbers to the operations (addition, subtraction, multiplication, and division). This again introduces students to new ways of thinking about the relationships between number and operations. Pattern finding is also considered in Chapter 5.

4.7 Using Technology to Teach Number and Operations

Calculators are commonplace today. Newer models boast features that supercomputers of the past could not dream of doing. With the advent of the graphing calculator in 1986, students routinely produce complex symbolic, tabular, and graphical representations. Although the implications for calculator use have changed greatly in the past decade, the question of what role calculators should play in mathematics classrooms remains a pertinent one, especially with respect to the teaching of number and operations.

Hembree and Dessart's seminal meta-analysis of calculator use and its effect on student learning (Hembree & Dessart, 1986), led to widespread calculator use for computation and problem solving at most grade levels. At the time of their study, graphing calculators were almost unheard of, so they naturally didn't study possible impacts of these calculators. However, they did find that calculator use in the mathematics classroom has a positive effect on student attitudes toward mathematics, and on their problem-solving and computation skills.

One concern held by teachers and parents alike is that students who rely heavily on calculators don't acquire proficiency in basic computation. For instance, if students use calculators, one might argue that many standard algorithms, such as adding fractions or even multiplying polynomials, don't need to be taught. The National Assessment of Educational Progress (NAEP) has attempted to address this issue in their 1996 report on calculator use. Data gathered during the administration of the 1996 NAEP examinations indicate that students who use calculators on a regular basis in mathematics class perform equally well or better on the NAEP examinations than students who are not exposed to calculators in their mathematics classes (Hawkins, Stancavage, & Dossey, 1998). This finding seems counterintuitive. How can calculators improve students' ability to solve math problems? One possible answer may be that when students are freed from tedious hand calculations, they have more time to focus on other aspects of mathematical problems and to connect concepts.

Calculators, and technology such as spreadsheet programs, can significantly improve math comprehension, particularly for students who have difficulties with routine calculations. Mental computation, most long-hand calculations, and estimation continue to be important skills that students should learn, but calculators can be effectively employed for time-consuming or difficult calculations that otherwise limit student work on problem solving, statistics, and probability. Certain important conventions and basic skills cannot be learned with a calculator, but some conventions, such as order of operations, can be reinforced by calculator use.

We all rely on our calculators for difficult and simple calculations. But when should we encourage and discourage calculator use in math class? Do we lose our ability to compute if we rely too heavily on our calculators? Appropriate uses can be encouraged as students learn to use calculators. Doing multidigit multiplication with manipulatives or with the standard algorithm helps students understand how our base-10 system works. Once students have a firm grasp of the base-10 system, they can check the reasonableness of calculator answers. Calculators provide quick results, but only users can determine the validity of the result. Validity is never automatic because a calculator's answer is only as good as the user's input. Hence, estimating skills for checking calculator answers will always be important.

Scientific calculators and graphing calculators perform many more operations than in the past. And each new version of calculator performs yet another set of functions that were once the exclusive domain of paper and pencil. Most scientific calculators, for instance, compute combinations and permutations. Graphing calculators and spreadsheet programs display sequences and series, and in a variety of representations. Sequence mode on some graphing calculators can build tables of values and draw graphs using discrete dots. Sequence mode can also investigate closed-form sequences because it can produce and list differences between successive terms. Here, students must have an understanding of both sequences and ways of investigating them, in order to use the calculator effectively. Again, technology should be used to enhance student understanding of mathematical concepts, not to replace it.

Exercise 4.7

The yellowbellied gnat problem presented in Section 4.3 is easily modeled and solved using matrix multiplication. What are the advantages of students using graphing calculators to perform the matrix multiplication for this problem? What are the disadvantages?

PROJECT Find Out for Yourself

Observe a high school math class and note how the teacher uses technology. (You may need to interview the teacher before or after class.)

1. In what situations are calculators used in this class? Who is using calculators? What is the nature of the problem being discussed or worked on? How are the calculators being used?
2. Does the teacher actively encourage and direct the use of calculators? For what purposes? Or is there just an expectation that students will use calculators?
3. Do you agree with how calculators were used in this classroom? What suggestions would you make?

4.8 Number and Operations Lesson Plans

Lesson 1: Number and Operation (Middle School Level)

Objectives: Use the locker problem to develop student understanding of the properties of positive integers. In the process of solving the problem and sharing solutions, students will

- make discoveries about prime numbers, perfect squares, and the number of unique factors for each type of integer;
- learn to use a simpler but similar problem as a problem-solving technique;
- make connections between this and other problems we have discussed that deal with factors and multiples;

- see the usefulness of various problem-solving strategies in making new and different discoveries.

Warm-Up: Review factors and multiples (worksheet); share various strategies for finding unique factors of a number greater than 500 (refer to yesterday's activity sheet).

Activities:

1. (5–10 minutes) Opening—after discussing and collecting responses to the warm-up, explain that students may want to keep multiples and factors in mind when they work on today's activity. Be sure students are comfortable with these ideas and with finding factors and multiples of integers.
2. (2–3 minutes) Organize students into groups (prearranged) and distribute scrap paper. Make available a variety of manipulatives such as logical blocks, pattern blocks, tiles, and stacking cubes.
3. (20 minutes) Distribute the locker problem and give students about 20 minutes to work together. Ask one member of each group to record the strategy or strategies used to find a solution.

The Locker Problem

On graduation day, 1000 seniors line up outside the school. As they enter the school, they pass by the school lockers, aptly numbered 1–1000. The first student opens all of the lockers. The second student closes every other locker beginning with the second locker. The third student changes the status of every third locker beginning with the third locker (if opened, the student closes it; if closed, the student opens it). The fourth student changes the status of every fourth locker, and so on. Which lockers remain open after all 1000 students enter the school?

4. (10–12 minutes) Bring students together and have groups share solution strategies. Be sure various strategies are discussed (they will likely have different approaches). If all strategies are similar, brainstorm other approaches as part of a whole-class discussion.
5. (2–3 minutes) Follow-up questions. Once a solution is found, ask some or all of the following questions:
   - What is unique about the number values of the lockers that remain open at the end? What characteristics do these numbers share?
   - What is it about these numbers (their properties) that determined that they would be left open?
   - If there are 5000 lockers and 5000 students perform this end-of-year ritual, how can you quickly determine which lockers would remain open? Explain why you know this strategy will work.

6. (5–10 minutes) Conduct a whole-class discussion of the first set of questions and be sure all students understand the questions.
7. (Remaining time) Allow time for students to rejoin groups and begin discussion of the next question. If time allows, groups should work on an algorithm for the final question. Inform students that they will present their results to the class. If no time is left, assign this for homework. Students or groups will present strategies and solutions at the start of next class.

Assessment:

- Warm-up activity—to give me a sense of where students are coming from, what they learned from previous lessons, and who might need extra guidance in today's lesson.
- Group work—watch students as they work to see who is contributing ideas and how group members are responding to and working with these ideas. Question students on occasion to get a better sense of their understanding of the problem, the solution strategies being used, their confidence in the solution, their ideas about why the solution has to do with perfect squares.
- Whole-class discussion—although I can only really assess the understanding of those who contribute to the discussion, I'll try to get a sense of what all students are thinking by listening and looking carefully.
- End of class—I'll ask each student to write a couple of sentences about their current thoughts and ideas about the problem and a justification of the solution. This will let me know where we need to begin tomorrow (More whole-class discussion of patterns within numbers or properties of numbers? More time in small groups to develop a justification? Sharing of final solutions and justifications?)

Closing: No closing activity planned for today unless students are ready to share justifications and solution strategies. The end-of-class assessment is a good way to wrap up the class.

Lesson 2: Number and Operation (Secondary Level)

Objectives: Students will work in groups and contribute to whole-class discussions to discover the difference between problems that involve permutations and ones that involve combinations. More specifically, students will

- work in groups to solve several number theory problems using strategies and techniques developed over several lessons;
- discuss strategies used and justify strategies;

- learn the differences between problems that can be solved as permutations and those that can be solved as combinations;
- discover the relationships between permutations and combinations and write a general formula for combinations in terms of permutations;
- apply this understanding to develop strategies for classifying similar problems.

Warm-Up: Ask students to write, *but not solve*, two different problems, one that uses permutations and one that uses combinations. The problems may use the same context in different ways. They should not label the problems in any way since these problems will be solved by others in the class, and part of the solution process will be to decide which technique to use.

Activities:

1. (1 minute) Collect problems written by students during warm-up. Arrange students into groups (preassigned).
2. (2–3 minutes) Explain that students will work together to solve problems developed by their classmates. The purpose of this exercise is to help them understand the differences between permutations and combinations. While working on the problems, they should consider and discuss the following (post these questions on the chalkboard or the overhead so they are visible while students work):
   - Do two different orderings of the same elements constitute two different solutions?
   - Are elements restricted in any way (such as, you have to choose a pair of socks after choosing pants since the pants and socks must match)?
   - What techniques learned earlier are most appropriate for solving the problem? Why?
   - Can you generalize the relationship between permutations and combinations by writing one in terms of the other? (see below)

Combinations in Relationship to Permutations

If r elements are collected or arranged from a set of n elements, then the number of combinations of n elements taken r at a time, $C(n, r)$, related to the number of permutations of n elements taken r at a time, $P(n, r)$, is

$$C(n, r) = \frac{P(n, r)}{r!} = \frac{P(n, r)}{P(r, r)}$$

3. (15–20 minutes) Give students time to work in groups on the problems and discuss the questions listed above. Distribute transparencies for students to record

the problems and their solutions. Let them know they should be prepared to discuss their responses to the questions.

4. (10–15 minutes) Bring class together for a discussion of various problems. Take volunteers to share the problems they solved and why they solved them in that way. Ask others in the class (other than the presenting group) to consider the questions and how they would have responded, in terms of differences or similarities between the problems. Group members who solved the problems or the person who wrote the problems should be given a chance to respond as well. If students don't see the generalized relationship between permutations and combinations, I will help them develop the equation through questioning (What is the difference between the result $C(n, r)$ and $P(n, r)$? How can we write that? What does $r!$ in the denominator represent? Why?)
5. (10 minutes) Provide groups with a list of 10 more problems (some of greater difficulty) and ask them to use what they learned in the discussion to quickly decide whether techniques of permutations or combinations should be used to solve each problem. Discussion of this with whole class will follow. If time allows, assign extension problems.
6. Assign these 10 problems for homework. Extension problems should be completed by the date of the unit test. If time allows, they can begin writing up solutions to the problems.

Assessment:

- Warm-up—these problems will be collected at the end of class. These will give me a sense of student understanding of the various techniques we have discussed in class during this unit on combinatorics.
- Group work—watch students as they work to see who is contributing ideas and how other group members are responding to and working with these ideas. Question students on occasion to get a better sense of their understanding of the problems, the techniques being used, their confidence in the solution, their ideas about when the different techniques should be used.
- Whole-class discussion—student presentations and the accompanying discussion will give me a sense of what ideas students are picking up on.
- Homework—student work on the follow-up problems and extensions will demonstrate their ability to use what was learned and apply it to new situations.
- Unit test—students will have the opportunity to put all techniques and strategies to work on the test. They will also be asked to share their reasoning for using a particular strategy. This should get at procedural and conceptual understanding.

Closing: Discussion/summary of the problems solved today and what they learned about the differences between permutations and combinations. Students will be asked to summarize what they learned during today's lesson and to make connections to earlier lessons on counting techniques and combinatorics. I can record a list on the chalkboard. This will help them study for the unit test.

Sample Extension Problems (circular permutations):

1. In how many different ways can you arrange 5 guests at a round table?
2. What if you want to choose 5 of 7 guests? How many different arrangements are there at a round table?
3. What if you want to arrange an r-element subset from an n-element set in a circle? How many different arrangements are there? Can you summarize this with a rule or equation?
4. In how many different ways can 3 men and 3 women be arranged at a round table?

References

Ashlock, R. B. (1982). *Error patterns in computation*. Columbus, OH: Merrill.

Hart, E. W. (1998). Algorithmic problem solving in discrete mathematics. *In* L. J. Morrow (ed.), *The teaching and learning of algorithms in school mathematics: NCTM 1998 yearbook* (pp. 251–267). Reston, VA: NCTM.

Hawkins, E. F., Stancavage, F. B., & Dossey, J. A. (1998). *School policies and practices affecting instruction in mathematics*. Washington, D.C.: National Center for Education Statistics.

Hembree, R., & Dessart, D. J. (1986). Effects of hand-held calculators in precollege mathematics education: A meta-analysis. *Journal for Research in Mathematics Education*, **17**(2):83–99.

Howden, H. (1989). Teaching number sense. *Arithmetic Teacher*, **36**(6):6–11.

Kamii, C. (1985). *Young children reinvent arithmetic*. New York: Teachers College Press.

Madell, R. (1985). Children's natural processes. *Arithmetic Teacher*, **32**(3):20–22.

Morrow, L. J. (ed.). (1998). *The teaching and learning of algorithms in school mathematics: NCTM 1998 yearbook*. Reston, VA: NCTM.

National Council of Teachers of Mathematics. (2000). *Principles and standards for school mathematics*. Reston, VA: NCTM.

Schoen, H. L. (ed.) (1986). *Estimation and mental computation: NCTM 1986 yearbook*. Reston, VA: NCTM.

Suydam, M. N. (ed.) (1978). *Developing computational skills: NCTM 1978 yearbook*. Reston, VA: NCTM.

Trafton, P. R. (1978). Estimation and mental arithmetic: Important components of computation. *In* M. N. Suydam (ed.), *Developing computational skills: NCTM 1978 yearbook* (pp. 196–213). Reston, VA: NCTM.

Usiskin, Z. (1986). Reasons for estimating. *In* H. L. Schoen (ed.), *Estimation and mental computation: NCTM 1986 yearbook* (pp. 1–15). Reston, VA: NCTM.

5
Patterns, Functions, and Algebra

"The science of algebra, independent of any of its uses, has all the advantages which belong to mathematics in general as an object of study, and which it is not necessary to enumerate. Viewed either as a science of quantity, or as a language of symbols, it may be made of the greatest service to those who are sufficiently acquainted with arithmetic, and who have sufficient power of comprehension to enter fairly upon its difficulties."

August DeMorgan

Overview

Algebra is the content area we most closely identify with high school mathematics. Yet few students completing secondary school mathematics programs can describe or define what algebra really is. In this chapter we examine algebra as a way to think about and solve problems in our world. Historically, algebra as a subject matter area grew from formalizing the algorithms needed to perform calculations with Hindu-Arabic numerals. Sometime around A.D. 850, the Arabic scholar Mohammed ibn-Musa al-Khowarizmi wrote his seminal text *Al-jabr wa'l muqabalah*. This text set forth the procedures and rules for solving linear and some second-degree equations. al-Khowarizmi's name gives us the word *algorithm* and the title of his book became our word *algebra*.

The conception and study of algebra has changed from its original focus on solving equations to a much broader one. Today, algebra can be viewed as a set of skills for manipulating equations and expressions; as the study of operational structures such as groups, rings, fields, and vector spaces; or as the study of the patterns of quantities and the changes in those quantities in real-world settings. In this chapter we touch on all of these interpretations as we examine the concepts, skills, and principles that comprise the study of algebra.

Focus on the NCTM *Standards*

The standard for the learning of algebra reflects its various interpretations. Quality curricula see that students have opportunities to study these varying conceptions and to make the connections between them. Algebra can be a dynamic study if viewed from the appropriate vantage point, one that views it as a study of patterns in data and as a new way to think about functional and structural relationships. Students should encounter algebraic ideas across their K–12 education, but, in particular, algebra plays a major role in their math studies from middle school onward. The NCTM standard for Algebra calls for students to develop four different areas of understanding.

Algebra Standard

- Understand patterns, relations, and functions
- Represent and analyze mathematical situations and structures using algebraic symbols
- Use mathematical models to represent and understand quantitative relationships
- Analyze change in various contexts (NCTM, 2000, p. 296)

Examine these four parts of the Algebra standard. What do they mean in terms of your own experience? How do they relate to your school's mathematics program and real-world demands in the workplace? How do patterns and functions relate to symbolic forms and to modeling? The answers to these questions will help us understand what algebra is all about.

Focus on the Classroom

The teaching of algebra is changing. Historically, schools have focused on equations and their symbolic manipulations. Any remaining time went into graphing algebraic relationships and solutions to equations. In the upper grades, teachers focused on the various forms of functions, their graphs, and applications. Little connection was made between functions and variables and equations.

Current recommendations for the study of algebra call for changing this focus. The change centers on the idea of quantity. In almost all algebraic applications, we describe some quantity—the number of tickets, the mass of an object, or the time required to accomplish a task. The idea of quantity, and changes in quantity, is central to modern conceptions of how to make algebra meaningful to students.

If we think of data from an experiment, we generally have two related sets of data, one that involves the time or number of observations and one that gives the quantity observed at some point. As an example, let's examine the population of the world at various points in history (see Table 5.1).

Can you discern a pattern? If we graph the data to see what happens (see Figure 5.1), we see that the world population has changed over time. We can also state this in mathematical symbols. In the language of algebra, variables are symbols for quantities that vary. In our example, the variable p (for population) varies over time. We can formalize this relationship by expressing the population as a function of the year. To do this, we write $p(x)$, where x is a variable representing the year at which we are interested in describing the population.

From these beginnings we can go on to find and describe a relationship, a function, that allows us to make predictions and analyze trends in population data. For example, we can use the pattern to predict future population values. Perhaps we want to examine the impact of the Black Death in the 1300s on the population of the world today. We can also find functions for pricing multiple purchases, calculating interest on loans, determining oxygen requirements for deep-sea divers at different depths, and a myriad of other situations.

Table 5.1 World population (in millions) across time.

Year	Population	Year	Population
0	170	1550	480
500	190	1650	550
1000	265	1750	725
1200	360	1850	1,175
1350	440	1950	2,556
1450	380	2000	6,001

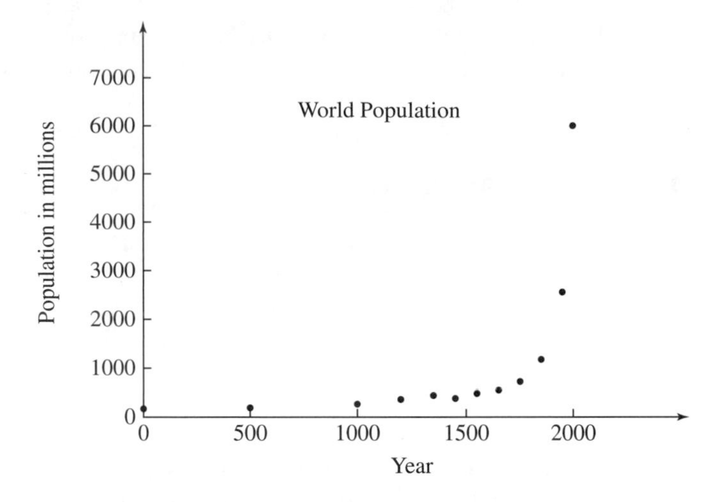

Figure 5.1 World population across time.

In other settings, we may examine properties of operations. For example, what number systems have the property that for all nonzero elements x in the system there is an element in the system, $1/x$, such that the product of x and $1/x$ results in the multiplicative identity 1 for that system? Such questions address the structure of the operations, in this case multiplication, in an algebraic system. Here x stands for an arbitrary element in a set of numbers, a slightly different conception of a variable. Such questions lead to another meaning of algebra—that of describing operational systems and their properties. It is the results of such study that give us the properties and theorems that we use to solve equations. At the middle grades these investigations play a smaller role in algebra, but this role increases as students move through the high school years.

At yet another level, we are interested in finding all numbers satisfying the equation

$$x^3 + 7x^2 + 14x + 8 = 0$$

To do this, we can draw a graph (Figure 5.2) and look for the zeros of the function shown, we can make a table of values (Table 5.2) and note where the expression on the left takes on the value 0, or we can use the rational root theorem or some other form of factoring (Table 5.3). In each case, we will find that the values that satisfy the equation are $x = -4, -2$, and -1.

The preceding example shows the connections between the tabular or numerical approach, the graphical approach, and the symbolic approach to doing algebra. We'll examine these three modes of representation as we discuss many of the different aspects of algebra in this chapter. Each representation mode requires varied levels of formalization and abstraction. It is this power and flexibility of algebra that we hope mathematics students in grades 7–12 come to appreciate and enjoy.

Table 5.2

x	y
−5	−12
−4	0
−3	2
−2	0
−1	0
0	8
1	30
2	72
3	140
4	240
5	378

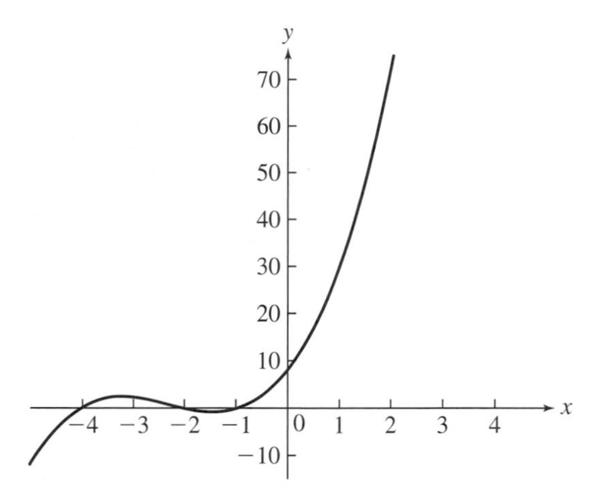

Figure 5.2 Graph of $x^3 + 7x^2 + 14x + 8 = 0$.

Table 5.3 Solution by factoring.

$$x^3 + 7x^2 + 14x = 0$$
$$(x + 4)(x^2 + 3x + 1) = 0$$
$$(x + 4)(x + 2)(x + 1) = 0$$
$$x = -4, \quad \text{or} \quad -2 \quad \text{or} \quad -1$$

5.1 Patterns, Functions, and Algebra in the Curriculum

The study of patterns takes many forms in school mathematics. In the early grades, students examine situations like 1, 3, 5, 7, __, __, __ and $\leftrightarrow$, $\updownarrow$, $\leftrightarrow$, $\leftrightarrow$, $\updownarrow$, $\updownarrow$, $\leftrightarrow$, $\leftrightarrow$, $\leftrightarrow$ __, __, __. In each situation, students learn to abstract the pattern, generalize their pattern into a rule, and use the rule to give the following terms in the sequence. Other examples from the early grades involve the number of sides in a polygon, turns or rotations of geometric figures, and growth patterns such as 1, 4, 16, 64, 256, __, __, __. Sometimes the answer students find is unique; in other cases there may be several defensible responses. Initially, we want students to analyze the changes and describe them. Later, as they advance, the focus shifts to identifying functions that represent specified patterns.

The first pattern above (1, 3, 5, 7, __, __, __), and the last pattern (1, 4, 16, 64, 256, __, __, __), exemplify very important patterns. The first shows the growth that occurs from a first term of 1 and has a constant difference between successive terms of 2. This is known as an *arithmetic sequence*. The second pattern has a first term of 1 and a constant ratio of 4 between successive terms. This is known as a *geometric sequence*. Their respective graphs are shown in Figures 5.3 and 5.4.

Comparing these graphs shows that the values of the first sequence result in the straight-line relationship $y = 2x + 1$ (Figure 5.3), while the values of the second se-

Table 5.4

n	a_n
0	1
1	3
2	5
3	7
⋮	⋮

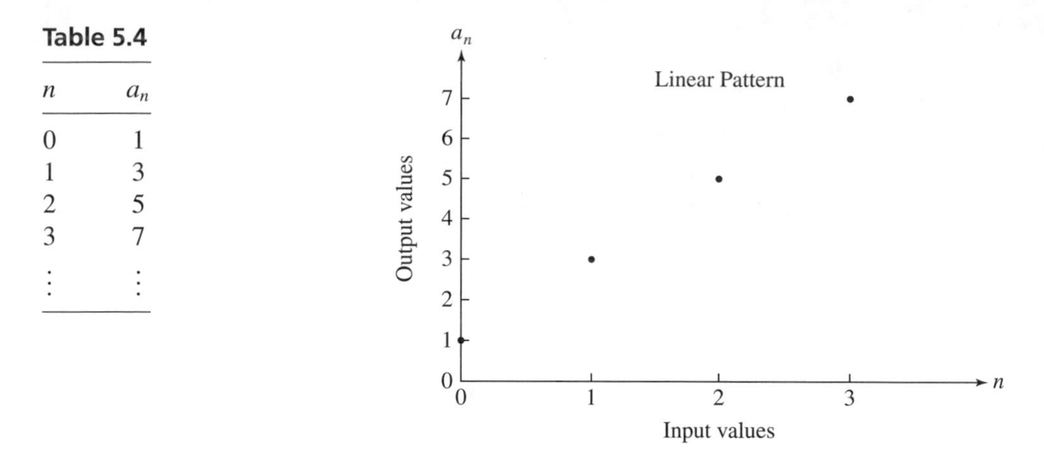

Figure 5.3 Graph of linear pattern 1, 3, 5, 7, ...

Table 5.5

n	a_n
0	1
1	4
2	16
3	64
⋮	⋮

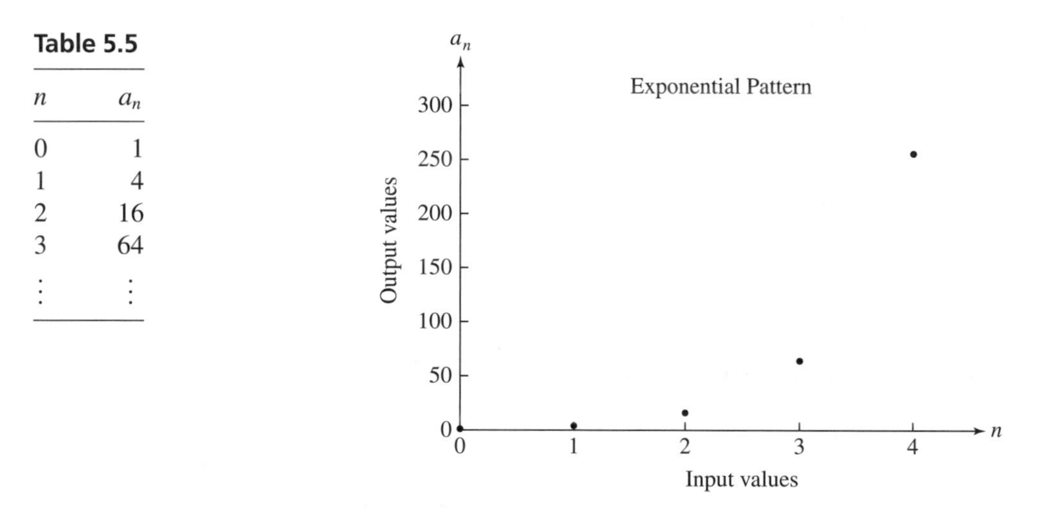

Figure 5.4 Graph of exponential pattern 1, 4, 16, 64, 265, ...

quence result in the exponential curve $y = 4^x$ (Figure 5.4). These sequences could have been described recursively as $a_0 = 1 =$ and $a_n = a_{n-1} + 2$ for $n \geq 1$ and by $a_0 = 1$ and $a_n = 4a_{n-1}$ for $n \geq 1$, respectively. These relationships, along with the change between successive values, become more obvious if we construct tables of their values (Tables 5.4 and 5.5).

To abstract the relationships shown in the data tables, we must recognize the structure of the change from one case to the next. This is pretty easy to do with values that have an ordered input pattern like those shown: 0, 1, 2, 3, ... Thus, it follows from Table 5.4 that if the input value 0 gives the output value 1, the output is $1 + 2$ for input 2, the output is $1 + 2 + 2$ or $1 + 2(2)$ for input 3. From this, it is a small step to guess that input x leads to output $1 + 2x$ or $2x + 1$. We can check this pattern by substituting in some of the later inputs in the table and seeing whether the outputs match the pattern. Then we can posit that the equation $y = 2x + 1$ matches the input values x with the output values y. The input values are sometimes called independent variables and the output values are dependent variables, because they depend on the values of the independent quantities selected.

From this, we can find the output (variables) associated with an input of 100. Using the expression $2(100) + 1$, we get 201. Evaluating expressions thus follows as a natural procedure, making sense in light of the idea that expressions represent arbitrary members of the sequence of values starting with input 0, which corresponds to the graph's y-intercept. We can also ask if the value of 405 appears anywhere in the list. Setting up the equation $2x + 1 = 405$, we solve to find that input $x = 202$ gives output 405. In a like manner, we find that no whole number input results in output 406.

Curricula need to provide students with activities that help them learn to abstract algebraic representations from data and patterns, and model real-world math problems.

Exercises 5.1

1. Quantities whose values don't change are called constants; quantities whose values do change are called variables. Which of the following quantities are constants and which are variables?

a. the length of a television show in minutes
b. the number of eggs in a gross of eggs
c. the speed of wind at the surface of the earth
d. the speed of light

2. Seventh grader Jerome measures the length of a board as 4 feet 6 inches. His friend Les says the board is 1.37 meters long. Since their values are different, Les says the length of the board is a variable. Is he correct? Why or why not? How would you respond?

3. Consider the following tables. Look for a pattern and complete the table. Then write an expression relating the input and output values.

a.

input x	-1	0	1	2	3
output y	3	1	-1		

b.

input r	0	1	1.5	2	2.5
output C	0	6.28	9.42		

c.

input x	0	1	2	3	4
output y	1	$\frac{1}{3}$	$\frac{1}{9}$		

d.

input x	1	2	3	4	5
output y	6	10	14		

4. A solid cube with edges three units long is made up of 27 individual unit cubes with white surfaces. Imagine that the exterior surface of the cube is sprayed with red paint. Consider the number of faces of each individual cube making up the $3 \times 3 \times 3$ cube that are now painted red. Use this information to provide answers for the blanks in the table below. What changes will result if we use a $4 \times 4 \times 4$ cube, a $5 \times 5 \times 5$ cube, and, finally, an $n \times n \times n$ cube?

Edges of large cube	Cubes with 3 red faces	2 red faces	1 red face	0 red faces	Total number of cubes
3	______	______	______	______	______
4	______	______	______	______	______
5	______	______	______	______	______
n	______	______	______	______	______

5. If we know the first three or four values in a sequence, can we make a rule for the pattern? Can we determine whether a pattern even exists? Illustrate your answer with some examples.

6. The following pair of recursive equations describe the location of a point in miles in the plane relative to the origin over time measured in hours.

$$x_t = x_{t-1} + 3,\ x_0 = 5 \qquad \text{and} \qquad y_t = y_{t-1} - 2,\ y_0 = 15$$

a. Complete the table and show the motion of the point in a graph.

Time	Horizontal location	Vertical location
0		
1		
2		
3		
4		

b. Find expressions describing the locations as a function of time.
c. Write an equation in y and x describing the path of the object.

7. The interior of a circle is divided into the maximum number of regions possible by drawing successive chords of the circle. No chords results in 1 region. One chord forms 2 regions. Two chords form 4 regions. What is the pattern here?

PROJECTS — Find Out for Yourself

1. Interview a middle school student, and show the student examples of arithmetic and geometric sequences. How does the student go about examining the values shown and describing the pattern(s) they see?

2. Ask several fellow math students to describe the difference between an independent and dependent variable. How do they answer your question? Do they do it verbally, graphically, symbolically? To what degree are they correct, or even close? Can you discern how they were introduced to these concepts from their explanations?

5.2 Patterns as the Basis of Mathematics

Pattern analysis helps us learn how to describe relationships. In many cases, we derive our descriptions of patterns from the verbal descriptions we are given. In other cases, like in the last section, our descriptions result from an analysis of data. It is important that students experience developing expressions and equations through these activities. This is good "doing of mathematics."

Teachers should be careful that students don't impose patterns where they do not exist. In many cases, patterns appear to be operating, but on closer examination, other forms of behavior become apparent. Describing patterns graphically or symbolically does a great deal to help students really understand the story told by the data.

Graphs help us explain trends and behavior patterns. Think of how we talk about values increasing or decreasing with time. Think about what it means to say something changes linearly or like a negative exponential. Such understandings grow from experiences where students gain a grasp of how variation in the data relates to standard patterns of change.

Symbols allow us to evaluate the pattern for a variety of values or to combine the pattern with other patterns to produce a new pattern. For example, if we know that a factory produces pencils, p, at a rate of $p = 1728h$ for each hour, h, a machine runs, then a factory with m machines will produce mpq/h pencils in q hours.

Perhaps the most important patterns for students to grasp initially are those of constant additive and multiplicative growth. Consider the pattern: 4, 7, 10, 13, 16, ... Viewing this pattern in terms of a table, we get Table 5.6.

Table 5.6

x	y	Δ
0	4	
		3
1	7	
		3
2	10	
		3
3	13	
		3
4	16	
⋮	⋮	⋮

The Δ column provides information on the rate of change relative to each of the unit-sized changes in the x-column. Here we see that the rate of change in the y-column is a constant addition of 3 units for each upward change of 1 unit in the x-column. If we think of the x-column as our input column and the y-column as our output column, we can say that increasing input by 1 causes a related increase of 3 in the output. Looking at this graphically, we would have something like that shown in Figure 5.5.

Alternatively, we can say that for every shift, or increase, of 1 in the independent variable, x, a corresponding shift of 3 occurs in the dependent variable, y. This constant growth pattern (the change always occurs in increments of 3) is the *slope* of the line associated with the output values. In statistics, the 3 is called the regression coefficient for the trend line fitting the data.

When we look at a pattern of data with a growth pattern modeled by repeated multiplication, like 2, 6, 18, 54, 162, . . . , the Δ column in the table looks different, as shown in Table 5.7. Here we see that the rate of change is not constant, although it does

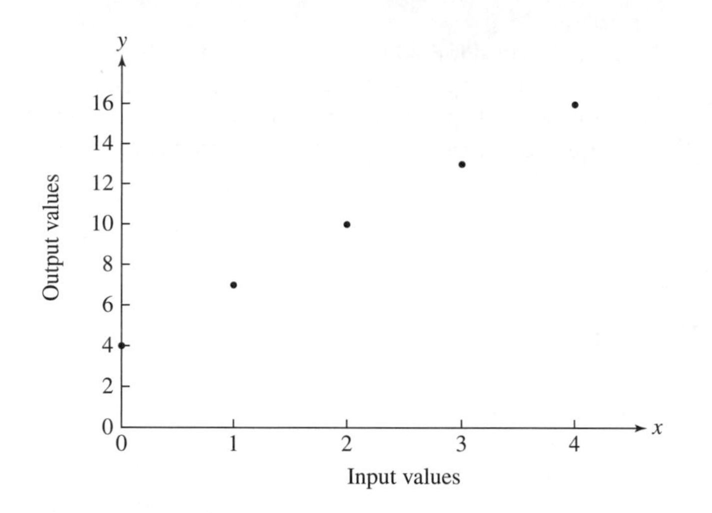

Figure 5.5 Graph of additive growth pattern 4, 7, 10, 13, 16. . . .

Table 5.7

x	y	Δ
0	2	
		4
1	6	
		12
2	18	
		36
3	54	
		108
4	112	
⋮	⋮	⋮

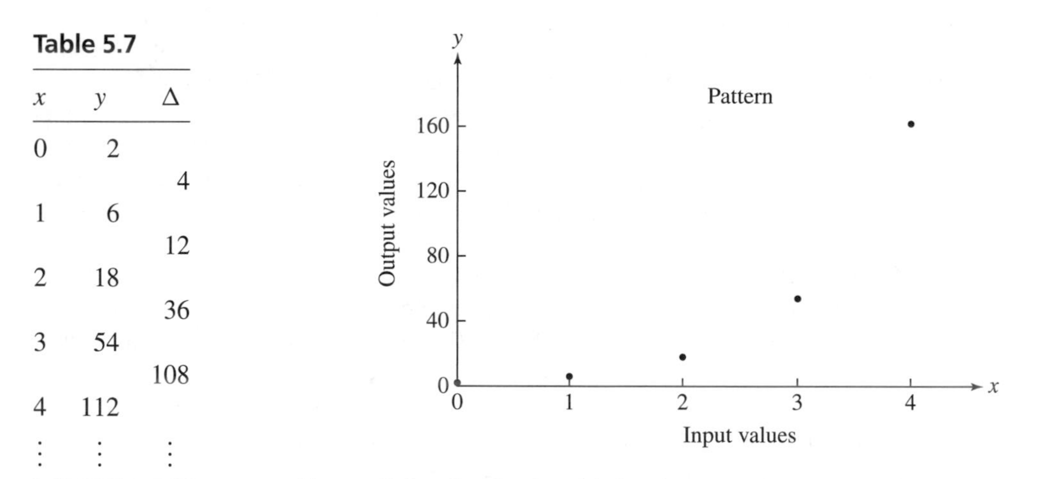

Figure 5.6 Graph of multiplicative growth pattern 2, 6, 18, 54, 162. . . .

grow as the values of the independent variable grow. Graphically, this pattern looks like Figure 5.6.

Examining patterns of change and the tables and graphs associated with them plays a central role in algebra, calculus, and statistics. The following exercises introduce you to certain aspects of the study of change.

Exercises 5.2

Consider the data in Exercises 1–4. Look for a pattern in the rate of change and use it to complete the tables. Then write an expression that relates the input and output values.

How are the numerical parameters in the expression related to the pattern noted in the changes?

1.

x	y	Δ
0	0	
1	0.5	
2	1.0	
3	1.5	
4	2.0	
⋮	⋮	

2.

x	y	Δ
0	6	
1	3	
2	0	
3	−3	
4	−6	
⋮	⋮	

3.

x	y	Δ
0	1	
1	7	
2	49	
3	343	
4	2401	
⋮	⋮	

4.

x	y	Δ
0	2	
1	5	
2	12.5	
3	31.25	
4	78.125	
⋮	⋮	

5. Latasha is comparing the prices of two film developing studios. Studio A has a \$4 service charge per order and an additional charge of 25¢ per picture while Studio B charges \$7.50 for a roll of 24 pictures. Compare these two rates graphically using

24-exposure rolls. Write a paragraph explaining which studio Latasha should use depending on the number of 24-picture rolls involved. What are the rates of change for these two options?

6. The population of metropolitan Las Vegas was approximately 1.2 million in 1998. The current rate of population increase for the area is 6.7% annually. Given that the groundwater and Colorado River allotments for Las Vegas can support 1.5 million people, how long will it be before the metropolitan area expands beyond its ability to provide water for its residents? Justify your response with a graph and give the expression you used to model the situation.

PROJECT Find Out for Yourself

No one knows how long it took to build the Great Pyramid at Giza. However, suppose that the project went somewhat according to the timeline shown here. Show this graph to a high school student and to a college student. Ask each one the following questions about the "story" told by the graph.

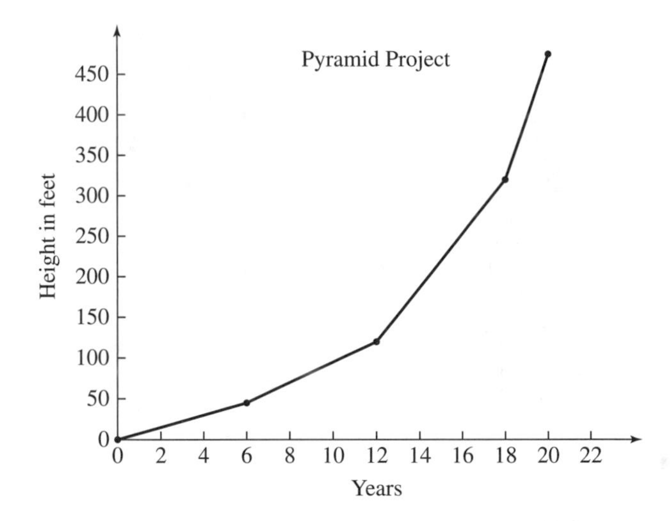

1. How much time has been allotted to each phase of the project?
2. In which phase does the height of the pyramid increase most rapidly? Least rapidly?
3. What is the approximate increase in the height per year during phase 3?
4. Calculate a rate of change for each period of the pyramid's construction.

What did you notice about each student's understanding of rate change and their ability to link that information with the data shown in the graph? How are the stages of this interview format related to knowledge about linear equations?

5.3 Variables, Expressions, and Equations

Developing a strong sense of what variables are and how to use them requires more than merely viewing variables as symbols for quantities, as we showed in the preceding sections (Usiskin, 1988). As students progress in their study of algebra, the role of variables in each of the following settings should be carefully introduced, discussed, and examined with students, and then contrasted.

$$P = 2l + 2w \quad (1)$$

$$A = \pi r^2 \quad (2)$$

$$4x = 50 \quad (3)$$

$$\text{For all real } \theta,\ \sin^2\theta + \cos^2\theta = 1 \quad (4)$$

$$\text{For all integers } x,\ 0 = x + -x \quad (5)$$

$$y = 3x + 2 \quad (6)$$

In each of these algebraic "sentences," variables take on different roles. The first two sentences talk about measures of geometric forms. In sentence (1), l and w represent the length and width of a rectangle. The P stands for the perimeter of the rectangle. In sentence (2), A stands for the area of a circle having radius r. Students often think the symbol π is a variable; however, it is a constant. Remember, it stands for 3.14159 . . . , which is the ratio of the circumference to the diameter for a circle. We usually express constants in numerical forms, but π, e, i, c, and a few others use letters.

In sentence (3), the x is a placeholder for an unknown value. This is probably the most frequent symbolic representation in algebraic equations. The θ in sentence (4) represents any element of some set in an identity that holds for all set elements. In this case, θ represents any real number. In a like manner, the x in (5) defines a property that holds for all integers, the additive inverse property for integers.

Sentence (6) has the variables x and y acting as covarying variables; that is, y's values change as x's values change, and vice versa. We see this use when equations describe the relationship between two, or more, quantities in a pattern or model.

In each sentence, the variables are used in combination with an operation. In the case of sentence (1), the formula for perimeter, the expression on the right side of the equation consists of two terms connected by the operation of addition. The first term expresses the portion of the perimeter that comes from combining the measures of the two lengths. The second term combines the measures of the two widths. In sentence (2), the formula for the area of the circle, the right-hand side of the equation is a single expression.

Developing such expressions is at the heart of translating patterns or verbal statements into mathematical symbols. Historically, students identified the variable that described the target in a word problem and then looked for key words that would help them translate the situation into an algebraic expression and/or equation. Take this problem, for example.

> Joan found one pair of jeans for \$31.25, and she noticed the same pair in the store two days later for \$32.50. How much had the price increased?

The word "increased" indicates a growth and hence signals "addition" under the key word approach. However, when we think through the problem and the operations involved, the equation of $x = 32.50 - 31.25$ gives the solution; this expression involves subtraction. Hence, the key word approach, while working in a number of cases, can cause confusion in certain settings.

Students are far better off if they think through situations in terms of the actions involved in the context described. In doing so, it is important that they get a firm grasp of the numerical operations required in the situation. Here operation sense comes into play. Is it a situation where groups are being combined and the total number is the desired goal? If it is two groups of different sizes, then addition may be the operation to consider. If it is a multiple combining groups of the same size, then multiplication may be called for. If the question is how many groups of a specified size are contained in total or how many each person gets when a total is shared, then division is the operation to focus on. If the question removes one group from another group and asks for the remaining group size, subtraction may be the key. Developing an operation sense and the ability to target the variable of interest, puts students well on their way to being able to represent problems in terms of expressions and equations.

Other knowledge that may help are students' overall vocabulary, their ability to visualize situations, and their grasp of necessary and sufficient factors. Table 5.8 lists key words that help students identify operations and describe relationships.

Example 1 ***Steve's Collection of Sports Cards***

Find an equation whose solution answers the question posed: Steve has 4 times as many baseball cards as he has football cards. He has 120 cards in all. How many does he have of each?

Let x be the number of football cards. Then $4x$ represents the number of baseball cards. Adding these two quantities gives the equation $x + 4x = 120$. ■

Table 5.8 Key words in word problems.

at most, at least	how many times more
as many as, times as many as	how many times less (more) than
consecutive	less than, is less than
difference	more than, is more than
divisible by	older than, younger than
even, odd	product
greater than, less than	quotient
greatest, least	sum
half as many as	times
how many more	equal shares

An equation is a statement of equality concerning two constants, variables, or expressions. The two sides of the equation represent the same quantities. To help students bridge the gap to writing equations, have them write the equation in words and pseudo-symbolic form. For example, have the students compare and contrast the meanings of

$$4 + (\text{number of football cards}) = (\text{number of baseball cards})$$

$$4 - (\text{number of football cards}) = (\text{number of baseball cards})$$

$$4 \times (\text{number of football cards}) = (\text{number of baseball cards})$$

$$4 \div (\text{number of football cards}) = (\text{number of baseball cards})$$

$$4 + (\text{number of baseball cards}) = (\text{number of football cards})$$

$$4 - (\text{number of baseball cards}) = (\text{number of football cards})$$

$$4 \times (\text{number of baseball cards}) = (\text{number of football cards})$$

$$4 \div (\text{number of baseball cards}) = (\text{number of football cards})$$

Note that the comparison has to do with determining (1) which groups of cards has the smaller amount, (2) what operation is involved in comparing the sizes of the two groups, and (3) which group needs to be multiplied to get to the other group.

One rather infamous study found (at least the results are infamous!) that many college students translated the statement that there are 4 times as many students as professors as $4S = P$ with the obvious assignment of the variables. Of course, this is the exact opposite of the correct relation! Perhaps the students weren't taught in high school to think this through in terms of physical representations. If they had, they would have noted that for each professor there must be 4 students, like the pairings shown in Figure 5.7. Continuing this pattern, we see that the correct equation should be $4P = S$. We can also envision this relationship as the ratio of $S{:}P = 4{:}1$.

Example 2 *Javier's Money*

Consider the following situation. Define the variables involved and develop an equation to solve the problem.

Javier has \$4 at the beginning of the day and he makes \$8 for each hour that he works. He starts work at 8 A.M. and works until 3 P.M. How much does Javier have at the end of the day? ■

In order to solve this problem, students must first determine the nature of the answer. Is a number of hours, an amount of money, or a day of the week called for by the problem? Once students target that the answer must be an amount of dollars, they can then recognize that the problem can be restated as

$$\begin{array}{lll} P \Rightarrow SSSS & P \Rightarrow SSSS & P \Rightarrow SSSS \\ & P \Rightarrow SSSS & P \Rightarrow SSSS \\ & & P \Rightarrow SSSS \end{array}$$

Figure 5.7 Visualizing ratio of students to professors.

Dollars at end of day = 4 ($ Javier had at the start) + Dollars he earns during the day

If we let D = number of dollars Javier has at the end of the day and h = number of hours that he works, we can translate the word equation into symbols:

$$D = 4 + 8h$$

To answer the original question, we need to find the number of hours worked. Working from 8 A.M. until noon is 4 hours; from noon until 3 P.M. is another 3 hours, so Javier worked 7 hours in all. Now, evaluating $D = 4 + 8h$ for $h = 7$, we have $D = 4 + 8(7)$, or $D = 60$. Thus, Javier had $60 at the end of the day.

As students work word problems and begin to develop expressions and equations, they are less likely to think of equations, especially the equals sign, as signifying "do something" or the right-hand side as "the answer" (Wagner & Parker, 1993).

Exercises 5.3

1. Write an expression that gives the value of each quantity:
   a. the number of yards of material purchased for r dollars at $3.50 per yard
   b. the number of sheets of paper left in a copy machine from a ream of 500 sheets after making 3 copies of a document that is x sheets long

2. A firework is launched at an angle of 70° with a velocity, v, of 240 ft/s. Its fuse is set to go off at 2.5 s after the launch. The expression for height, h, of a firework as a function of time is given by $h = 0.94vt + 0.5gt^2$ where t is the number of seconds after the launch and g is the constant acceleration due to gravity, which is -32 ft/s^2. At what height will the firework explode?

3. A rancher has four fields with the same area and a 2-acre homestead. If she divides her land equally among her three children, how much land will each child inherit? Write two equivalent expressions for your answer.

4. The target recommended heart rate for exercise, in beats per minute, is given by the expression $0.8(200 - a)$ where a stands for age.
   a. What is the target heart rate for a 14-year-old?
   b. If a 45-year-old's heart rate is 132 beats per second, are they over, at, or under their recommended heart rate?

PROJECTS Find Out for Yourself

1. Ask a student to explain the meaning of the formula for the surface area of a rectangular box—that is, $A = 2lw + 2lh + 2wh$, where A is the area, l the length, w the width, and h the height.

2. Ask a student to write expressions for the following. Then ask how they developed the expressions they did.

a. 3 more than a
b. the cube of the sum of x and y
c. 7 less than the product of w and h
d. the width of a rectangle with area 72 cm^2 and height h cm

5.4 Functions: Their Representations—Their Properties

Key to student understanding of algebra and modeling is understanding what constitutes a function and the properties of functions. Most students first see functions in terms of graphs that satisfy the vertical line test. They next see them in terms of the $f(x)$ notation, while, in fact, this is not the function, but only a symbol for the value of the function at a point x. Later, they may see functions discussed in a more abstract form. Such discussions usually deal with arbitrary functions f and g and involve the sum, difference, product, quotient, and composition of two or more functions. This is most common in precalculus $(f \circ g)$ or trigonometry $(\sin\theta^2 + \cos\theta^2)$. The development of how to think about functions and how to interpret and operate with them in each of their representations requires a strong conceptual understanding.

The concept of function grew out of an attempt to capture the dynamics of motion and quantity through notation and the written word (Kliener, 1989). Only recently has function notation in terms of ordered pairs appeared on the scene. This notation, while clean and abstract, does little to help students understand underlying connections between related independent and dependent variables.

Developing the idea of a function as a rule that explains the relationship between changes in independent and dependent variables or between input and output, is not an easy thing to pull off. You will find that multiple representations of the concept come in handy. When we graph data like $1, 3, 5, 7, \ldots$, it is easy to move from the graph to a table like Table 5.9.

Table 5.9

x	y
0	1
1	3
2	5
3	7
4	9
⋮	⋮

Students are generally then ready to move the symbolic form: $y = 2x + 1$ or $f(x) = 2x + 1$. However, they may not fully understand what this symbolism means. For example, how is the change of 2 units with each step captured by the expression or formula, what is the name given to this rate change in interpreting linear equations or functions? In making a table of values, should the listing for x start with 0 or with 1? Each of these decisions helps the student focus on subtle nuances in the concept of function.

Functions are usually defined as a relationship between two sets of numbers that links each number in the first set with a unique number in the second set. This special relationship is characterized as a graph where a vertical line intersects the graph at a single point. If the line intersects the graph in two or more points for some value x along the vertical axis, then the graph is that of an arbitrary relation, not a function. See, for example, the graph of the parabola $x = y^2$ opening to the right shown in Figure 5.8. Here we see that the number 4 on the x-axis, is associated with both $+2$ and -2 along the y-axis. Thus, the vertical line at $x = 4$ intersects the graph in 2 points. Such an equation is said to represent a relation, but not a function.

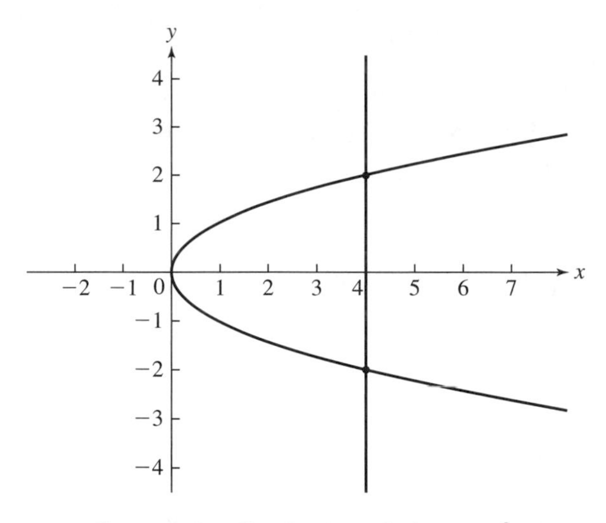

Figure 5.8 Graph of parabola $x = y^2$.

Often, textbooks define "function" in terms of a set of ordered pairs. That is,

> A function is a correspondence of items in one set A, called the *domain*, with items in another set B, called the *range*, such that
>
> 1. Every item in set A is corresponds with an item in set B.
> 2. No item in set A corresponds with more than one item in set B.

Property 1 says that every element in the first set gets mapped (or corresponded) with some item in set B. Thus, every element, a, in the domain set A gets matched up with some value, b, in the range set B. We show this in a variety of ways in Figure 5.9.

While this definition has a great deal to say in terms of developing the structure of abstract algebra and establishing theorems concerning functions, it offers little in terms of developing initial student understanding (Bednarz, Kieren, & Lee, 1996).

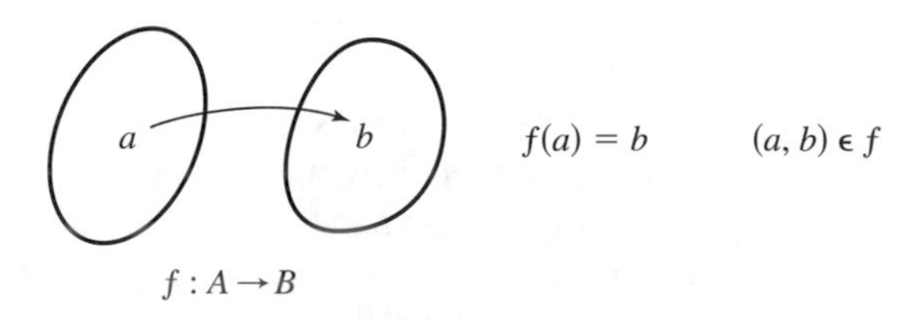

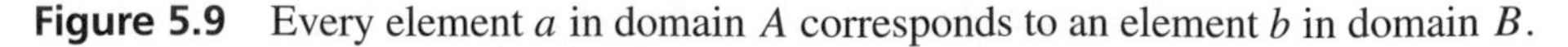

Figure 5.9 Every element a in domain A corresponds to an element b in domain B.

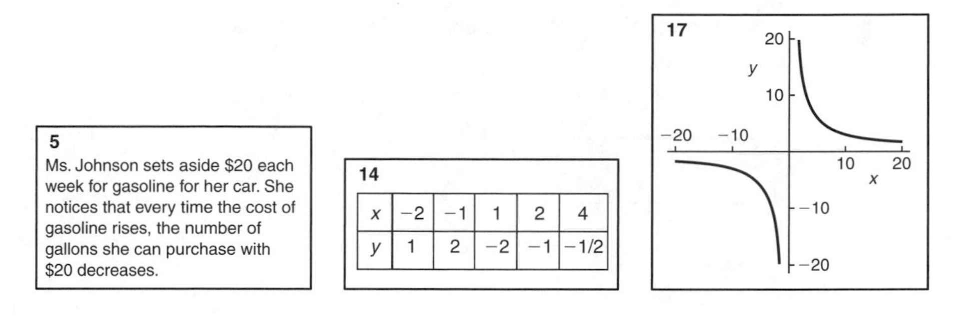

Figure 5.10 Practice cards for working with functions.

Besides developing a graphical understanding of the function concept that is rooted in its representation in tabular and symbolic formats, we also want students to begin to make connections about various classes of functions. In particular, we want students to have a good grasp of representations that show linear, exponential, quadratic, and periodic relationships. Cooney (1996) suggests that these ties can be built by having students work with sets of cards that show functions in graphical, verbal, symbolic, and tabular formats. Students classify the cards according to the type of representation, those that signify the same kind of function, those that bear a given analogical relationship to one another, or those having some other particular property. Some example cards are shown in Figure 5.10.

In another approach to examining functions and their representations, students accumulate a "tool kit" of functions and experiment with each change in "tool" parameters. For example, students identify and create examples of linear, exponential, quadratic, and periodic functions. In addition, they should have command of graphs of the square root function, inverses of familiar functions (especially logarithmic and circular functions), and inverse variation.

We also want students to be able to shift between different representations as they manipulate expressions and equations. For example, they should be able to graph zeros of a function and equation solutions. They should be able to explain how a system of equations can be solved by examining the graph of the equations.

Exercises 5.4

1. Which of the following graphs is a graph of a function? Explain why or why not.

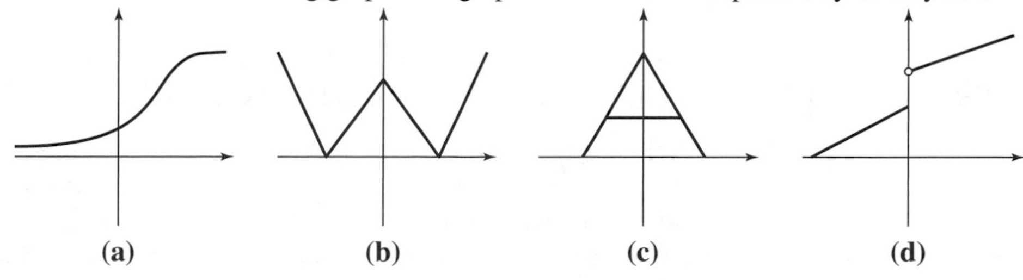

2. Which of the relations depicted below are functions?

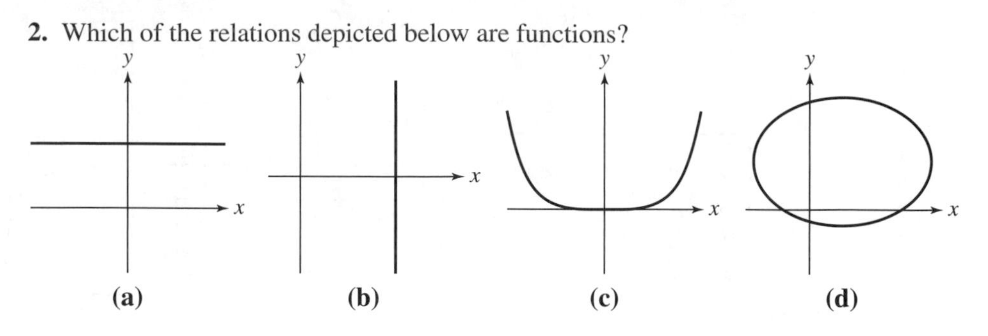

3. What do we mean when we say that a function is one-to-one? How can this property get confused with the definition of a function in terms of ordered pairs? What misconceptions can it cause? What is unique about the graph of a one-to-one function?

4. What is a constant function? How can its definition be confused with the definition of a general function? How can you prevent this in class?

5. Consider the graphs of the two functions shown below. How are they related to one another? How are they related and what role does $1/(x+2)$ play in that relationship?

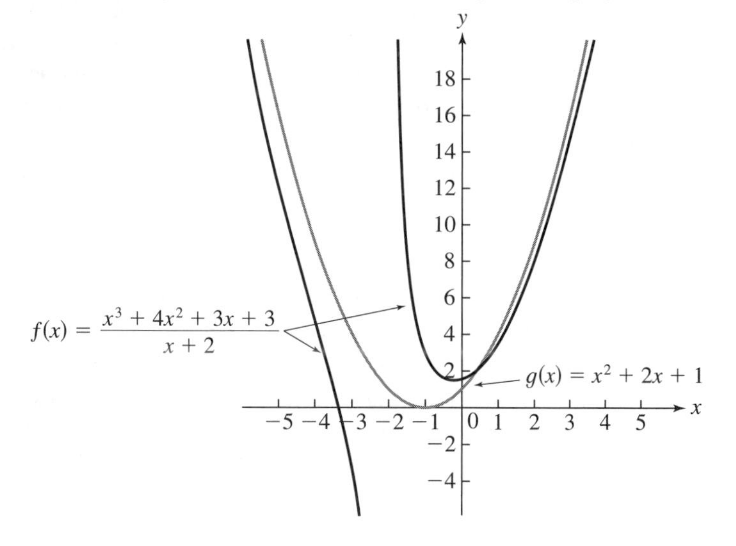

6. Explain how you would use a graphing calculator to describe the shape, behavior, and zeros of the function $f(x) = \ln x - x - 1$. Describe what such an investigation might entail and discover. What is the significance of using a calculator to approach this task?

PROJECTS ||||||||||| Find Out for Yourself |||||||||||

1. Susan is working with the function $f(x) = -x+7$. She claims that $f(2+3) = f(2)+f(3)$ for this function. Is she right? Can this property be generalized to any group of functions? Is it always true for functions? Answer these questions yourself and then ask some of your classmates in another math class about their views on this.

2. Explore the history of the function concept in the Kleiner reference (see Chapter References), or on the Internet (see http://www.groups.dcs.st=andrews.ac.uk/~history). Which mathematicians played significant roles in the development of this mathematical concept, and what were their contributions?

5.5 Solving Equations, Inequalities, and Systems of Equations

Solving Equations

Solving equations, or alternatively finding outputs for specific inputs—this is what most people think algebra "does." That is, most people think algebra is a set of disconnected rote rules for manipulating abstract algebraic expressions "with x's and y's in them." Dreyfus and Eisenberg (1987) found that many students who can solve traditional algebra exercises often have only minimal understanding of the algebraic or graphical *meaning* of the problems. Other research indicates that when meaning does exist, it exists in terms of graphical interpretations. This may be because the graphical representation takes the "x's and y's" and translates them into a less abstract format—a visual format (Markovits et al., 1988).

Solving equations and inequalities is at the heart of the eighth and ninth grade algebra programs. However, much of the concept development supporting this activity is already in place through algebraic reasoning activities in the earlier grades. For example, students in the fourth grade can reason through to the solution of $3(x + 5) = 30$ by "unfolding" the equation in this manner:

> What I need to find is some number when multiplied by 3 gives me 30. That number must be 10. But this number is really some number plus 5. What number plus 5 gives me 10? The answer is 5. Let me check it. 3 times the quantity 5 plus 5 is the same as 3 times 10, or 30. So the desired number must be 5.

Other students choose to graph the line $y = 3(x + 5)$ and the line $y = 30$ on the same set of coordinate axes, as shown in Figure 5.11, then look for the x-coordinate of the point where the two lines cross. Here that point appears to be the point $(5, 30)$. The value of x appears to be 5 and it checks. Hence, $x = 5$.

Yet other students attack the problem by constructing a table like Table 5.10. As successive values are substituted for x, the value of the expression $3x + 15$ gets closer and closer to 30, until it equals 30 when $x = 5$.

Finally, students may apply the traditional algorithm developed in almost all algebra classes. This approach makes use of the equal additions property of addition. This property says when the same number is added to both sides of an equation, the resulting equation is equivalent to the former equation—that is, has the same solution as the former equation. Carrying out the steps and noting the property of algebraic operations that justifies the equivalence, students follow the process shown in Table 5.11.

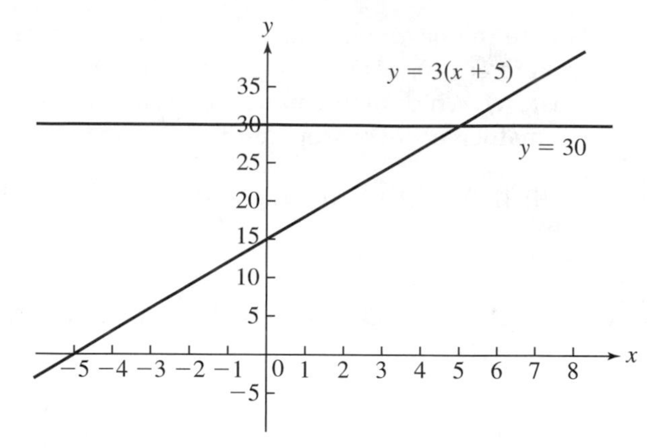

Figure 5.11 Graphs of $y = 3(x + 5)$ and $y = 30$.

Table 5.10

x	$3x + 15$
0	15
1	18
2	21
3	24
4	27
5	30
6	33

Table 5.11

$3(x + 5) = 30$	
$3x + 15 = 30$	Distributive property of multiplication over addition
$(3x + 15) - 15 = 30 + -15$	Equal addition property
$3x + (15 + -15) = 30 + -15$	Associative property of addition
$3x + 0 = +15$	Additive inverse property and addition fact
$3x = +15$	Additive identity property
$\frac{1}{3}(3x) = \frac{1}{3}(+15)$	Equal multiplication property
$(\frac{1}{3} \cdot 3)x = \frac{+15}{3}$	Associative property of multiplication
$1x = +5$	Multiplication inverse property
$x = +5$	Multiplicative identity property

A major part of the algebra teacher's task is to bring meaning to the algorithms that underlie the seemingly mystical manipulations that make up the operational side of algebra. A first step in doing this is getting students to tackle situations that are full of meaning—that is, when the variables represent quantities that students can visualize (Schifter, 1999). First, ask students to think about what a reasonable answer might be. Next, have them "unfold" the equation in linear situations. A good bridge here is a balance scale analogy where "what is done to one side of an equation must be done to the other side." The example in Figure 5.12 moves through such an analysis with the equation $4x + 7 = 31$.

Such an approach helps students bridge to manipulating algebraic transformation rules working first with a concrete reference, one of those connections we want them to make (English & Halford, 1995). Note that this is not simply "transposing" values from one side of the equation to the other by changing sign or operation as some people teach. This process, although it results in equivalent expressions, is not viewed in the same

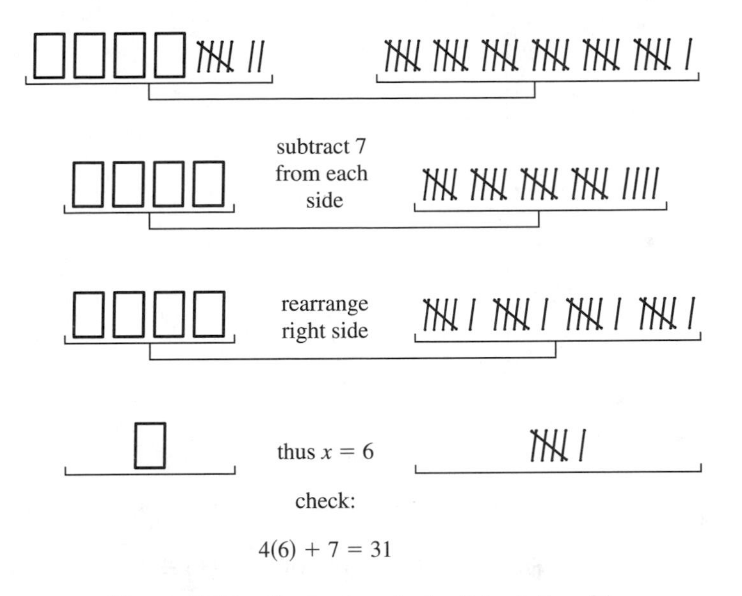

Figure 5.12 Student analysis of $4x + 7 = 31$.

manner by learners (Kieran, 1989). Equal treatment of both sides of the equation has a symmetry to it that students can grasp easily. In building such understandings, don't forget to help students see that the process generates equivalent equations, equations that have the same solution (Steinberg et al., 1990).

Students need to realize that in dealing with an equation of the form $y =$ (some expression in x), the point or points where the graph of the equation crosses the x-axis are the solutions to the equation. That is, they are the values for x, such that when substituted into the equation will result in a y-value of 0. Make sure students understand that solutions are also called roots. Should the relationship be expressed in functional form as $f(x) =$ (some expression in x), then these values are also known as the zeros of the function, because $f(x) = 0$ for these values of x.

Example 1 ***Graphing Functions***

Find the zeros of the following functions graphically.

a. $f(x) = 4x - 22$

b. $g(x) = e^x - 3$ ■

Graphing these functions we get the graphs shown in Figures 5.13a and b. The roots appear to be 5.5 and something around 1.1, respectively. The solution to part a is easy to establish through formal manipulations. The solution to part b is more difficult because a graph only gives an approximation. Solving it analytically gives us the value of $\ln x = 3$. Checking this value sharpens our solution a bit to $1.0986\ldots$. However, an exact numerical solution is beyond our reach. Zooming with graphic calculators gives closer and closer estimates of the value.

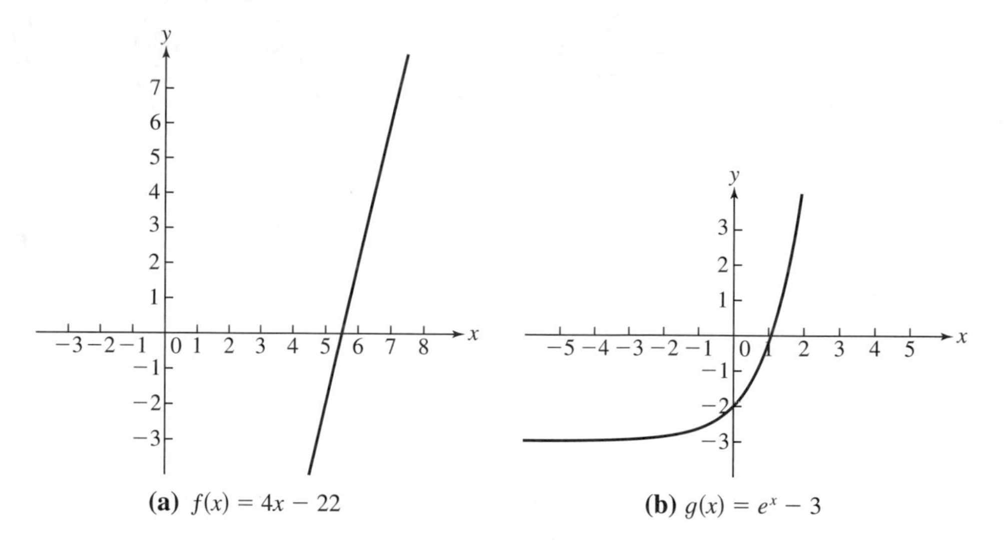

Figure 5.13 Graphs of (a) $f(x) = 4x - 22$, and (b) $g(x) = e^x - 3$.

Solving Inequalities

Solving inequalities is much like solving equations. The only difference is that in almost all cases solutions to inequalities result in solution regions, rather than discrete sets of points. For example, the solution to $x - 3 > 5$ is an infinite set, the set of all real numbers x such that $x > 8$. In a like manner, the inequality $y > x - 3$ is an infinite region in the plane consisting of all points (x, y) such that they lie above the line $y = x - 3$, as shown in Figure 5.14.

Students quickly learn to graph the equations that serve as region boundaries, like $y = x - 3$ for Figure 5.14 and then select a point to determine which side of the boundary should be shaded. In the case of $y > x - 3$, the origin $(0, 0)$ shows that 0 is greater than $0 - 3$; hence, the side that contains the origin is the solution set. The

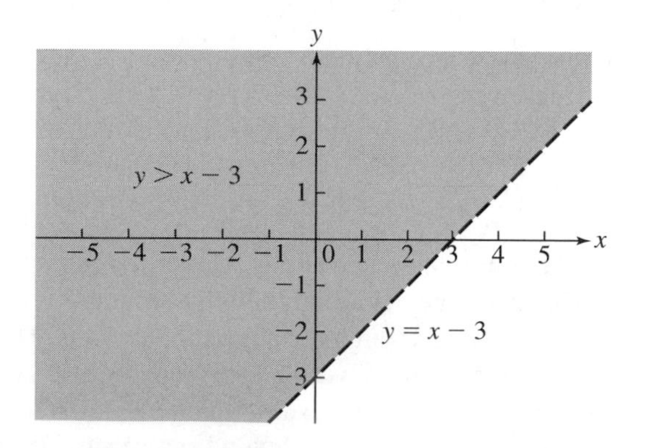

Figure 5.14 Solution region of the strict inequality $y > x - 3$.

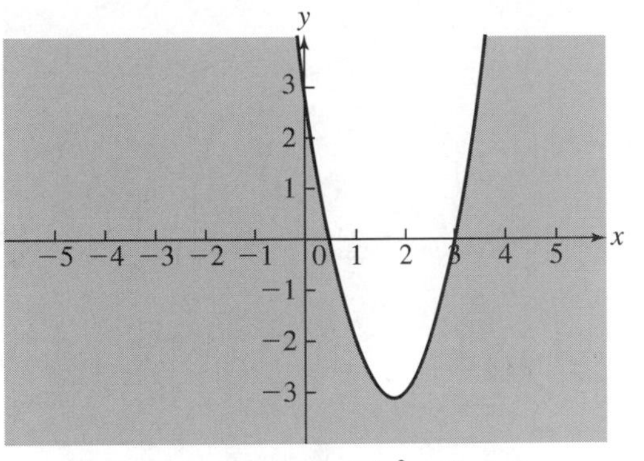

Solution region for $y \le 2x^2 - 7x + 3$.

Figure 5.15 Solution region of the inequality $y \le 2x^2 - 7x + 3$.

line $y = x - 3$ is dashed because it is not part of the solution for the strict inequality. (Don't forget to remind students that $>$ signals a *strict* inequality, whereas $\ge$ does not.) Students thus learn when to include the boundary. Figure 5.15 graphs $y \le 2x^2 - 7x + 3$. Here the boundary is solid because this is not a strict inequality. The shaded region is the portion of the plane that contains the origin, because $0 \le 2(0)^2 - 7(0) + 3$.

Solving a System of Equations

The solution of a set of simultaneous equations, usually referred to as a system of equations, is a value, or a set of values, that satisfies each equation in the system. For example, the system of equations

$$x + y = 2 \tag{1}$$

$$x + 2y = 3 \tag{2}$$

has the solution $(1, 1)$. When the substitutions $x = 1$ and $y = 1$ are made, the two original equations are satisfied. Such systems can be solved in several ways. Figure 5.16 shows a graphical solution. The point where the lines cross satisfies both equations.

The system can also be solved by substitution. To do this, we solve equation (1) for y in terms of x, and get $y = 2 - x$. Substituting $2 - x$ for y into equation (2) gives $x + 2(x - 3) = 3$. We then find $x = 1$. Substituting 1 into $x + y = 2$ gives $y = 1$. Checking the proposed solution $(1, 1)$ in $x + 2y = 3$ shows that $(1, 1)$ is indeed the solution.

We can also solve the system with matrices. To do this, we rewrite the system

$$x + y = 2$$

$$x + 2y = 3$$

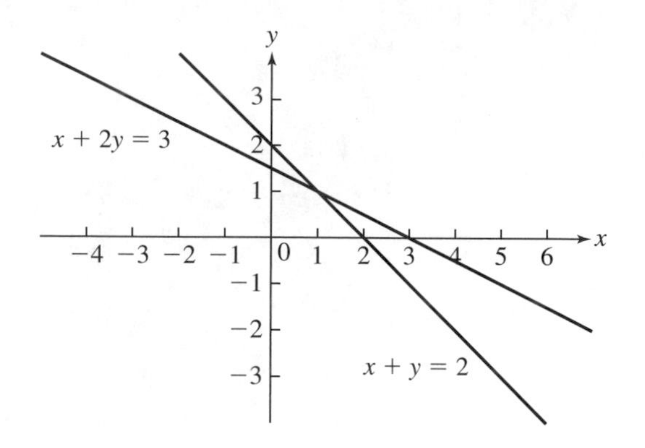

Figure 5.16 Graphs of the system of equations $x + y = 2$ and $x + 2y = 3$.

as the matrix product

$$\begin{bmatrix} 1 & 1 \\ 1 & 2 \end{bmatrix} \begin{bmatrix} x \\ y \end{bmatrix} = \begin{bmatrix} 2 \\ 3 \end{bmatrix}$$

where the first matrix is the coefficient matrix **A**, the second one is the variable matrix **X**, and the third one is the constant matrix **B**. This product can be represented as $\mathbf{AX} = \mathbf{B}$. Solving by matrix algebra, we get $\mathbf{X} = \mathbf{A}^{-1}\mathbf{B}$, recalling that we have to multiply on the left by $\mathbf{A}^{-1}$ because multiplication of matrices is not commutative. This product is easily carried out on calculators using the matrix calculation operations.

The discussion of solution methods for systems of equations is a good place to demonstrate the efficiency of algorithms. As the number of variables in the equations increases—3 equations in 3 unknowns, 4 equations in 4 unknowns, and so on—some interesting solution methods come into play. Some programs stress Gauss-Jordan elimination; others stress Cramer's rule. Gauss-Jordan elimination reduces the system of equations through elementary row operations to the triangular form shown in Table 5.12. Then solving the final equation for z, we back-substitute into equation (2) to solve for y. Finally, using these values for y and z in equation (1), we find x. Our solution can be checked in the original equations.

Cramer's rule sets up the system of equations in matrix form $\mathbf{AX} = \mathbf{B}$, checks to see if the determinant of **A** is nonzero, and then solves for each variable by forming a quotient for each variable. The numerator of the quotient that gives the value for x is the determinant of the coefficient matrix **A** where the column of coefficients associated with x in the three equations is replaced by the corresponding equation constants from the same rows in matrix **B**. The denominator of the quotient is the determinant of matrix **B**. This process is then repeated for the other variables, appropriately substituting constants in the coefficient matrix for each variable's column.

When we examine the computations required by these two approaches for an arbitrary system of n equations in n unknowns, we see the superiority of Gauss-Jordan elimination over Cramer's rule (Steinberg, 1974). As n increases, Gauss-Jordan elimination requires about

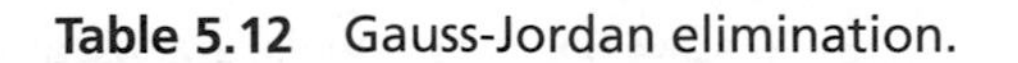

Table 5.12 Gauss-Jordan elimination.

Solve the system of equations

$$2x + 4y + 2z = 6 \quad (1)$$

$$-2x + 1y + 3z = 4 \quad (2)$$

$$3x + 2y + 1z = 5 \quad (3)$$

Construct tableau

$$\begin{bmatrix} 2 & 4 & 2 & 6 \\ -2 & 1 & 3 & 4 \\ 3 & 2 & 1 & 5 \end{bmatrix} \xrightarrow{\frac{1}{2}R_1} \begin{bmatrix} 1 & 2 & 1 & 3 \\ -2 & 1 & 3 & 4 \\ 3 & 2 & 1 & 5 \end{bmatrix} \xrightarrow{2R_1+R_2} \begin{bmatrix} 1 & 2 & 1 & 3 \\ 0 & 5 & 5 & 10 \\ 3 & 2 & 1 & 5 \end{bmatrix}$$

$$\xrightarrow{R_3+-3R_1} \begin{bmatrix} 1 & 2 & 1 & 3 \\ 0 & 5 & 5 & 10 \\ 0 & -4 & -2 & -4 \end{bmatrix} \xrightarrow{\frac{1}{5}R_2} \begin{bmatrix} 1 & 2 & 1 & 3 \\ 0 & 1 & 1 & 2 \\ 0 & -4 & -2 & -4 \end{bmatrix}$$

$$\xrightarrow[R_3+4R_2]{R_1+(-2R_2)} \begin{bmatrix} 1 & 0 & -1 & -1 \\ 0 & 1 & 1 & 2 \\ 0 & 0 & 2 & 4 \end{bmatrix} \xrightarrow{\frac{1}{2}R_3} \begin{bmatrix} 1 & 0 & -1 & -1 \\ 0 & 1 & 1 & 2 \\ 0 & 0 & 1 & 2 \end{bmatrix}$$

$$\xrightarrow[R_2+(-R_3)]{R_1+R_3} \left[\begin{array}{ccc|c} 1 & 0 & 0 & 1 \\ 0 & 1 & 0 & 0 \\ 0 & 0 & 1 & 2 \end{array}\right]$$

Hence $x = 1$, $y = 0$, and $z = 2$ is the solution to the system.

$$\frac{n^3}{3} + \frac{n^2}{2} - \frac{5n}{6} \text{ additions} \quad \text{and} \quad \frac{n^3}{3} + n^2 - \frac{n}{3} \text{ multiplications}$$

Cramer's rule, on the other hand, requires about

$$\frac{n^4}{3} \text{ multiplications} \quad \text{and} \quad \frac{n^4}{3} \text{ additions}$$

Given this superiority and the fact that Cramer's rule introduces determinants into the curriculum as well, it is hard to make an argument for solving a system of equations using Cramer's approach.

Exercises 5.5

1. Provide a solution, with justification for your actions, for the equation

$$\frac{3}{7}x - 12 = \frac{-8}{3}x + 5\frac{1}{2}$$

2. Explain how to solve $x^2 - 13x + 42 = 0$ using symbols, graphs, and tables.

3. A system of 3 equations in 3 unknowns can have a unique solution, an infinite number of solutions, or no solution.
   a. Explain geometrically how each of these cases can exist.
   b. Explain how you can tell from the Gauss-Jordan elimination process which of these cases holds for a particular problem.

4. Find the solution, in real numbers, to the inequality $x^2 + x - 16 < 4$. Justify your steps. Also graph the solution set for the inequality.

5. What points (x, y) in the plane satisfy $3y \geq 2x - 7$ and $y < x^2 + 2x + 7$? Illustrate your answer.

6. Solve the following systems of equations:

a. $$\begin{cases} x + 2y + 3z = 1 \\ x + y - z = 0 \\ x + 2y + z = 3 \end{cases}$$

b. $$\begin{cases} x + y + z = 3 \\ x - y + z = 3 \\ x + z = 2 \end{cases}$$

c. $$\begin{cases} x - z + w = -1 \\ x + y + z - w = 2 \\ -y - 2z + 3w = -3 \\ 5x + 2y - z + 4w = 1 \\ -x + 2y + 5z - 8w = 7 \end{cases}$$

7. We often use the terms *inconsistent*, *consistent*, *independent*, and *dependent* when we discuss systems of equations. What do they mean and what are their ramifications in terms of numbers of solutions?

8. Solve the system of equations $\begin{cases} 41x + 40y = 81 \\ 40x + 39y = 79 \end{cases}$

Suppose this system arose in a research application and that the researcher found the two constants to be 80.99 and 79.01. How does this change of 0.01 in both constants impact the solution? This system is an example of an ill-conditioned system. Find out what this means and explain the shift in solutions graphically.

PROJECT Find Out for Yourself

Present the following application (Dossey et al., 2000) to an Algebra II student and ask them to solve it. What responses do you expect? What did you find out in giving the problem to the student?

Compact discs (CDs) can be purchased through CD clubs. Some clubs charge you a membership fee, then a certain amount for each CD purchased. Other clubs only charge you for each CD you purchase. The graph shows the total cost for purchasing

CDs from the Blues, Jazz, and Rock CD clubs. Use the information in the graph below to answer these questions:

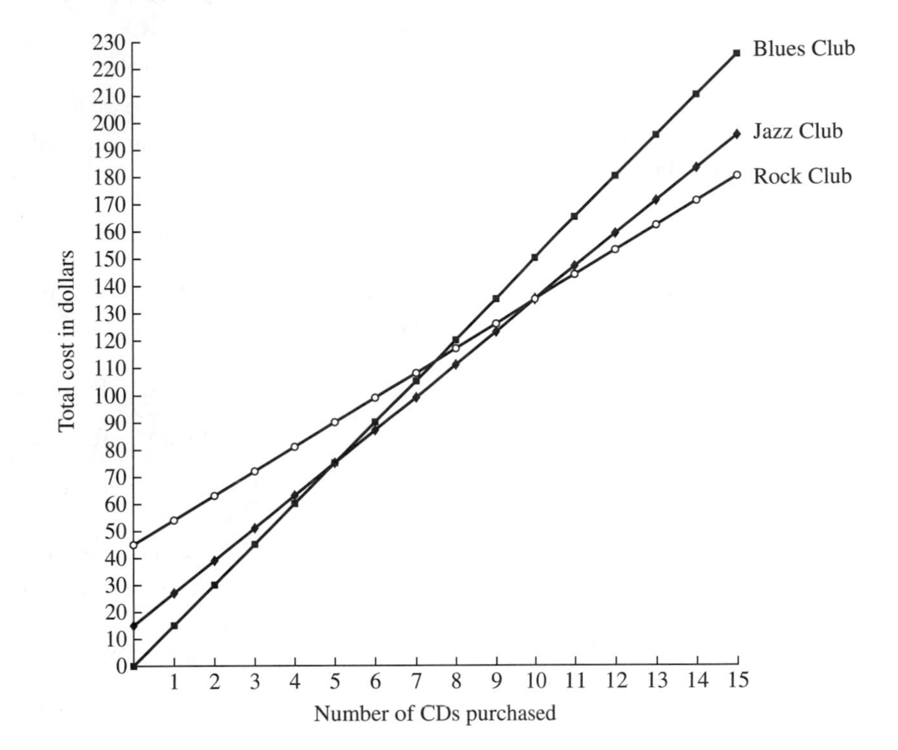

a. Which club has the largest membership fee, and what is it? Where is such information indicated in the graphs?

b. Which club charges the most for each CD purchased, and what is it? How did you determine this answer?

c. If any recording you might desire is available through any of the clubs, describe the ranges in which each club might give the best deal if you compare purchases of CDs in ranges of numbers of CDs purchased per year.

5.6 Modeling Using Proportion: Transformations That "Linearize the Data"

In this section we formalize the concept of proportion and use it to uncover relationships among variables. We then examine the technique of geometric similarity, a powerful concept for uncovering relationships among variables. We'll also see how proportion can be used to fit models to data.

Proportion

Two positive quantities x and y are proportional if one quantity is a constant positive multiple of the other; that is, if

$$y = kx$$

for some positive constant k. We write $y \propto x$ in this situation, and say that y "is proportional to" x. Thus,

$$y \propto x \quad \text{if and only if} \quad y = kx \quad \text{for some constant } k > 0 \tag{1}$$

Of course, if $y \propto x$, then $x \propto y$ also, because the constant k in proportion (1) is greater than zero and then $x = \frac{1}{k}y$. Other examples of proportions include

$$y \propto x^2 \quad \text{if and only if} \quad y = k_1 x^2 \quad \text{for } k_1 \text{ a constant} \tag{2}$$

$$y \propto \ln x \quad \text{if and only if} \quad y = k_2 \ln x \quad \text{for } k_2 \text{ a constant} \tag{3}$$

$$y \propto e^x \quad \text{if and only if} \quad y = k_3 e^x \quad \text{for } k_3 \text{ a constant} \tag{4}$$

In proportion (2), $y = kx^2$, $k > 0$ so we also have $x \propto y^{1/2}$ because $x = \left(\frac{1}{\sqrt{k}}\right) y^{1/2}$. We can now link proportion through a transitive rule for proportionality:

$$\text{if } y \propto x \text{ and } x \propto z, \quad \text{then } y \propto z$$

Thus, any variables proportional to the same variables are proportional to each other.

Now let's explore a proportion geometrically. In proportion (1), $y = kx$ yields $k = y/x$. Thus, k may be interpreted as the tangent of the angle θ depicted in Figure 5.17, and the relation $y \propto x$ defines a set of points along a line in the plane with angle of inclination θ.

If we compare the general form of a proportionality relationship $y = kx$ with the equation for a straight line $y = mx + b$, we can see that the graph of the proportion is a line (possibly extended) passing through the origin. If we plot the variables for proportions (2), (3), and (4), we obtain the straight-line graphs in Figure 5.18.

The graphing principle used in Figures 5.17 and 5.18 may be new to you. To plot y versus x^2 from a table of values relating y and x, first calculate the value of x^2 for each value of x. Next draw a coordinate system in which the horizontal axis represents x^2

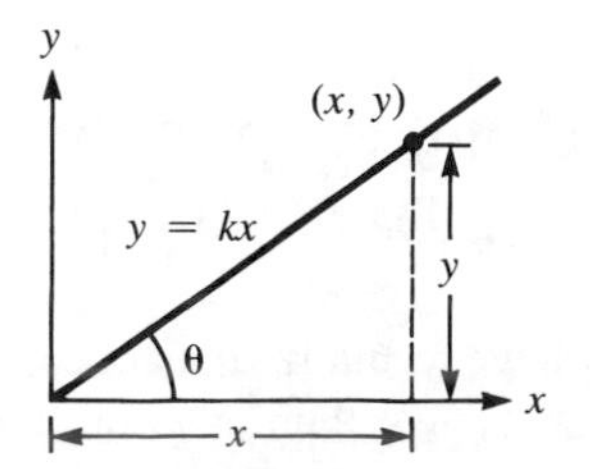

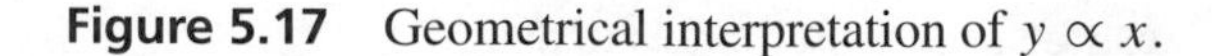
Figure 5.17 Geometrical interpretation of $y \propto x$.

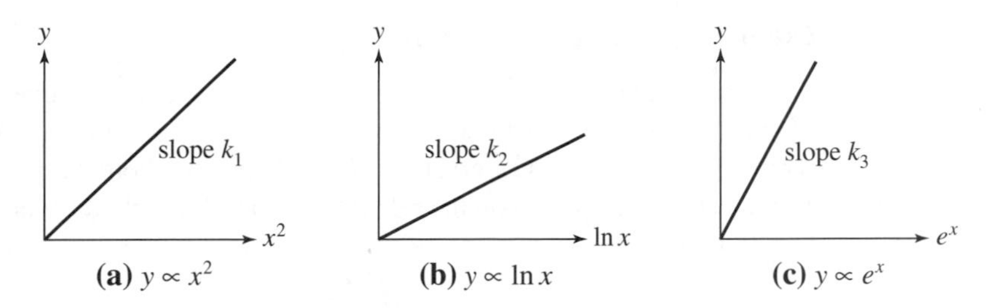

Figure 5.18 Geometric interpretations: (a) proportion (2); (b) proportion (3); (c) proportion (4).

values (not x values). Finally, plot the ordered pairs (x^2, y) in this rectangular coordinate system for each x and y related in your original table. For instance, the original table

y	3	12	27	48	75
x	1	2	3	4	5

produces the table

y	3	12	27	48	75
x^2	1	4	9	16	25

and graph in Figure 5.19. The graph is a straight line with slope 3. We can follow a similar procedure to plot y versus $\ln x$ and y versus $f(x)$ for any function of x.

It is important to note that not every line represents a proportionality relationship: The y-intercept *must* be zero; that is, the line must pass through the origin.

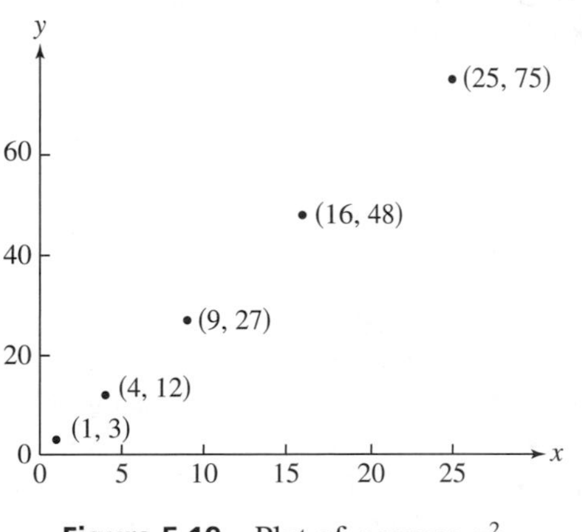

Figure 5.19 Plot of y versus x^2.

Example 1 ***Kepler's Third Law***

To further understand the idea of proportion, let's examine one of the famous proportionalities: Kepler's third law. In 1601, the German astronomer Johannes Kepler became director of the Prague Observatory. Kepler had been helping Tycho Brahe in collecting 13 years of observations on the relative motion of the planet Mars. By 1609, Kepler had formulated his first two laws:

1. Each planet moves along an ellipse with the sun at one focus.
2. For each planet, the line from the sun to the planet sweeps out equal areas in equal times.

Kepler spent many years verifying these laws and formulating the third law

$$T = cR^{3/2}$$

where T is the period in days, R is the mean distance to the sun, and c is a constant. The data in Table 5.13 is taken from the 1993 World Almanac.

In Figure 5.20, we plot the period versus the mean distance to the $\frac{3}{2}$ power. The plot approximates a line that projects through the origin. We can easily estimate the slope (constant of proportionality) by picking any two points that lie on the line passing through the origin:

$$\text{slope} = \frac{90,466.8 - 88}{220,869.1 - 216} \approx 0.410$$

We estimate the model to be $T = 0.410R^{3/2}$. ■

Table 5.13 Orbital periods and mean distances of planets from the sun.

Planet	Period (days)	Mean Distance (millions of miles)
Mercury	88.0	36
Venus	224.7	67.25
Earth	365.3	93
Mars	687.0	141.75
Jupiter	4,331.8	483.80
Saturn	10,760.0	887.97
Uranus	30,684.0	1,764.50
Neptune	60,188.3	2,791.05
Pluto	90,466.8	3,653.90

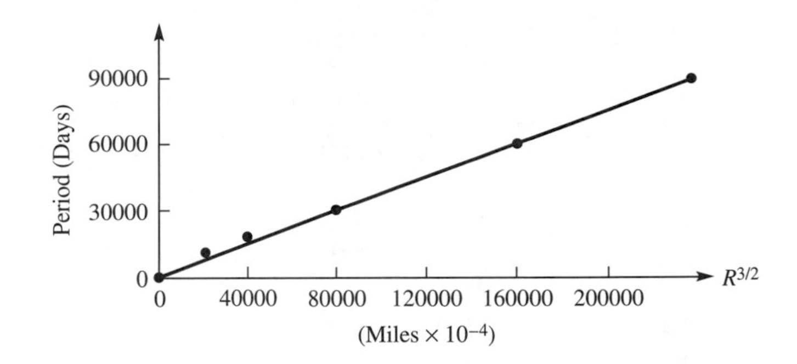

Figure 5.20 Graph of Kepler's third law as a proportion.

Example 2 ***A Spring-Mass System***

Consider a spring-mass system, such as the one shown in Figure 5.21. An experiment is conducted to measure the stretch of the spring as a function of the mass (measured as weight) placed on the spring. The data collected for this experiment is displayed in Table 5.14. The graph of the data in Figure 5.22 shows an approximate straight line (passing through the origin).

The data appear to follow the proportionality rule that elongation (e) is proportional to the mass (m), $e \propto m$. The straight line appears to pass through the origin. Understanding this system geometrically helps us determine whether proportionality is a reasonable assumption for estimating the slope, k. In this case, the assumption appears valid, so we estimate the constant of proportionality by picking any two points that lie on our straight line as

$$\text{slope} = \frac{\Delta \text{ elongation}}{\Delta \text{ mass}} = \frac{3.25 - 1.875}{200 - 100} \approx 0.0135$$

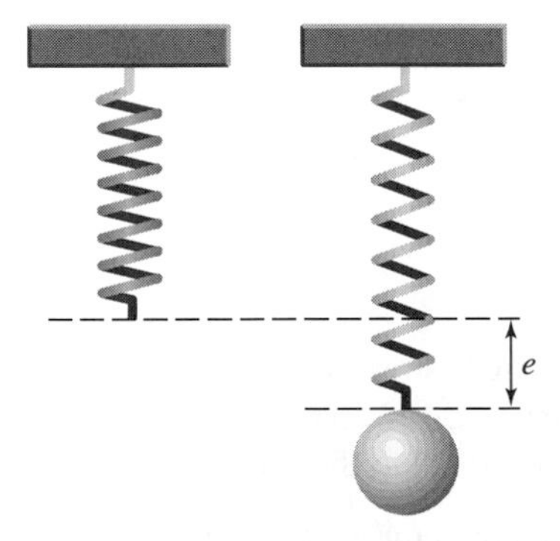

Figure 5.21 Spring-mass system for Example 2.

Table 5.14 Spring-mass system.

Mass	Elongation
50	1.000
100	1.875
150	2.750
200	3.250
250	4.375
300	4.875
350	5.675
400	6.500
450	7.250
500	8.000
550	8.750

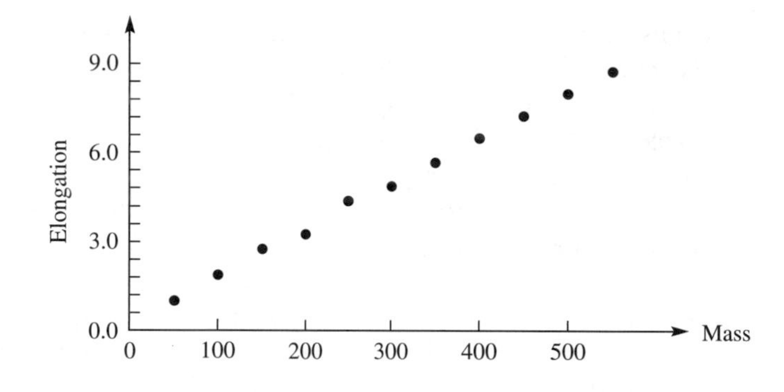

Figure 5.22 Graph of data for the spring-mass system.

We estimate our model as

$$e = 0.0135m$$

We then examine how close our model fits the data (see Figure 5.23) and conclude that it is reasonable. ■

Exercises 5.6

1. If a spring is stretched 0.37 in. by a 14-lb force, what stretch will be produced by a 9-lb force? By a 22-lb force? Assume Hooke's law, which asserts the distance stretched is proportional to the force applied.
2. If an architectural drawing is scaled so that 0.75 in. represents 4 ft, what length represents 27 ft?

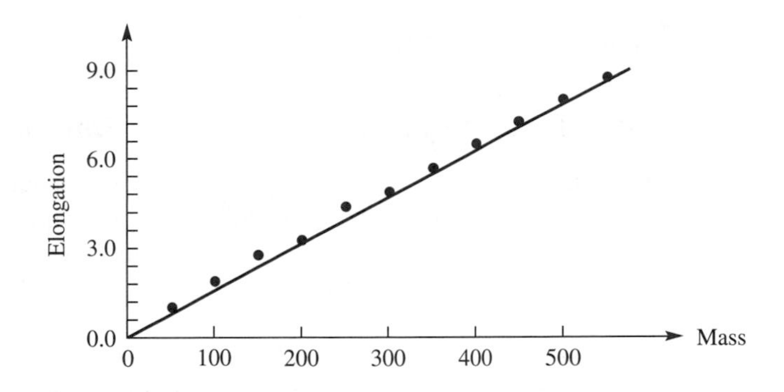

Figure 5.23 Graph of the spring-mass system data with proportionality line.

3. Determine if the following data support a proportionality argument for $y \propto z^{1/2}$. If so, estimate the slope.

y	3.5	5	6	7	8
z	3	6	9	12	15

4. A new planet is discovered beyond Pluto at a mean distance to the sun of 4004 million miles. Using Kepler's third law, estimate for time T of the new planet's orbit around the sun.

5. Use a graph to explain the meaning of the proportion $y \propto \frac{u}{v}$.

6. In Example 2, if a 1000-g mass is added to the spring, how far would the spring stretch? If a 1,000,000-g mass is added, how far would the spring stretch? Is the model valid for any added mass?

7. In Example 2, convert the weights from grams into pounds. Find the model that now relates elongation to weight in pounds.

PROJECTS Modeling Explorations

1. Consider an automobile suspension system. Build a model that relates the stretch (or compression) of the spring to the mass it supports. If available, obtain a car spring or bungi cord and measure the stretch for different masses supported by the spring or bungi cord. Test your proportionality argument graphically. If it is reasonable, find the constant of proportionality.

2. Write and present to class a 10-minute report on Hooke's law.

5.7 Algebra Lesson Plans

Lesson 3: Solving a Linear Equation (Middle School Level)[1]

Objectives: Students will work with a variety of models to examine procedures for solving linear equations. The focus will be on pulling together the various ways discussed in previous work and developing a rich and deep understanding of the processes available. As part of this, students will

1. Review the various models that they have used to solve linear equations to date—tabular, concrete materials, balance beam, unfolding, and symbolic.
2. Look for connections between equivalent aspects of each approach to see how they relate.
3. Compare and contrast the advantages and disadvantages of each solution approach.
4. Translate a verbal situation into expressions, form an equation for the situation, solve it, and write a sentence explaining the answer.

Activities:

1. (5–10 minutes) Warm-Up Monitor the pairs working in the groups, assessing how they progress through the problems in the following overhead:

Solve each equation using the methods indicated in the table.

a. $3x - 6 = 15$

b. $4x + 3 = 25$

c. $7 = x/5 - 1$

d. $0.4 - 3x = 5x + 2$

e. $3x + 15 = 45$

[1] Based on lessons from McDougal Littell's *Math thematics 3*, and Scott Foresman-Addison Wesley's *Middle school math course 3.*

Methods to Use By Partner Pairs

Problem/ method	Table	Algebra tiles	Balance	Unfolding	Symbolic
a	Pairs 1, 2	Pairs 9, 10	Pairs 7, 8	Pairs 5, 6	Pairs 3, 4
b	Pairs 3, 4	Pairs 1, 2	Pairs 9, 10	Pairs 7, 8	Pairs 5, 6
c	Pairs 5, 6	Pairs 3, 4	Pairs 3, 4	Pairs 9, 10	Pairs 7, 8
d	Pairs 7, 8	Pairs 5, 6	Pairs 5, 6	Pairs 1, 2	Pairs 9, 10
e	Pairs 9, 10	Pairs 7, 8	Pairs 5, 6	Pairs 3, 4	Pairs 1, 2

2. (10–15 minutes) Opening Remarks Indicate that over the past week, we have been looking at a variety of ways to solve equations. In essence these all give the same result, but they represent the process in different ways. Today, we'll tie them together by seeing how their steps are similar. Review the steps to solving (a), (c), and (d). Ask them first how these are alike and then how they are different.

   Review solving (a) and (c) in each of the five ways. Note the steps that students use. How do they relate across the different approaches? How does adding an inverse to both sides relate to adding or removing a weight from the scales or balance beam? How does moving up or down in input values relate to adding or removing weights? How does doing the opposite operation in unfolding relate to adding the inverse in the symbolic approach? In this part of the lesson, focus on questioning the students and expecting them to give the responses.

3. (5–10 minutes) Whole-Class Discussion Here is an opportunity to get the students involved in assessing the advantages and beginning to move to the unfolding or symbolic approaches as effective ways to solve equations. The focus might be on adopting these because they are efficient and make it easier to talk about our solutions.

4. (10–12 minutes) Setting Up and Solving Linear Equations Present students with the following:

   (a) Mrs. Algebrite pays an electric bill that has a special service fee of $5 and electric charges of $18. If electricity costs $0.10 per kilowatt hour, how many kilowatt hours did she use? If h is the number of hours, set up an equation to represent this situation, solve it, and write a response to the question posed.

   (b) A company's retirement program gives yearly retirement benefits based on a formula. The formula for an 18-year employee is given by $R = (18s)/60$, where s is the person's annual salary at the time of retirement.

   Ask students to set up equations for, then solve, each problem.

5. (Remaining Time) Discuss solutions and assign the homework for the following class period. Remind students to write a full sentence as part of their answer.

Assessment:

1. Warm-Up Circulate through the room observing the students' work using the various approaches. Ask them to compare the ease of each method as you visit with small groups. Ask them to show how a specific action using the various materials (balance beam, unfolding, and so on) can be stated symbolically.
2. Opening Remarks Watch to see the various techniques students use to solve the equations. Which approaches cause the most problems? Why? Can students make the connections between related parts of each solution method? Question and display examples to ensure that all students have a solid grasp of the material before continuing.
3. Whole-Class Discussion Sharing techniques for finding area will let me know how flexible students are and how well they understand the concept of area of a nonstandard region.
4. Setting and Solving Circulate about the class observing student work. Especially observe those that seemed a bit timid or confused in the whole-class discussion. Stop by their desks and make sure they understand.
5. Remaining Time Ask students if there are any questions. Pull together the points made about the various approaches, note why the symbolic approach gets so much emphasis, note that it is important to have a strong conceptual feel for what is happening as one moves to interpreting and writing equations and solving problems.

Lesson 4: Locating Solutions for Linear Programming Problems (Secondary Level)[2]

Objectives: Students will develop an intuitive understanding of why maximum (minimum) values for the objective function for a linear programming problem occur at one vertex or along an edge of the feasible region defined by the constraints.

- Work in pairs to explore a model for a particular linear programming problem—its constraints, its feasible region, its objective function, and the evaluation of its objective function.
- Build and interpret a three-dimensional model to interpret the objective function for a linear programming problem whose feasible region is two-dimensional.

[2] Built from linear programming section in Dossey, J. A., et al. (1996). *Addison-Wesley secondary math: Focus on advanced algebra*. Menlo Park, CA: Addison-Wesley.

- Develop an intuitive understanding of why the maximum (minimum) values for the objective function are associated with corners or edges of the feasible region.
- Appreciate the need for mathematical proof.
- Illustrate the nature of a function, $o(x, y)$, whose values $z = o(x, y)$ represent the values of an objective function for points in the feasible region of a linear programming problem.

Warm-Up: Students will work in pairs to determine the feasible region and objective function for a linear programming problem of the type they have been considering. This problem will serve as the basis for the intuitive development of the theorem that the maximum (minimum) values of the objective function occur at vertices or along an edge of the feasible region.

Activities:

1. (10 minutes) Warm Up Students enter the classroom and begin work with their partners on solving the linear programming problem, which is presented on an overhead transparency.

> Solve the following problem with your partner. Check your solution with your neighbor. Make sure that you graph the feasible set, find the coordinates of its corners, and note the value of the objective function at each corner.
>
> An automobile manufacturer makes cars and trucks in a factory that is divided into two shops. The first shop, which does the basic assembly, needs 5 worker-days per truck and 2 worker-days per car. The second shop, which does the finishing touches, needs 3 worker-days for both cars and trucks. The first shop has 180 worker-days available per week and the second shop has 125 worker-days available. The profit is \$500 per car and \$700 per truck. How many of each type of vehicle should the manufacturer produce each week to maximize profit? What is the maximum profit?

2. (5 minutes) Presentation of Solution Select one student to show his or her solution at the board. Make sure that all of the parts of the solution required are presented. 15 cars and 30 trucks for \$28,500.

3. (20–25 minutes) Examination of Objective Function Discuss the solution and note that the objective function can be represented as a graph. The feasible set might look like this:

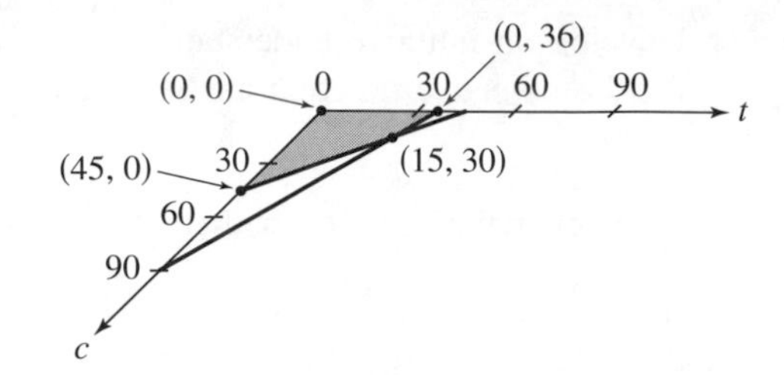

Note that at each corner we can construct segments whose height is equal to the value of the objective function for that particular corner of the feasible region. For example, the height of the segment at the origin is $500(0) + 700(0)$, or 0. In a like manner, the other segments are $(0, 36) \rightarrow 25{,}200$, $(15, 30) \rightarrow 28{,}500$, and $(0, 45) \rightarrow 22{,}500$. This would appear like this:

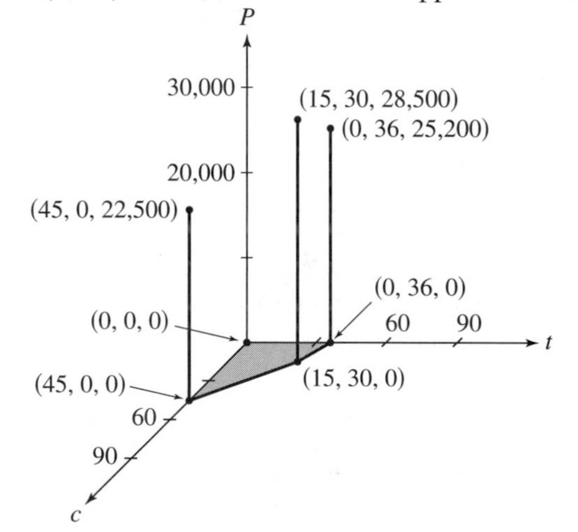

Now, place a sheet of posterboard across the tops of the segments. Note that they all fall in the same plane. Where must the maximal value be?

The vertices are in a plane; therefore, they are either (1) at the same height (and all feasible solutions produce the same result); or (2) one is a different height; or (3) all three are different heights. Pick one vertex. Is it higher, equal to, or lower than its neighboring vertex? A little geometry quickly convinces us that the highest point is at a boundary of the region above the feasible set. This may be a value associated with a point above a vertex or a segment of points above one boundary segment, or all points in the plane above the feasible set. The last situation happens only if the objective function is a constant. Hence, we only consider the first two options.

Note that the maximal value can be found by visiting successive vertices of the feasible set and noting the value of their objective functions. Either one of them will be a maximum (minimum), or two consecutive vertices will have the same maximal (minimal) value. Thus, all of the point segments connecting them will have this maximal (minimal) value. Note that the equation of the plane is

$$z = 500x + 700y$$

like a linear equation through the origin, but one dimension higher. Note the similarities in the equation formats: $ax + by + cz + d = 0$ with a z-intercept of 0 when $x = 0$ and $y = 0$ in $500x + 700y - z = 0$.

4. (5 minutes) Multivariate Function Notation Introduce the $o(c, d) = p$ notation to tie to the graph and talk about functions of two variables. Note that each point in the feasible region (c, d) has a unique value p associated with it and that no point (c, d) is associated with more than one point. Hence, the objective function is really a function; it simply maps ordered pairs into single values. Talk a little about $f(x, y) = z$ in general. Note operations like addition are functions $(2, 3) \rightarrow 5$, but also examine functions like $f(x, y) = x^2 + y^2$.
5. (5 minutes) (Remaining Time) Ask pairs to begin work on writing out the justification. Encourage students to share ideas, calling on volunteers. Assign students to work on one example from the text using the function notation, drawing a picture of the towers, and examining that the points do lie in a plane.

Assessment:

1. Warm-Up Watch student command of parts of solving the linear programming problem. This work should be almost automatic at this point. Select a pair of volunteers and ask them to share their solution on the board.
2. Group Work (informal assessment) Watch student solutions and ask questions about values in the solution, especially values of the objective function at the feasible region's boundary points. Question students to get a sense of their understanding of the problem, of the techniques used, of their confidence in their solution, and their ideas about why the process works.
3. Whole-Class Discussion Make sure that students are following the development of the plane and boundary points by asking questions and probing their understandings.
4. Homework Draw out student connections of the multivariate representations with the development of functions for a single variable; that is, $f : x \rightarrow y$ with values $f(x) = y$.
5. Extension Problem This is a good way to see if students can apply what they learned to a new situation.

References

Bednarz, N., Kieran, C., & Lee, L. (1996). *Approaches to algebra: Perspectives for research and teaching*. Dordrecht, The Netherlands: Kluwer.

Dossey, J., Jones, C. O., Klag, P., Kennedy, D., Kilpatrick, J., Lappan, G., Silver, E., & Smith, M. (2000). *Quantitative literacy and the middle school curriculum*. Unpublished manuscript. New York: The College Board.

Dreyfus, T., & Eisenburg, T. (1987). On the deep structure of functions. *In* J. C. Bergeron, N. Herscovics, & C. Kieran (eds.). *Proceedings of the 11th International Conference for the Psychology of Mathematics Education* (Vol. 1, pp. 190–196). Montreal: University of Laval.

English, L., & Halford, G. (1995). *Mathematics education: Models and processes*. Mahwah, NJ: Erlbaum.

Kieran, C. (1989). The early learning of algebra: A structural perspective. *In* S. Wagner and C. Kieran (eds.). *Research issues in the learning and teaching of algebra*. Reston, VA: NCTM.

Kleiner, I. (1989). Evolution of the function concept: A brief survey. *College Mathematics Journal* **20**(4):282–300.

NCTM (2000). *Principles and standards for school mathematics*. Reston, VA: NCTM.

Schifter, D. (1999). Reasoning about operations: Early algebraic thinking in grades K–6. *In* L. Stiff (ed.). *Developing mathematics reasoning in grades K–12, 1999 yearbook of the NCTM*. Reston, VA: NCTM.

Steinberg, R. (1974). *Computational matrix algebra*. New York: McGraw Hill.

Steinberg, R., Sleeman, D., & Ktorza, D. (1990). Algebra students' knowledge of equivalence of equations. *Journal for Research in Mathematics Education* **22**(2):112–121.

Wagner, S., & Parker, S. (1993). Advancing algebra. *In* P. W. Wilson (ed.). *Research ideas for the classroom: High school mathematics* (pp. 119–139) Reston, VA: NCTM.

Usiskin, Z. (1988). Conceptions of school algebra and uses of variables. *In* A. F. Coxford, & A. P. Shulte (eds.). *The ideas of algebra, K–12, 1988 yearbook of the NCTM* (pp. 8–19). Reston, VA: NCTM.

6
Geometry and Measurement

Alas, in our schools identification and classification of shapes usually stop just at the point where they can begin to be really interesting—where they begin to explore structures in three-dimensional space. How many people realize that even polygons that are not flat can be interesting and important? Many molecules have polygonal shapes, but often these polygons are crumpled and their conformations are the key to their chemical properties. Besides finite polygons and polygons whose edges don't cross, there are zig-zag, star, and helical polygons. . . . Soap bubbles, soap films, and froths are also endless sources of fascinating geometrical principles.

Marjorie Senechal (1990)

Overview

The study of geometry at the middle and high school levels is the study of two- and three-dimensional shapes and their properties and relationships. Geometric shapes model our world, in a sense determine our world, so knowledge of geometry is both useful and important. The study of geometry has always been a major part of the high school curriculum, but it has not always been included in the middle school curriculum. Geometry should play an important role at all levels of school mathematics. Through geometry students learn to reason about, and compare, objects; to measure attributes of objects; to represent objects through drawings or with coordinates; and to visualize and mentally manipulate geometric objects.

In this chapter we investigate the teaching of geometry and measurement at the middle school and secondary levels. The chapter focuses on topics appropriate for students in grades 6–12, as well as issues related to the learning of geometry. In particular, we will see what research on geometric reasoning and the measurement of objects has to tell us about teaching these subjects. As in Chapters 4 and 5, this chapter gives sample lesson plans that exemplify geometry teaching as described in the NCTM *Principles and Standards*.

Focus on the NCTM *Standards*

The NCTM *Principles and Standards* includes two standards in the areas of geometry and measurement. The first of these two, Geometry, focuses on geometric objects, their characteristics, and various representations. The second, Measurement, focuses on systems of measurement and techniques for measurement. These systems and techniques encompass the measurement of geometric objects as well as other mathematical objects and phenomena such as data sets and measurable attributes such as the slope of a line and velocity. In this chapter we consider measurement with respect to geometric objects.

As noted in the Geometry standard, the study of geometry should include a focus on the various aspects of geometric objects, such as relationships between objects, characteristics of objects under transformations, and the connections between geometric objects and the physical world. This standard also highlights visualization of geometric objects as a problem-solving tool.

Geometry Standard

Instructional programs from prekindergarten through grade 12 should enable all students to

- Analyze characteristics and properties of two- and three-dimensional geometric shapes and develop mathematical arguments about geometric relationships;

- Specify locations and describe spatial relationships using coordinate geometry and other representational systems;
- Apply transformations and use symmetry to analyze mathematical situations;
- Use visualization, spatial reasoning, and geometric modeling to solve problems. (NCTM, 2000, p. 41)

The Measurement standard focuses on what attributes of geometric objects can be measured and techniques for measuring them, and also addresses precision in measurement. In this chapter we summarize the details of both the Geometry and Measurement standards and discuss implications for the middle school and high school curriculum.

Measurement Standard

Instructional programs from prekindergarten through grade 12 should enable all students to

- Understand measurable attributes of objects and the units, systems, and processes of measurement;
- Apply appropriate techniques, tools, and formulas to determine measurements. (NCTM, 2000, p. 44)

Focus on the Classroom

Abstract geometric objects are used to describe and approximate our physical world. Hence, if we can reason about these objects and work with them, whether abstractly or otherwise, we can make sense of the physical world. The study of geometry and measurement prepares students to interact with the geometric objects and relationships that they will encounter throughout their lives. As noted above, a student's ability to understand geometric relationships is also a valuable problem-solving tool. And as students learn to make sense of geometric objects, they develop reasoning skills that will serve them well, not only in mathematics but also in their everyday activities.

6.1 Geometry and Measurement in the Curriculum

Students learn to classify geometric shapes at an early age. As early as preschool, teachers help students to distinguish between circles and squares, and between rectangles and triangles. By middle school, students are familiar with most common geometric shapes—triangles, quadrilaterals, and circles—and many of their properties, as well as much of the terminology that relates the shapes and their properties. Thus, even though some middle school students may not have been introduced to a hierarchical classification system for quadrilaterals, they most likely can identify, and distinguish between, general rhombuses and squares. They can probably also identify characteristics that quadrilaterals have in common (for example, the sum of the measures of the interior angles of a quadrilateral is 360).

The study of geometry in middle school and high school goes beyond recognition and classification of basic shapes, however. Objects and relationships studied in grades 6–12 include the component parts of various figures (such as the angle sum of a polygon), comparisons among types of figures, as well as analysis of two- and three-dimensional figures. In middle school and high school, students can expect to work with geometric objects in various representations. By high school, students should recognize situations when reasoning within coordinate systems is better than taking a nonmetric or synthetic approach.

Geometry and Spatial Sense

The number of topics and concepts available to middle school and high school students is large. With a strong background knowledge of geometric figures and their characteristics, students build a solid understanding of the fundamentals of geometry through their middle school and high school years. For example, using their understanding of circles and triangles, middle school students can identify properties and characteristics of spheres, cylinders, and cones. Recognizing this connection between two- and three-dimensional objects is an important step toward understanding the objects that make up our world.

Students should also be expected to recognize, analyze, and justify relationships among two-dimensional geometric objects, such as the equal measure of the base angles of an isosceles triangle. And they should learn to use various tools that help them reason geometrically. For example, using a coordinate plane, slope, and distance, they can show that the two pairs of opposite sides of a parallelogram are parallel and congruent. Transformational geometry is another area of two-dimensional geometry that fits well in the middle school and secondary math curriculum. Topics in transformational geometry include describing symmetries of objects, position and orientation of objects in a plane, and using these characteristics to determine the congruence and similarity of figures. In addition, students in grades 9–12 should have opportunities to explore the geometry of polar coordinates and vectors, and to explore connections to trigonometry, algebra, and functions.

Measurement

When students search for relationships or compare figures or components of figures, they often rely on a numerical relationship, a measurement. Numbers are often an obvious choice for younger student comparisons because numbers are familiar to them and easy to work with. However, in order to relate numbers to geometric objects, students must understand what attributes are measurable, what units are appropriate for measuring the figure, and how to accurately measure the attributes. These issues and the features of the measurement systems we use are all important concepts to be grappled with in the teaching and learning of geometry.

Results of national tests show that most students in middle school and high school have a good sense of the appropriate units for measuring objects and they are able to use measurement instruments. However, these test results also indicate that students at the middle school level have an incomplete conceptual understanding of area. High school students, on the other hand, have difficulty with the concepts of volume and surface area (Kenney & Kouba, 1997). Physical models and applications of these concepts can help students make sense of the various measurable attributes of geometric figures.

Teaching and Learning Geometry and Measurement

Mathematics educators and learning theorists such as Piaget and the van Hieles have suggested that student geometric reasoning skills progress through stages in the ways they reason about geometric objects (Piaget, 1970; Senk, 1989). Table 6.1 lists the levels of geometric thinking proposed by the van Hieles (1959), as well as a more basic level, level 0, proposed by Clements and Battista (1992). According to this model, student thinking progresses sequentially through levels, but not all students progress at the same rate. Although there is some controversy over the validity of these theories, they have been useful for monitoring growth of knowledge and, over several decades, have been used to develop appropriate geometry curricula.

As students progress through school, we assume they are also progressing from one level of geometric understanding to another. There is, however, some evidence to support the claim that students move backward through the levels when new geometric concepts are introduced. For instance, when students at level 2 first encounter a new object such as a dodecahedron, they may recognize it by its overall appearance on several

Table 6.1 The van Hiele levels of geometric thinking.[1]

Level	Description
Level 0—Pre-recognition	Students unable to identify many common shapes.
Level 1—Visual	Students identify shapes according to appearance.
Level 2—Analytic	Students characterize shapes by their properties.
Level 3—Abstract/Relational	Students can form abstract definitions.
Level 4—Formal Deduction	Students can establish geometric theorems.
Level 5—Rigor	Students can reason formally about mathematics.

[1] As described by Clements and Battista, 1992.

occasions yet cannot name its properties, such as the number of vertices or the number of faces that meet at a vertex. Thus, the student drops from level 2 back to level 1 until more familiar with the new object. This suggests a wavering between levels that can occur as students continue to build understanding at a particular level. Whatever the direction of movement, research shows that forward progression through the van Hiele levels is slow (Clements & Battista, 1992; Senk, 1989). In fact, by the time they graduate from high school, most students reach only level 2 or level 3 (Senk, 1985, 1989).

It is clear that students need help in developing geometric understanding. Is it possible to move students beyond the van Hiele's level 3 of geometric thinking during high school? If not, can we at least help all students achieve level 3? If we can move students beyond level 3, how do we do it?

Technology enhances geometric thinking. Dynamic geometry software packages allow students to easily visualize and manipulate geometric objects. This enables them to investigate properties of objects and form conjectures based on their explorations. It is likely that these types of investigations increase student *knowledge* of geometric objects, but more work has to be done in order to increase student *understanding* of these aspects. This brings to light an important progression of understanding not alluded to in the van Hiele model.

Many mathematics educators support a progression of teaching and learning that allows students to more fully develop their understanding of concepts (Hodgson & Morandi, 1996; NCTM, 2000; Reid, 1998). This progression consists of (1) time for exploration, (2) the expectation that students will develop conjectures based on findings from explorations, (3) searching for a justification of the conjecture, and (4) developing a formal argument or proof to validate the conjecture. These elements appear to develop student thinking well beyond level 2 of the van Hiele model. This model of teaching and learning is quite time intensive for the classroom teacher and for the students. Several mathematics curricula developed in response to the NCTM *Standards* (1989) follow such a model and will be highlighted in this chapter.

6.2 Exploring Geometric Relationships

Relationships Among Shapes

As early as preschool, children can identify and classify the basic geometric shapes. Through classification activities, students begin to understand the relationships among geometric figures. Notions of common characteristics are important building blocks for more advanced geometric reasoning. By the time students reach middle school and certainly throughout high school, they use these understandings and relationships to develop more sophisticated classification systems, to compare two- and three-dimensional objects, to explore complex relationships, and to formulate and justify conjectures.

By middle school, students should be able to classify triangles and quadrilaterals. Triangles can be classified according to the relationships of their sides as equilateral, scalene, or isosceles. We also classify them by describing their angles, such as right, acute, or obtuse triangles. How do acute triangles differ from obtuse triangles or

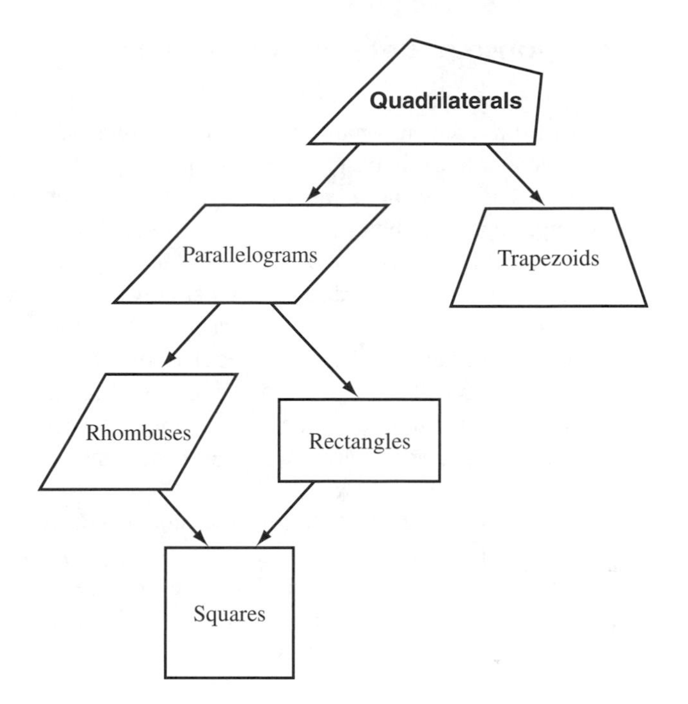

Figure 6.1 The hierarchy of quadrilaterals.

right triangles? Can a triangle be obtuse and isosceles? Can a triangle be obtuse and equilateral? Once students become familiar with the features and the vocabulary associated with triangles, they can explore questions such as these. Being able to justify their conclusions is just as important as coming up with the correct conclusion. Similar investigations with quadrilaterals lead to further classification schemes. In the case of the quadrilateral, a hierarchical map helps illustrate the classification system (see Figure 6.1). Students should be encouraged to continue these types of investigations with other common geometric figures. For example, can we classify pentagons in a way similar to quadrilaterals?

Problem to Investigate: Classifying Triangles

Create a visual aid, using Figure 6.1 as a reference, for classifying triangles. Be sure to include the six properties noted above: equilateral, isosceles, scalene, right, acute, and obtuse. What other properties should be included? Where do the classifications overlap? Is any set of triangles a subset of another? Is this evident in your drawing?

Relationships Between Dimensions

Fostering explorations of two-dimensional geometric figures leads nicely into investigations of three-dimensional objects. Once again, students should be encouraged to look for characteristics that can be used to identify a class of similar objects, such as the shape of faces and the number of edges that meet at a vertex. One place to start such investigations is in the two-dimensional plane. What we know about two-dimensional objects can often be used to make sense of three-dimensional counterparts. For example, to investigate the classification of regular polyhedra (How many are there? What shapes are their faces?), record the characteristics of specific regular polygons and use this information to determine whether a regular polyhedron can be constructed from a set of regular polygons. This exploration leads to several discoveries: (1) There are only a finite number of regular polyhedra; (2) regular polyhedra can only be constructed using equilateral triangles, squares, or regular pentagons.

Definition 1 A **regular polyhedron** is a three-dimensional geometric figure whose faces are all one type of regular polygon and whose vertices all contain the same configuration of faces. For example, the tetrahedron has four faces that are all equilateral triangles. Three equilateral triangles meet at each of the four vertices. *Note*: A regular polygon has equal edges and equal angle measures. ■

Research on student spatial abilities in geometry indicates that many students have difficulties visualizing geometric figures (Banchoff, 1990; Clements & Battista, 1992; Balemenos, Ferrini-Mundy, & Dick, 1987). Yet, contrary to the beliefs of many, spatial abilities can be developed through practice. Math class is a logical place for students to investigate two- and three-dimensional geometry, and to learn and practice spatial visualization skills. Physical manipulatives are often helpful for students to mentally and physically investigate the objects they are trying to understand. Computer models of three-dimensional figures can also be helpful for visualizing objects and their characteristics.

Whether working in two or three dimensions, teachers should encourage students to investigate and justify less obvious relationships within and among geometric objects. Example 1 describes an investigative activity to help students recognize the relationship between the sides of a right triangle.

Example 1 *Exploring Right Triangles*

The freshman geometry teacher, Mr. Roberts, has created many precut right triangles, enough for each group of three students to investigate at least two triangles. As is typical in his class, students enter class and are immediately assigned work groups. He distributes the triangles and asks each group to try to find a relationship between the lengths of the triangle's sides. A couple of groups decide to chart the lengths of the sides of each triangle. They then use the chart to find patterns and relationships. ■

The Pythagorean relationship for right triangles is the most obvious relationship that students will find. However, more advanced students might be asked to use their knowledge of trigonometry and the Pythagorean theorem to find other relationships, such as ratios of sides and the sine of each corresponding angle to arrive at the law of sines. Similar activities lead students to more relationships, such as the ratio of sides in an isosceles right triangle or in a triangle with 30-, 60-, and 90-degree angles. Example 2 provides another idea for helping students investigate geometric relationships.

Example 2 ***Exploring Geometric Relationships***

Mrs. Mickelson's advanced geometry class is studying rotational and reflectional symmetry of various two-dimensional objects. She now asks them to consider the five regular polyhedra. She poses the question, "What does it mean for a polyhedron to have rotational symmetry?" In two dimensions, her students locate a point that acts as the center of rotation. But does a point work for the three-dimensional model? This is the first issue students must consider. Mrs. Mickelson wants them to realize that by stepping up one dimension (from two to three), they must also increase the dimension of the center of rotation (from a point to a line). In addition, she encourages them to use what they know about rotational symmetry in two dimensions. She asks, "Where did centers of rotation exist for the regular polygons? What does that tell you about potential places to look for rotational symmetry in regular polyhedra?" After about 15 minutes of small-group discussions on rotational symmetry of tetrahedra and cubes, she asks students to consider the reflectional symmetry of these objects: "Again, think about this—what relationships will exist when we move from two dimensions to three dimensions?" ■

Problems to Investigate

1. Describe the lines or axes of rotational symmetry in a tetrahedron. How many different kinds of axes are there? How many of each type are there?
2. Describe the planes of reflectional symmetry in a tetrahedron. How many different types of planes are there? How many of each type are there?

Transformational Relationships

The concepts of congruence and similarity arise naturally when students investigate relationships among triangles or other polygons. These relationships also lead to discussions of geometric transformations. One transformation of a figure is a rotation about a point. Whether the axis of rotation is within, on, or outside of the object, the act of rotating the object changes its position. What else changes after the object is rotated? What characteristics of a figure are preserved when the figure is rotated? More generally, students can investigate and identify transformations that result in a figure congruent to the original figure. Of those transformations that produce a figure congruent to the original—reflections, rotations, translations, and glide reflections—what relationships do they hold? These transformations are known as isometries. Isometries preserve distance between points. Besides preserving distance and hence producing congruent triangles, isometries hold many other relationships. For instance, all isometries also preserve angle measure and betweenness of points. We discuss the relationships between isometries in greater detail in Section 6.4.

Not all transformations are isometries. Many transformations, such as those that shrink or stretch an object, don't preserve the size and shape of the original object. If the shape but not the size is preserved, the transformation is a similitude because the figure and its image under the transformation are similar figures. One type of similitude is a dilation. Dilations also preserve slope. That is, if a dilation is performed on a triangle in the coordinate plane, each side of a triangle and its image under the dilation will have the same slope (the lines containing the sides will be parallel).

Relationships to Other Topics

The concepts of congruence and similarity connect nicely to other math topics as well as to art and architecture. Congruence and aspects of symmetry appear often in art and designs. Decorators talk about balance or symmetry of the decorations. Have students examine a design. Can they identify the symmetries that exist? What knowledge of congruence and transformations do they need? Likewise, similarity is often used in the design world, particularly in architecture. Architects create and read scale drawings of the proposed constructions. They often translate these drawings into scale models. This requires a good understanding of geometric shapes and an ability to visualize the relationship between the two-dimensional drawing and the three-dimensional object. The proportions between measurements in a model and the real object to be constructed must also be understood. Thus, the relationship between measurements and the geometric or physical objects is a crucial one.

Explorations in geometry lead to the discovery of geometric relationships that build on prior understandings. As students come to recognize new relationships, they intuitively formulate conjectures, or beliefs, about the way things work. To help students develop a deeper understanding of their conjectures, and to foster the mental habit of justifying results, students should be encouraged to construct arguments that validate relationships. Usually, the process of exploring and discovering the relationship provides insights into how to construct the argument. Even so, students need to learn to organize this information and present it in such a way that they can answer not only their own questions, but also the questions of others. Justifying conclusions is an excellent way to learn logical reasoning, a valuable skill in all areas of mathematics and in everyday life.

Exercises 6.2

1. What important features are used to classify polygons? To classify polyhedra?
2. What characteristics of circles might be used to classify them?
3. Develop a classification scheme for (a) polyhedra and (b) prisms. If possible, use a drawing or tree to indicate relationships and subsets among the classifications in your scheme.

4. Describe the three different axes of rotational symmetry in a cube. How are these axes of rotation related to the points of rotational symmetry of the square? To the lines of reflectional symmetry of the square?

5. What axes of rotation exist for the regular square pyramid? How do you know where to look to identify axes of rotation in polyhedra?

6. Use the Internet to research polyhedra in nature. Here are two good web sites to start with:

   - www.forum.swarthmore.edu
   - www.georgehart.com.virtual-polyhedra/vp.html

   a. Which polyhedra occur naturally in our world? Explain.
   b. Write up a short investigation (that might take up to 15 minutes for students to complete) of geometry in nature that would be appropriate for high school students.

PROJECTS Find Out for Yourself

1. Visit a high school algebra class.
   a. When do students use geometric reasoning or other ideas from geometry?
   b. Describe ways in which geometry or geometric reasoning could be incorporated to enhance student learning of the algebra lesson you observed.

2. Interview a middle school student.
   a. Before the interview, develop a sequence of questions that will probe the student's understanding of the relationships between two- and three-dimensional objects.
   b. Write a couple of paragraphs to describe the student's understanding of two- and three-dimensional relationships. Be sure to support your claims with evidence from the interview.

6.3 Geometric Similarity

Geometric similarity is a concept related to proportionality and can be useful to simplify the modeling process.

Definition Two objects are said to be **geometrically similar** if there is a one-to-one correspondence between points of the objects such that the ratio of distances between corresponding points is constant for all possible pairs of points. ■

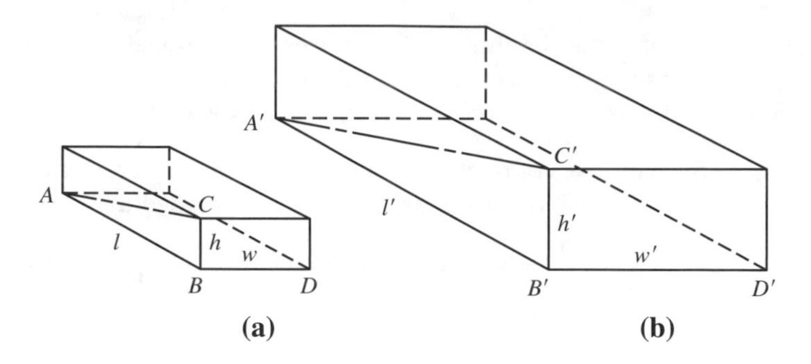

Figure 6.2 Two geometrically similar objects.

For example, consider the two boxes depicted in Figure 6.2. Let l denote the distance between the points A and B in Figure 6.2a, and l' the distance between the corresponding points A' and B' in Figure 6.2b. Other corresponding points in the two figures, and the associated distances between the points, are marked in the same way. For the boxes to be geometrically similar, it must be true that

$$\frac{l}{l'} = \frac{w}{w'} = \frac{h}{h'} = k \quad \text{for some constant } k > 0$$

Let's interpret this result geometrically. In Figure 6.2, consider the triangles ABC and $A'B'C'$. If the two boxes are geometrically similar, these two triangles must be similar. The same argument can be applied to any corresponding pair of triangles, such as CBD and $C'B'D'$. Thus, *corresponding angles are congruent for objects that are geometrically similar*. In other words, the shape is the same for two geometrically similar objects, and one object is simply an enlarged copy of the other. Think of geometrically similar objects as scaled replicas of one another, as in an architectural drawing where all the dimensions are simply scaled by some constant positive factor k.

One resulting advantage for two geometrically similar objects is a simplification in certain computations, such as volume and surface area. For the boxes depicted in Figure 6.2, consider the following argument for the ratio of the volumes V and V':

$$\frac{V}{V'} = \frac{lwh}{l'w'h'} = k^3 \tag{1}$$

Similarly, the ratio of their total surface areas S and S' is given by

$$\frac{S}{S'} = \frac{2lh + 2wh + 2wl}{2l'h' + 2w'h' + 2w'l'} = k^2 \tag{2}$$

Not only are these ratios immediately known once the scaling factor k is specified, the surface area and volume can be expressed as a proportionality in terms of some selected **characteristic dimension**. For example with $l/l' = k$, we have

$$\frac{S}{S'} = k^2 = \frac{l^2}{l'^2}$$

Therefore,

$$\frac{S}{l^2} = \frac{S'}{l'^2} = \text{constant}$$

holds for any two geometrically similar objects. That is, surface area is always proportional to the square of the characteristic dimension length:

$$S \propto l^2$$

Likewise, volume is proportional to the length cubed:

$$V \propto l^3$$

Thus, if you are interested in some function depending on an object's length, surface area, and volume, for example,

$$y = f(l, S, V)$$

you can express all the function arguments in terms of some selected characteristic dimension, such as length, giving

$$y = g(l, l^2, l^3)$$

The characteristic dimension can be any convenient one-dimensional measurement.

Geometric similarity is a powerful simplifying assumption. To illustrate the concepts of both proportionality and geometric similarity, we present the following example.

Example *A Bass Fishing Derby*

Consider a sport fishing club that, for conservation purposes, encourages its members to release their fish immediately after catching them. The club also wants to award members at their annual derby for total weight of fish caught by individuals: honorary membership in The 1000 Pound Club, Greatest Total Weight Caught During a Derby, and so forth. But how do members determine the weight of a fish they've caught? People have suggested carrying portable scales; however, portable scales tend to be inconvenient and inaccurate, especially for smaller fish. So, the club has asked you to come up with a way to estimate the weight of their freshly caught bass in terms of some easily measurable dimensions.

Many factors that affect fish weight can easily be identified. Different species have different shapes and different average weights per unit volume (weight density) based on the proportions and densities of meat, bone, and so on. Gender also plays a role, especially during spawning season. The various seasons themselves probably have a considerable effect on weight.

Since we are seeking a general rule, let's make a simplifying assumption that the average weight density for bass is constant. Later, it may be desirable to refine our model if the results prove unsatisfactory or if we determine that considerable variability in density does exist. Furthermore, let's also neglect gender and season. Thus, initially, we will predict weight as a function of size (volume) and constant average weight density.

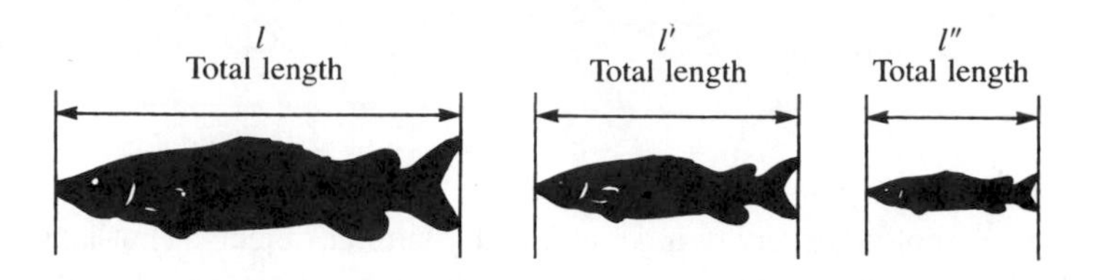

Figure 6.3 Fish that are geometrically similar are simply scaled models of one another.

Assuming that all bass are geometrically similar, the volume of any one bass is proportional to the cube of some characteristic dimension. Note that we are not assuming any particular shape, but only that the bass are "scaled models" of one another. The basic shape can be quite irregular so long as the ratio between corresponding pairs of points in two distinct bass remains constant for all possible pairs of points. This idea is illustrated in Figure 6.3.

Next, we'll choose the length l of the fish as the characteristic dimension. This choice is depicted in Figure 6.3. Thus, the volume of a bass satisfies the proportionality

$$V \propto l^3$$

Because weight W is volume times average weight density and a constant average density is being assumed, it follows immediately that

$$W \propto l^3$$

Let's test our model. Consider the following data collected during a recent fishing derby:

Length, l (in.)	14.5	12.5	17.25	14.5	12.625	17.75	14.125	12.625
Weight, W (oz)	27	17	41	26	17	49	23	16

If our model is correct, then the graph of W versus l^3 should be a straight line passing through the origin. The graph, which approximates a straight line, is presented in Figure 6.4. (Note that our judgment here is qualitative. In Chapter 9, we develop analytic methods to determine a "best-fitting" model for collected data.)

Let's accept the model, at least for further testing, based on the small amount of data presented so far. Since the data point $((14.5)^3, 26)$ happens to lie along the line we've drawn in Figure 6.4, we can estimate the slope of the line as $26/3049 = 0.00853$, yielding the model

$$W = 0.00853 l^3 \tag{3}$$

Of course, if we had drawn our line a little differently, we would obtain a slightly different slope. In Chapter 9, you will be asked to show analytically that the coefficients that minimize the sum of squared deviations between the model $W = kl^3$ and the given data points is $k = 0.008437$. A graph of the model in (3) is presented in Figure 6.5, which also plots the original data points. ■

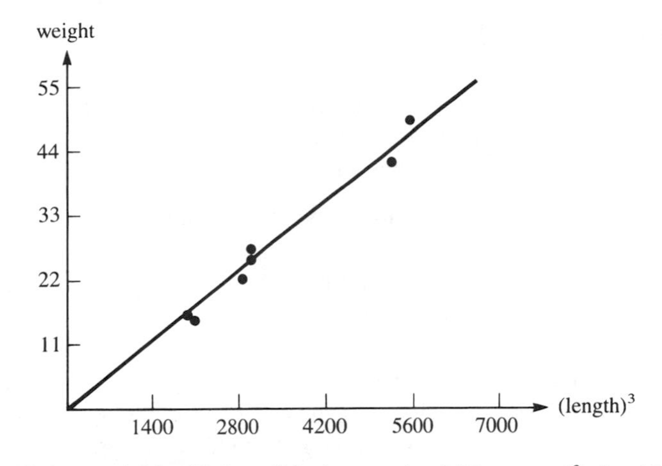

Figure 6.4 If the model in (3) is valid, the graph of W versus l^3 should be a straight line passing through the origin.

The model in (3) provides a convenient rule of thumb. For example, from Figure 6.5 we might estimate that a 12-in. bass weighs approximately 1 lb. This means that an 18-in. bass should weigh approximately $(1.5)^3 = 3.4$ lb and a 24-in. bass approximately $2^3 = 8$ lb. For the fishing derby, a card converting the length of a caught fish to its weight in ounces or pounds could be given to each fisher, or a retractable metal tape marked with a conversion scale if the use of the rule becomes popular enough. A conversion scale for the model in (3) is as follows:

Length (in.)	12	13	14	15	16	17	18	19	20	21	22	23	24	25	26
Weight (oz.)	15	19	23	29	35	42	50	59	68	79	91	104	118	133	150
Weight (lb)	0.9	1.2	1.5	1.8	2.2	2.6	3.1	3.7	4.3	4.9	5.7	6.5	7.4	8.3	9.4

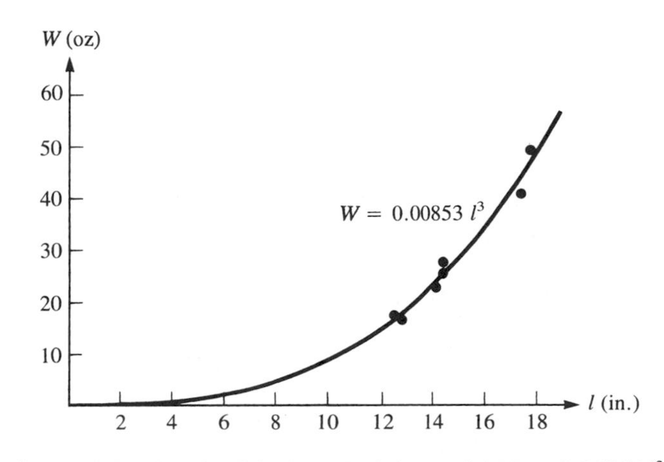

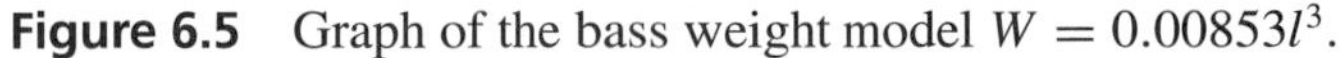
Figure 6.5 Graph of the bass weight model $W = 0.00853 l^3$.

Exercises 6.3

1. Consider a 20-lb pink flamingo that stands 3 ft tall and has 2-ft legs. Model the height and leg length of a 100-lb ostrich. What assumptions are necessary? Are they reasonable assumptions?

2. An object is sliding down a ramp inclined at an angle of θ radians and attains a terminal velocity before reaching the bottom. Assume that the drag force caused by the air is proportional to Sv^2, where S is the cross-sectional area perpendicular to the direction of motion and v is the speed. Further assume that the sliding friction between the object and the ramp is proportional to the normal weight of the object. Determine the relationship between the terminal velocity and the mass of the object. If two different boxes, weighing 600 and 800 lb, are slid down the ramp, find the relationship between their terminal velocities.

3. Assume that under certain conditions the heat loss of an object is proportional to the exposed surface area. Relate the heat loss of a cubic object with side length 6 in. to one with a side length of 12 in. Now, consider two irregularly shaped objects, such as two submarines. Relate the heat loss of a 70-ft submarine to a 7-ft scale model. Suppose you are interested in the amount of energy needed to maintain a constant internal temperature in the submarine. Relate the energy needed in the actual submarine to that required by the scaled model. Specify the assumptions you have made.

4. Consider the situation of two warm-blooded adult animals essentially at rest and under the same conditions (as in a zoo). Assume the animals maintain the same body temperature and that the energy available to maintain this temperature is proportional to the amount of food provided to them. Challenge this assumption. If you are willing to assume that the animals are geometrically similar, relate the amounts of food necessary to maintain their body temperatures to their lengths and volumes. (*Hint:* See Exercise 3.) List any assumptions you have made. What additional assumptions are necessary to relate the amount of food necessary with their body weight?

PROJECTS Exploring Modeling

1. *Superstars* In the TV show "Superstars," the top athletes from various sports compete against one another in a variety of events. The athletes vary considerably in height and weight. To compensate for this in the weight-lifting competition, the body weight of the athlete is subtracted from his lift. What kind of relationship does this suggest? Use the following table, which displays the winning lifts at the 1976 Montreal Olympic Games (weights are in pounds) to show this relationship:

Class	Max weight (lb)	Snatch (lb)	Jerk (lb)	Total lift (lb)
Flyweight	114.5	231.5	303.1	534.6
Bantamweight	123.5	259.0	319.7	578.7

(continued)

Class	Max weight (lb)	Snatch (lb)	Jerk (lb)	Total lift (lb)
Featherweight	132.5	275.6	352.7	628.3
Lightweight	149.0	297.6	380.3	677.9
Middleweight	165.5	319.7	418.9	738.5
Light heavyweight	182.0	358.3	446.4	804.7
Middle heavyweight	198.5	374.8	468.5	843.3
Heavyweight	242.5	385.8	496.0	881.8

Researchers suggest that the strength of a muscle is proportional to its cross-sectional area. Using this model for strength, construct a model relating lifting ability and body weight. List all assumptions. Do you have to assume that all weight lifters are geometrically similar? Test your model with the data provided.

Now consider a refinement to your model. Suppose a certain amount of body weight is independent of size in adults. Suggest a model that incorporates this refinement and test it against the data provided.

Criticize the use of the preceding data. What data would you really like in order to handicap weight lifters more fairly? Who is the best weight lifter according to your models? Suggest a handicapping rule of thumb for the "Superstars" show.

2. *Heart Rate of Birds* Warm-blooded animals use large quantities of energy to maintain body temperature because of the heat loss through the body surface. In fact, biologists believe that the primary energy drain on a resting warm-blooded animal is maintenance of body temperature.
   a. Construct a model relating blood flow through the heart to body weight for warm-blooded animals. Assume that the amount of energy available is proportional to the blood flow through the lungs, which is the source of oxygen. Assuming the least amount of blood needed to circulate, the available energy will equal the energy used to maintain the body temperature.
   b. The following data relate weights of some types of birds to their heart rate measured in beats per minute. Construct a model that relates bird heart rate to body weight. Discuss the assumptions of your model. Use the data provided to check your model.

Bird	Body weight (g)	Pulse rate (beats/min)
Canary	20	1,000
Pigeon	300	185
Crow	341	378
Buzzard	658	300
Duck	1,100	190
Hen	2,000	312
Goose	2,300	240
Turkey	8,750	193
Ostrich	71,000	60–70

Source: Data from A. J. Clark, *Comparative physiology of the heart*, New York: Macmillan, 1977, p. 99.

3. *Heart Rate of Mammals* The following data relate the weights of some mammals to their heart rate in beats per minute. Refer to Project 2a to construct a model that relates heart rate to body weight. Discuss the assumptions of your model. Use the data provided to check your model.

Mammal	Body weight (g)	Pulse rate (beats/min)
Vespergo pipistrellas	4	660
Mouse	25	670
Rat	200	420
Guinea pig	300	300
Rabbit	2,000	205
Little dog	5,000	120
Big dog	30,000	85
Sheep	50,000	70
Man	70,000	72
Horse	450,000	38
Ox	500,000	40
Elephant	3,000,000	48

Source: Data from A. J. Clark, *Comparative physiology of the heart*, New York: Macmillan, 1977, p. 99.

4. *Loggers* Loggers like to use readily available measurements to estimate the number of board feet of lumber in a tree. (A board foot of lumber is $\frac{1}{12}$ of a cubic foot, equivalent to a board 12 in. wide, 12 in. long, and 1 in. deep.) Assume the loggers measure tree diameters in inches at (their) waist height.

   Develop a model that predicts board feet as a function of diameter in inches. Use the following data for your test:

x	17	19	20	23	25	28	32	38	39	41
y	19	25	32	57	71	113	123	252	259	294

The variable x is the diameter of a ponderosa pine in inches and y is the number of board feet divided by 10.

a. Consider two separate assumptions, allowing each to lead to a model. Completely analyze each model.

- All trees are right circular cylinders and all trees are about the same height.
- All trees are right circular cylinders and the height of the tree is proportional to the diameter.

b. Which model appears better and why? Justify your conclusions.

5. *Racing Shells* If you have been to a rowing regatta, you might have observed that the more oarsmen in a boat, the faster the boat travels. It is therefore reasonable to investigate whether a mathematical relationship exists between the

speed of a boat and the number of crew members. Consider the following assumptions (partial list) in formulating a model:

a. The total force exerted by the crew is constant for a particular crew throughout the race.

b. The drag force experienced by the boat as it moves through the water is proportional to the square of the velocity times the wetted surface area of the hull.

c. Work is defined as force times distance. Power is defined as work per unit time.

Number of crew	Race time (s)	
	Race 1	Race 2
1	429.6	430.2
2	412.2	406.2
4	379.8	367.8
8	346.8	343.8

[*Hint:* When additional oarsman are added to a shell, it isn't obvious whether the *force* is proportional to the number in the crew, or the *power* is proportional to the number in the crew. Which assumption appears the most reasonable? Which yields a more accurate model?]

6. *Scaling a Braking System* Suppose after years of experience your auto company has designed an "optimum braking system" for its prestigious full-sized car. That is, the distance required to brake the car is the best in its weight class, and the occupants feel the system is very smooth. Your firm decides to build cars in the lighter weight classes. Discuss how you would "scale" the braking system of your present car in order to obtain the same performance in the smaller versions. Be sure to consider the hydraulic system and the size of the brake pads. Does a simple geometric similarity suffice? Suppose that the wheels are scaled in such a manner that the pressure (at rest) on the tires is constant in all car models. Will the brake pads seem proportionally larger or smaller in the scaled-down cars?

6.4 Transformational Geometry

Slides, flips, turns, shrinking, and stretching are the major topics of transformational geometry at the middle school and high school levels. This area of geometry is especially interesting to students since it is dynamic and the concepts are relatively easy to delve into at an introductory level. Students may be familiar with the motions associated with transformations through experiences in everyday life and can use this knowledge to make sense of the more mathematical aspects. Transformational geometry also helps students build connections between various geometric and nongeometric concepts.

In the 1800s, a mathematician named Felix Klein promoted the development of all of geometry through transformations. He defined geometry as the study of properties that remain unchanged under different sets of transformations. This view of geometry ties it closely to the study of algebra.

Isometries

Transformational geometry is dynamic in the sense that it describes motions in space in a mathematical way. The original figure and the result of a transformation are static geometric objects, but a movement of some kind (whether purely analytic or physical) has transformed the original figure into the resulting or *image figure*. Slides (translations), flips (reflections), and turns (rotations) all exemplify basic rigid motions on the surface of a plane that transform geometric objects. The function that transforms an object by way of a translation, reflection, or rotation is known as an *isometry*. A glide reflection—the composition of a translation and a reflection—is the fourth isometry. As noted earlier, isometries preserve the shape and size of objects; hence the image of a figure under an isometry is congruent to the original figure. Knowing this helps students realize that isometries preserve angle measure, length of segments, and "betweenness" of points. Students can work with compositions of isometries to investigate some of the following questions:

- What common isometry is equivalent to the composition of two reflections over parallel lines?
- What common isometry is equivalent to the composition of two reflections over intersecting lines?
- Which isometries or compositions of isometries preserve slope of a line?

Isometries describe movement; thus, it is most helpful for students to experience these transformations in a dynamic way. Since all isometries are compositions of at most three reflections, they can all be performed using a mirror. Many types of manipulatives are available that allow students to perform the movements either physically or with the help of technology, but any kind of mirrored object can be used to help them create reflections of objects over lines. Vendors of educational materials frequently sell transparent plastic manipulatives such as MIRAs. A typical MIRA is shown in Figure 6.6.

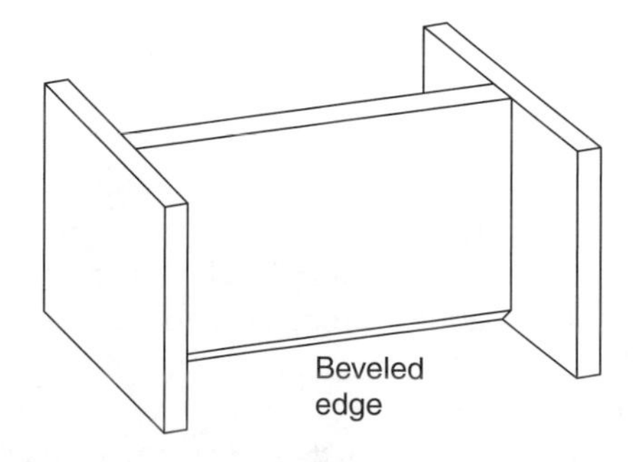

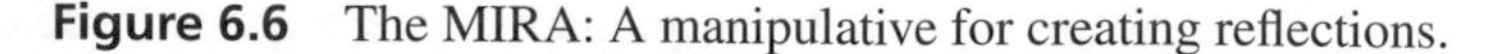

Figure 6.6 The MIRA: A manipulative for creating reflections.

Dynamic geometry software packages, such as the Geometer's Sketchpad, or sophisticated graphing calculators, such as the Texas Instruments TI-89, perform many transformations with a single command. For instance, select a geometric figure, and the program allows you to translate the object along a specified vector (alternately, for a given distance and angle of rotation). The computer then displays the original object and its translated image. Since the image depends on the original object, its shape won't vary unless the shape of the original object is changed. Note that, in general, the manipulation of geometric objects on the computer screen doesn't maintain the shape and size of the object. Hence, these manipulations—such as "dragging" a vertex—are not isometries.

Other Transformations

In the coordinate plane, a transformation is a function that is a bijection (one-to-one and onto), and hence there are many transformations of the coordinate plane, most of which do not preserve distance and angle measure, as an isometry does. Linear transformations, those that map lines to lines, are easier to get a handle on, and thus are most useful and more easily understood by students in class. A linear transformation may stretch a figure or compress it. It may also stretch a figure in one direction and compress it in another, as shown in Figure 6.7. Such a transformation can be described by the following function of two variables: $f(x, y) = (\frac{1}{2}x, 3y)$. This function clearly shows that under the transformation, all x-coordinates of the original figure will shrink by a factor of 2, and the y-coordinates of the original figure will stretch by a factor of 3.

Other transformations of the coordinate plane include similitudes and dilations. Both of these transformations preserve the shape of a figure (it either shrinks or stretches proportionally, but not both). Hence, the image of a figure under a similitude or dilation is similar to the original figure. Although dilation can be thought of as a special case of similitude (a similitude in which slope of lines is preserved under the transformation), it is sometimes more helpful to portray the similitude as a composition of a dilation and an isometry. Figure 6.8 shows a similitude that results from a dilation of factor 2 composed with a 45 degree counterclockwise rotation about the origin.

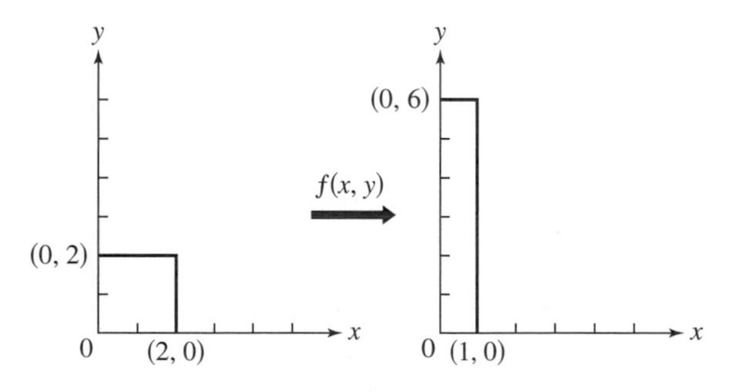

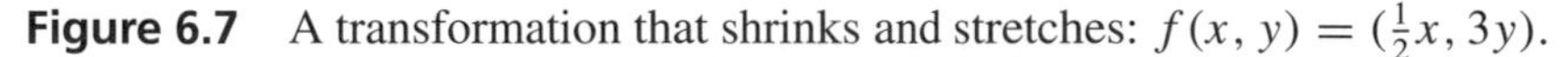

Figure 6.7 A transformation that shrinks and stretches: $f(x, y) = (\frac{1}{2}x, 3y)$.

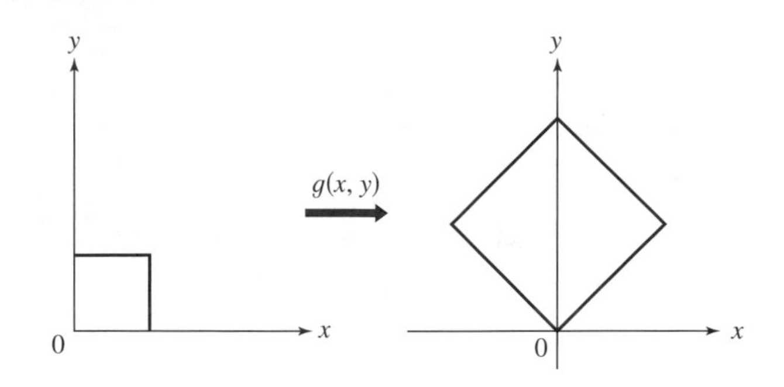

Figure 6.8 A similitude as a composition of a dilation and a rotation: $g(x, y) = \left(\frac{3\sqrt{2}}{2}x - \frac{3\sqrt{2}}{2}y, \frac{3\sqrt{2}}{2}x + \frac{3\sqrt{2}}{2}y\right)$.

Problem to Investigate

Decide whether the composition of a dilation and a translation is a dilation or a more general similitude. Be sure to justify your response.

By the middle school years, students should be expected to recognize and describe some of the basic isometries, as well as dilations and similitudes. Recognizing examples and nonexamples is key to building a solid understanding. By the end of high school, students should understand and represent various transformations using drawings and function notation. This connection to the coordinate plane is essential to exploring more sophisticated transformations and other representations, such as matrices.

Connections to Other Concepts

The study of transformations involves an understanding of many other mathematical concepts. Transformational geometry can be used either to build on other concepts or as a stepping stone to introduce these same concepts. For example, congruence and similarity are connected to isometries (they preserve size and shape of figures) and dilations (preserve shape but not size). An early introduction to isometries and dilations leads nicely into the study of congruence and similarity.

Students can study transformations on the coordinate plane and use functions or compositions of functions to describe the motion of the transformation. Also using a coordinate system, they can check properties of transformations. For example, distance is preserved under any isometry (using the distance formula), but slope of lines is preserved only under translations and dilations (or compositions of these two transformations). Hence, a solid understanding of coordinate geometry and of functions is very useful in studying transformations.

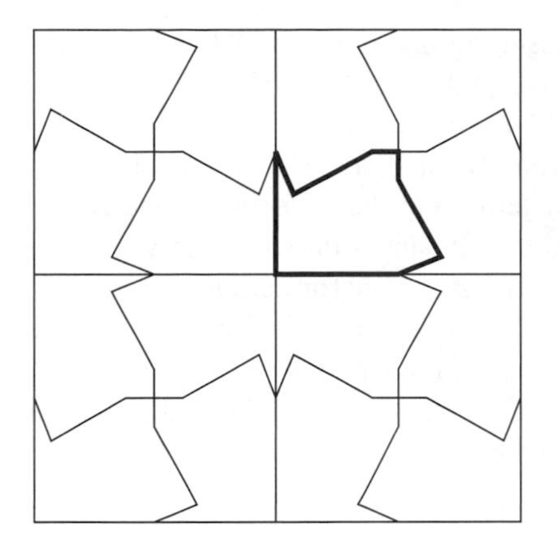

Figure 6.9 A tessellation created through reflecting and rotating the base object.

In addition, transformations segue into the study of tessellations or tilings of the plane, and symmetry. Students create tessellations by transforming a single geometric figure. By repeatedly translating (as well as rotating or reflecting) the figure, the entire plane can be tiled. In Figure 6.9, a tessellation is formed by reflecting and rotating the base figure shown in bold. Tessellations often possess either reflectional or rotational symmetry (or both). The names *reflectional* and *rotational* highlight the connection between symmetries and transformational geometry.

Geometric objects frequently have reflectional or rotational symmetry. If a two-dimensional figure has reflectional symmetry, then there exists a line in the plane such that when the figure is reflected over the line, the image is exactly the same as the original figure in shape, size, and location (see Figure 6.10a). In Figure 6.10b, the figure

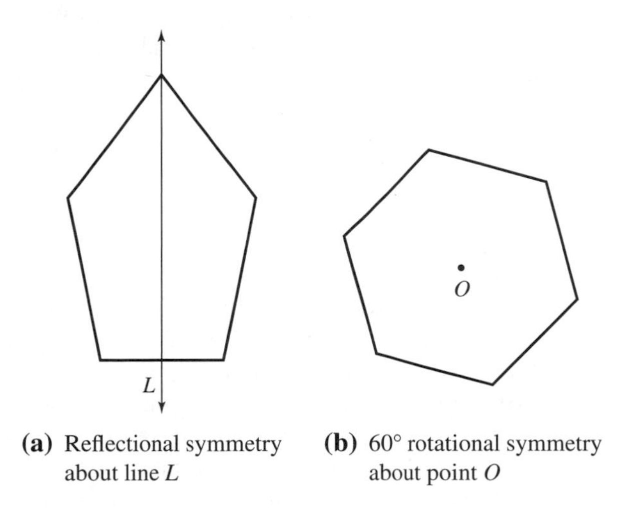

L
O
(a) Reflectional symmetry about line L
(b) 60° rotational symmetry about point O

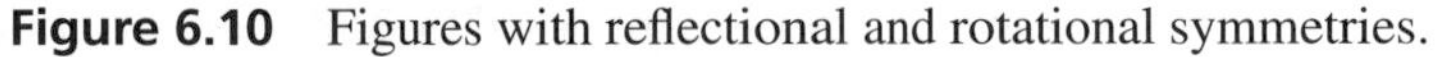

Figure 6.10 Figures with reflectional and rotational symmetries.

exhibits rotational symmetry. For this figure to have rotational symmetry, there must be a point on the figure for which, if the figure is rotated about the point for an amount less than 360°, the image will be exactly the same as the original figure in shape, size, and location. Thus, students with knowledge of isometries have a basis for understanding and determining the reflectional and rotational symmetry of geometric objects. Conversely, if students study symmetry first, they can apply their knowledge of reflections and rotations to transformational geometry.

As noted earlier, ideas of transformation help students make sense of reflectional and rotational symmetry in two dimensions. Introducing students to analogous three-dimensional ideas develops spatial visualization skills and strengthens their conceptions of symmetry.

Problem to Investigate

Determine all lines of reflectional symmetry for the square, and use these to determine the planes of symmetry for the cube. Which planes of reflectional symmetry for the cube correspond to a line of symmetry for the square? Which do not?

Applications of Transformational Geometry

Real-world applications of transformational geometry offer a plethora of topics for the middle school and high school curricula. One common connection is, again, symmetry. Many artistic and functional objects from ancient civilizations demonstrate symmetries, such as strip patterns on pottery and patterns woven into clothing, blankets, or baskets. And, of course, today symmetry and other patterns created through transformations surround us. Figure 6.11 shows examples of such designs from ancient and modern times.

Although they probably haven't thought of it this way, dilations are common transformations in the lives of high school students. They are no doubt familiar with magnifications and reductions in photo enlargements or photocopy reductions. These are examples of dilations. The object maintains its shape (the picture does not look distorted), but the size is changed. Students will recognize other size-changing transformations in maps and in translating map images to the path or road they're on. In a more analytical sense, transformations also describe the change that takes place in the slope of a plank when a weight is placed on the plank.

By high school, students can use their knowledge of other mathematical concepts and symbols to describe transformations in various ways. Thus far, we've given a few examples of transformations as functions of two variables or described in words such as "counterclockwise rotation of 90°." The *counterclockwise* rotation of 90° is a rotation in the positive direction—as opposed to a *clockwise* rotation of 90°, which is a negative

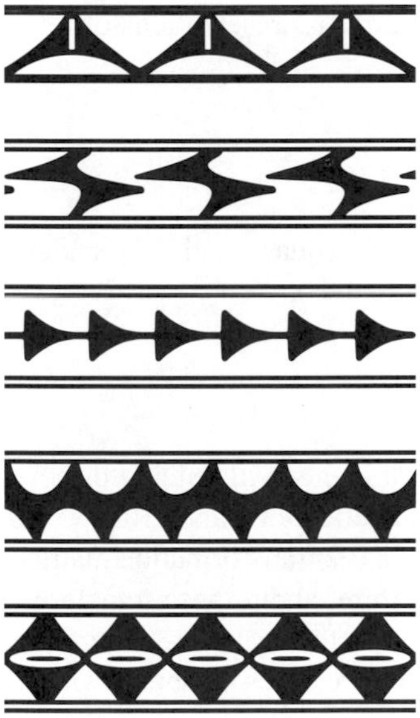

(a) Examples of strip patterns found on pottery in the Southwestern United States.

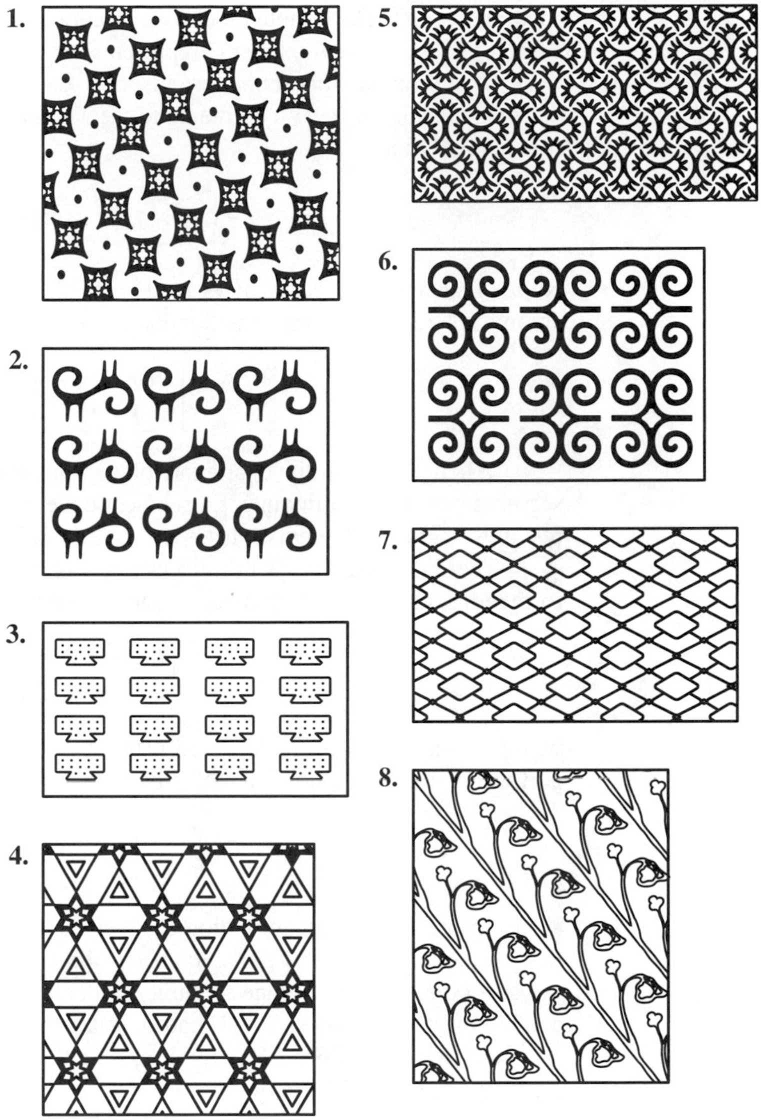

(b) Examples of two-dimensional patterns from (1) painted ceiling, ancient Egypt; (2) stamped cloth, Ghana; (3) woven bag, Native Americans; (4) window lattice, Persia; (5) cotton printer's block, India; (6) stamped cloth, Ghana; (7) window lattice, China; (8) cotton printer's block, India.

Figure 6.11

rotation (−90°). The counterclockwise rotation of 90° can also be represented by the symbol $R_{90°,C}$, where the letter C represents the center of rotation.

As a function, this rotation sends the x-coordinate of a point to the y-coordinate and the opposite of the y-coordinate to the x-coordinate, which can be represented in function notation as

$$f(x, y) = (-y, x), \quad \text{or} \quad \text{as a matrix} \begin{bmatrix} 0 & -1 \\ 1 & 0 \end{bmatrix}$$

Notice that when the coordinates (1, 1) and (2, 3), displayed vertically in the matrix, are multiplied by the transformation matrix, we get the following result:

$$\begin{bmatrix} 0 & -1 \\ 1 & 0 \end{bmatrix} \cdot \begin{bmatrix} 1 & 2 \\ 1 & 3 \end{bmatrix} = \begin{bmatrix} -1 & -3 \\ 1 & 2 \end{bmatrix}$$

When plotted on the coordinate plane (Figure 6.12), we see that the image of the segment with given endpoints is a segment under a 90° counterclockwise rotation. The resulting segment has endpoints (−1, 1) and (−3, 2).

There are many ways of introducing students to matrix representations of transformations. One way is to provide a matrix such as

$$\begin{bmatrix} 0 & 2 & 2 & 0 \\ 0 & 0 & 2 & 2 \end{bmatrix}$$

with columns representing the coordinates of the vertices of a square, and have students perform matrix multiplication with a given transformation matrix

$$\begin{bmatrix} a & b \\ c & d \end{bmatrix}$$

to discover how the transformation changes the square. Students will quickly discover that rotations of 90°, 180°, or 270° about the origin and reflections about the x- and y-axes of the coordinate plane are quite similar in that the 2×2 transformation matrices contain only the numbers 1, −1, and 0. Ask students to think about these transformations and why they only allow for 1, −1, or 0 as entries in the matrices. After some practice, students will discover a variety of transformation matrices on their own. For

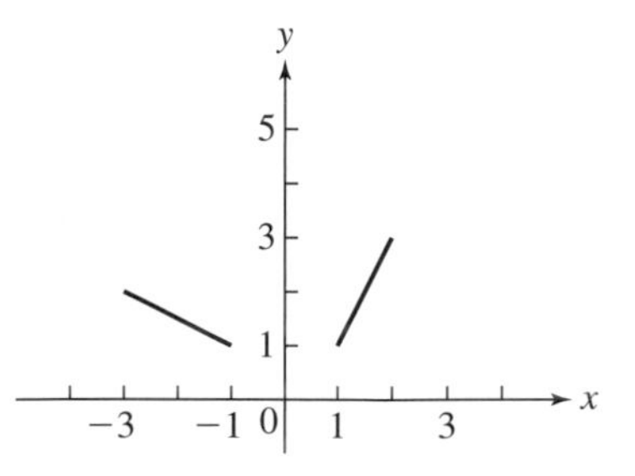

Figure 6.12 90° rotation of a segment about the origin.

instance, ask students to find the matrix that transforms the given square to its image under a dilation of factor 3 centered at the origin. Using the coordinates of the given and transformed figures, students perform matrix operations to find the values of a, b, c, and d in the transformation matrix, as shown in the following Problem to Investigate.

Problem to Investigate

Use the equation to determine the transformation matrix for a dilation of factor 3 centered at the origin:

$$\begin{bmatrix} a & b \\ c & d \end{bmatrix} \cdot \begin{bmatrix} 0 & 2 & 2 & 0 \\ 0 & 0 & 2 & 2 \end{bmatrix} = \begin{bmatrix} 0 & 6 & 6 & 0 \\ 0 & 0 & 6 & 6 \end{bmatrix}$$

In general, the study of transformations at the high school level is useful for representing the movement and distortion of geometric objects in the plane. As demonstrated, transformations easily connect to various other geometry topics at both the middle school and high school levels. In addition, transformations help students make sense of the fundamental concepts of congruence and similarity. Stressing the use of transformations in mathematical and real-world problem solving such as map reading and archeology is enriching and helps to solidify understanding.

Exercises 6.4

1. Reflections are often thought of as the most basic isometry because they can be used to derive any other isometry. Show that a translation or a rotation is a composition of at most two reflections.

2. Use the ideas from Exercise 1 to show that any isometry is the composition of at most three reflections.

3. A glide reflection is a fourth isometry that is a composition of two other isometries (a translation and a reflection over a line perpendicular to the line of translation). How else can the glide reflection be represented? Be specific.

4. What matrices describe the following transformations? Explain.
  a. reflection over the x-axis
  b. 90° clockwise rotation
  c. reflection over the line $y = x$

5. Find an example of symmetry in art or nature.
  a. Describe the symmetry you found.
  b. How can you use this example and others as part of a lesson on symmetry at the middle school or high school level?

6. John copies a picture on the library's copy machine. He doesn't notice that someone left the machine set at 80% of the original size (setting: 0.80). He needs a copy in the original size, but he no longer has the original picture.

a. What setting should John use on the copier to obtain an original-size picture from his reduced copy? Explain your reasoning.

b. What setting should he use if he wants a copy 20% larger than the original? Explain.

6.5 Geometric Representations

Through geometry and other mathematical experiences, students will learn that geometric ideas are useful for problem solving in innumerable areas. For example, how do we know the heights of mountains we've never climbed? How can we find the best location for a new airport in a state? Geometric models of numerical and algebraic situations are especially helpful in these types of investigations and analyses. For instance, drawing or constructing a physical model of an abstract geometric figure helps middle school students explain to classmates why the "flip-and-multiply" method works for dividing fractions. Coordinate geometry allows students to visually represent a function in order to analyze its behavior. Students should also be encouraged to use geometric representations to make sense of data, algebraic and numeric procedures, and other mathematical phenomena. Geometry allows learners to visualize situations in ways that provide key insights into concepts and problem situations.

Representing Algebra, Data, and Functions

Although manipulatives don't play a large role in the mathematics curriculum above grades 5 or 6, perhaps they should. There are many hands-on tools that students can use in problem solving. Many of them give students having trouble with a concept yet another avenue to understanding. Pattern blocks and a variety of similar objects can represent sequences or series to give different insights into patterns or closed forms of series. Algebra tiles, unlike pattern blocks, were created with a specific purpose in mind—to represent such algebraic concepts as the multiplication of binomials visually. When beginning algebra students can visualize an operation, they can perform it more easily. For example, here is a way students can investigate the distributive property. To see the result of $3(x + y)$, students place three sets of x and y tiles together (shown in Figure 6.13). By looking at this arrangement, they can easily see that this is the same as three x tiles and three y tiles, or $3x + 3y$. The drawback of algebra tiles, however, is that the values of x and y remain constant rather than variable. This may confuse students who struggle with the concept and various uses of variables. Thus, it is crucial that students also work with the more abstract notions of variables and algebraic manipulations on a daily basis.

Representations of data can also take geometric forms such as bar graphs or other two- and three-dimensional objects. In a typical bar graph, for instance, the width of the bars remains constant while bar height varies to represent different quantities. In a

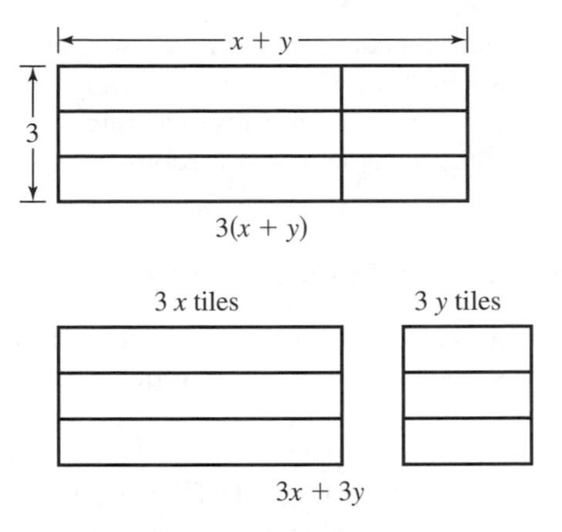

Figure 6.13 Algebra tiles used to investigate the distributive property.

sense, it is the area of the bar that represents the quantity, but students focus on only the height of the bar. The idea of area as the measurement of interest escapes them. Yet students in their own graphs frequently vary both height and width, so this is an important concept. Students need to recognize that when they double the length and width of a square for a data display, the new square has an area 4 times as large as the original square (Figure 6.14a). Students frequently have even more difficulty realizing that in three dimensions, the scale factor increases by a power of 3. Thus, a sphere whose radius is twice as large as another sphere has a volume that is 8 times (2^3) greater than the smaller sphere, as shown in Figure 6.14b. More about working with students to develop data displays can be found in Chapter 7.

Knowledge of geometry is also useful for problem solving in more advanced classes such as advanced algebra, precalculus, and calculus. Coordinate geometry is essential for exploring properties of functions and for portraying mathematical situations. For instance, graphical representations of functions allow learners to "see" how the function behaves at certain crucial points such as where local maximums or asymptotes exist. Whether students create their own graphs or use graphing calculators to display the graph of a function, they will become familiar with basic properties of families of functions. Carefully designed activities make students aware of the basic shapes

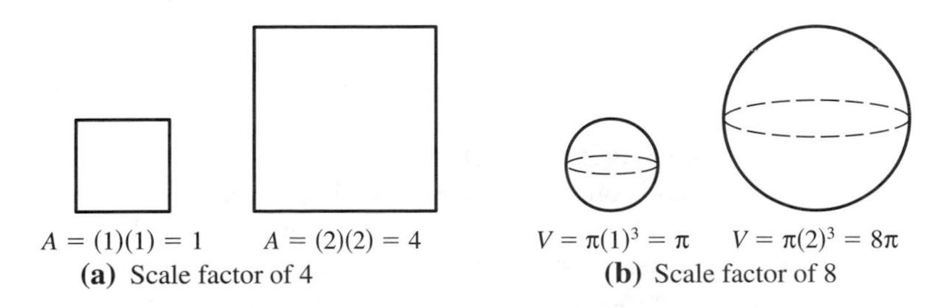

Figure 6.14 Representing data with geometric figures.

of quadratic functions or of the relationship between the location of an asymptote and the equation of a rational function. Graphs of functions are also quite useful for pre-calculus and calculus students exploring the idea of limits, particularly for exponential, logarithmic and hyperbolic functions.

Fractals and Graph Theory

Many areas of geometric reasoning and representations that are not part of the traditional middle school and high school curriculum can prove enriching for students of all ability levels. Fractal geometry is one such area. In a geometric sense, fractals are created through an iterative process of adding to or subtracting from a basic figure. Figure 6.15 shows the first three stages of a famous fractal known as the Sierpinski triangle or gasket. This fractal is created by connecting the midpoints of the sides of the dark triangles to form a smaller triangle in the center. This center triangle is then pulled from the original to get the next stage. As students will notice, the creation of a fractal is an infinite process.

Fractal geometry excites students at all levels because it is dynamic. Students at the middle school and high school level get keenly interested in building fractals and make numerous discoveries along the way as they measure various aspects of fractals. Using the first three stages of the equilateral Sierpinski triangle depicted in Figure 6.15, students can compare perimeter and area and get a feel for what happens to area when sizes vary. Students use both geometry and measurement in such comparisons. They can also explore other concepts of dimension with fractals. Martin (1991) gives an interesting investigation of the fractional dimensions of fractals.

Problem to Investigate

Using Figure 6.15, assume the triangle used to create the Sierpinski triangle is equilateral and each side has a length of 2. Compare the perimeter and area of the triangle at various stages. What do you notice? What would be the perimeter and area for stage 4 of this fractal? Can you predict the perimeter and area for other stages of the fractal?

Be aware that the perimeter of the Sierpinski triangle includes the perimeter of all the darkened triangles. Similarly, only the areas of the darkened triangles need to be calculated and summed to find the area of the Sierpinski triangle.

Graph theory is another area that can enrich the geometry curriculum. Graph theory, which is often introduced with discrete mathematics, is the study of graphs or networks. Networks are used to plot shortest routes. These routes can be as simple as finding the shortest time for a mail route or as complex as routing millions of telephone calls in the shortest time, something we take for granted but that is no mean feat. Some of the most formidable mathematicians of our time work on graph theory.

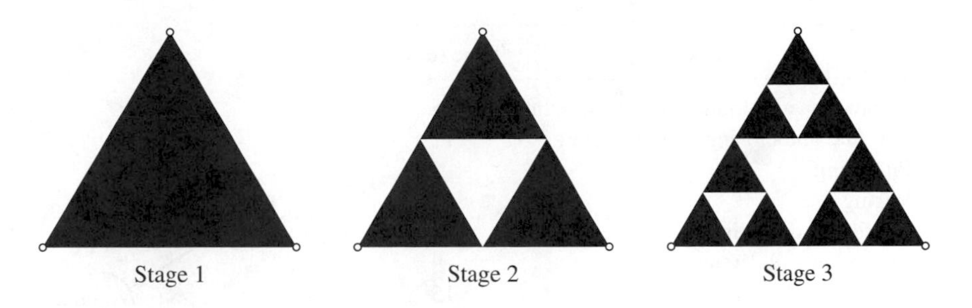

Figure 6.15 First three stages of the Sierpinski triangle.

So how does the U.S. Post Office figure out routes that allow mail to be delivered to each mailbox in the shortest amount of time? The route can be represented as a network of vertices (mailboxes) and edges (streets). The solution then involves determining the shortest path that includes all vertices (but not necessarily all edges). Note that traveling down each street at least once in the most efficient way appears on the surface to be equivalent to the mail route problem, but it is not. Give students opportunities to experiment and discover the differences between the two problems, then brainstorm lists of common applications of graph theory.

Graphs or networks can be used to represent a wide variety of problem situations. The well-known handshake problem, found in the exercises, is a classic example for which graph theory provides a simple visual representation of the problem solution. The snow plow problem is an example of an *Euler circuit*. Suppose a city needs to plow its streets in a particular area in the most efficient way. That is, the plow driver traverses each street no more than one time, beginning and ending at the city dump. Consider the two configurations of city streets shown in Figure 6.16. For which configurations of streets is this possible? A configuration of vertices and edges for which it is possible to traverse each edge exactly once while beginning and ending in the same spot is called

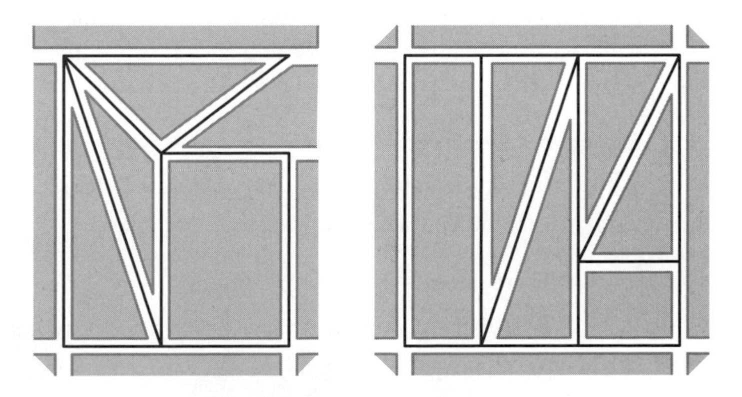

Figure 6.16 The snow plow problem.

Code	Task	Time (days)
A	Excavation and foundation construction	5
B	Framing and closing main structure	12
C	Plumbing	5
D	Wiring	3
E	Heating and cooling installation	7
F	Insulation and dry wall	9
G	Exterior siding, trim, painting	15
H	Interior finishing and painting	7
I	Carpeting	3
J	Landscaping	4

Figure 6.17 Digraph for building a house. *Source:* Adapted from Peressini et al. (1992), *Precalculus and discrete mathematics*, Glenview, IL: Scott Foresman.

an *Euler circuit*. With some practice, students can develop methods for locating Euler circuits and even identifying properties of Euler circuits.

Graphs can also be used to map a series of functions or tasks that lead to more efficient production. The edges of these graphs can only be traveled in one direction and are known as *digraphs*. Digraphs help to coordinate the work of a team so that it is easy to see which tasks must be completed before others and which can be completed simultaneously. Imagine how useful this is in constructing a football stadium, or the space station, or planning 12-hour surgeries.

Figure 6.17 shows a digraph for construction of a house. The various branches indicate jobs or parts of jobs that can be completed simultaneously. Those tasks required to be completed before progressing are indicated by the linear arrangement. From this graph, can you predict the number of days for completing the house?

A Different Look at Geometry

Beyond Euclidean geometry of the traditional high school curricula, many different geometric representations or geometric systems allow students to investigate the geometric foundations of systems that are familiar to them. Finite geometries, for one, provide students with a simplified axiom system and, hence, a means to understand how to model systems. The following example is a nice place to start with high school students.

Example *Smalltown's Committees*[2]

In the village of Smalltown there live only a few people. People in Smalltown like to work on committees. The following facts are given:

- There are exactly four committees.
- Two distinct committees have exactly one person in common.
- Each person in Smalltown is on exactly two committees.

[2] Thanks to Tom Shilgalis for suggesting this example.

From this information, decide how many people live in Smalltown. Does each committee contain the same number of persons? ■

Many variations on this idea exist. As noted, since the set of axioms or "facts" is limited, students can easily experiment to discover a possible model and other conditions of the model.

Another area of interest for students at the middle school and high school levels is non-Euclidean geometry. As with finite geometries, non-Euclidean geometries give students the opportunity to work with an unfamiliar axiom system, which helps them notice differences between the new system and the all-familiar Euclidean one. Taxicab geometry is an example of a non-Euclidean geometry. Taxicab geometry restricts movement to a grid (see Figure 3.2, Gridville), hence the distance between two points must be measured along the "streets" or the horizontal and vertical segments of the grid. Challenge students to explore other differences between taxicab geometry and Euclidean geometry. How are lines defined in taxicab geometry? Do two points determine a unique line? In other words, is there exactly one path along a line that leads from one point to another? A nice book that explores taxicab geometry in some depth is *Taxicab Geometry*, by Eugene Krause (1975).

Spherical geometry is another non-Euclidean system that is very accessible to students in grades 6–12. Although this geometric system is quite different from the Euclidean system, students often have an intuitive sense about spherical geometry because they are familiar with spheres. Physical models (balls) and some string can provide lots of food for thought as students work within this system. Again, ask students how lines should be defined and whether two points will determine a unique line. Lenart (1996) provides a wonderful set of spherical geometry activities for students from middle school through high school and college. The exercise set here gives a taste of possible student activities.

Exercises 6.5

1. Ten people are at a party. If each person shakes hands with every other person, how many handshakes are required? Use a graph to model this situation and to find the solution. How can you use the graph to predict the number of handshakes if there are n people at the party?

2. Here is an extension to the handshake problem. The party has grown. Now there are 47 people. Over the course of the evening, various people shake hands. Is it possible for each person at the party to shake hands with exactly 9 others? [A graph is useful here too.]

3. Use the following set of axioms to create a drawing or model of this finite geometry system. Then use the model and the axioms to develop at least two conjectures about the system.
   a. There is at least one line.
   b. Every line contains at least three points.

c. Two distinct points are on one and only one line.
d. Not all points are on the same line.

4. In Euclidean geometry, a circle is defined as the set of all points that are equidistant from a given point. Use this definition in a taxicab geometry setting. What shape results?

5. You are given two points on the sphere. Describe how you would find the shortest path between the two points. If you extend this line segment beyond the two points for as long as you can, what does the path look like now? Use this discovery to define lines on the sphere.

6. Is it possible to draw a triangle on a sphere? Explain how you would do so. What about a parallelogram? Explain. Which other regions or polygons from Euclidean geometry are possible or impossible to draw on a sphere?

6.6 The Importance of Justification

Earlier sections of this chapter suggest teaching geometry through student exploration (of relationships among objects, of motions in the plane and in space, through modeling phenomena with geometric representations, and so on). Explorations help students formulate conjectures and build a sense of geometric structures. But how can students be sure their conjectures are valid? Will ten examples be enough to convince a classmate that a property holds? Justification in geometry and in all areas of mathematics is important whether students are generating their own conjectures or you are providing them with theorems or ideas. A justification not only proves that a statement is true, it also helps students solidify ideas. Justifications answer the question, Why is this so? And students who ask why and then seek the answer are developing a healthy mathematical curiosity.

New mathematics is most often first established empirically, by developing and refining conjectures. Mathematicians then ascertain the validity, or truth, of the conjecture by proving it with deductive reasoning, and a theorem is established. Students should be encouraged to follow such progressions of problem posing and proving in geometry. Since so much of the content of geometry in the middle school and high school can be investigated with paper-and-pencil constructions or with the aid of dynamic geometry utilities, students can play mathematician, developing and refining conjectures. As they hone their reasoning skills, they can also polish their communication skills; learning to write and present arguments to deductively support their claims will stand them in good stead in many areas.

Howe (1998) claims that in using formal reasoning to establish the truth of their conjectures, students are making sense of mathematics. But what should geometry teachers expect from their students in terms of deductive arguments? Should all students learn to write two-column proofs, such as those that have become synonymous with high school geometry classes? At what level can we expect students to perform in proof writing? If we subscribe to van Hiele's theory of the development of geometric thinking, and if we believe the research in this area (Clements & Battista, 1992;

Senk, 1989), then by the time students reach high school, most are functioning at either level 2 (analytic) or level 3 (abstract/relational). At these levels, it is difficult for students to construct valid deductive arguments. If this is so, should proofs be omitted all together? Many researchers and mathematicians would disagree with this solution (Battista & Clements, 1995; Howe, 1998; Senk, 1989).

The NCTM, for one, calls for introducing reasoning and proof in earlier grades with progressively more attention to proofs as students progress through high school. Others support this stance, but note that many misconceptions about proofs need to be addressed (Dreyfus & Hadas, 1987; Usiskin, 1997). Dreyfus and Hadas suggest that students frequently form misconceptions about these six principles of geometric proofs:

- A theorem has no exceptions. A mathematical statement is said to be correct only if it is correct in every conceivable instance.
- Even "obvious" statements have to be proved. In particular, a proof may not be built on the apparent features of a figure.
- A proof must be general. One or more particular cases cannot prove a general statement. However, one counterexample is sufficient to refute it.
- The assumptions of a theorem must be clearly identified and distinguished from the conclusion.
- The converse of a correct statement is not necessarily correct.
- Complex figures consist of basic components whose identification may be indispensable in a proof (Dreyfus & Hadas, 1987, pp. 48–49).

These principles are specific to geometric proofs, but can be revised to apply to mathematical reasoning and proof in general. If teachers, aware of these basic principles, can develop classroom activities that introduce students to the principles, then students will have the building blocks necessary to develop deductive arguments. The road to developing good arguments and to making sense of arguments is not a smooth one for students, but when they are expected to justify their work on a daily basis, they do learn what constitutes proofs and how to construct valid arguments.

PROJECT Find Out for Yourself

Interview a high school student to determine her or his understanding of proof and justification based on the six principles listed above. Use Exercises 6.8 (1 and 2) and the accompanying figure on page 241 to develop your interview.

(a) Before the interview, write a series of questions that will probe the student's understanding of, and beliefs about, the situations presented. Be sure to ask the student to support any claims.

(b) After the interview, write a couple of paragraphs to describe the student's understanding of, and beliefs about, proof and justification.

6.7 Using Technology to Understand Geometry

If exploration and formulating conjectures are two important components for doing and learning about geometry, then dynamic geometry software packages can enhance geometry learning (Lehrer & Chazan, 1998). Dynamic software packages enable students to quickly and accurately construct and investigate entire classes of figures. This is advantageous if the goal of geometry learning is to understand the relationships of geometric objects. Many software packages work in either a coordinate plane or a more general Euclidean plane; this allows students to translate among representations. In short, a wide variety of useful features help students construct basic shapes and objects, perform transformations, track the steps in creating a particular figure, and use animated objects to depict the transformation process.

When students explore geometry on the computer, they are drawn into deeper and more complex investigations. However, both teachers and students need to be aware of the appropriate uses of technology. For example, simply producing a multitude of examples in support of a conjecture does not constitute a formal justification of the conjecture. With some guidance and suggestions, students can develop the disposition to question and justify what they witness on the computer screen. In general, the curiosity sparked by the ease of exploration on the computer motivates students to make new discoveries and to make sense of these discoveries. Hopefully, this leads to making connections among other concepts and other areas of mathematics.

Problem to Investigate

Construct the following figures using a dynamic geometry software package. Be sure to "drag" each vertex to guarantee that the figure retains its essential components (that is, the equilateral triangle that you create is still equilateral when you drag any vertex).

- An equilateral triangle
- A square
- A parallelogram
- An isosceles triangle
- An isosceles trapezoid

"Constructing" Figures with Dynamic Software

In dynamic software (as the preceding problem alludes to), it is possible to draw a geometric figure, such as an equilateral triangle, that doesn't remain equilateral once you begin to explore by "dragging" the vertices. In other words, there is a difference between drawing a figure and *constructing* a figure with dynamic geometry software. This is analogous to drawing an equilateral triangle without a ruler and compass. By simply "eye-balling" it, you probably come pretty close to an equilateral triangle, but when carefully measured, you'll find that it isn't equilateral. However, if you use your construction tools (ruler and compass) precisely, you can create a triangle with three congruent sides.

The constructed equilateral triangle proves to be important for further explorations. For instance, if you want students to discover that the altitudes, the angle bisectors, and the medians of any equilateral triangle always coincide, they must begin with an accurate equilateral triangle. Even more important, if you want students to use the power of dynamic software—dragging vertices to explore a multitude of equilateral triangles without continually redrawing them—then students must construct the triangle in such a way that it maintains its three congruent sides when a vertex is dragged.

In order to construct a geometric figure with a software package, students must understand some basic constructions as well as the properties of the figure they are constructing. Students who know how to construct angles with compass and straightedge can use this skill on the computer. For instance, when swinging an arc with a compass for a paper-and-pencil construction, students need to know where to place the point of the compass and how wide to open the compass. On the computer, this translates to knowing where to place the center of a circle and how to measure the radius of the circle. The tools in software packages provide a variety of ways to create circles. Students should remember, however, that the entire circle will be drawn, not merely the small arc needed to complete the construction. Hence, many extra lines on the computer screen need to be dealt with during construction and hidden from the viewing screen when the construction is completed.

Knowing the properties of a figure is also helpful in constructing figures with dynamic geometry software. When constructing a square, for example, students use the facts that all angles are right angles and all sides are congruent. Thus, students start by drawing a segment to represent the first side. Knowing that there should be two right angles at the ends of the segment, students use the software to construct lines perpendicular to the segment that pass through its endpoints. Note that this also involves some knowledge of constructions because a perpendicular line can only be constructed if you identify (1) the line or segment to which you wish it to be perpendicular, and (2) the point through which you want the perpendicular to pass. At this point, the figure will look like the one shown in Figure 6.18a.

Two questions may now arise. How do you construct a segment congruent to the base segment so the sides all have the same length? If you mark off these sides so they are congruent to the base and then draw a fourth side, how do you know the last two angles created will always be right angles? As the teacher, it is your job to help students realize or search for the answers to these questions. Figure 6.18b answers the first question. Just as there are a variety of construction tools, so there are a variety of ways to construct a square. Some will be valid methods (that provide a square that remains a square when dragged), while others will not. Another valid method begins with the segment and uses the transformation tools to rotate the segment 90° clockwise and then counterclockwise about the appropriate endpoints of the segment.

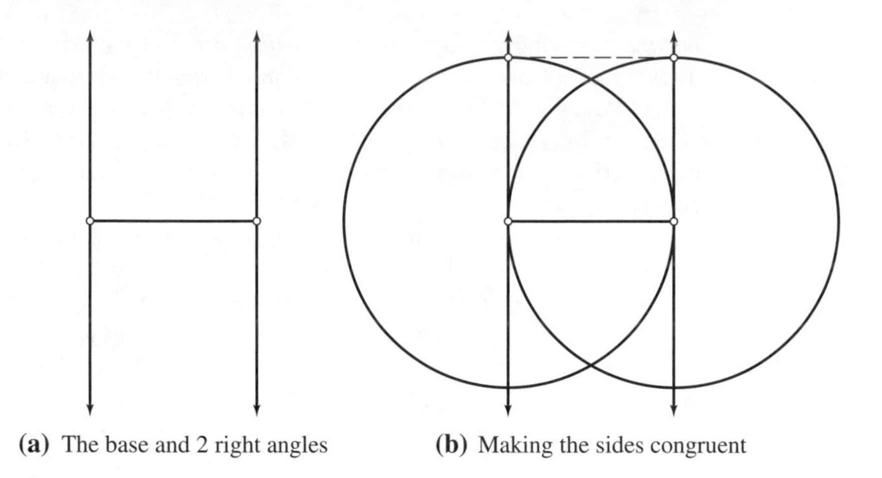

(a) The base and 2 right angles (b) Making the sides congruent

Figure 6.18 Construction of a square with dynamic geometry software.

Explorations and Problem Posing with Dynamic Software

As noted above, dynamic geometry software packages offer a unique opportunity for students to investigate multiple cases of a particular situation or relationship. In this way, students develop, test, and refine conjectures about the characteristics, properties, and relationships of geometric figures. Students thus begin to reason inductively. Inductive reasoning leads to many new discoveries and serves to motivate students to explore and question relationships beyond what they see on the computer screen.

Problem posing is an important part of continued growth for students and mathematicians alike. Mathematicians continually question the discoveries they make and wonder where these new discoveries will lead next. In an environment of exploration and conjecturing, students also learn to pose questions and seek answers, and to investigate new ideas. Dynamic geometry software, as a demonstration tool in class or in a computer lab, helps create an environment that motivates students to explore and question. But software cannot create the environment alone. Teachers and students must endeavor to create a culture of inquiry. To do this, tasks must be appropriate, and teachers must model the questioning and problem posing that is so central to successful explorations in mathematics.

Keep in mind that exploration, problem posing, and conjecturing do not take the place of deductive reasoning. Inductive exploration should be followed by deductive reasoning; give students time to pull ideas together and justify or prove the results they obtain on the computer. Software tools often allow students to investigate the elements of potential deductive arguments, as illustrated in the following example.

Example *Exploring Deductive Arguments*

A couple of students in a high school geometry class claim that the perpendicular bisector of the base of an isosceles triangle is also the bisector of the opposite angle. Mr. Hernandez decides to require all students to either construct a valid argument or find a counterexample for this claim. Most students decide to develop an argument to prove the claim. Danielle and Maya use a software package to construct their proof. They first construct an isosceles triangle and the perpendicular bisector of the base, then notice that the perpendicular bisector appears to contain the opposite vertex. They wonder if this is always the case. By dragging vertices, they quickly create many more isosceles triangles. In all cases, the perpendicular bisector of the base contains the opposite vertex. They measure the two angles created at the vertex, and find that they are indeed congruent, and remain congruent as the isosceles triangle is dragged. Maya wonders why the perpendicular bisector of the base always contains the opposite vertex. The two ponder the properties of perpendicular bisectors as they endeavor to answer Maya's question. Later, Danielle decides to find congruent triangles in the figure in order to prove that the two angles at the vertex are congruent. The software helps them test this possibility. ■

In this example, Danielle and Maya used various features of the software to perform important tasks in the proof process. They first explored the situation to convince themselves of the potential truth of the statement. Some of their explorations were attempts to find counterexamples, which led Maya to an important piece of the proof—the need to show why the perpendicular bisector of the base will contain the vertex of an isosceles triangle. Danielle used the software to explore further in the hope that they could find congruent triangles and the appropriate congruent angles. Through this investigation the software gave them visual feedback and quick (and relatively accurate) measurements. They also used the software to convince themselves, and to find ways of developing their deductive argument.

Other Technology Tools for Geometry

Besides the various dynamic software utilities available (and the many supplemental programs), a variety of other hi-tech tools also enhance geometric understanding. For example, the turtle geometry packages such as LOGO continue to be viable tools for beginning geometry students. These programs require students to provide commands or rules that will lead the "turtle" to create a figure or design. The basic commands are easy for students to learn, but it takes more thought and an understanding of geometric relationships to create a desired figure. For example, students need to know that the interior angles of a regular pentagon measure 108 so they can enter the appropriate commands. The dynamic capabilities of the turtle geometry programs are limited, but the programs are useful for basic explorations of shapes, and for helping students understand how the commands they type move the turtle. In a sense, the turtle itself proves the validity or invalidity of the commands.

Another useful tool for enhancing geometry understanding is the graphing calculator. Many of these now provide features similar to dynamic software packages. They also provide students with the means of computing with matrices—which we saw as connected to transformational geometry in Section 6.2—and graphing the results of these computations on the coordinate plane. A possible investigation of this sort is presented in the exercises.

In general, a wide range of technology is useful for demonstrating geometric properties and relationships, as well as for allowing students to explore, investigate, and justify ideas. As with any technological tool, it is not the tools themselves that increase understanding but what the students do with the tools and how they interact with them. This requires much forethought and careful planning, and attention to the nature of the classroom culture that the teacher and students establish. Some great resources are listed in Section 6.9, the high school technology-based lesson.

Exercises 6.7

1. You are given an equilateral triangle with sides of length 10. Inside the triangle is a point labeled D. Find the sum of the distances from point D to each side of the triangle. Vary the position of point D using a dynamic geometry software package. As you investigate this problem, be sure to justify your results.
2. What 2×2 matrix describes the reflection of an object over the line $y = x$? Use this matrix and a graphing calculator to find the coordinates of the image of the triangle whose vertices are $A(2, 1)$, $B(-1, 3)$, and $C(0, 5)$. Plot the original triangle and its reflected image on the graphing calculator.
3. Use a dynamic geometry program to justify that angle-angle-side (AAS) is a valid theorem for proving that two triangles are congruent.
4. You have two noncongruent quadrilaterals. Can they have five pairs of corresponding congruent parts? Can they have six pairs of corresponding congruent parts? Use a dynamic geometry software program to investigate.
5. You work for the World Food Organization (WFO) and have been asked to locate a new reservoir that three villages will use. Optimally you want to locate the reservoir so that it is equidistant from the three villages. Use a dynamic geometry software package to locate a point equidistant from three villages. What do you notice about the relationship between this point and the perpendicular bisectors of the sides of the triangle formed by the three villages?
6. You are also helping the WFO to locate another reservoir, but this time four villages will use the reservoir. Describe how you will locate the best spot for the reservoir.
7. For a given triangle, how can you locate a square such that it is inscribed in the triangle? Use a dynamic geometry software package to investigate this question.

PROJECT Find Out for Yourself

How might dynamic geometry software be beneficial for students who are struggling with the properties of parallelograms (such as which parallelograms have congruent or perpendicular diagonals and why)?

1. Develop a short activity module involving the properties of parallelograms that you can use with middle school or high school students. The module should in-

clude some introductory explorations, some deeper investigations, and a request for justifying results. Be sure to provide a closing activity or further questions to explore.

2. If possible, implement the module as part of a clinical experience or present your ideas to a teacher and ask to have at least some of them used with the students. Discuss the results of implementing the activities. Did the activities make sense to students? In what ways did the activities enhance student understanding?

6.8 Spatial Reasoning

Spatial visualization involves creating and manipulating mental representations of objects, both abstract objects and real-world objects. We commonly use drawings in mental manipulation, like the drawing of six squares in Figure 6.19 and its accompanying question. At other times, actual objects facilitate the visualization process. In addition, a wealth of mental imagery helps us spatially visualize and manipulate objects that we cannot physically see or hold. Spatial reasoning relies, at least partially, on spatial visualization.

Spatial reasoning abilities are essential to creative thought in mathematics (Clements & Battista, 1992). In addition, spatial reasoning provides yet more tools for problem solving in geometry and in other areas of mathematics, and gives students a better sense of the geometric structures they've worked with both in and out of class. In the middle school curriculum, spatial reasoning abilities help students visualize transformations under an isometry and help them visualize three-dimensional objects in two-dimensional pictures. At the high school level, spatial reasoning is useful when working with three-dimensional objects, transformations, and functions. For instance, in a typical integral calculus problem, students calculate the volume of an object created by revolving a portion of the graph of a function about the x-axis. This is easier when they can visualize the shape of the object after rotating the graph of its function.

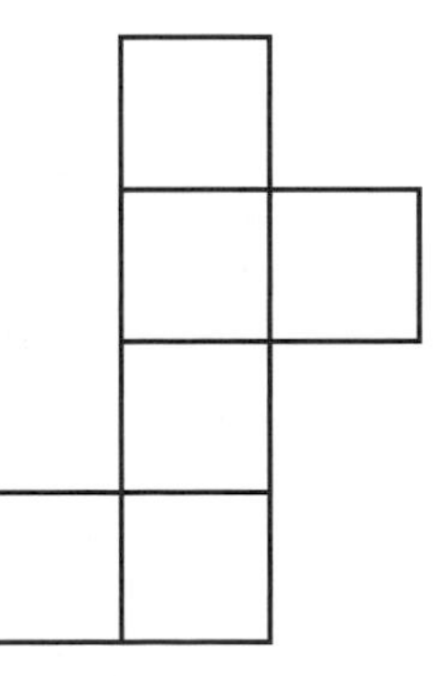

Figure 6.19 Can this figure be folded to form a closed cube?

We live in a three-dimensional world, yet most of the mathematics, and in particular the geometry, that students study in school is presented in two dimensions. Visualizing in three dimensions is important for understanding our world, and it is imperative that students develop this crucial skill. This is not inherent in many students, and should be a part of the mathematics curriculum at all levels (Banchoff, 1990; Clements & Battista, 1992). In fact, research indicates that when students are given opportunities to develop their spatial reasoning skills, they are more likely to be successful in mathematics at all grade levels (Balomenos, Ferrini-Mundy, & Dick, 1987; Clements & Battista, 1992; Senechal, 1990).

Spatial Visualization in the Classroom

Activities that develop spatial visualization take many forms. It's easy to integrate them into the curriculum using the topic at hand or by making them the focus of a unit or two. There are many ways to integrate visualization activities, and textbook series often take advantage of a particular concept or topic to introduce problem situations that involve spatial reasoning. For instance, a unit on polygons can be extended to consider how to use a cube and a piece of paper to create polygons with 3, 4, 5, 6, or 7 sides. Is it possible to pass a plane (piece of paper) through a cube in such a way as to create a 5-sided polygon? Is it possible for the five-sided polygon to be a regular pentagon? Geometry projects also easily accommodate spatial reasoning components. A unit on scale factors and drawing or creating scale models nicely incorporates spatial reasoning as students visualize and draw two-dimensional models of three-dimensional objects. The reverse of this is to ask students to create a three-dimensional model of a city or neighborhood from a two-dimensional sketch.

Any unit on three-dimensional geometry necessarily involves spatial visualization. Students must translate back and forth from three-dimensional objects to two-dimensional representations. At times students may have physical models available. Even so, the mental images of three-dimensional objects can go a long way in helping students to reason geometrically in three dimensions. Some possible investigations are provided here and in the exercises that follow this section.

Problem to Investigate

If you take a regular tetrahedron (polyhedron with four equilateral triangular faces) and slice the vertices off as far as the halfway marks along the edges of the tetrahedron, what geometric figure remains?

1. Describe the new figure by the number of faces and the shapes of its faces. How do you know that this is the case?
2. How does the surface area of the new figure compare with the surface area of the original tetrahedron? Explain.

3. How does the volume of the new figure compare with the volume of the original tetrahedron? Explain.

Building Spatial Abilities

> [Visualization] is also important for all of mathematics. To study change, we need to see it; to study data, we examine various graphical representations. We try to grasp the concept of higher dimension by drawing pictures and making models. . . . But it is not true that we instinctively know how to "see" any more than we know how to swim.
>
> (Senechal, 1990, p. 168)

As is noted in the quote, spatial reasoning abilities must be cultivated. With physical models and computer-generated visual representations as aids, students learn to build their own visual models, but they need practice and encouragement to move beyond models. They must also learn to distinguish key features in a given figure, to visually isolate these features, and to use them in the reasoning process. This occurred naturally for Danielle and Maya in the example in Section 6.7 when they decided to focus on the two small triangles created by the perpendicular bisector. Drawings and mental images help us analyze relationships that aren't obvious, and this builds our spatial reasoning abilities. Resources that provide a variety of research and activities on spatial reasoning include Senechal (1990), Geometry's Future (1990), and the visualization units from the Connected Mathematics middle school series (Lappan et al., 1996).

Exercises 6.8

1. You are given the two drawings shown below, which represent a front view and a side view of a block structure. What is the least number of blocks that can be used to build a structure that fits these two conditions?

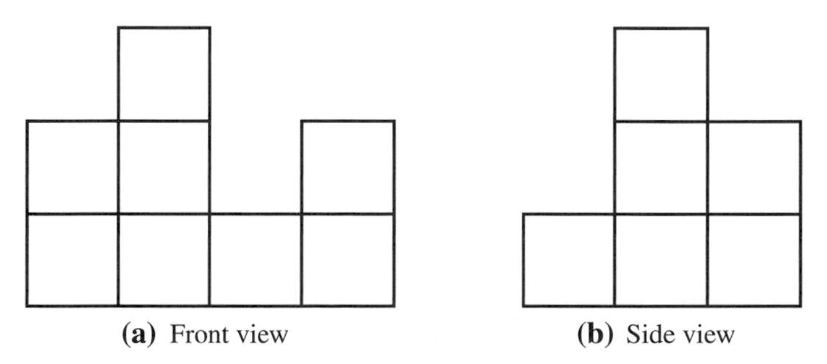

(a) Front view **(b)** Side view

2. How can you modify part b of this drawing so that it represents a possible top view of the block structure represented in part a?

3. What two two-dimensional objects can be used to create a closed cylinder? Use these objects to find the surface area of a closed cylinder with radius 3 units and height 8 units.

4. Imagine a cube (six square faces).
   a. If you slice off the vertices of the cube as far as the halfway marks on each of its edges, what figure remains? Describe the figure by the number of faces and the shapes of its faces.
   b. How do the surface area and volume of the new object compare to the surface area and volume of the original cube? Explain your results.
   c. Now take the same original cube and slice off the vertices as far as the $\frac{1}{3}$ marks on each of its edges. What figure remains? How do you know? How do the surface area and volume of the new object compare with the original cube?

6.9 Geometry and Measurement Lesson Plans

Lesson 5: Geometry and Measurement (Middle School Level)

Objectives: Students will work independently or together on geoboard activities to get a sense of the board's power and to discover an interesting geometric relationship (Pick's theorem). In the process of working through this discovery lesson, students will

1. Make discoveries about the geoboard and its relationships to the real number system.

2. Refresh their memories on certain geometric relationships such as the lengths of diagonals of various rectangles.

3. Use and appreciate a variety of techniques for finding the area of shapes on the geoboard.

4. Discover a relationship between the area of geoboard figures and the other characteristics of the figure (number of border pegs, number of interior dots).

Pick's Theorem

$$\text{Area of region} = \frac{\text{interior pegs}}{2} + \text{boundary pegs} - 1$$

Warm-Up: Have geoboards, square dot paper, and rubber bands waiting. Pair desks so each student works with a partner. Have students use rubber bands to create as many different (noncongruent) triangles on the geoboard as possible. Ask them to justify that no two triangles are congruent. After 5–7 minutes, have student pairs share their findings with other pairs nearby.

Activities:

1. (3 minutes) Opening How easy or difficult was the warm-up assignment? Did you learn anything from your neighbors? How many different triangles are there?
2. (30–35 minutes) Distribute discovery worksheet (see below) and let students work in pairs. Encourage them to share solutions or strategies with other pairs nearby.
3. (10 minutes) Begin sharing findings from initial problems on the discovery worksheet. There will not be enough time for discussing Pick's theorem, but students should be encouraged to compare responses and strategies for finding area of figures. Begin class tomorrow with follow-up to today's work and discussion of Pick's theorem.

Assessment:

1. Warm-Up Introduce students to the geoboard and get a sense of how they organize their findings.
2. Group Work Note the various techniques students employ for finding the area of figures. Which figures cause the most problems? Why? Question students about alternative ways to find area. Question them about patterns in the data.
3. Whole-Class Discussion Sharing of techniques for finding area will let me know how flexible students are and how well they understand the concept of area of a nonstandard region.
4. Homework Students will complete the worksheets and look for patterns hopefully to find the desired generalization.

Closing: Summarize various ways of organizing data (for finding number of different lengths and squares) and techniques for finding area of regions on the geoboard. Encourage students to look for patterns in the area data and a generalization based on the data (formula for finding area of any geoboard region).

Discovery Worksheet

- Materials needed: 1 geoboard or dot paper
- Variety of rubber bands

Directions: Work with your partner to complete the parts of the worksheet below. Base your answers on the results you get from working with the geoboard or the dot paper. Be sure to justify your responses. The distance between two vertically or horizontally adjacent pegs is considered 1 geo. The area of a small square of vertical and horizontal pegs is 1 square geo.

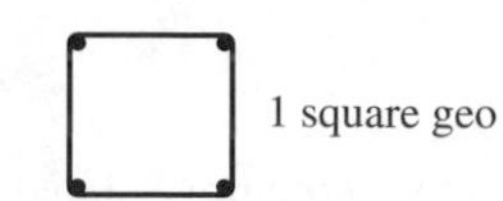

1. How many different segments of 1 geo can you construct on a geoboard? Explain how you arrived at your answer.
2. How many segments of different lengths is it possible to construct on a geoboard? What is the length of each segment? How do you know you have found all possible answers?
3. How many different-size squares can you construct on a geoboard each having its vertices at pegs? What is the area of each? How do you know?
4. How are questions 2 and 3 related to one another?
5. Look at Figures A–D in the figure shown. What is the area of each figure in square geos? How can you justify each of your answers?

   Figure A_______________

   Figure B_______________

   Figure C_______________

   Figure D_______________

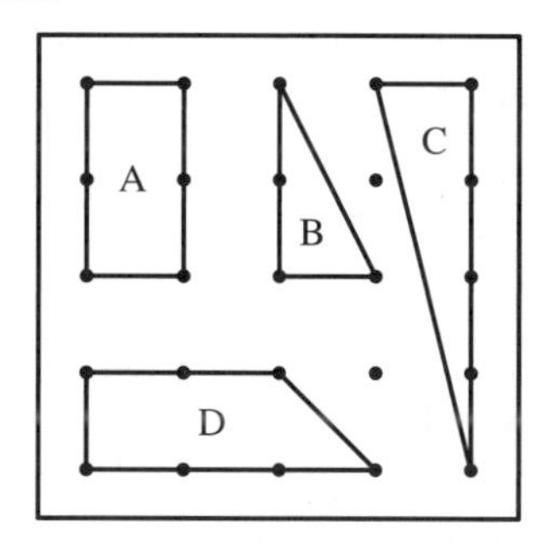

6. Look at Figures E and F in the figure shown. What is the area of each figure in square geos? How can you justify each of your answers?

   Figure E_______________

   Figure F_______________

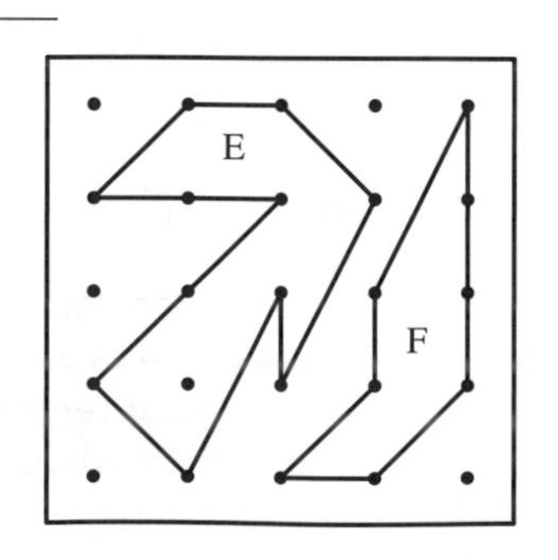

7. Look at Figure G in the figure shown. What is the area of the figure in square geos? How can you justify your answer?

   Figure G_______________

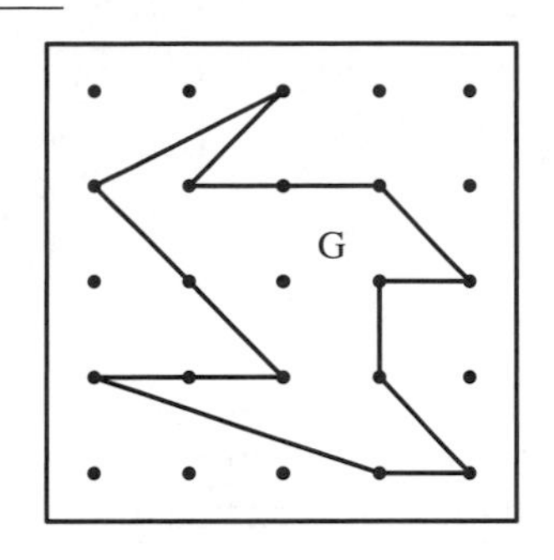

8. Describe the general methods that you have used to find the areas of figures on the geoboard. Do they remind you of any postulates that you studied in Euclidean geometry? If so, what postulates?

9. Look at Figures A–E in the figure shown. What is the area of each figure in square geos? How can you justify each of your answers?

   Figure A_______________

   Figure B_______________

   Figure C_______________

   Figure D_______________

   Figure E_______________

   What property do these five figures have in common?

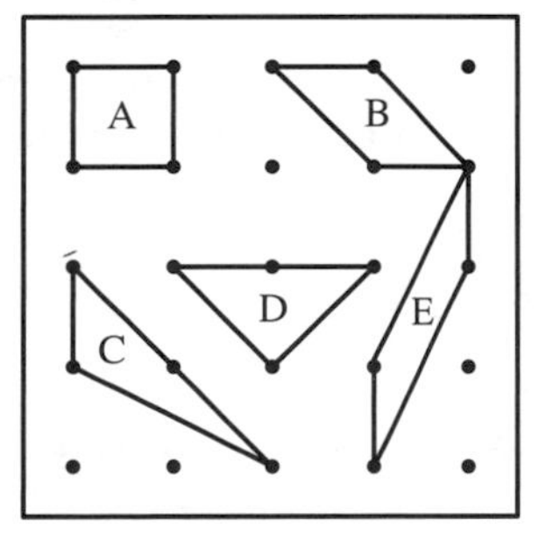

10. Look at Figures A–D in the figure shown. What is the area of each figure in square geos? How can you justify each of your answers?

    Figure A_______________

    Figure B_______________

    Figure C_______________

    Figure D_______________

What property do these four figures have in common?

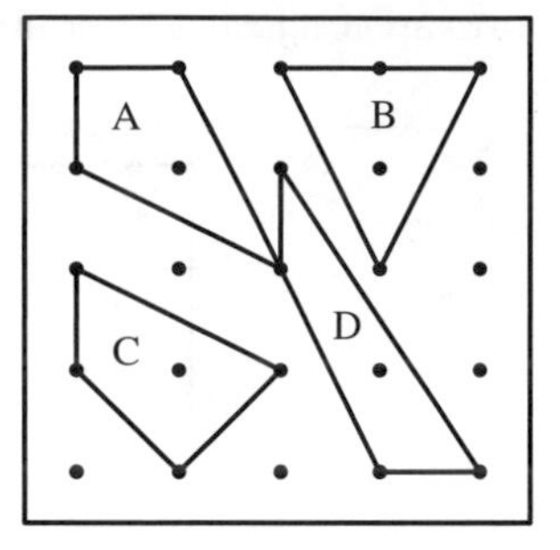

11. Look at Figures A–C in the figure shown. What is the area of each figure in square geos? How can you justify each of your answers?

    Figure A_______________

    Figure B_______________

    Figure C_______________

    What property do these three figures have in common?

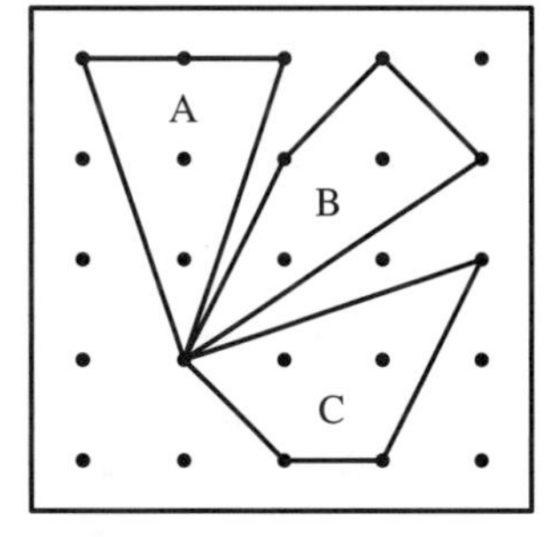

12. Look at Figures A–C in the figure shown. What is the area of each figure in square geos? How can you justify each of your answers?

    Figure A_______________

    Figure B_______________

    Figure C_______________

    What property do these three figures have in common?

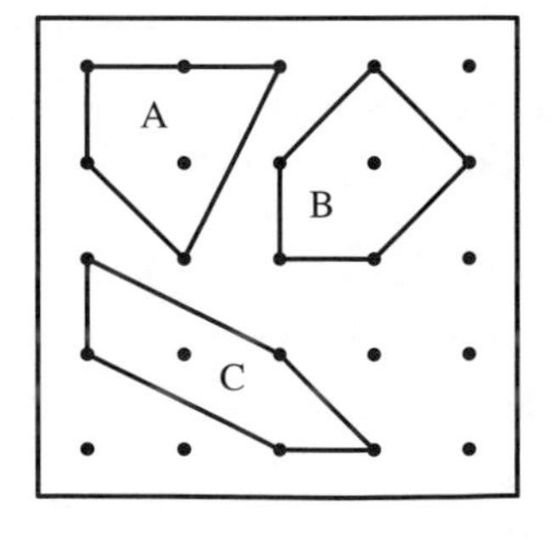

13. Look at Figures D–F in the figure shown. What is the area of each figure in square geos? How can you justify each of your answers?

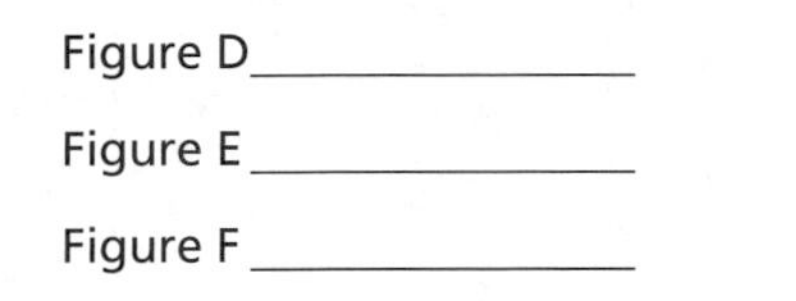

Figure D_______________

Figure E_______________

Figure F_______________

What property do these three figures have in common?

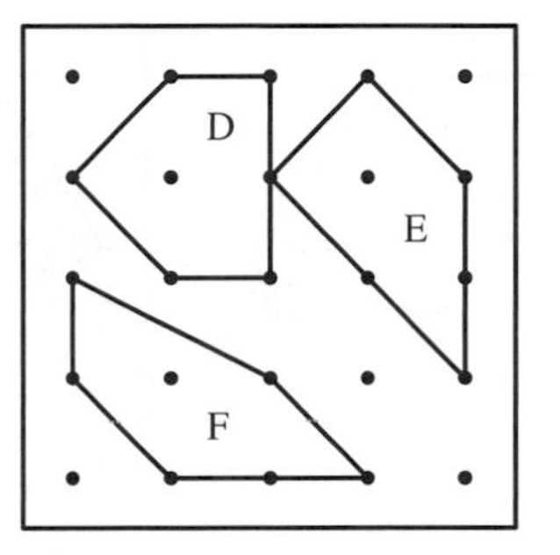

14. Look at Figures G–H in the figure shown. What is the area of each figure in square geos? How can you justify each of your answers?

Figure G_______________

Figure H_______________

What property do these two figures have in common?

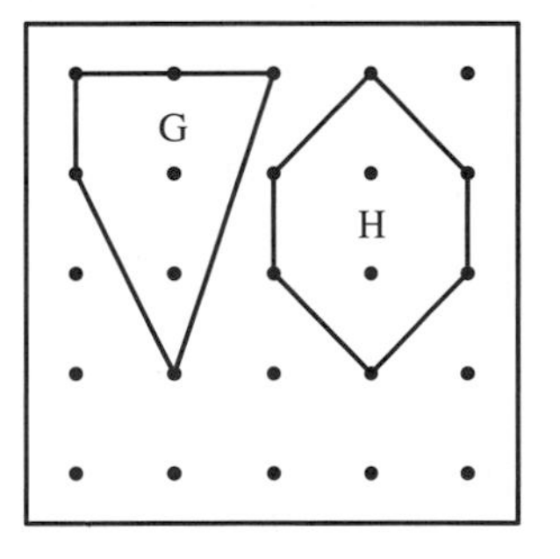

15. Using the information that you have derived in completing Problems 9–14, can you write a generalization about the area of figures on the geoboards that have all of their vertices at one of the nails on a geoboard? How might you justify the property?

16. Test out the generalization that you formed above in Problem 15 on the following shapes. Use your generalization to find the areas first and then check your answers with the methods that you discussed in Problem 8.

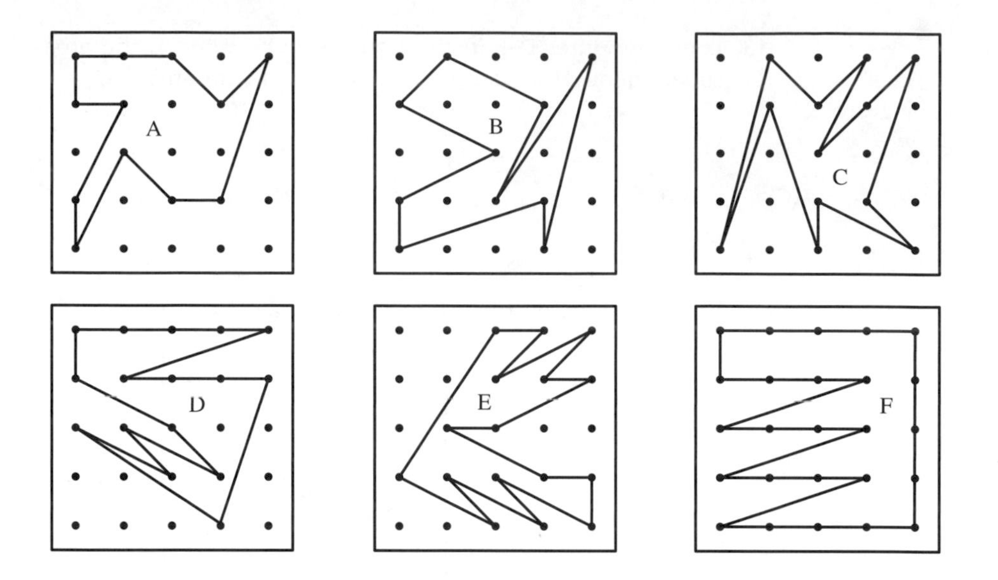

Lesson 6: Geometry and Measurement (Secondary Level)[3]

Objectives: Students will work with a geometry software package to explore, make conjectures, and develop a justification for conjectures based on distances in an equilateral triangle. More specifically, students will

- Work in pairs to explore the sum of the distances from a point inside an equilateral triangle to each side of the triangle.
- Build and test conjectures related to the sum of the distances.
- Develop ways of justifying these conjectures based on further explorations.
- Write an explanatory proof of at least one conjecture to be shared with classmates.
- Appreciate the need for mathematical proof.

Warm-Up: Students will work in pairs to construct an equilateral triangle with the geometry software package in the computer lab. They might recall constructions done previously with straightedge and compass and use these techniques to "construct" an equilateral triangle on the computer. Alternatively, they may use transformations (60° rotations) to construct the triangle.

[3] These activities adapted from M. D. de Villiers (1999). *Rethinking proof with the Geometer's Sketchpad.* Emeryville, CA: Key Curriculum Press.

Activities:

1. (2–3 minutes) Share construction techniques of equilateral triangle.

2. (2–3 minutes) Explain that students will use knowledge of equilateral triangles, perpendicular distances, and the geometry software to explore and write conjectures. Then let students know they will be expected to develop a good justification or proof of at least one conjecture (conjectures will be shared and discussed as a whole class prior to writing up justifications). The purpose of this lesson is to use previously learned knowledge to make new discoveries and to develop a valid justification that explains the new discovery.

3. (20–25 minutes) Distribute worksheet (below) and give students time to explore, develop conjectures, and test conjectures. Visit groups as they work in order to check understanding and to guide them if they are not finding anything significant. Sample probing questions:

   - How can you find the distance from a point to a line?
   - How do you use the software package to calculate a sum of the distances?
   - What do you notice about the sum when point P is exactly in the center of the triangle?
   - What do you notice about the sum when point P is close to a vertex of the triangle?
   - Does it make sense that the sum is always the same?
   - How can you check that this is true for any equilateral triangle?
   - Why is it true for every equilateral triangle?
   - What does the sum correspond to? Why? (length of an altitude of the triangle)

4. (10 minutes) Bring class together to discuss findings. Ask students to share various conjectures with the group as well as why they are convinced the conjecture is true. Brainstorm some initial ideas for justifying the conjectures—for example, draw auxiliary lines (which would be helpful?), check various properties of smaller triangles or other figures within the larger equilateral triangle (perimeter, area, etc.), investigate properties of equilateral triangle that would help explain. Once all significant conjectures and ideas are recorded, ask pairs to choose one and develop an explanation for why the conjecture is true.

5. (Remaining time) Ask pairs to begin work on justification. This may require more exploration on the computer. Encourage groups who have chosen the same conjecture to share ideas. Visit groups to help them consider ideas and perhaps ask questions to lead them in a particular direction.

Assessment:

1. Warm-Up The process of constructing the equilateral triangle will help me assess student knowledge of properties of equilateral triangles. Selected volunteers will be asked to share their construction techniques.
2. Group Work (informal assessment) Watch students as they work to see who is contributing ideas and how other group members are responding to, and working with, these ideas. Question students on occasion to get a better sense of their understanding of the problems, the techniques being used, their confidence in the conjectures, their ideas about why the conjectures are true (see probing questions above).
3. Whole-Class Discussion Hopefully many students will share conjectures and ideas about why they are true. Call on a variety of students.
4. Homework Students work on the justifications and turn these in to be graded.
5. Extension Problems This is a good way to see if students can apply what they learned to a new situation.

Closing: No closing activity, but have students continue to think about and build the justification for their chosen conjecture. If some groups are close to being done, have them consider extension problems.

Possible Extension Problems:

- What can you say about the sum of the distances if point P is not inside the triangle?
- What can you say about the sum of the distances if the triangle is not equilateral?
- Where should you locate point P in an arbitrary triangle so the sum of the distances is as small as possible? Explain.
- Now try a rhombus. Where should you locate point P in a rhombus so that the sum of the distances from point P to each side of the rhombus is minimal? Explain.
- What happens if you try this for a more general quadrilateral?

Discovery Worksheet

Using the equilateral triangle you have already constructed, follow the directions below and respond to the questions. Keep in mind that your conjectures should be supported by data and reasoning.

Directions:

- Label your equilateral triangle *ABC*.
- Place a point *P* in the interior of triangle *ABC* (it doesn't matter where).
- Measure the distance from point *P* to each of the sides *AB*, *BC*, and *AC* of triangle *ABC*. (How do you measure distance from a point to a line?)
- Calculate the sum of the three distances.

Questions:

1. As you move/drag point *P* around the interior of the triangle, what do you notice about the sum of the distances?
2. Now drag a vertex to change the size of your triangle. Again, drag point *P* around the interior of the triangle. What do you notice about the sum of the distances?
3. Write one or more conjectures about what you have discovered. Be sure to use complete sentences when writing your conjectures.
4. Why are you convinced that your conjecture is true?
5. How might you verify the conjecture?
6. Construct an altitude in your triangle. What is an altitude? How can you construct one?
7. What do you notice about the length of the altitude in comparison with the sum of the distances from point *P* to each of the sides of the triangle? Write your finding as a conjecture.
8. Why do you think this relationship holds?
9. Does this relationship hold no matter what size equilateral triangle you have?
10. How might you verify this relationship?

Further Investigations:

11. Highlight triangle *ABP* and find its area. Write an expression for the area of triangle *ABP* in terms of the length of the side *AB* (call its length l) and its height (call it h_1).
12. Highlight triangle *ACP* and find its area. Write an expression for the area of triangle *ACP* in terms of side *AC* and its height (call the height h_2). What is the length of side *AC*?
13. Highlight triangle *BCP* and find its area. Write an expression for the area of triangle *BCP* in terms of side *BC* and its height (call the height h_3).

14. Add the three areas and simplify the expression (taking out any common factors).

15. Now write an expression for the area of the entire equilateral triangle. How does this area compare to the sum of the areas of the other three triangles? Write an equation that shows this relationship.

Conclusions:

16. Choose one of your conjectures from your previous work.

17. Use these discoveries to help you justify the conjecture you chose from above.

18. Prepare a clear and explanatory way of presenting your conjecture and justification to the class.

References

Balomenos, R. H., Ferrini-Mundy, J., & Dick, T. (1987). Geometry for calculus readiness. *In* M. M. Lindquist & A. P. Shulte (eds.) *Learning and teaching geometry, K–12: 1987 Yearbook* (195–209). Reston, VA: National Council of Teachers of Mathematics.

Banchoff, T. (1990). Dimensions. *In* L. A. Steen (ed.), *On the shoulders of giants: New approaches to numeracy* (11–59). Washington, D.C.: National Academy Press.

Battista, M. T., & Clements, D. H. (1995). Geometry and proof. *Mathematics Teacher* **88**(1):48–54.

Clements, D. H., & Battista, M. T. (1992). Geometry and spatial reasoning. *In* D. A. Grouws (ed.), *Handbook of research on mathematics teaching and learning* (420–464). New York: Macmillan.

Crisler, N. (1995). *Symmetry and patterns*. Lexington, MA: COMAP.

Dreyfus, T., & Hadas, N. (1987). Euclid may stay—and even be taught. *In* M. M. Lindquist & A. P. Shulte (eds.), *Learning and teaching geometry, K–12* (47–58). Reston, VA: The National Council of Teachers of Mathematics.

Hodgson, T., & Morandi, P. (1996). Exploration, explanation, formalization: A three-step approach to proof. *Primus* **6**:49–57.

Howe, R. (1998). The AMS and mathematics education: The revision of the "NCTM Standards." *Notices of the American Mathematical Society* **45**(2):243–247.

Kenney, P. A., & Kouba, V. L. (1997). What do students know about measurement? *In* P. A. Kenney & E. A. Silver (eds.), *Results from the sixth mathematics assessment of the National Assessment of Educational Progress* (141–163). Reston, VA: National Council of Teachers of Mathematics.

Krause, E. F. (1975). *Taxicab geometry*. Reading, MA: Addison Wesley.

Lappan, G., Fey, J. T., Fitzgerald, W. M., Friel, S. N., & Phillips, E. D. (1996). *Connected mathematics*. Palo Alto, CA: Dale Seymour.

Lehrer, R., & Chazan, D. (1998). *Designing learning environments for developing understanding of space and geometry*. Mahwah, NJ: Erlbaum.

Malkevitch, J., Banchoff, T., Crowe, D., Garfunkel, S., Klee, V., Meyer, W., Senechal, M., Thurston, W., Conway, J., & Doyle, P. (1990). *Geometry's future*. Arlington, MA: COMAP.

Martin, T. (1991). Fracturing our ideas about dimension. *In* C. Findell (ed.), *NCTM student math notes*. Supplement to *NCTM News Bulletin* **28**(2):1–4.

National Council of Teachers of Mathematics (2000). *Principles and standards for school mathematics*. Reston, VA: National Council of Teachers of Mathematics.

Piaget, J. (1970). *Genetic epistemology*. New York: Columbia University Press.

Reid, D. A. (1998). Sharing ideas about teaching proving. *Mathematics Teacher* **91**:704–706.

Senechal, M. (1990). Shape. *In* L. A. Steen (ed.), *On the shoulders of giants: New approaches to numeracy* (139–181). Washington, D.C.: National Academy Press.

Senk, S. L. (1985). How well do students write geometry proofs? *Mathematics Teacher* **78**(6):448–456.

Senk, S. L. (1989). Van Hiele levels and achievement in writing geometry proofs. *Journal for Research in Mathematics Education* **20**(3):309–321.

Strutchens, M. E., & Blume, G. W. (1997). What do students know about geometry? *In* P. A. Kenney & E. A. Silver (eds.), *Results from the sixth mathematics assessment of the National Assessment of Educational Progress* (165–193). Reston, VA: National Council of Teachers of Mathematics.

Usiskin, Z. (1997). Applications in the secondary school mathematics curriculum: A generation of change. *American Journal of Education* **106**:62–84.

Van Hiele, P. M. (1959). *Development and learning process: A study of some aspects of Piaget's psychology in relation with the didactics of mathematics*. Groningen, The Netherlands: Wolters.

7
Data Analysis and Probability

Statistics has some claim to being a fundamental method of inquiry, a general way of thinking that is more important than any of the specific facts or techniques that make up the discipline. If the purpose of education is to develop broad intellectual skills, statistics merits an essential place in teaching and learning.

David S. Moore (1990)

Overview

Data analysis and probability were singled out for significant roles in K–12 mathematics programs by the original NCTM *Standards* (1989). The *Principles and Standards for School Mathematics* (NCTM, 2000) merges these two areas into one standard. In this chapter we examine various methods to present data and how to choose a method and then interpret it. We look at ways to organize, interpret, and discuss data drawn from one- and two-variable settings, with particular emphasis given to interpreting graphs and other data-based displays. We will see how the study of trend lines can be connected with related algebra topics. We then shift our focus to probability and chance.

The chapter closes with an examination of simulations and how we use them to model events in our world. Special emphasis will be given to modeling both deterministic (there is an answer) and stochastic (the answer depends on chance) contexts.

Focus on the NCTM *Standards*

The NCTM standard for Data Analysis and Probability lists four specific outcomes. They are that students learn to

- Formulate questions that can be addressed with data and collect, organize, and display relevant data to answer them.
- Select and use appropriate statistical methods to analyze data.
- Develop and evaluate inferences and predictions that are based on data.
- Understand and apply basic concepts of probability. (NCTM, 2000, p. 324)

We live in a data-rich society; therefore, the ability to comprehend the meanings embedded in data and to work with data to answer specific questions is an important educational outcome. Even more important, students need to be able to discern what questions are amenable to investigating with data and how to collect, organize, and interpret that data. The *Standards* calls for an emphasis on exploratory data analysis. Examining data sets provides students with concrete understanding of the meanings of central tendency, or average, and variability, or spread, in such sets. As students manipulate data sets and examine how to represent relationships through a variety of graphical forms, they develop strong conceptual understanding of the foundations required for further study of statistics.

As part of their study of analytic methods, students develop an understanding of when these methods are appropriate and how to conduct trend analyses. For example, given a set of data reporting behaviors over time or the relationship between two variables, students learn to develop trend lines and work to ferret out information about correlation between variables. Students also confront questions about the degree to which they can interpolate or extrapolate from their analyses.

A third part of the study related to this standard is the consideration of probability. Probability can be described as the study of chance. The *Standards* recommends that students be given a background in probability and experience using it in common settings. More formal study is left for the collegiate level.

Focus on the Classroom

The teaching of data analysis and probability can be one of the most interesting and exciting areas of the middle and secondary school curriculum. The range of real-world situations that students can apply their classroom learning to is enormous. And each analysis they perform will be very like what professionals do: Professionals formulate questions, collect data, and make judgments about the meaning of the data. Students are quick to make this connection.

As students begin to examine chance, they can simulate outcomes on the computer of situations that would be all but impossible to compute by hand. Laboratory lessons involving chance help students begin to estimate intuitively. As students make the transition from hands-on to formal conceptions of probability, great care must be taken to help them avoid common misconceptions.

7.1 Exploring Data

Data Analysis and Probability made its first significant entry into the middle and high school curriculum during the 1980s through the joint American Statistical Association/NCTM Quantitative Literacy Project. This project and its associated publications dealing with data exploration, sampling, survey methods, and probability, provided teachers with the first well-developed and coordinated materials for teaching data analysis and probability in the spirit of the *Standards*. The materials and the professional development workshops that accompanied them helped teachers move from a study of statistics (measures of central tendency, standard deviation, and hypothesis testing) to a study of data, its organization and representation, and its interpretation. This move was accompanied by a parallel shift in the study of probability from classical models based on permutations and combinations to a study of chance based on fundamental principles and simulations. As such, the study of classical probability and statistics gives students an intuitive basis that later helps them understand data prior to summarizing and testing it with the tools of classical probability and statistics.

At the heart of this change is the placement of evidence and data at center stage. Students from the intermediate grades forward formulate questions about situations, collect relevant data, and learn to tell the story with data. They make decisions about what data to collect, how to avoid biases in the collection, and learn about sampling and the role of randomness. They then must organize their data, and represent and interpret it. Today students, even at the middle school level, routinely choose among dot plots, stem-and-leaf plots, box plots, and a multitude of other formats in order to present univariate data appropriately. At the next level, students shift to more involved data analyses, including various forms of bivariate data and their associated scatterplots, and they consider questions of correlation, trend, distribution, and covariation. The concept of "best fit" leads naturally to ways of describing trend lines, their rates of change, what the intercepts on their graphs mean, and the domain over which their fits may be interpretable.

Exploratory Data Analysis

Exploratory Data Analysis, or EDA for short, is perhaps the most modern piece of the mathematical sciences in the school curriculum. This field of interactive, exploratory data investigation springs from the work of John Tukey of Princeton University and his colleagues (Tukey, 1977; Landwehr & Watkins, 1987). At the heart of EDA is the discovery of meaning in data sets. Rather than merely develop statistics, or compute such parameters as averages or spreads, EDA develops graphical displays that reveal patterns and identify special subgroups within a data set.

These methods include simple approaches that apply to single-variable (univariate) situations such as measuring one aspect of human performance, or determining the judgments of a panel or the responses to a single survey question. However, these methods also can be used in parallel analyses of data sets. We will look at three EDA techniques, the dot plot, the stem-and-leaf plot, and the box plot.

The *dot plot* is perhaps the first EDA method students use to analyze data. In it, numerical data are plotted as dots above a number line that reflects the span of the data. Such plots, sometimes called line plots, help students identify the extremes (points at the ends of the data set), outliers (data points that fall outside the other data), and clusters (data points grouped close together). For example, consider the dot plot in Figure 7.1, in which scores from the Third International Mathematics and Science Study's mathematics examination are given for eighth graders from various nations (Beaton et al., 1996). The data range from 354 (South Africa) to 643 (Singapore). Further analyses reveal that Singapore, Korea (607), Japan (605), Hong Kong (588), Flemish-speaking Belgium (565), and the Czech Republic (564) are outliers at the top end of the scale. South Africa, Columbia (385), Kuwait (392), Iran (428), and Portugal (454) are outliers at the low end of the scale. In the middle of the distribution, there are two clusters of scores separated at about 500. One cluster runs from slightly below 480 to 500 and the other from slightly above 500 to around 550. The groupings of Japan and Korea and of Belgium and the Czech Republic might also be considered as two small clusters. Such displays of data are easy ways to get a feel for the data, its visual spread, its outliers, its clusters, and its major gaps, if any.

Historically, in the classical approach, the arithmetic mean would be found by summing the scale scores and dividing by 41, a process that would have resulted in the parameter known as the mean. In this case, the *mean* of the 41 national scores is 513

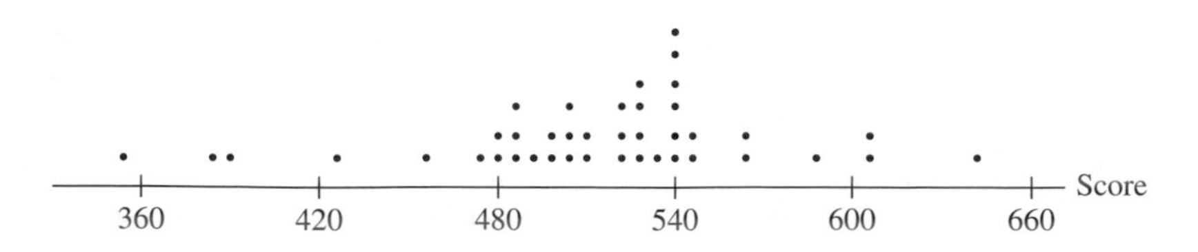

Figure 7.1 National achievement scores in mathematics for various nations.

1970s	|
1980s	𝍸 𝍸 𝍸 |||
1990s	𝍸 𝍸 𝍸 𝍸 𝍸 𝍸 ||
2000s	|

Figure 7.2 Frequency plot of year of manufacture of 52 used cars in the classifieds (197|3 means 1973).

when rounded to the nearest integer. The *median*, or the score point above which and below which 50% of the scores fall is 522. The median can be thought of the point that divides the data set into two halves, each having the same number of data points in it. When the number of score points in the set is odd, the median is one of the score points. When the number of score points is even, we find the arithmetic mean of the two mid-most points and list it as the median. For example, 3 is the median of 1, 2, 4, 5 as it is the average of 2 and 4 and has two scores above it and two scores below it. The third historical measure of central tendency was the *mode*. The mode is the most frequently occurring score in the data set. An examination of the 41 national scores reveals that there are three points along the score scale with two countries associated. These are at 541, with the Netherlands and Slovenia, 527, with Ireland and Canada, and 522, with Thailand and Israel. Thus, we would say that this set of data is tri-modal. Finding the difference between the extreme scores of 643 and 354, we see that the *range* of the data is 289. Using technology, we calculate the *standard deviation* as 56.37 points. These last two measures are measures of the spread of the data.

A second EDA method for exploring the structure of a data set is the *stem-and-leaf plot*. But first let's look at the frequency diagram shown in Figure 7.2, which groups data on used cars by decades. Each tally mark represents the year of manufacture of 52 used cars advertised in the classifieds. Examining the diagram, we can easily see that most of these cars were made in the 1980s and 1990s, but we cannot break the data down into smaller increments.

The stem-and-leaf plot in Figure 7.3 represents the same data set but groups the data by "stems." The stems represent decades like the tally rows in the frequency diagram. The vertical bar separates the year of manufacture into decades. That is, 198| represents 1980–89. The numbers to the right of the bar each represent the year of manufacture for each used car. These digits are the "leaves." For example,

Year of Used Cars

197	3
198	244556677778889999
199	00000000011225566777777778888999
200	0

Figure 7.3 Stem-and-leaf plot of year of manufacture of 52 used cars in the classifieds (197|3 means 1973).

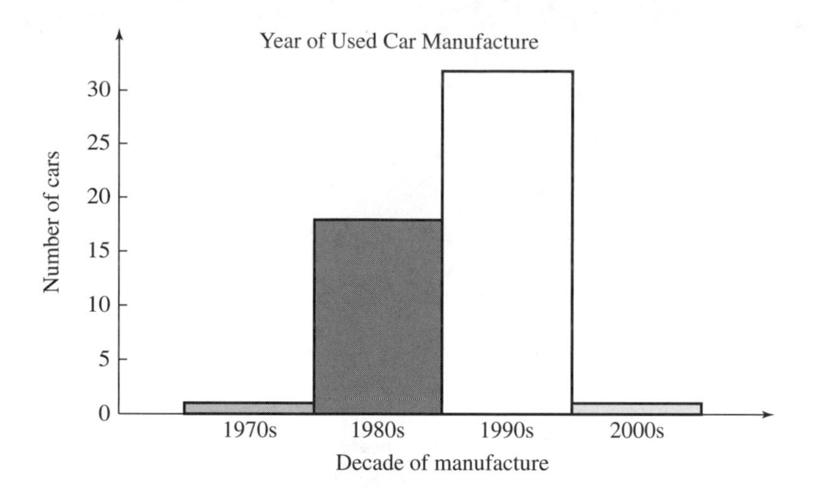

Figure 7.4 Histogram of year of manufacture of 52 used cars in the classifieds.

198|244556677778889999 groups used car ads for cars manufactured in 1982, 1984, 1985, 1986, 1987, 1988, and 1989. Thus, the stem-and-leaf plot shows the decade data like the frequency diagram but also shows the additional breakdown of number of car ads by year.

Note that the leaves are ordered in increasing fashion for each stem. This makes the plot an *ordered* stem-and-leaf plot. If we rotate the stem-and-leaf plot 90 degrees counterclockwise, we get something like a *histogram*, which is a classical approach to displaying data. A histogram for the used car data is shown in Figure 7.4. Note that we lose detail—the data is once again grouped only by decade, and now we only know an approximate number manufactured in each decade.

When a comparison between two groups is desired, a *back-to-back* stem-and-leaf plot can be used, which facilitates comparing the extremes, outliers, clusters, and overall shape of the two groups. The back-to-back stem-and-leaf plot in Figure 7.5 shows the year of manufacture for Fords and Chevrolets in used car ads. Here the stem runs up the center and the leaves to each side. The data reflect little difference in the years of manufacture for the two makes, with the possible exception of slightly more Chevys from the early 1980s.

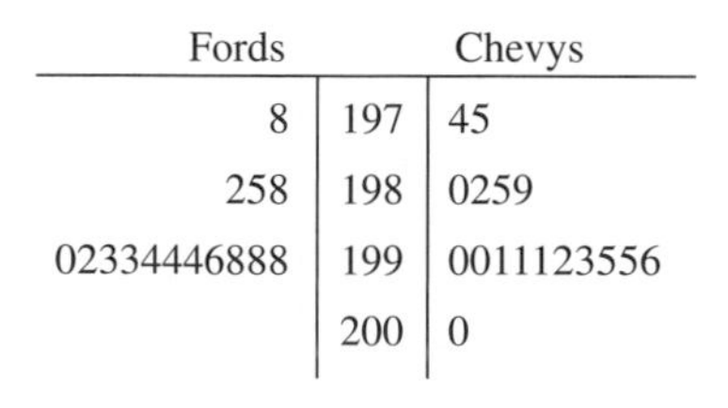

Fords		Chevys
8	197	45
258	198	0259
02334446888	199	0011123556
	200	0

Figure 7.5 Back-to-back stem-and-leaf plot of used car ads for Fords and Chevys (197|4 means 1974; 8|197 means 1978).

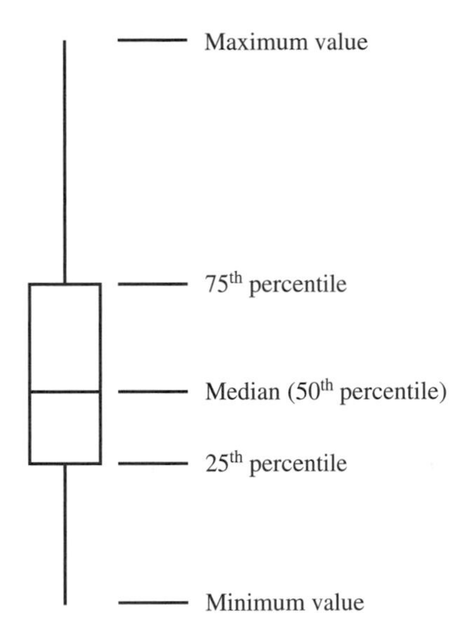

Figure 7.6 Values needed to construct box-and-whisker plots.

The third major EDA approach to examining data is the *box plot.* This format is sometimes referred to as a box-and-whisker plot. Before we look at a box plot, let's consider what information is needed in order to construct one.

As Figure 7.6 shows, students need to understand percentiles, median, and minimum and maximum data values to create box plots. Tukey (1977) suggests that using a five-number summary table makes building box plots simpler. Such a table contains the following values for a given data set:

	Median	
Lower Hinge		Upper Hinge
Minimum Value		Maximum Value

Hinges are values that are halfway from one extreme to the median.

Consider the data set in Table 7.1, which lists the 30 largest cities in 2000 and their populations. Using Tukey's methods, we count in from each extreme of the data set to the middle, to find the *depth* of each data point. The median is found by taking the total count of data points and dividing it in half if the count is odd and dividing the (count plus 1) in half if the count is even. The data set shown in Table 7.1 gives a median value of depth 11.55 and an average depth of 15.5.

To find the hinges, we find the median of data points that are halfway between the extremes and the median. Starting at the median, 11.55, we discard half values, and add 1. Half of this value gives us the depths of the hinges, we discard the 0.5 in 15.5 to get 15, add 1 and divide by 2: depth of 8 for the hinges. Counting in 8 data points from each extreme, the upper hinge is 14.1 and the lower hinge is 9.3. Finally, we note that the

Table 7.1 The 30 largest cities and their populations in the year 2000.

City	Population (millions)	Depth of data	
Tokyo, Japan	27.9	1	
Bombay, India	18.1	2	
São Paulo, Brazil	17.8	3	
Shanghai, China	17.2	4	
New York, U.S.	16.6	5	
Mexico City, Mexico	16.4	6	
Beijing, China	14.2	7	
Jakarta, Indonesia	14.1	8	Upper hinge = 14.1
Lagos, Nigeria	13.5	9	
Los Angeles, U.S.	13.1	10	
Calcutta, India	12.7	11	Median depth = 11.55
Tianjin, China	12.4	12	
Seoul, South Korea	12.3	13	
Karachi, Pakistan	12.1	14	Upper hinge = 14.1
Delhi, India	11.7	15	Median = 11.55
Buenos Aires, Argentina	11.4	15	
Manila, Philippines	10.8	14	
Cairo, Egypt	10.7	13	
Osaka, Japan	10.6	12	
Rio de Janero, Brazil	10.2	11	
Dhaka, Bangladesh	10.2	10	
Paris, France	9.6	9	
Istanbul, Turkey	9.3	8	Lower hinge = 9.3
Moscow, Russia	9.3	7	
Lima, Peru	8.4	6	
Teheran, Iran	7.3	5	
London, United Kingdom	7.3	4	
Bangkok, Thailand	7.3	3	
Chicago, U.S.	7.0	2	
Hyderabad, India	6.7	1	

two extremes are 6.7 and 27.9. These values together give us our five-number summary table for the 30 cities of

	11.55	
9.3		14.1
6.7		27.9

Using these values, we can now use a software package to construct the box plot shown in Figure 7.7 for the city populations. The hinges correspond essentially to the 25th and 75th percentiles, that is, the points below which lie 25 and 75 percent of the data points in the data set. In a like manner, the median is the 50th percentile. These

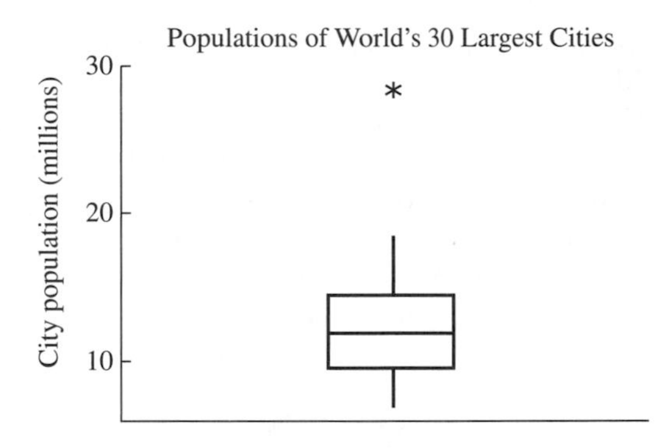

Figure 7.7 Box plot of the populations in 2000 of the world's 30 largest cities.

three percentile points are often referred to as Q_1, Q_2, and Q_3. The distance between Q_1 and Q_3 is known as the *interquartile range*.

The box plot for this data shows what happens when they are made by software packages. The upper whisker extends out only to 18.1 and places a dot at 27.9 for Tokyo. The dot tells us that the value for Tokyo is an outlier. If we skipped this qualification, the upper whisker would extend to 27.9.

Figure 7.8 shows *parallel* box plots, which enable us to compare two or more groups using the same scale. Here we compare the populations of cities from our original list but break them out by continent, in this case, the Americas and Asia. The box plot for the Americas uses 7 cities, while that for Asia uses 15 cities. Here we see a tighter interquartile range for the Asian cities, meaning that Asian city populations cluster closer to the median, but a greater overall variability because of Tokyo. The medians for the two groups are comparable.

Once students can both construct and interpret dot, stem-and-leaf, and box plots, they have experienced much of the content of classical summary statistics. Ours is a statistics-driven society, and students will find unending numbers of real-world exam-

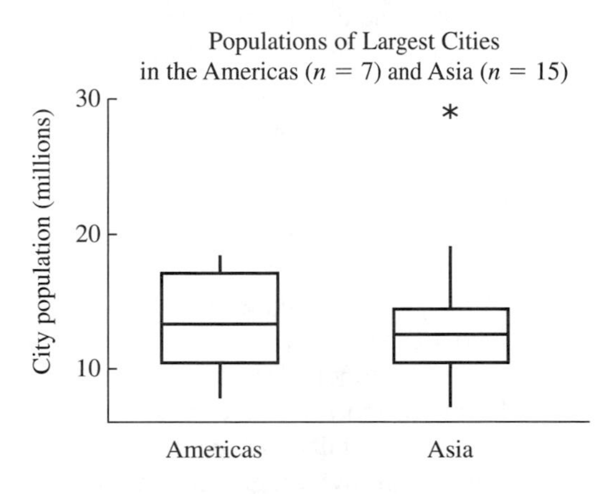

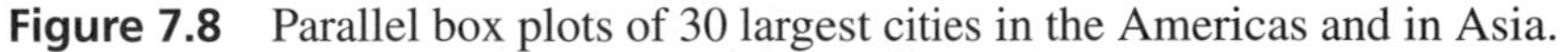
Figure 7.8 Parallel box plots of 30 largest cities in the Americas and in Asia.

ples, from extremely pertinent ones to ones they will find humorous, to work with and critique as they endeavor to understand what these statistics mean and how they affect our lives. That said, students should be given lots of opportunities to calculate and distinguish between measures of central tendency (mean, median, and mode), and between measures of spread (range, standard deviation, variance, and interquartile range). These values take on meaning in terms of the experience students have had in studying and comparing different data sets. Absent this experience, data and their related graphics have little meaning for students.

Thinking About Collecting Data

Of course, the real purpose of studying data, their graphical representation, and the associated statistics, is to enable students to investigate and make decisions about questions that are important to their lives. Central to this objective is helping students determine how to frame good questions and design ways to collect data that will ferret out the pertinent information. In general, data evolves from one of three types of collection: surveys, sets of observations, and experiments (NCTM, 2000; Landwehr, Swift, & Watkins, 1987; Burrill et al., 1992; Zawojewski et al., 1991).

Surveys give us responses from a representative sample (of individuals or objects) that answer a specific set of questions. Carefully framed questions allow for sets of data to be collected that can then be organized, represented, analyzed, and interpreted in a valid manner to answer the questions asked. In order for the results to be generalized, the sample of individuals or objects must be representative of the group or population about which we desire generalizations. This has required the development of careful ways to draw samples so that they are representative of the larger group or population. For example, if we want to say something about the opinions of students at East High School, then the sample we survey must reflect the characteristics of the total student body.

Developing a sampling plan can be quite involved, depending on whether we want to collect data for the entire group or by subgroups of the population of interest. In the first case, we simply need a simple random sample, but in the latter, we need a stratified random sample, where individuals are randomly selected from each subgroup's membership on a proportional basis. Entire graduate courses are devoted to the methods for selecting random samples and processing the resulting data.

A second source of data is from observational studies. Such data are tallies made by observers watching a particular phenomenon. For example, we might tally the number of women versus men who purchase popcorn at a movie. The results might be used to determine how much popcorn to make at various times or with different types of movies. Another observational study might investigate traffic safety by counting the number of people who come to full stops behind the white line at the stop sign near a grade school crossing. Such observations could be part of a study to increase traffic safety around the school. Here the emphasis is on determining what event needs to be observed, what categories of data should be collected, and how they should be reported.

A third source of data is experimental studies. For example, to compare the impact of two fertilizers on soybean growth, we might subdivide a field with homogeneous soil and water conditions into a matrix of 3×2 plots arranged as shown in Figure 7.9. Each

Fertilizer A	No Fertilizer	Fertilizer B
No fertilizer with irrigation	Fertilizer B with irrigation	Fertilizer A with irrigation

Figure 7.9 Experimental planting plan for soybean growth study.

row of plots in the field is arranged randomly and the two rows also differ in that one row is irrigated. This experimental design allows us to compare fertilizer/no fertilizer, fertilizer A/fertilizer B, and irrigation/no irrigation.

Student facility in these skills increases only when they are involved in the decision making that accompanies planning the study, and collecting and interpreting the data from surveys, observational studies, and experiments. Such activities are easily incorporated into lesson plans that involve social studies, physical education, the sciences, or other curricular topic areas (Burrill, Scheaffer, & Rowe, 1991).

Exercises 7.1

Use dot, stem-and-leaf, and box plots to develop graphical displays for the data sets in Exercises 1–4. Then interpret the story told by the data.

1. Density of state populations (people per square mile) in 1990 (*World Almanac Books*, 2000)

AL	79.6	KY	92.8	ND	9.3
AK	1.0	LA	96.9	OH	264.9
AZ	32.3	ME	39.8	OK	45.8
AR	45.1	MD	489.2	OR	29.6
CA	190.8	MA	767.6	PA	265.1
CO	31.8	MI	163.6	RI	960.3
CT	678.4	MN	55.0	SC	115.8
DE	340.8	MS	54.9	SD	9.2
DC	9882.8	MO	74.3	TN	118.3
FL	239.6	MT	5.5	TX	64.9
GA	111.9	NE	20.5	UT	21.0
HI	172.5	NV	10.9	VT	60.8
IL	12.2	NH	123.7	VA	158.3
ID	205.6	NJ	1042.0	WA	73.1
IN	154.6	NM	12.5	WV	74.5
IA	49.7	NY	381.0	WI	90.1
KS	30.3	NC	136.1	WY	4.7

2. Acreage per farm by state (*New York Times*, 2000)

AL	194	KY	154	ND	1274
AK	1625	LA	273	OH	186
AZ	3582	ME	186	OK	dna
AR	298	MD	168	OR	435
CA	320	MA	95	PA	128
CO	1092	MI	200	RI	87
CT	93	MN	361	SC	196
DE	215	MS	276	SD	1354
DC	dna	MO	274	TN	131
FL	236	MT	2091	TX	582
GA	226	NE	844	UT	773
HI	262	NV	2300	VT	200
IL	490	NH	135	VA	180
ID	352	NJ	86	WA	393
IN	236	NM	2831	WV	176
IA	340	NY	205	WI	210
KS	731	NC	162	WY	3761

3. Number of home runs hit by National and American League champions for 1970–1999 (*New York Times*, 2000)

Year	National League	American League	Year	National League	American League	Year	National League	American League
1970	45	44	1980	48	41	1990	40	51
1971	48	33	1981	31	22	1991	38	44
1972	40	37	1982	37	39	1992	35	43
1973	44	32	1983	40	39	1993	46	46
1974	36	32	1984	36	50	1994	43	40
1975	38	36	1985	37	40	1995	40	50
1976	38	32	1986	37	40	1996	47	52
1977	52	39	1987	49	49	1997	49	53
1978	40	46	1988	39	42	1998	70	56
1979	48	45	1989	47	36	1999	65	48

4. Ten tallest buildings in feet (and number of floors in each) for Chicago and New York (*World Almanac*, 2000)

Chicago: 1450 (110); 1136 (80); 1127 (100); 1007 (61); 995 (62); 961 (65); 871 (66); 859 (74); 853 (67); 850 (60)

New York: 1368 (110); 1362 (110); 1250 (102); 1046 (77); 951 (67); 927 (71); 915 (59); 866 (48); 861 (72); 850 (70)

5. Describe an experiment that would test whether students and teachers differ in their preferences for soft drink A or soft drink B.

6. How would you conduct a survey to decide the mascot for a grade 5–8 middle school?

PROJECTS Find Out for Yourself

1. Ask three middle school students to describe the story told by a graph involving the comparison of two data sets by box plots. One data set has a higher mean, but the other data set has less variability.

2. Find an article in the local newspaper that uses statistics to defend a stance. Is their use of statistics appropriate?

7.2 Exploring Bivariate Data

In a manner similar to that for univariate data, student introduction to bivariate and multivariate data needs to progress from representation and interpretation to more classical statistical treatments. Students first need to recognize the nature of, and some sources of, such data. Data sets like hours worked and paycheck total, miles driven and gallons of gasoline used, and amount of fat and protein in common fast-food sandwiches make good examples.

Scatterplots

Bivariate data (data that relates the values of two variables) is usually graphed as sets of ordered pairs. Such graphs are called *scatterplots*, because they show how the data is "scattered" relative to the two axes, which represent the two variables. Table 7.2 provides such an example; it lists the lengths of antennae and wings, measured in centimeters, for two species of small insects known as midges. The data is displayed in a scatterplot in Figure 7.10. The Af midges, marked by ♦ and the Apf midges, marked by ○, differ in their value to humankind. The Af midges are valuable pollinators of crops in Brazil, while the Apf midges are carriers of a debilitating brainstem disease. How would you use the data in Figure 7.10 to begin to describe the difference in the two midges (Fusaro, 1989)?

A scatterplot should always have labeled axes, a title, and a legend identifying points, such as shown in Figure 7.10, if the two variables involved are special data

Table 7.2 Wing and antenna lengths of midge flies.

	Apf Midges								
Wing length (cm)	1.78	1.86	1.96	2.00	2.00	1.96			
Antenna length (cm)	1.14	1.20	1.30	1.26	1.28	1.18			
	Af Midges								
Wing length (cm)	1.72	1.64	1.74	1.70	1.82	1.82	1.90	1.82	2.08
Antenna length (cm)	1.24	1.38	1.36	1.40	1.38	1.48	1.38	1.54	1.56

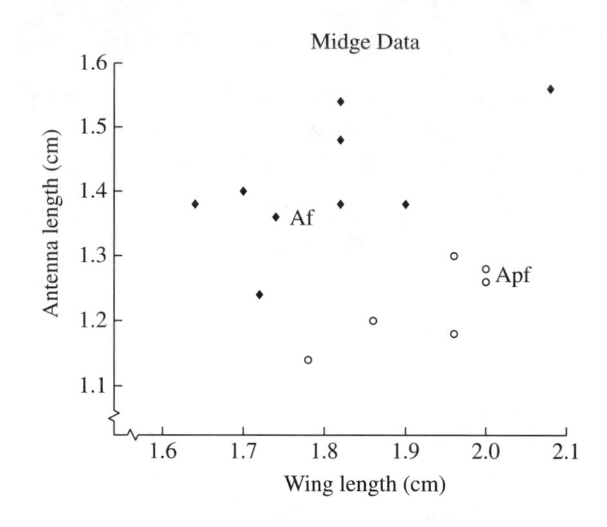

Figure 7.10 Scatterplot of midge data.

from a specific category, individual, or country. Such labeling facilitates interpreting the information. In other cases, when the variables are entirely captured by the labels on the axes, additional labeling is not necessary. The actual building of a scatterplot can be tied directly to students' experience with graphing ordered pairs of numbers.

Once students understand how to construct scatterplots, they can be asked to write paragraphs interpreting scatterplot data. For example, consider the scatterplot in Figure 7.11, which contains information on IQ and reading scores of 18 students. How would you describe the relationship between the IQ scores and reading scores shown in the scatterplot?

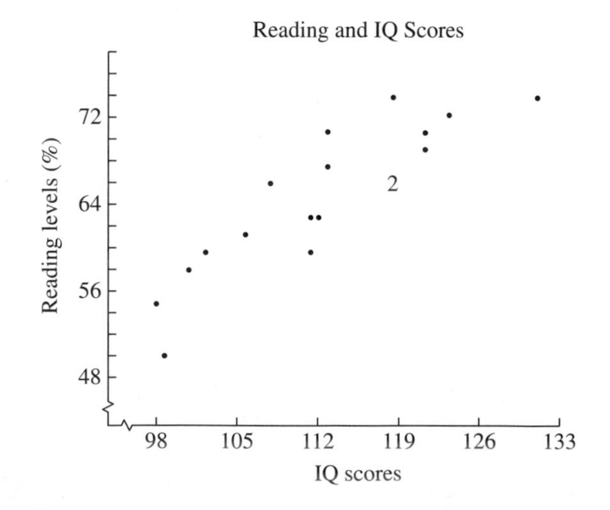

Figure 7.11 Scatterplot of IQ and reading score data.

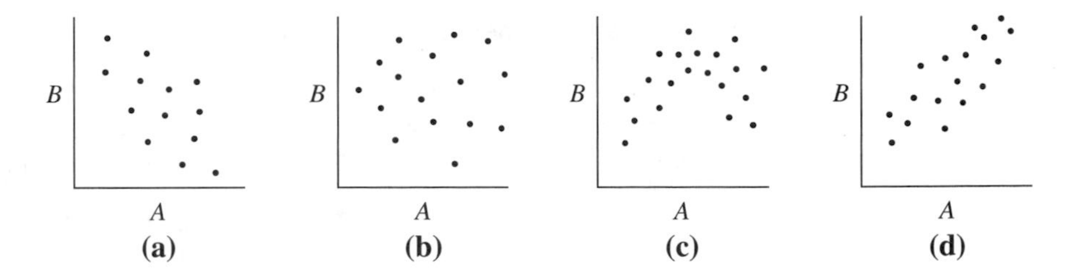

Figure 7.12 "Clouds" of data in scatterplots.

Correlation

Another feature described in scatterplots is the shape of the "clouds" of data collected in surveys. For example, consider the scatterplots in Figure 7.12. In scatterplot (a), the data seems to fall along a broad "line" running from the upper left to lower right. This pattern indicates that increasing scores on factor A are associated with decreasing scores on factor B. That is, as one score increases, the other decreases. In scatterplot (b), the data shows no discernible pattern relating factors A and B. In scatterplot (c), the data demonstrates a *curvilinear* pattern. If, for example, factor A is time spent on homework, and factor B is test scores, this pattern implies that up to a point, increased study time results in higher scores, but beyond that point, increased study appears to result in decreasing test scores. This point is sometimes called the *point of diminishing returns*. The fourth set of data, scatterplot (d), exhibits a linear pattern like plot (a), but here increased scores on factor A are associated with increased scores on factor B. That is, as one score increases, so does the other.

A statistic, called the *correlation coefficient*, measures the degree of linear relationship present in the scatterplots in Figure 7.12. The correlation coefficient r, runs from -1 to $+1$. Bivariate data sets, where the values of the coordinates either increase or decrease together, and whose "clouds" cluster along a diagonal from lower left to upper right are positively correlated. Negative correlations show the opposite, diagonals that run from upper left to lower right. Curvilinear data will also have r values between -1 and $+1$, but the pattern is only obvious when we look at the data in a scatterplot.

Trend Lines

Policy makers often want to identify trends in data. That is, is there a relationship between two or more variables in a set of data? For example, consider the data listed in Table 7.3 and graphed in Figure 7.13, which shows percentage of students living in poverty and the overall state mean achievement scores in mathematics for the District of Columbia and 40 states that participated in the National Assessment of Education Progress in 1996 (Reese et al., 1997). We can see a definite pattern: Achievement scores correlate negatively with poverty; that is, scores decrease as poverty increases.

Finding "best fit" is a key manipulation in identifying trend lines. The "line of best fit" for a data set is a line that "best" represents the data in terms of some form of accounting for the error between the predicted values associated with the line and

Table 7.3 Math achievement/poverty data.

State	Mathematics achievement score	Percent of students living in poverty	State	Mathematics achievement score	Percent of students living in poverty
AL	257	19.5	MS	250	28.2
AK	278	11.7	MO	273	23.6
AZ	268	23.4	MT	283	12.3
AR	262	20.4	NE	283	12.5
CA	263	25.3	NM	262	29.2
CO	276	9.9	NY	270	23.5
CT	280	18.6	NC	268	18.4
DE	267	9.8	ND	284	11.6
DC	233	30.0	OR	276	13.7
FL	264	22.1	RI	269	13.3
GA	262	18.5	SC	261	18.7
HI	262	12.0	TN	263	20.1
IN	276	13.7	TX	270	26.8
IA	284	13.5	UT	277	9.9
KY	267	13.6	VT	279	7.0
LA	252	36.8	VA	270	12.6
ME	284	9.6	WA	276	14.6
MD	270	17.2	WV	265	22.0
MA	278	12.2	WI	283	12.1
MI	277	17.9	WY	275	10.7
MN	284	13.7			

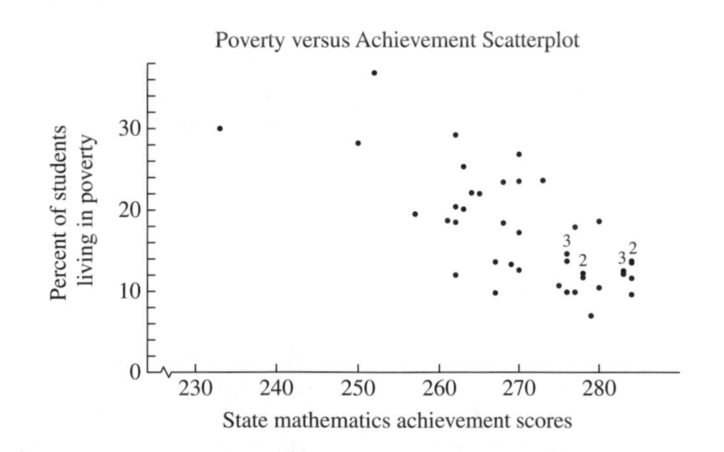

Figure 7.13 Scatterplot showing a negative correlation of the two variables.

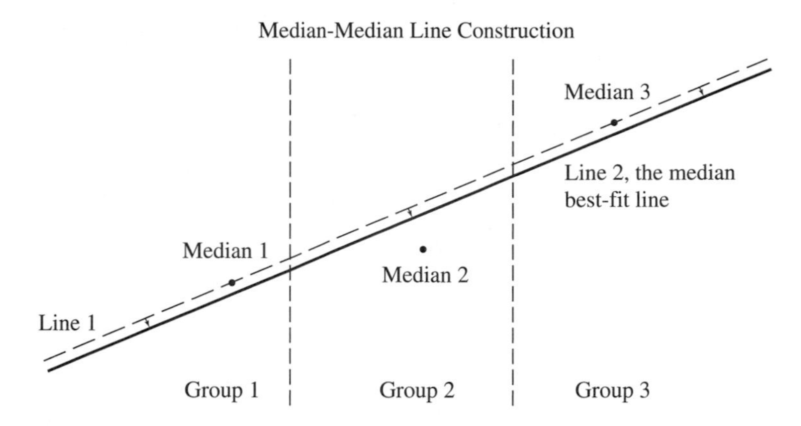

Figure 7.14 Constructing a best fit using the median-median line method.

the actual observed data values. A number of different "best" lines can be drawn. The *median-median line* fit is constructed by dividing the data into three equal groups and finding the ordered pair in each group that represents the medians of the x- and y-coordinates for the ordered pairs in that third of the data. The two ordered pairs from groups 1 and 3 are then connected by a line. The line l is then moved one-third of the way toward group 2's median point, as shown in Figure 7.14. The median-median line fit represents an easy way to find the best fit. Using the y-intercept and the slope, we can easily write the equation of line l; in this case, it is

$$y = -0.47x + 142.74 \tag{1}$$

The median-median line is especially useful for data with outliers.

The *line of least squares* gives us another way to fit data. Statistics courses often refer to this line as the linear regression line. It minimizes the sum of squared distance from the points in the data set to the line itself, as shown in Figure 7.15.

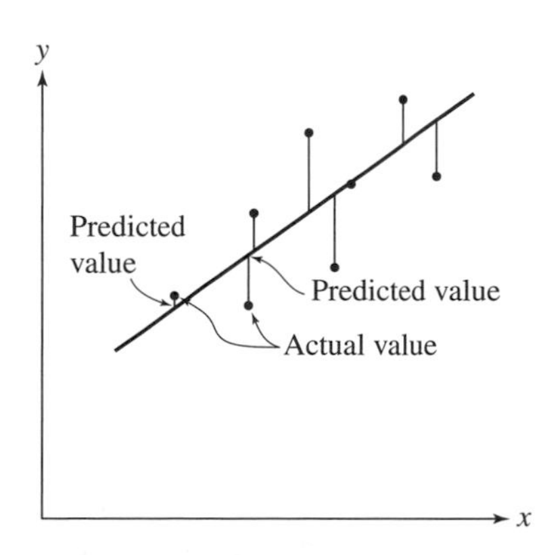

Figure 7.15 Constructing a line of least squares.

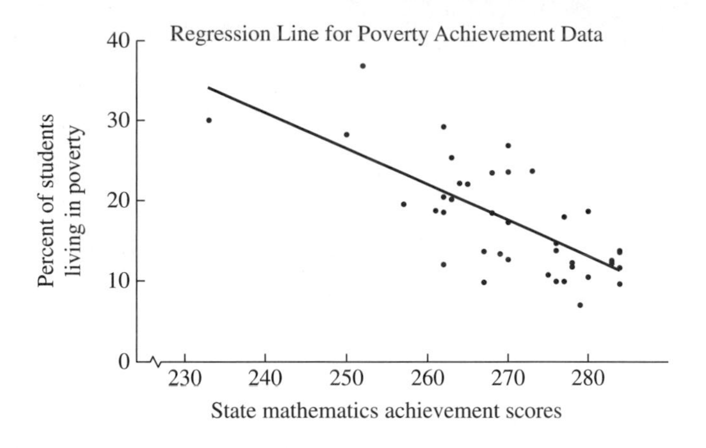

Figure 7.16 Graph of the linear regression line $y = -0.45x + 138.54$ for the math achievement/poverty data.

The equation for the linear regression line k for the math achievement/poverty data shown in Figure 7.16 is

$$y = -0.45x + 138.54 \tag{2}$$

Figure 7.17 graphs the best fit line and the linear regression line for the math achievement score/poverty data and shows that the difference between the two lines is very small.

Now let's look at another valuable measure, the amount of error. If we use the equations to predict values, then compare the predictions against actual observations, we can determine how closely the equations model reality, that is, how well they fit the actual situation.

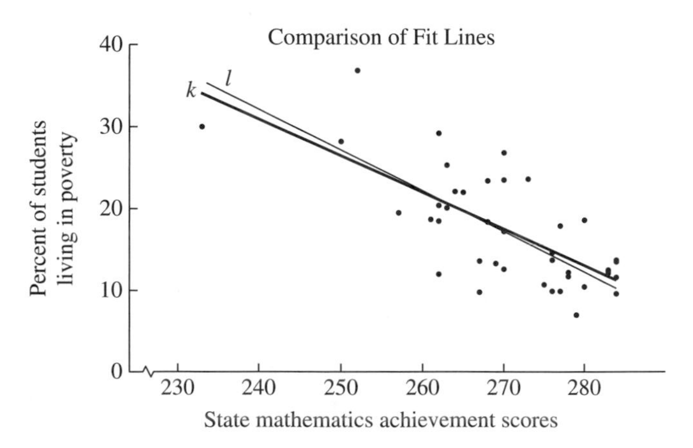

Figure 7.17 Comparing the median-median line l and the linear regression line k for the math achievment/poverty data.

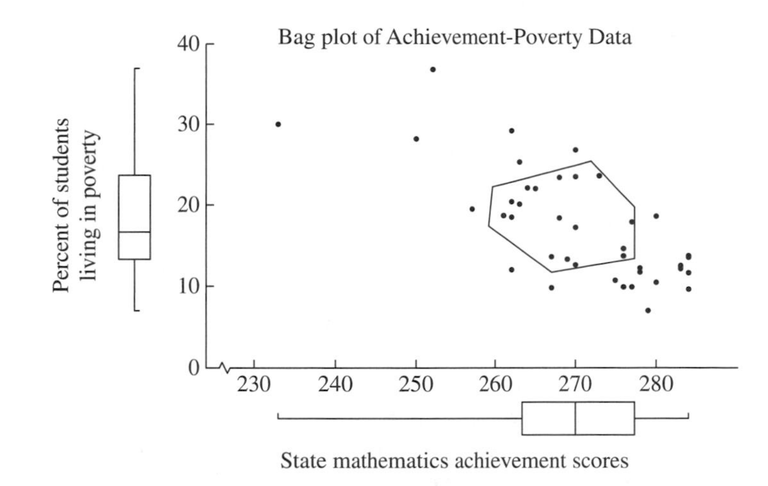

Figure 7.18 Bag plot of math achievement/poverty data.

Substituting values associated with a state's percent of students living in poverty into equations (1) and (2), we can find the achievement values the lines would predict. These values can then be compared with the actual values observed. The difference is the amount of error. Plotting the actual values against the predicted values, we can find the degree to which they are correlated. The closer this value is to $+1$, the better the fit. Using median-median line (1) and linear regression line (2) determined above, we see that both fits have a correlation of 0.72; that is, the lines fit equally well.

Another way to analyze bivariate data such as this is through the bag plot, another EDA approach developed by John Tukey and co-workers (Rousseeuw, Ruts, & Tukey, 1999.) This approach is shown in Figure 7.18. Here a box plot is built for each variable's data along their respective axes. Then a "bag" is drawn around all the points whose coordinates fall in the region described by the box plots. This is the "data bag." These points are in the middle 50 percentile for both variables, and hence are quite representative of each variable. This approach also identifies potential outliers.

Exercises 7.2

1. Consider the following data. Make a scatterplot, and on it draw an estimated line of best fit, then find both the median-median line and the line of linear regression. (Use appropriate sofware.) How close was your estimated line to the computed lines?

x	y	x	y	x	y
25	23	38	47	18	23
14	20	38	50	16	26
16	24	5	9	15	18
5	8	3	7	18	22

(continued)

x	y	x	y	x	y
18	28	15	22	17	25
12	20	16	23	9	8
16	26	4	8	14	24
25	37	11	14	1	1
11	20	12	20	14	14
6	15	49	55	7	12
1	1	10	15	12	20
18	25	6	15	12	22

2. Analyze the following four sets of data. You will find that they have similar means and standard deviations. What did you note about their data patterns? Compare the lines of fit given by the median-median approach and the linear regression approach. Which seems to work best (Anscombe, 1973)?

x_1	y_1	x_2	y_2	x_3	y_3	x_4	y_4
10	8.04	10	9.14	10	7.46	10	6.58
8	6.95	8	8.14	8	6.77	8	5.76
13	7.58	13	8.74	13	12.74	13	7.71
8	8.81	9	8.77	9	7.11	9	8.84
11	8.33	11	9.26	11	7.81	11	8.47
14	9.96	14	8.10	14	8.84	14	7.04
6	7.24	6	6.13	6	6.08	6	5.25
4	4.26	4	3.10	4	5.39	4	12.50
12	10.84	12	9.13	12	8.15	12	5.56
7	4.82	7	7.26	7	6.42	7	7.91
5	5.68	5	4.74	5	5.73	5	6.89

3. The table lists the U.S. population from 1800 to the present. Can you identify a trend? Is it linear or curvilinear? Predict the population for 2000 from the trend line. Estimates for 2000 put our population at 274.27 million people. How well does the predicted value match the estimate given?

Date	U.S. population	Date	U.S. population
1800	5.31	1900	76.21
1810	7.24	1910	92.23
1820	9.64	1920	106.02
1830	12.87	1930	123.20
1840	17.07	1940	132.16
1850	23.19	1950	151.33
1860	31.44	1960	179.32
1870	38.56	1970	203.30
1880	50.19	1980	226.54
1890	62.98	1990	248.77

4. Choose two characteristics about your classmates that you would like to investigate. (Select some characteristics that may have policy implications relative to undergraduate study.) Collect data on these characteristics and make a data display showing the relationship between these variables.

5. Consider the following data relative to the winning times at the last ten Olympic Games for men and women's 400-meter freestyle swimming. What trends do you notice in the data?

	400-Meter freestyle times (sec)	
Year	Men's	Women's
1964	4:12.2	4:43.3
1968	4:09.0	4:31.8
1972	4:00.3	4:19.4
1976	3:51.9	4:09.9
1980	3:51.3	4:08.8
1984	3:51.2	4:07.1
1988	3:47.0	4:03.9
1992	3:45.0	4:07.2
1996	3:48.0	4:07.3
2000	3:40.6	4:05.8

PROJECT Find Out for Yourself

Examine a number of high school algebra textbooks. How many of them develop lines of best fit? How many of them relate the coefficient of the x term to the rate of change in the dependent variable for each unit of change in the value of the independent variable?

7.3 Chance and Probability

The consideration of chance in the classroom moves to a new level as students enter middle school. Notions of simple probability associated with single-stage events, such as rolling a die, are extended to compound events with two or more steps. Students ponder questions that involve finding the probability of making both shots in a one-and-one basketball game, knowing the probability associated with a single shot. Students routinely use tree diagrams, contingency tables, and geometric models to determine the chances of an event occurring. Along the way, they learn to distinguish between mutually exclusive and independent events.

Unlike their parents, today's students encounter chance and probability by doing and experiencing, not by memorizing formulas. Technology plays a greater role in high school as students begin to run simulations.

Basic Probability

As students move from the consideration of simple events in the intermediate grades, their study of probability becomes a bit more formal as they learn to list elements in a sample space and use the ratio model to compute the probability of joint events. In particular, they experience situations that extend the rolling of dice and the spinning of spinners. The middle grades see students beginning to consider such problems as the following.

A bag has five red cubes and five green cubes. What is the probability that you draw two green cubes from the bag in a row, if the first cube drawn is not replaced?

A fly lands on a dart board. What is the probability that the fly lands in the shaded region?

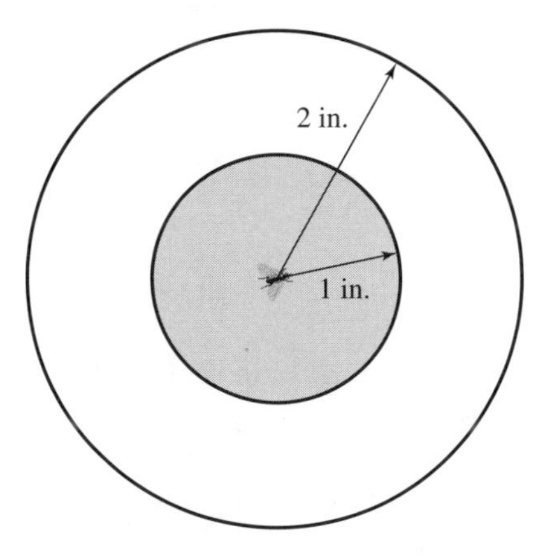

The Judi and Anne Show provides weekly movie reviews. Judi and Anne signify their choices by giving films a "thumbs up" or a "thumbs down." The table shows their votes for 220 movies over the past year. What is the probability that Anne will give a movie a "thumbs up" if we know that Judi has given the movie a "thumbs down"?

Anne

Judi	👍	👎	Total
👍	80	27	107
👎	36	77	113
Total	116	104	220

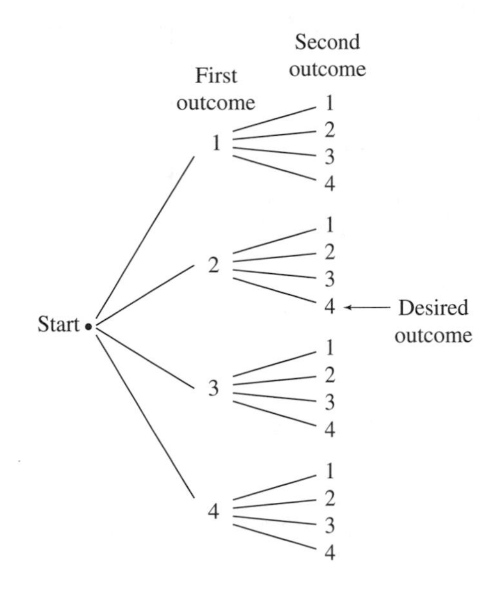

Figure 7.19 Using a tree diagram to find the sample space of an event.

The ideas of experiments, outcomes, and events are central to developing a basic knowledge of probability in the middle school years. In order to understand the role of the sample space of an event, students write exhaustive lists and construct tree diagrams. For example, to determine the probability of spinning a 2 and then a 4 on a spinner with four congruent regions numbered 1, 2, 3, and 4, students might construct the tree shown in Figure 7.19. There are 16 possible outcomes; only one gives a spin of 2 followed by a spin of 4. Thus, the probability of this event is $\frac{1}{16}$.

Students apply the ideas of probability to a variety of games to determine whether they are "fair" or "unfair." For example, a die is thrown and player 1 wins if the number rolled is less than 3. Otherwise, player 2 wins. Is this game fair? Games extend easily into simulating the probability of an event. Consider the situation in the following example.

Example ***Covering the dot***

What is probability that a disk with a 3-in. diameter lands entirely within the 7-in^2 target shown in Figure 7.20 in such a manner that it covers the dot at the target's center? ■

To solve this problem, we can consider the ratio of successes to total tries. The maximum area that can be covered by a disk falling within the boundaries of the square is $40 + (1.5)^2$ in^2, or about 47.068 in^2; however, to be a success, the disk has to cover the dot in the center of the square. Hence, the circle must land in one of the positions where its 3-in. diameter touches or covers the dot. This is the circle of radius 3 in. shown in Figure 7.21. Thus, 9π in^2 is the target success area. Since the total area that can be covered is 47.068 in^2, the probability of success is approximately 0.6007 . . . .

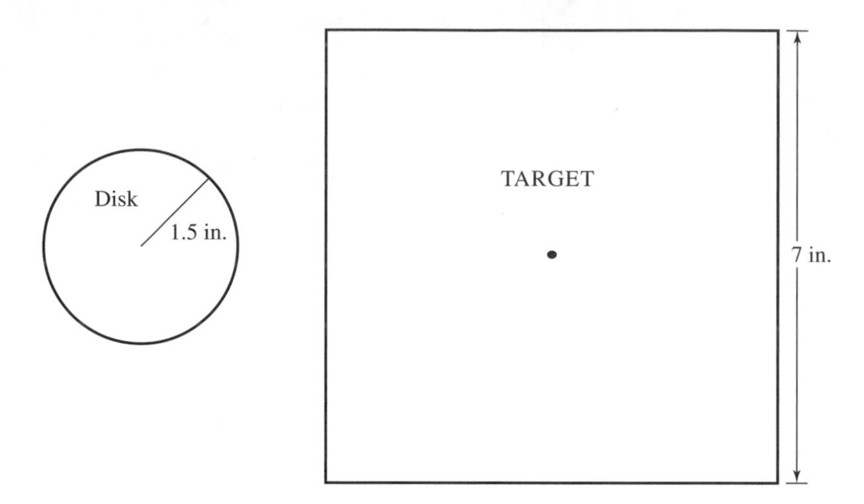

Figure 7.20 The target for the covering the dot example.

We can also find the probability of success for this event by actually tossing a disk onto such a target many, many times, keeping track of the number of tosses and the number of successes. Or, alternately, we can use our calculators to select a random ordered pair of numbers from within the square with corners (0, 0), (0, 7), (7, 7), and (7, 0). If the distance from the point (3.5, 3.5) to the ordered pair (x, y) selected is less than or equal to 3 units, then the toss is a success. When this process is completed a large number of times, say 1000, the ratio of successes to total tries will approximate the theoretical probability of 0.6007 . . . .

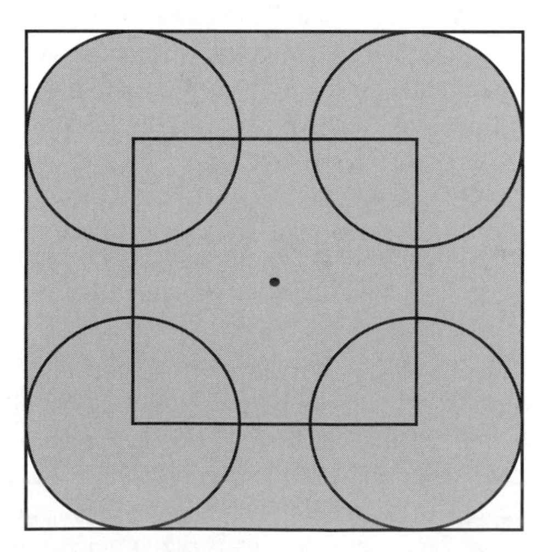

Figure 7.21 Target success area for the example.

Simulations and experimental probability derivations gives students a strong understanding of chance. When students experience only formal, theoretical probability situations, their intuitive understandings are limited and their abilities to estimate probabilities are impaired.

Independence and Conditional Probability

As students enter high school, they begin to consider independent and mutually exclusive events. Determining whether one outcome affects another event's outcome gives students opportunities to wrestle with the nature of independence. In doing so, students must understand the notions of "independence" and "mutually exclusive."

For example, consider the tossing of a pair of colored dice. Let event A be the red die showing a 1 and event B be the green die showing a 2. These two events are independent, as the outcome of one event has no effect on the outcome of the other event. Hence, these two events are independent of one another. However, mutually exclusive events are two events that cannot happen at the same time. For example, let event A be that the sum of the faces of the two dice is odd and let event B be that the two faces showing are themselves odd. Two odd faces would sum to an even number; hence, these two events cannot happen at the same time—that is, they have no outcomes in common.

From here students learn that the probability of a compound event composed of independent events is found by multiplying the probabilities of the two events. On the other hand, the probability of a compound event made up of mutually exclusive events is found by summing the probabilities of the two individual, mutually exclusive events. These are two of the first theorems of probability that students encounter.

Students also study complementary events. Two events are complementary if they partition the sample space for an experiment into two disjoint sets whose union is the entire sample space for the experiment. For example, the events of "heads" and "tails" are complementary events with respect to flipping a coin, and the sum of their probabilities is 1. Two events A and B are complementary if $P(A) = 1 - P(B)$ and the union of the sample spaces for A and B constitutes the set of all possible outcomes for the original experiment.

Exercises 7.3

1. What are the outcomes and their probabilities for the experiment consisting of rolling a pair of fair dodecahedral die and finding the sum of the two numbers showing?

2. Which sequence of coin flips is more likely, or do the two sequences have the same probability? Explain your answer carefully in words.

   Sequence A: HHHHHHTTTTHT

   Sequence B: HTHTHHTHTHHT

3. In a playoff game, a player is fouled in the act of shooting. Her shooting average for free throws is 78%. What is the probability that she will make both free throws?

4. If Jansen throws two darts at the target shown, what is the probability that the sum of the numbers hit is even?

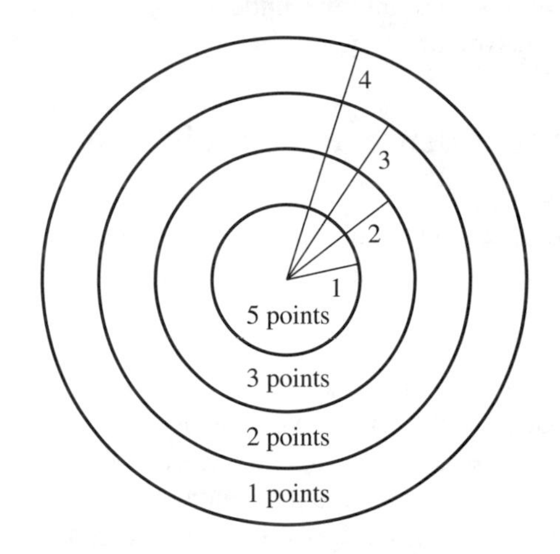

5. Break a piece of thin spaghetti into three separate pieces. What is the probability that the three pieces can be arranged to form a triangle?

6. A test for a medical condition that is found in approximately 1% of the population is believed to be about 97% accurate. If Laura reads positive on the test, what is the probability that she has the disease?

PROJECTS — Find Out for Yourself

1. Interview a number of students, and ask them the following questions:
   a. Two bags have black and white counters. Bag A has 3 black counters and 1 white one. Bag B has 6 black counters and 2 white ones. Which bag gives the better chance of drawing a black counter on a single draw?
   b. If the chance of having a boy or a girl were the same, in which type of hospital would you expect there to be more days on which at least 60% of the babies born were boys?

      - A large hospital
      - A small hospital
      - No difference

2. Read "Thinking About Uncertainty: Probability and Statistics" by Shaughnessy and Bergman in *Research Ideas for the Classroom: High School Mathematics*. Reston, VA: NCTM (Wilson, 1993). Were the responses to your questions in project 1 representative of the findings that Shaughnessy and Bergman report?

7.4 Simulating Deterministic Behavior

Introduction

In many situations, because of complexity or other factors, we cannot construct analytic (symbolic) models that adequately explain observed behaviors. In such situations, we can usually conduct experiments to investigate the relationship between the dependent variable(s) and selected values of the independent variable(s) within some range. To collect the data, we may be able to observe the behavior directly. In other instances, we may have to duplicate the behavior (possibly in a scaled-down version) under controlled conditions. Suppose we need to determine the drag force on a proposed submarine. Since it is not feasible to build a prototype, we build a scaled model to *simulate* the behavior of the actual submarine. Another example of this type of simulation is using a scaled model of a jet airplane in a wind tunnel to estimate the effects of very high speeds for various designs of the aircraft.

In other instances where the behavior cannot be explained analytically, or data collected directly, the modeler might *simulate* the behavior *indirectly* in some manner, and then test the various alternative models being considered to estimate how each represents the behavior. Data can then be collected to determine which alternative works best. We now study one method for simulating behavior, called **Monte Carlo simulation**, which is typically accomplished with the aid of computers.

Suppose we are investigating the service provided by a small harbor with ship unloading facilities. Monte Carlo simulation allows us to duplicate the arrival of ships at the harbor and the time it takes to service each ship. We can portray the distribution of arrival times and of service times in various possible daily patterns. After we simulate many trials, we want the daily distribution of arrivals and service times to mimic their real-world counterparts. Once we are satisfied that the behavior is adequately duplicated, we can then investigate alternative ways to provide additional unloading facilities, or reduce service times. Using a large number of trials as our data base, we compute appropriate statistics, such as the average total processing time, or the length of the longest queue. These statistics help us determine the best strategies for operating the harbor system. We model a harbor system and investigate alternative strategies for improving service in Section 7.7.

Monte Carlo simulations use random numbers, and we discuss random-number generation in Section 7.5. Loosely speaking, a sequence of random numbers uniformly distributed in an interval m to n is a set of numbers with no apparent pattern, where each number between m and n can appear with equal likelihood. For example, if you toss a six-sided die 100 times and record the number showing each time, you will have written down a sequence of 100 random integers approximately uniformly distributed over the interval 1 to 6. The tossing of a coin can be duplicated by generating a random number and assigning it a head if the random number is even and a tail if the random number is odd. If this trial is replicated a large number of times, you would expect heads to occur about 50% of the time. However, an element of chance is involved. A run of 100 trials could produce 51 heads and the next 10 trials could produce all heads (although this is not very likely). Thus, the 110-trial experiment is actually worse than the 100-trial experiment. Processes with an element of chance involved are called **probabilistic**, as

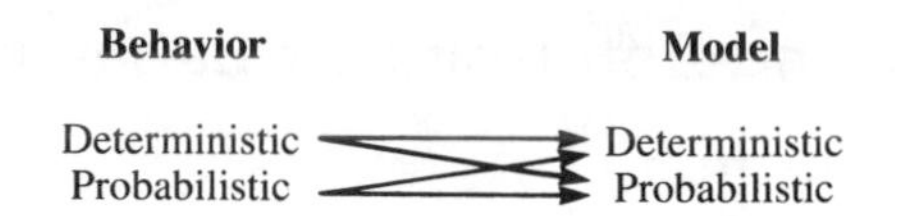

Figure 7.22 Behaviors and models can be deterministic or probabilistic.

opposed to **deterministic**, processes. Monte Carlo simulation is therefore a probabilistic process.

The behavior being modeled may be either deterministic or probabilistic. For instance, the area under a curve is deterministic (even though it may be impossible to find it precisely). On the other hand, the time between arrivals of ships at the harbor on a particular day is probabilistic behavior. In Figure 7.22, deterministic models can be used to approximate either deterministic or probabilistic behavior and, likewise, Monte Carlo simulations can be used to approximate deterministic behavior (as you will see with a Monte Carlo approximation to an area under a curve) or a probabilistic one. However, the real power of Monte Carlo simulation lies in modeling probabilistic behavior.

A principal advantage of Monte Carlo simulation is the relative ease with which it can approximate complex probabilistic systems. Additionally, Monte Carlo simulation estimates performance over a wide range of conditions, rather than a very restricted range as often required by analytic models. Furthermore, because a particular submodel can be changed rather easily in Monte Carlo simulations, sensitivity analysis becomes a viable option. Analysts also have control over the level of detail. For example, a very long time frame can be compressed or a small time frame expanded; this is a great advantage over experimental models. Finally, there are very powerful, high-level simulation languages available (such as GPSS, GASP, PROLOG, SIMAN, SLAM, and DYNAMO) that eliminate much of the tedious labor in constructing simulation models.

Area Under a Curve

Let's see how Monte Carlo simulation can be used to model a deterministic behavior, the area under a curve. We first find an approximate value to the area under a nonnegative curve. Suppose $y = f(x)$ is some given continuous function satisfying $0 \leq f(x) \leq M$ over the closed interval $a \leq x \leq b$. Here the number M is simply some constant that *bounds* the function. This situation is shown in Figure 7.23. Notice that the area we seek is wholly contained within the rectangular region of height M and length $b - a$ (the length of the interval over which f is defined).

Now we select a point $P(x, y)$ at random from within the rectangular region. To do so, you must generate two random numbers, x and y, satisfying $a \leq x \leq b$ and $0 \leq y \leq M$, and interpret them as a point P with coordinates x and y. Once $P(x, y)$ is selected, ask yourself if it lies within the region below the curve. That is, does the y-coordinate satisfy $0 \leq y \leq f(x)$? If the answer is yes, then count the point P by adding one to some counter. Two counters will be necessary; one to count the total points generated and a second to count those points that lie below the curve (see Figure 7.23). You can then calculate an approximate value for the area under the curve using the following formula:

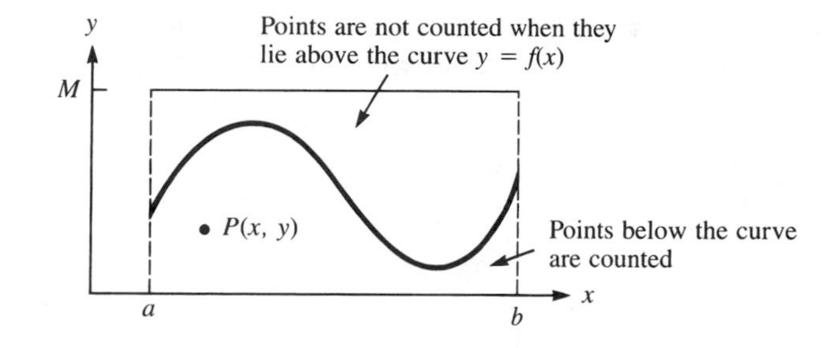

Figure 7.23 The area under the nonnegative curve $y = f(x)$ over $a \leq x \leq b$ is contained within the rectangle of height M and base length $b - a$.

$$\frac{\text{area under curve}}{\text{area of rectangle}} \approx \frac{\text{number of points counted below curve}}{\text{total number of random points generated}}$$

Monte Carlo simulations typically require a large number of trials before the deviation between the predicted and true values becomes small. A discussion of the number of trials needed to ensure a predetermined level of confidence in the final estimate requires a background in statistics. However, as a general rule, in order to double the accuracy of the result (that is, cut the expected error in half) about four times as many experiments are necessary.

The following algorithm gives the sequence of calculations needed for a computer Monte Carlo simulation for finding the area under a curve.

MONTE CARLO AREA COMPUTER ALGORITHM

Input Total number n of random points to be generated in the simulation.

Output AREA = approximate area under the specified curve $y = f(x)$ over the given interval $a \leq x \leq b$, where $0 \leq f(x) < M$.

Step 1 Initialize: COUNTER = 0.

Step 2 For $i = 1, 2, \ldots, n$, do steps 3–5.

Step 3 Calculate random coordinates x_i and y_i, satisfying $a \leq x_i \leq b$ and $0 \leq y_i < M$.

Step 4 Calculate $f(x_i)$ for the random x_i coordinate.

Step 5 If $y_i \leq f(x_i)$, then increment the COUNTER by 1. Otherwise, leave COUNTER as is.

Step 6 Calculate AREA $= M(b - a)$ COUNTER$/n$.

Step 7 OUTPUT (AREA)
STOP

Table 7.4 Monte Carlo approximation to the area under the curve $y = \cos x$ over the interval $-\pi/2 \le x \le \pi/2$.

Number of points	Approximation to area	Number of points	Approximation to area
100	2.07345	2000	1.94465
200	2.13628	3000	1.97711
300	2.01064	4000	1.99962
400	2.12058	5000	2.01429
500	2.04832	6000	2.02319
600	2.09440	8000	2.00669
700	2.02857	10000	2.00873
800	1.99491	15000	2.00978
900	1.99666	20000	2.01093
1000	1.96664	30000	2.01186

Table 7.4 gives the results of several different simulations to obtain the area beneath the curve $y = \cos x$ over the interval $-\pi/2 \le x \le \pi/2$, where $0 \le \cos x < 2$.

The actual area under the curve $y = \cos x$ over the given interval is 2 square units. Note that even with the relatively large number of points generated, the error is significant. (Try an upper bound of $0 \le \cos x \le 1$.) For functions of one variable, the Monte Carlo technique is generally not competitive with quadrature techniques that you learn in numerical analysis. The lack of an error bound and the difficulty in finding an upper bound M are disadvantages as well. Nevertheless, the Monte Carlo technique can be extended to functions of several variables and becomes more practical in that situation.

Exercises 7.4

1. Each ticket in a lottery contains a single hidden number according to the following scheme: 55% of the tickets contain a 1, 35% contain a 2, and 10% contain a 3. A participant in the lottery wins a prize by obtaining all three numbers—1, 2, and 3. Describe an experiment that could be used to determine how many tickets you would expect to buy in order to win a prize.

2. Record companies A and B both produce classical music recordings. Label A is a budget label and 5% of A's new CDs exhibit a significant degree of warpage. Label B is manufactured under tighter quality control and is consequently more expensive than A. Only 2% of B's CDs are warped. You purchase one label A and one label B recording at your local store on a regular basis. Describe an experiment that could be used to determine how many times you would expect to make such a purchase before buying two warped CDs in the same purchase.

3. Using Monte Carlo simulation, write an algorithm to approximate π by considering the number of random points selected inside the quarter circle:

$$Q : x^2 + y^2 = 1, \quad x \geq 0, \quad y \geq 0,$$

where the quarter circle is taken to be inside the square:

$$S : 0 \leq x \leq 1 \text{ and } 0 \leq y \leq 1$$

Use the equation that $\pi/4 = \text{area } Q/\text{area } S$.

4. Using Monte Carlo simulation, write an algorithm to calculate the area of the ellipse:

$$\frac{x^2}{2} + \frac{y^2}{4} \leq 16$$

that lies in the first quadrant, $x > 0$, $y > 0$.

7.5 Generating Random Numbers

In the preceding section, we developed algorithms for Monte Carlo simulations to find areas and volumes. A key ingredient common to each algorithm is the need for random numbers. Random numbers have a variety of applications. These include gambling problems, problems similar to finding the area and volume previously presented, and in modeling larger complex systems such as air traffic control situations.

In some sense, computers do not generate random numbers, because they employ deterministic algorithms. Computers really generate sequences of pseudo-random numbers that, for all practical purposes, may be considered random. There is no single best random number generator or best test to ensure randomness.

There are complete courses of study for random numbers and simulations that give in-depth coverage of the methods and tests for pseudo-random number generators. Our purpose here is to introduce a few random number methods that can be used to generate sequences of numbers that are nearly random.

Many programming languages, such as PASCAL, BASIC, and other software (MINITAB, Quattro Pro, and EXCEL, to name a few) have built-in random number generators.

Middle-Square Method

The middle-square method was developed in 1946 by John Von Neuman, Stan Ulam, and Nicholas Metropolis at Los Alamos Laboratories to simulate neutron collisions as part of the Manhattan Project. Their middle-square method works as follows:

1. Start with a 4-digit number x_0, called the *seed*.
2. Square it to obtain an 8-digit number (add leading zeros if necessary).
3. Take the middle 4 digits as the next random number.

Continuing in this manner, we obtain a sequence that appears to be random over the integers from 0 to 9999. These integers can then be scaled to any interval a to b.

For example, if we want numbers from 0 to 1, we divide the 4-digit numbers by 10,000. Let's illustrate the middle-square method.

Pick a seed, say $x_0 = 2041$, and square it (adding leading zeros) to get 04165681. The middle 4 digits give the next random number, 1656. Generating nine random numbers in this way yields

n	0	1	2	3	4	5	6	7	8	9
x_n	2041	1656	7423	1009	0180	0324	1049	1004	80	64

We can use more than 4 digits if we wish, but we must always take the middle number of digits equal to the number of digits in the seed. For example, if $x_0 = 653217$ (6 digits), its square 426,692,449,089 has 12 digits. Thus, take the middle 6 digits as the random number, namely, 692449.

The middle-square method is reasonable, but it has a major drawback in its tendency to degenerate to zero (where it will stay forever). With the seed 2041, the random sequence does seem to be approaching zero. How many numbers can be generated until we are (almost) at zero?

Linear Congruence

The linear congruence method was introduced by D. H. Lehmer in 1951, and a great majority of the pseudo-random numbers used today are based on this method. Its main advantage over other methods is that seeds can be selected that generate patterns that eventually cycle (discussed below). However, the length of the cycle is so large that the pattern doesn't repeat itself on large computers for most applications. The method requires the choice of three integers: a, b, and c. Given some initial seed, say x_0, we generate a sequence by the rule

$$x_{n+1} = (a \times x_n + b) \bmod(c)$$

where c is the modulus, a the multiplier, and b the increment. The qualifier mod(c) in the equation means to obtain the remainder after dividing the quantity $(a \times x_n + b)$ by c. For example, with $a = 1$, $b = 7$, and $c = 10$,

$$x_{n+1} = (1 \times x_n + 7) \bmod(10)$$

means x_{n+1} is the integer remainder upon dividing $x_n + 7$ by 10. Thus, if $x_n = 115$, $x_{n+1} = \textit{remainder}\left(\frac{122}{10}\right) = 2$.

Before investigating the linear congruence methodology, we need to discuss **cycling**. This is a major problem with random number generation. Cycling means the sequence repeats itself, and although undesirable, it is also unavoidable. At some point, all pseudo-random number generators begin to cycle. Cycling is easier to understand with an example.

If we set our seed at $x_0 = 7$, we find $x_1 = (1 \times 7 + 7) \bmod(10)$ or $14 \bmod(10)$, which is 4. Repeating this same procedure, we obtain the following sequence:

$$7, 4, 1, 8, 5, 2, 9, 6, 3, 0, 7, 4, \ldots$$

and the original sequence repeats again and again. Note that cycling occurs after 10 numbers. The methodology produces a sequence of integers between 0 and $c - 1$ inclusively before cycling (which includes the most possible remainders after dividing the integers by c). Cycling is guaranteed with at most c numbers in the random number sequence. Nevertheless, c can be chosen to be very large, and a and b chosen in such a way to obtain a full set of c numbers before cycling begins to occur. Many computers use $c = 2^{31}$ for the large value of c. Again, we can scale the random numbers to obtain a sequence between any limits a and b, as required.

A second problem that can occur with the linear congruence method is a lack of statistical independence among the members in the list of random numbers. Any correlations between nearest neighbors, next-nearest neighbors, third-nearest neighbors, and so forth are generally unacceptable. (Because we live in a three-dimensional world, third-nearest neighbor correlations can be particularly damaging in physical applications.) Pseudo-random number sequences can never be completely statistically independent because they are generated by a mathematical formula or algorithm. Nevertheless, the sequence will appear (for practical purposes) independent when it is subjected to certain statistical tests. These concerns are best addressed in a course in statistics.

Exercises 7.5

1. Use the middle-square method to generate
a. 10 random numbers using $x_0 = 1009$
b. 20 random numbcrs using $x_0 = 653217$
c. 15 random numbers using $x_0 = 3043$
Comment about the results of each sequence. Was there cycling? Did each sequence degenerate rapidly?

2. Use the linear congruence method to generate
a. 10 random numbers using $a = 5$, $b = 1$, and $c = 8$
b. 15 random numbers using $a = 1$, $b = 7$, and $c = 10$
c. 20 random numbers using $a = 5$, $b = 3$, and $c = 16$
Comment about the results of each sequence. Was there cycling? If so, when did it occur?

PROJECTS Exploring Modeling

1. Complete the requirement for UMAP module 269, "Monte Carlo: The Use of Random Digits to Simulate Experiments," by Dale T. Hoffman. Hoffman presents, explains, and uses the Monte Carlo technique to find approximate solutions to several realistic problems. Simple experiments are included for student practice.

2. "Random Numbers" by Mark D. Myerson, UMAP 590. This module, in more depth than our presentation, discusses methods for generating random numbers, and presents tests for determining the randomness of a string of numbers. Complete this module and prepare a short report on testing for randomness.

3. Write a computer program to generate uniformly distributed random integers in the interval $m < x < n$, where m and n are integers, according to the following algorithm:

   Step 1 Let $d = 2^{31}$ and choose N (the number of random numbers to generate).

   Step 2 Choose any seed integer Y such that

$$999999 > Y > 100000$$

   Step 3 Let $i = 1$.

   Step 4 Let $Y = (15625Y + 22221) \operatorname{mod}(d)$.

   Step 5 Let $X_i = m + \operatorname{floor}((n - m + 1)Y/d)$.

   Step 6 Increment i by $1 : i = i + 1$.

   Step 7 Go to step 4 unless $i = N + 1$.

   Here $\operatorname{floor}(p)$ means the largest integer not exceeding p.

   For most choices of Y, the numbers $X_1, X_2, \ldots$ form a sequence of (pseudo) random integers as desired. One possible recommended choice is $Y = 568731$.

   To generate random numbers (not simply integers) in an interval a to b with $a < b$, use the preceding algorithm, replacing the formula in step 5 by letting

$$X_i = a + \frac{Y(b - a)}{d - 1}$$

   (Computed with the accuracy of your computer.)

4. Write a program to generate 1000 integers between 1 and 5 in a random fashion so that 1 occurs 22% of the time, 2 occurs 15% of the time, 3 occurs 31% of the time, 4 occurs 26% of the time, and 5 occurs 6% of the time. Over what interval would you generate the random numbers? How do you decide which integer from 1 to 5 has been generated according to its specified chance of selection?

5. Write a program or use a spreadsheet to find the approximate areas in Exercises 3 and 4 in Section 7.4.

7.6 Simulating Probabilistic Behavior

Key to good Monte Carlo simulation practices is an understanding of the axioms of probability. The term *probability* refers to the study of both randomness and uncertainty as well as the quantifying of the likelihoods associated with various outcomes. Probability can be seen as a long-term average. For example, if the probability of an event occurring is 1 out of 5, then in the long run, the chance of the event happening is $\frac{1}{5}$.

Over the long haul, the probability of an event can be thought of as the ratio of

$$\frac{\text{number of favorable events}}{\text{total number of events}}$$

In this section we model simple probabilistic behavior to build intuition and understanding. In Section 7.7, we develop submodels of probabilistic processes to incorporate in simulations.

We will examine three simple probabilistic models:

1. Flip of a fair coin
2. Roll of a fair die or pair of dice
3. Roll of an unfair die or pair of unfair dice

Example 1 *A Fair Coin*

Most people realize that the chance of obtaining a head or a tail on a coin is $\frac{1}{2}$. What happens if we actually start flipping a coin? Will one out of every two flips be a head? Probably not. Again, probability is a long-term average. Thus, in the long run, the ratio of heads to the number of flips approaches 0.5. Let's define $f(x)$ as follows, where x is a random number between [0, 1]:

$$f(x) = \begin{cases} \text{Head}, & 0 \le x \le 0.5 \\ \text{Tail}, & 0.5 < x \le 1 \end{cases}$$

Note that $f(x)$ assigns the outcome head or tail to a number between [0, 1]. We want to take advantage of the cumulative nature of this function as we make random assignments to numbers between [0, 1]. In the long run, we expect to find the following occurrences:

Random number interval	Occurrences	
	Proportion	Cumulative
$x < 0$	0.00	0.00
$0 \le x \le 0.5$	0.50	0.50
$0.5 < x \le 1.0$	0.50	1.00
$1 < x$	0.00	1.00

■

Let's illustrate this idea using the following algorithm.

MONTE CARLO FAIR COIN COMPUTER ALGORITHM

Input Total number n of random flips of a fair coin to be generated in the simulation.

(continued)

Output Probability of getting a head when we flip a fair coin.

Step 1 Initialize: COUNTER = 0.

Step 2 For $i = 1, 2, \ldots, n$, do steps 3 and 4.

Step 3 Obtain a random number x_i between 0 and 1.

Step 4 If $0 \leq x_i \leq 0.5$, then COUNTER = COUNTER + 1. Otherwise, leave COUNTER as is.

Step 5 Calculate $P(\text{head}) = \text{COUNTER}/n$.

Step 6 OUTPUT Probability of heads, $P(\text{head})$.
STOP

Table 7.5 lists our results for various choices n of the number of random x_i generated. Note that as n gets large, the probability of heads occurring is 0.5, or half the time.

Example 2 ***Roll of a Fair Die***

Rolling a fair die adds a new twist to the process. In the flip of a coin, only one event is assigned. Now we must devise a method to assign six events because a die has six faces, $\{1, 2, 3, 4, 5, 6\}$. The probability of each event occurring is $\frac{1}{6}$; that is, each number is equally likely to occur. As before, this probability of a particular number occurring is defined to be

$$\frac{\text{number of occurrences of a particular face from } \{1, 2, 3, 4, 5, 6\}}{\text{total number of trials}}$$

We can use the following algorithm to generate our experiment for a roll of a fair die.

Table 7.5 Results from flipping a fair coin.

Number of flips	Number of heads	Proportion of heads
100	49	0.49
200	102	0.51
500	252	0.504
1000	492	0.492
5000	2469	0.4930
10,000	4993	0.4993

MONTE CARLO ROLL OF A FAIR DIE ALGORITHM

Input Total number n of random rolls of a die in the simulation.

Output The percentage or probability for faces $\{1, 2, 3, 4, 5, 6\}$.

Step 1 Initialize COUNTER 1 through COUNTER 6 to zero.

Step 2 For $i = 1, 2, \ldots, n$, do steps 3 and 4.

Step 3 Obtain a random number satisfying $0 \leq x_i \leq 1$.

Step 4 If x_i belongs to these intervals, then increment the appropriate COUNTER.

$$\begin{array}{ll}
0 \leq x_i \leq \frac{1}{6} & \text{COUNTER } 1 = \text{COUNTER } 1 + 1 \\
\frac{1}{6} < x_i \leq \frac{2}{6} & \text{COUNTER } 2 = \text{COUNTER } 2 + 1 \\
\frac{2}{6} < x_i \leq \frac{3}{6} & \text{COUNTER } 3 = \text{COUNTER } 3 + 1 \\
\frac{3}{6} < x_i \leq \frac{4}{6} & \text{COUNTER } 4 = \text{COUNTER } 4 + 1 \\
\frac{4}{6} < x_i \leq \frac{5}{6} & \text{COUNTER } 5 = \text{COUNTER } 5 + 1 \\
\frac{5}{6} < x_i \leq 1 & \text{COUNTER } 6 = \text{COUNTER } 6 + 1
\end{array}$$

Step 5 Calculate probability of each face $j = \{1, 2, 3, 4, 5, 6\}$ by (COUNTER $j)/n$.

Step 6 OUTPUT probabilities.
STOP

We see from Table 7.6 that with 100,000 rolls (trials) we are close (for these trials) to the expected results.

Table 7.6 Results from a roll of a fair die.

	Number of rolls					
Face	10	100	1000	10,000	100,000	Expected results
1	0.300	0.190	0.152	0.1703	0.1652	0.1667
2	0.00	0.150	0.152	0.1652	0.1657	0.1667
3	0.100	0.190	0.157	0.1639	0.1685	0.1667
4	0.00	0.160	0.180	0.1653	0.1685	0.1667
5	0.400	0.150	0.174	0.1738	0.1676	0.1667
6	0.200	0.160	0.185	0.1615	0.1652	0.1667

■

Example 3 ***Roll of an Unfair Die***

Let's consider a probability model where each event is not equally likely. Assume the die is biased according to the following empirical distribution:

Face	P(roll)
1	0.1
2	0.1
3	0.2
4	0.3
5	0.2
6	0.1

The cumulative occurrences for the function to be used in our computer algorithm is

Value of x_i	Assignment
$[0, 0.1]$	ONE
$(0.1, 0.2]$	TWO
$(0.2, 0.4]$	THREE
$(0.4, 0.7]$	FOUR
$(0.7, 0.9]$	FIVE
$(0.9, 1]$	SIX

We model the roll of an unfair die using the following algorithm.

MONTE CARLO ROLL OF AN UNFAIR DIE ALGORITHM

Input Total number n of random rolls of a die in the simulation.

Output The percentage or probability for faces $\{1, 2, 3, 4, 5, 6\}$.

Step 1 Initialize COUNTER 1 through COUNTER 6 to zero.

Step 2 For $i = 1, 2, \ldots, n$, do steps 3 and 4.

Step 3 Obtain a random number satisfying $0 \leq x_i \leq 1$.

Step 4 If x_i belongs to these intervals, then increment the appropriate COUNTER.

$$
\begin{array}{ll}
0 \leq x_i \leq 0.1 & \text{COUNTER } 1 = \text{COUNTER } 1 + 1 \\
0.1 < x_i \leq 0.2 & \text{COUNTER } 2 = \text{COUNTER } 2 + 1 \\
0.2 < x_i \leq 0.4 & \text{COUNTER } 3 = \text{COUNTER } 3 + 1 \\
0.4 < x_i \leq 0.7 & \text{COUNTER } 4 = \text{COUNTER } 4 + 1 \\
0.7 < x_i \leq 0.9 & \text{COUNTER } 5 = \text{COUNTER } 5 + 1 \\
0.9 < x_i \leq 1 & \text{COUNTER } 6 = \text{COUNTER } 6 + 1
\end{array}
$$

Step 5 Calculate probability of each roll $y = \{1, 2, 3, 4, 5, 6\}$ by (COUNTER $j)/n$.

Step 6 OUTPUT probabilities.
STOP

The results are shown in Table 7.7. Note that a large number of rolls (trials) is required in order for the model to approach the expected results.

Table 7.7 Results from a roll of an unfair die.

	Number of rolls					
Face	100	1000	5000	10,000	40,000	Expected results
1	0.130	0.080	0.094	0.0948	0.0948	.1
2	0.08	0.099	0.099	0.0992	0.0992	.1
3	0.210	0.199	0.192	0.1962	0.1962	.2
4	0.35	0.320	0.308	0.3082	0.3081	.3
5	0.280	0.184	0.201	0.2012	0.2011	.2
6	0.150	0.120	0.104	0.1044	0.1045	.1

■

In the next section we use these ideas to simulate a real-world probabilistic situation.

Exercises 7.6

1. You arrive at the beach for a vacation and are dismayed to learn that the local weather station is predicting a 50% chance of rain every day. Using Monte Carlo simulation, predict the chance that it rains three consecutive days during your vacation.

2. Use Monte Carlo simulation to approximate the probability of three heads occurring when five fair coins are flipped.

3. Use Monte Carlo simulation to simulate the sum of 100 consecutive rolls of a fair die.

4. "Loaded" dice are known to follow the distribution shown on page 294. For example, die 1 will show a 4 30% of the time. Use Monte Carlo simulation to simulate the sum of 300 rolls of two unfair dice.

Face	Die 1	Die 2
1	0.1	0.3
2	0.1	0.1
3	0.2	0.2
4	0.3	0.1
5	0.2	0.05
6	0.1	0.25

5. Make up a game that uses a flip of a fair coin, and then use Monte Carlo simulation to predict the results of the game.

PROJECTS Exploring Modeling

1. *Blackjack*—Construct and perform a Monte Carlo simulation of Blackjack (also called Twenty-One). The rules of Blackjack are as follows:

Most casinos use six or eight decks of cards when playing this game to inhibit card counters. You will use two decks of cards in your simulation (104 cards total). There are only two players, you and the dealer. Each player receives two cards to begin play. The cards are worth their face value for 2–10, 10 points for face cards (Jack, Queen, and King), and either 1 or 11 points for aces. The object of the game is to obtain a total as close to 21 as possible without going over (called busted) 21.

If the first two cards total 21 (ace/10 or ace/face card), this is called blackjack and is an automatic winner (unless both you and the dealer have blackjack, in which case it is a tie, or push, and your bet remains on the table). Winning via blackjack pays you 3 to 2, or 1.5 to 1 (a $1 bet reaps $1.50 and you don't lose the $1 you bet).

If neither you nor the dealer has blackjack, you (the player) can take as many cards as you want, one at a time, to try to get as close to 21 as possible. If you go over 21, you lose and the game ends. When you are satisfied with your score, you stand. The dealer then draws cards according to the following rules:

The dealer stands on 17, 18, 19, 20, or 21. The dealer must draw a card if the total is 16 or less. The dealer always counts aces as 11 unless it causes him or her to bust, in which case it is counted as a 1. For example, an ace/6 combination for the dealer is 17, not 7 (the dealer has no option) and the dealer must stand on 17. However, if the dealer has an ace/4 (for 15) and draws a King, then the new total is 15, because the ace reverts to its value of 1 (so as not to go over 21). The dealer would then draw another card.

If the dealer goes over 21, you win (including your bet money; you gain $1 for every $1 you bet). If the dealer's total exceeds your total, you lose all the money you bet. If the dealer's total equals your total, it is a push (no money exchanges hands; you don't lose your bet, but neither do you gain any money).

What makes the game exciting in a casino, is that the dealer's original two cards are one up, one down, so you don't know the dealer's total and must play the odds based on the one card showing. You don't need to incorporate this twist into your simulation for this project. Here's what you are required to do:

Run through 12 sets of 2 decks playing the game. You have an unlimited bankroll (don't you wish!) and bet $2 on each hand. Each time the 2 decks run out, the hand in play continues with 2 fresh decks (104 cards). At that point, record your standing (plus or minus x dollars). Then start again at 0 for the next deck. So your output will be the 12 results from playing each of the 12 sets of decks, which you can then average or total to determine your overall performance.

What about your strategy? That's up to you! But here's the catch—you will assume that you can see neither of the dealer's cards (so you have no idea what cards the dealer has). Choose a strategy to play, and then play it throughout the entire simulation. (Blackjack enthusiasts can consider implementing doubling down and splitting pairs into their simulation, but this is not necessary.)

Provide your instructor with the simulation algorithm, computer code, and output results from each of the 12 sets of decks.

2. *Darts*—Construct and perform a Monte Carlo simulation of a darts game. Points are awarded as follows:

Dart board area	Points
Bulls-eye	50
Yellow ring	25
Blue ring	15
Red ring	10
White ring	5

From the origin (the center of the bulls-eye), the radius of each ring is as follows:

Ring	Thickness (in.)	Distance to outer ring edge from the origin (in.)
Bulls-eye	1.0	1.0
Yellow	1.5	2.5
Blue	2.5	5.0
Red	3.0	8.0
White	4.0	12.0

The board has a radius of 1 ft (12 in.).

Make an assumption about the distribution of how the darts hit on the board. Write an algorithm and code it in the computer language of your choice. Run 1000 simulations to determine the mean score for throwing five darts. Also determine which ring has the highest expected value (point value times the probability of hitting that ring).

3. *Craps*—Construct and perform a Monte Carlo simulation of the popular casino game of craps. The rules are as follows:

There are two basic bets in craps, pass and don't pass. In the *pass* bet, you wager that the shooter (the person throwing the dice) will win; in the *don't pass* bet, you wager that the shooter will lose. We will play by the rule that on an initial roll of 12 (boxcars), both pass and don't pass bets are losers. Both are even-money bets.

Conduct of the game:

- Roll a 7 or 11 on the first roll: Shooter wins (pass bets win and don't pass bets lose).
- Roll a 12 on the first roll: Shooter loses (boxcars, pass and don't pass bets lose).
- Roll a 2 or 3 on the first roll: Shooter loses (pass bets lose, don't pass bets win).
- Roll 4, 5, 6, 8, 9, 10 on the first roll: This becomes the point. The object then becomes to roll the point again before rolling a 7. The shooter continues to roll the dice until the point or a 7 appears. Pass bettors win if the shooter rolls the point again before rolling a 7. Don't pass bettors win if the shooter rolls a 7 before rolling the point again.

Write an algorithm and code it in the computer language of your choice. Run the simulation to estimate the probability of winning a pass bet and the probability of winning a don't pass bet. Which is the better bet? As the number of trials increases, to what do the probabilities converge?

4. *Horse Race*—Construct and perform a Monte Carlo simulation of a horse race. Simulate the Mathematical Derby with the following entries and estimated winning probabilities:

Mathematical Derby	
Entry's name	Probability of winning
Euler's Folly	0.13
Leapin' Leibniz	0.18
Newton Lobell	0.10
Count Cauchy	0.07
Pumped up Poisson	0.22
Loping L'Hopital	0.02
Steamin' Stokes	0.06
Dancin' Dantzig	0.22

Construct and perform a Monte Carlo simulation of 1000 horse races. Which horse won the most races? Which horse won the least races? Do these results surprise you? Provide the tallies of how many races each horse won with your output.

5. *Roulette*—In American roulette, there are 38 spaces on the wheel, 0, 00, and 1–36. Half the spaces numbered 1–36 are red, and half are black. The two spaces 0 and 00 are green.

Simulate the playing of 1000 games betting either red or black (which pay even money, 1:1). Bet $1 on each game and keep track of your earnings. What are the earnings per game betting red/black according to your simulation? What was your longest winning streak? Longest losing streak?

Simulate 1000 games betting green (pays 17:1—if you win, you add $17 to your kitty; if you lose, you lose $1). What are your earnings per game betting green according to your simulation? How does it differ from your earnings betting red/black? What was your longest winning streak betting green? Longest losing streak? Which strategy do you recommend using, and why?

7.7 Queuing: A Harbor System

In the preceding section, we began modeling probabilistic *behaviors* using Monte Carlo simulation. In this section you will learn a method to approximate more probabilistic *processes*. Additionally, a check is made to determine how well the simulation duplicates the process being studied. We consider a queuing problem.

Unloading Ships

Consider a small harbor with unloading facilities for ships. Only one ship can be unloaded at any one time. Ships queue up to unload cargo, and the time between ship arrivals varies from 15 to 145 minutes. Ship unloading times depend on the type and amount of cargo and vary from 45 to 90 minutes. We seek answers to the following questions:

1. What is the average and maximum time per ship in the harbor?
2. If a ship's *waiting time* is the time between its arrival and the start of unloading, what is the average and maximum waiting time per ship?
3. What percent of the time are the unloading facilities idle?
4. What is the length of the longest line (queue)?

To get some reasonable answers, we can simulate harbor activity using a computer or programmable calculator. Let's assume that the arrival times between successive ships and unloading time per ship are uniformly distributed over their respective time intervals. That is, the arrival time between ships can be any integer between 15 and 145, and any integer within that interval can appear with equal likelihood. Before giving a general computer algorithm to simulate the harbor system, let's consider a 5-ship situation.

We have the following data for each ship.

	Time in minutes				
	Ship 1	Ship 2	Ship 3	Ship 4	Ship 5
Time between arrivals	20	30	15	120	25
Unload time	55	45	60	75	80

Because Ship 1 arrives 20 min after the clock commences at $t = 0$ min, the harbor facilities are idle for 20 min at the start. Ship 1 immediately begins to unload, which takes 55 min. Meanwhile, Ship 2 arrives on the scene at $t = 20 + 30 = 50$ min after time 0. Ship 2 can't start to unload until ship 1 finishes, at $t = 20 + 55 = 75$ min. This means that Ship 2 must wait $75 - 50 = 25$ min before unloading begins. The situation can be depicted in timeline 1:

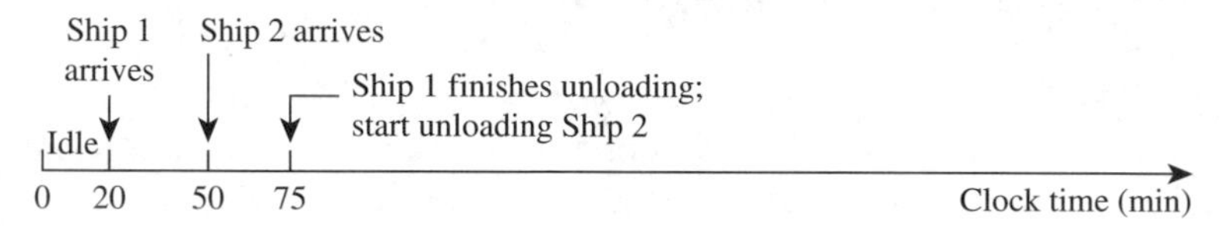

Now before Ship 2 starts to unload, Ship 3 arrives at time $t = 50 + 15 = 65$ min. Because Ship 2 starts to unload at $t = 75$ min and takes 45 min, Ship 3 cannot start unloading until $t = 75 + 45 = 120$ min, when Ship 2 is finished. Thus, Ship 3 must wait $120 - 65 = 55$ min. This situation is depicted in timeline 2:

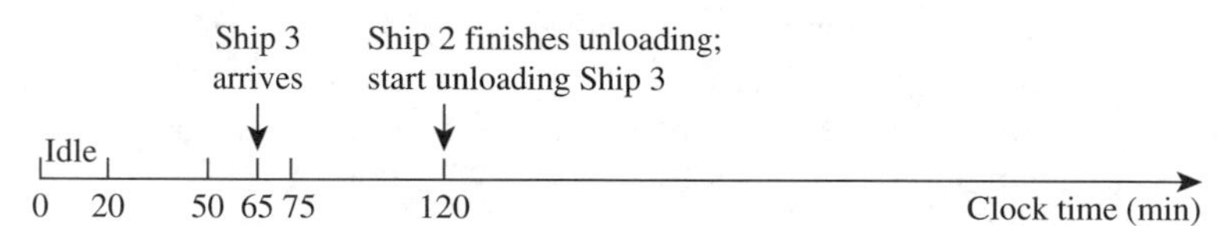

Ship 4 doesn't arrive in the harbor until $t = 65 + 120 = 185$ min. Ship 3 finishes unloading at $t = 120 + 60 = 180$ min, and the harbor facilities are idle for $185 - 180 = 5$ min. The unloading of Ship 4 commences immediately upon its arrival, as depicted in timeline 3:

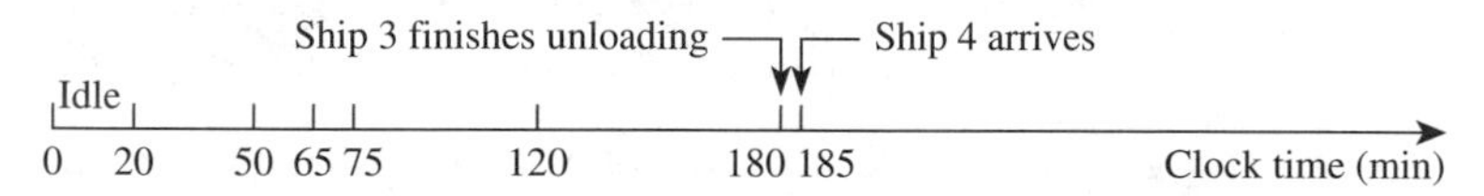

Finally, Ship 5 arrives at $t = 185 + 25 = 210$ min, before Ship 4 finishes unloading at $t = 185 + 75 = 260$ min. Thus, Ship 5 must wait $260 - 210 = 50$ min before it starts to unload. The simulation is complete when Ship 5 finishes at $t = 260 + 80 = 340$ min. The final situation is shown in timeline 4:

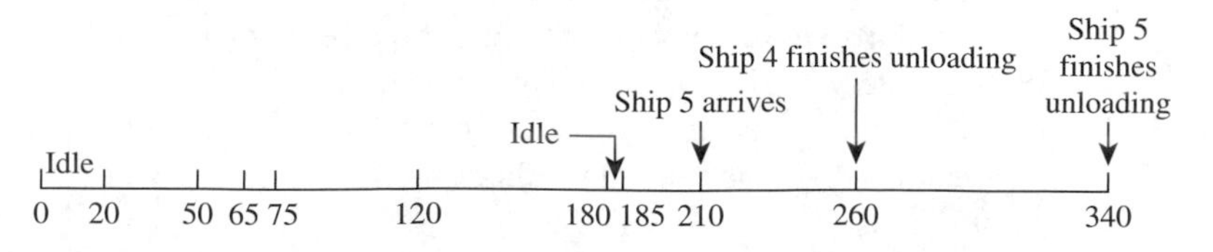

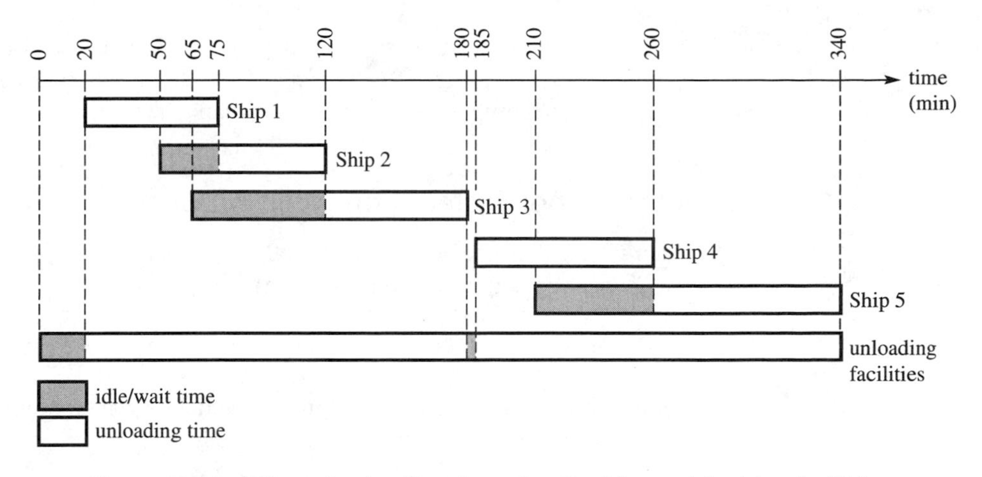

Figure 7.24 Idle and unloading times for the ships and docking facilities.

In Figure 7.24, we summarize the waiting and unloading times for the five ships. Table 7.8 summarizes the results of the entire simulation of the five hypothetical ships. Note the total waiting time spent by all five ships before unloading is 130 min. This waiting time represents a cost to the shipowners and is a source of customer dissatisfaction with the docking facilities. On the other hand, the docking facility has only 25 min of total idle time. It is in use 315 out of the total 340 min in the simulation, or approximately 93% of the time.

Suppose the owners of the docking facilities are concerned with the quality of service they are providing and want to evaluate various alternatives in order to see if improving service justifies the added cost. Several statistics can help them evaluate service quality. For example, the maximum time a ship spends in the harbor is 130 min (ship 5), whereas the average is 89 min. Generally customers are very sensitive to the time spent waiting. In this example, the maximum wait time is 55 min, while the aver-

Table 7.8 Summary of the harbor system simulation.

Ship	Random time between ship arrivals	Arrival time	Start service	Queue length at arrival	Wait time	Random unload time	Time in harbor	Dock idle time
1	20	20	20	0	0	55	55	20
2	30	50	75	1	25	45	70	0
3	15	65	120	2	55	60	115	0
4	120	185	185	0	0	75	75	5
5	25	210	260	1	50	80	130	0
Totals (if appropriate):					130			25
Averages (if appropriate):					26	63	89	

Note: All times are given in minutes after the start of the clock at time $t = 0$.

age wait time is 26 min. Some customers may take their business elsewhere if queues are too long. In this case, Ship 2 has the longest wait time. The following Monte Carlo simulation algorithm computes statistics for alternative systems.

Terms for Harbor System Computer Algorithm

***between**$_i$* Time between successive arrivals of ships i and $i - 1$ (a random integer varying between 15 and 145 min)

***arrive**$_i$* Time from start of clock at $t = 0$ when ship i arrives at the harbor

***unload**$_i$* Time required to unload ship i at the dock (a random integer varying between 45 and 90 min)

***start**$_i$* Time from start of clock at which ship i commences unloading

***idle**$_i$* Time dock facilities are idle immediately *before* commencement of unloading ship i

***wait**$_i$* Time ship i is idle before unloading commences

***finish**$_i$* Time from start of clock to completion of unloading for ship i

***harbor**$_i$* Total time ship i spends in the harbor

HARTIME Average time per ship in the harbor

MAXHAR Maximum time of ship in the harbor

WAITIME Average wait time per ship before unloading

MAXWAIT Maximum ship wait time

IDLETIME Percent of total simulation time unloading facilities are idle

HARBOR SYSTEM SIMULATION ALGORITHM

Input Total number n of ships for the simulation.

Output HARTIME, MAXHAR, WAITIME, MAXWAIT, and IDLETIME.

Step 1 Randomly generate between_1 and unload_1. Then set $\text{arrive}_1 = \text{between}_1$.

Step 2 Initialize all output values:

$$\text{HARTIME} = \text{unload}_1, \qquad \text{MAXHAR} = \text{unload}_1,$$

$$\text{WAITIME} = 0, \qquad \text{MAXWAIT} = 0, \qquad \text{IDLETIME} = \text{arrive}_1.$$

Step 3 Calculate finish time for unloading of ship_1:

$$\text{finish}_1 = \text{arrive}_1 + \text{unload}_1$$

Step 4 For $i = 2, 3, \ldots, n$, do steps 5–16.

Step 5 Generate the random pair of integers between_i and unload_i over their respective time intervals.

Step 6 Assuming the time clock begins at $t = 0$ minutes, calculate the time of arrival for ship_i:

$$\text{arrive}_i = \text{arrive}_{i-1} + \text{between}_i$$

Step 7 Calculate the time difference between the arrival of ship_i and the finish time for unloading the previous ship_{i-1}:

$$\text{timediff} = \text{arrive}_i - \text{finish}_{i-1}$$

Step 8 For nonnegative timediff, the unloading facilities are idle:

$$\text{idle}_i = \text{timediff} \quad \text{and} \quad \text{wait}_i = 0$$

For negative timediff, ship_i must wait before it can unload:

$$\text{wait}_i = -\text{timediff} \quad \text{and} \quad \text{idle}_i = 0$$

Step 9 Calculate the start time for unloading ship_i:

$$\text{start}_i = \text{arrive}_i + \text{wait}_i$$

Step 10 Calculate the finish time for unloading ship_i:

$$\text{finish}_i = \text{start}_i + \text{unload}_i$$

Step 11 Calculate the time in harbor for ship_i:

$$\text{harbor}_i = \text{wait}_i + \text{unload}_i$$

Step 12 Sum harbor_i into total harbor time HARTIME for averaging.

Step 13 If $\text{harbor}_i > \text{MAXHAR}$, then set $\text{MAXHAR} = \text{harbor}_i$. Otherwise leave MAXHAR as is.

Step 14 Sum wait_i into total waiting time WAITIME for averaging.

Step 15 Sum idle_i into total idle time IDLETIME.

(continued)

Step 16 If $\text{wait}_i > \text{MAXWAIT}$, then set $\text{MAXWAIT} = \text{wait}_i$. Otherwise leave MAXWAIT as is.

Step 17 Set $\text{HARTIME} = \text{HARTIME}/n$, $\text{WAITIME} = \text{WAITIME}/n$, and $\text{IDLETIME} = \text{IDLETIME}/\text{finish}_n$.

Step 18 OUTPUT (HARTIME, MAXHAR, WAITIME, MAXWAIT, IDLETIME)
STOP

Table 7.9 gives the results, according to the harbor system algorithm, of six independent simulation runs of 100 ships each.

Now suppose you are a consultant for the owners of the docking facilities. What is the effect of acquiring better equipment for unloading cargo so that unloading time is reduced to between 35 and 75 min per ship? Table 7.10 gives the results of our simulation algorithm.

You can see from Table 7.10 that reducing the unloading time per ship by 10–15 min decreases the times ships spend in the harbor, especially the wait times. However, the percent of the time during which the dock facilities are idle nearly doubles. The situation is favorable for shipowners because it increases the availability of each ship for hauling cargo. Thus, the traffic coming into the harbor is likely to increase. Table 7.11 gives the simulated results for the situation in which the traffic increases to the extent

Table 7.9 Harbor system simulation results for 100 ships.

Average time of a ship in the harbor	106	85	101	116	112	94
Maximum time of a ship in the harbor	287	180	233	280	234	264
Average ship wait time	39	20	35	50	44	27
Maximum ship wait time	213	118	172	203	167	184
Percentage of time dock facilities are idle	18%	17%	15%	20%	14%	21%

Note: All times are given in minutes. Time between successive ships is 15–145 min. Unloading time per ship varies from 45 to 90 min.

Table 7.10 Harbor system simulation results for 100 ships with better unloading equipment.

Average time of a ship in the harbor	74	62	64	67	67	73
Maximum time of a ship in the harbor	161	116	167	178	173	190
Average ship wait time	19	6	10	12	12	16
Maximum ship wait time	102	58	102	110	104	131
Percentage of time dock facilities are idle	25%	33%	32%	30%	31%	27%

Note: All times are given in minutes. Time between successive ships is 15–145 min. Unloading time per ship varies from 35 to 75 min.

Table 7.11 Harbor system simulation results for 100 ships with increased traffic.

Average time of a ship in the harbor	114	79	96	88	126	115
Maximum time of a ship in the harbor	248	224	205	171	371	223
Average ship wait time	57	24	41	35	71	61
Maximum ship wait time	175	152	155	122	309	173
Percentage of time dock facilities are idle	15%	19%	12%	14%	17%	6%

Note: All times are given in minutes. Time between successive ships is 10–120 min. Unloading time per ship varies from 35 to 75 min.

that the time between successive ships is reduced to between 10 and 120 min. You can see from this table that now the ships spend more time waiting with the increased traffic, but the harbor facilities are idle much less of the time. Moreover, both the shipowners and the dock owners are benefiting from the increased business.

Model Refinement

Now let's turn to the situation documented in Table 7.12. Suppose we are not satisfied with the assumption that the arrival time between ships (that is, their interarrival times) and the unloading time per ship are *uniformly distributed* over the time intervals $15 \leq \text{between}_i \leq 145$ and $45 \leq \text{unload}_i \leq 90$, respectively. So it is decided to collect

Table 7.12 Data collected for 1200 ships using the harbor system.

Time between arrivals	Number of occurrences	Probability of occurrence	Unloading time	Number of occurrences	Probability of occurrence
15–24	11	0.009			
25–34	35	0.029			
35–44	42	0.035	45–49	20	0.017
45–54	61	0.051	50–54	54	0.045
55–64	108	0.090	55–59	114	0.095
65–74	193	0.161	60–64	103	0.086
75–84	240	0.200	65–69	156	0.130
85–94	207	0.172	70–74	223	0.185
95–104	150	0.125	75–79	250	0.208
105–114	85	0.071	80–84	171	0.143
115–124	44	0.037	85–99	109	0.091
125–134	21	0.017		1200	1.000
135–145	3	0.003			
	1200	1.000			

Note: All times are given in minutes.

experimental data for the harbor system and incorporate the results in our model. We observe (hypothetically) 1200 ships using the harbor to unload their cargoes, and collect the data displayed in Table 7.12.

We add consecutively the probabilities of each time between arrivals, as well as probabilities of each individual unloading time. The results are shown in the cumulative histograms in Figure 7.25.

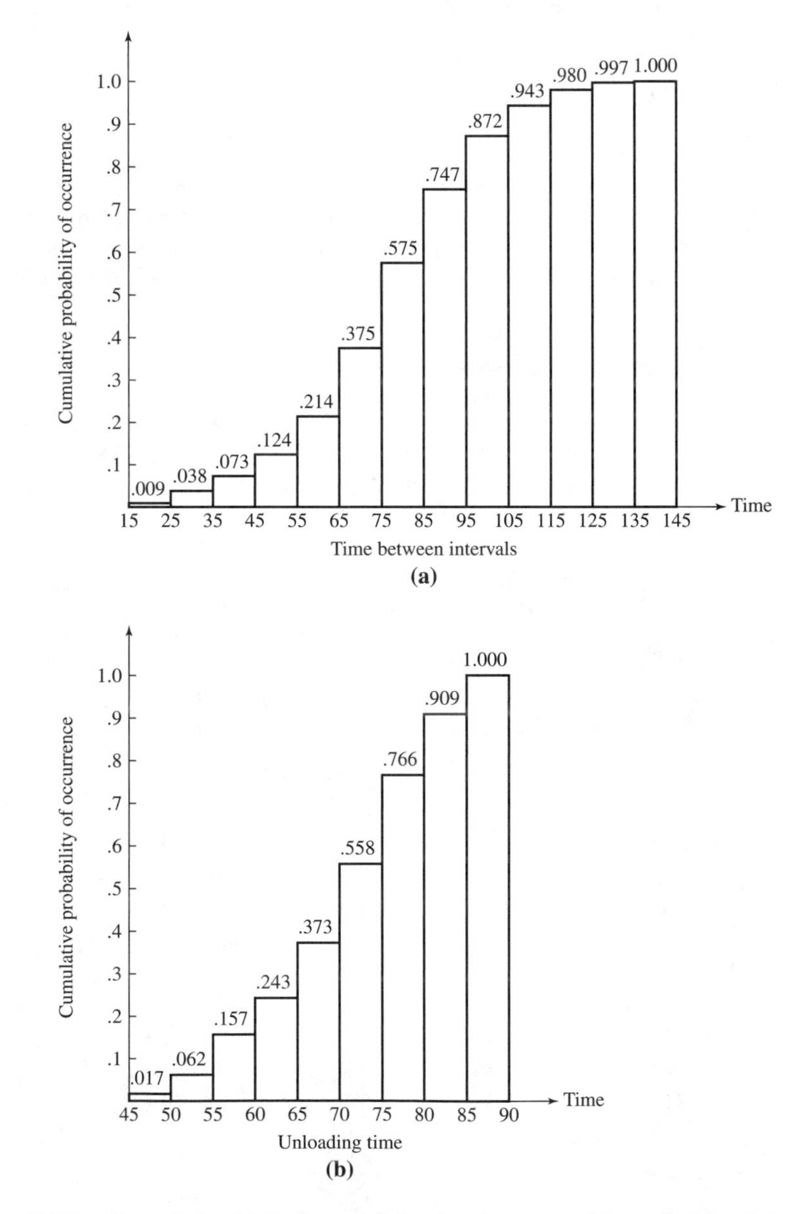

Figure 7.25 Cumulative histograms of the time between ship arrivals and the unloading times, from the data in Table 7.12.

Next we use random numbers uniformly distributed over the interval $0 \le x \le 1$ to duplicate the various interarrival times and unloading times based on the cumulative histograms. We construct line segments through the endpoints of each interval as illustrated in Figure 7.26 for the unloading time data. Let's find the line segments for the unloading time data.

In Figure 7.26 we show line segments connecting the endpoints of each interval for the unloading data. For each interval we find a line segment that passes through the following points:

Line segment	Interval	Left and right endpoints
$S_1(x)$	$0 \le x < 0.017$	$(45, 0), (50, 0.017)$
$S_2(x)$	$0.017 \le x < 0.062$	$(50, 0.017), (55, 0.062)$
$S_3(x)$	$0.062 \le x < 0.157$	$(55, 0.062), (60, 0.157)$
$S_4(x)$	$0.157 \le x < 0.243$	$(60, 0.157), (65, 0.243)$
$S_5(x)$	$0.243 \le x < 0.373$	$(65, 0.243), (70, 0.373)$
$S_6(x)$	$0.373 \le x < 0.558$	$(70, 0.373), (75, 0.558)$
$S_7(x)$	$0.558 \le x < 0.766$	$(75, 0.558), (80, 0.766)$
$S_8(x)$	$0.766 \le x < 0.909$	$(80, 0.766), (85, 0.909)$
$S_9(x)$	$0.909 \le x \le 1.000$	$(85, 0.909), (90, 1.000)$

Consider line segment $S_5(x)$. Letting u represent the unloading time, we wish to find the line segment $u = mx + d$ that passes through the points $(65, 0.243)$ and $(70, 0.373)$. Substituting,

$$65 = 0.243x + d$$

$$70 = 0.373x + d$$

The solution is $u = 38.462x + 55.654$. We solve for the other line segments in a similar manner. The results are summarized in Tables 7.13 and 7.14 on page 307.

Finally, we incorporate our linear segment submodels into the simulation model for the harbor system by generating between$_i$ and unload$_i$ for $i = 1, 2, \ldots, n$ in steps 1 and 5 of our computer algorithm, according to the rules displayed in Tables 7.13 and 7.14. Using these submodels, Table 7.15 gives the results of six independent simulation runs of 100 ships each.

Exercises 7.7

1. Using the data from Table 7.12 and the cumulative histograms of Figure 7.25, construct cumulative plots of the time between arrivals and unloading time submodels. Calculate equations for the linear segments over each random number interval. Compare your results with the linear segments given in Tables 7.13 and 7.14.

2. Use a smooth polynomial to fit the data in Table 7.12 to obtain arrivals and unloading times. Compare results to those in Tables 7.13 and 7.14.

3. Modify the harbor system computer algorithm to track the number of ships waiting in the queue.

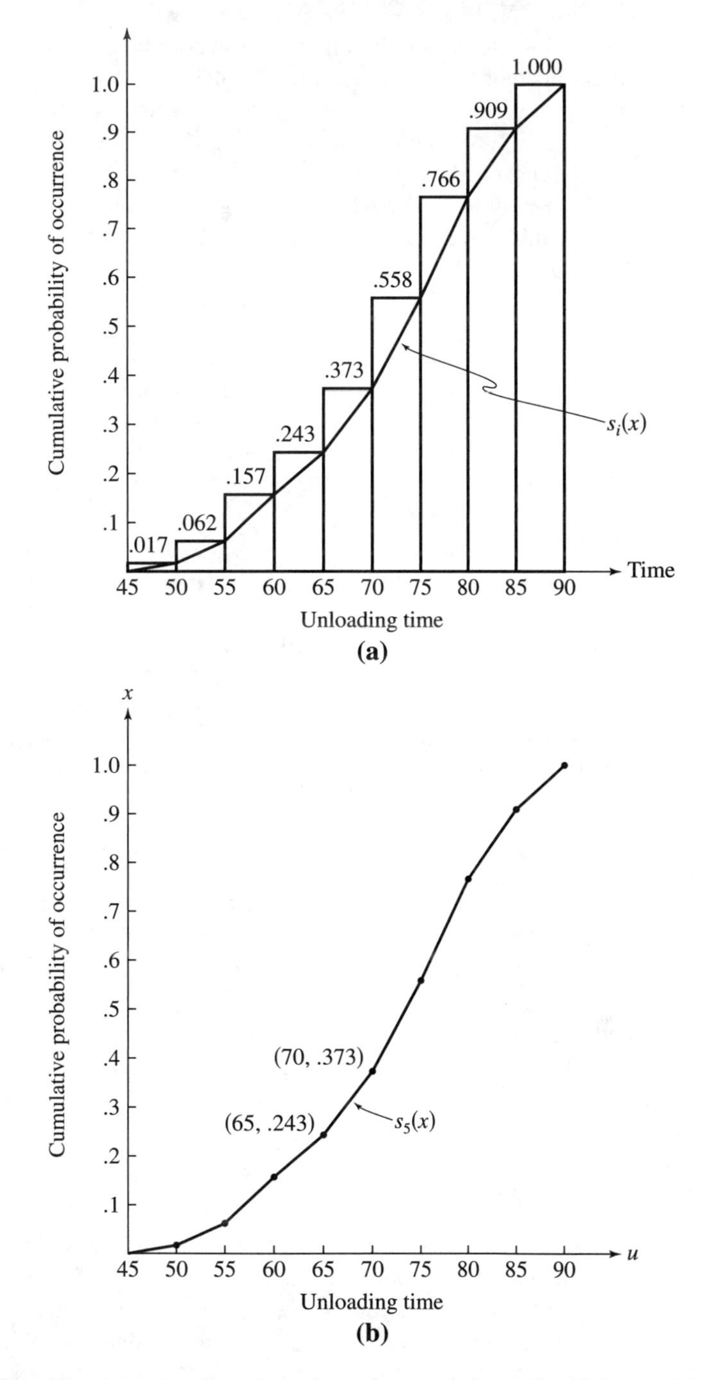

Figure 7.26 For the unloading time data, the endpoints of each interval in the cumulative histogram are connected, resulting in the model shown in Figure 7.26b.

Table 7.13 Linear segment submodels give the time between arrivals of successive ships as a function of a random number in the interval [0,1].

Random number interval	Corresponding arrival time	Linear segment
$0 \le x < 0.009$	$15 \le b < 25$	$b = 1111.1x + 15$
$0.009 \le x < 0.038$	$25 \le b < 35$	$b = 344.8x + 21.897$
$0.038 \le x < 0.073$	$35 \le b < 45$	$b = 285.7x + 24.143$
$0.073 \le x < 0.124$	$45 \le b < 55$	$b = 196.1x + 30.686$
$0.124 \le x < 0.214$	$55 \le b < 65$	$b = 111.1x + 41.222$
$0.214 \le x < 0.375$	$65 \le b < 75$	$b = 62.1x + 51.708$
$0.375 \le x < 0.575$	$75 \le b < 85$	$b = 50.0x + 56.25$
$0.575 \le x < 0.747$	$85 \le b < 95$	$b = 58.1x + 51.57$
$0.747 \le x < 0.872$	$95 \le b < 105$	$b = 80.0x + 35.24$
$0.872 \le x < 0.943$	$105 \le b < 115$	$b = 140.8x - 17.817$
$0.943 \le x < 0.980$	$115 \le b < 125$	$b = 270.3x - 139.86$
$0.980 \le x < 0.997$	$125 \le b < 135$	$b = 588.2x - 451.47$
$0.997 \le x \le 1.000$	$135 \le b \le 145$	$b = 3333.3x - 3188.3$

Table 7.14 Linear segment submodels give unloading times as a function of a random number in the interval [0,1].

Random number interval	Corresponding unloading time	Linear segment
$0 \le x < 0.017$	$45 \le u < 50$	$u = 294.12x + 45.000$
$0.017 \le x < 0.062$	$50 \le u < 55$	$u = 111.11x + 48.111$
$0.062 \le x < 0.157$	$55 \le u < 60$	$u = 52.632x + 51.737$
$0.157 \le x < 0.243$	$60 \le u < 65$	$u = 58.14x + 50.872$
$0.243 \le x < 0.373$	$65 \le u < 70$	$u = 38.462x + 55.654$
$0.373 \le x < 0.558$	$70 \le u < 75$	$u = 27.027x + 59.919$
$0.558 \le x < 0.766$	$75 \le u < 80$	$u = 24.038x + 61.587$
$0.766 \le x < 0.909$	$80 \le u < 85$	$u = 34.965x + 53.217$
$0.909 \le x \le 1.000$	$85 \le u \le 90$	$u = 54.945x + 35.055$

Table 7.15 Harbor system simulation results for six independent runs of 100 ships each.

Average time of a ship in the harbor	108	95	125	78	123	101
Maximum time of a ship in the harbor	237	188	218	133	250	191
Average ship wait time	38	25	54	9	53	31
Maximum ship wait time	156	118	137	65	167	124
Percentage of time dock facilities are idle	9%	9%	8%	12%	6%	10%

Note: Based on the data exhibited in Tables 7.13 and 7.14. All times are given in minutes.

4. Most small harbors have a maximum number of ships N_{max} that can be accommodated in the harbor area while they wait to be unloaded. If a ship cannot get into the harbor, assume it goes elsewhere to unload its cargo. Refine the harbor system computer algorithm to take these considerations into account.

5. Suppose the owners of the docking facilities decide to construct a second facility in order to accommodate more ships. When a ship enters the harbor, it goes to the next available facility, which is facility 1 if both facilities are available. Using the same assumption for interarrival times between successive ships and unloading times as in the initial text example, modify the computer algorithm for a system with two facilities.

6. Construct a Monte Carlo simulation of a baseball game. Use individual batting statistics to simulate the probability of a single, double, triple, home run, or an out. In a more refined model, how would you handle walks, hit batsman, steals, and double plays?

PROJECTS Exploring Modeling

1. Write a computer simulation to implement the harbor system algorithm.

2. Write a computer simulation to implement a baseball game between your two favorite teams (see Exercise 6).

3. Pick a traffic intersection with a traffic light. Collect data on vehicle arrival times and clearing times. Build a Monte Carlo simulation to model traffic flow at this intersection.

4. Complete the requirements of UMAP module 340, "The Poisson Random Process," by Carroll O. Wilde. Probability distributions are introduced to obtain practical information on random arrival patterns, interarrival times or gaps between arrivals, waiting line buildup, and service loss rates. The Poisson distribution, the exponential distribution, and Erlang's formulas are used. The module requires an introductory probability course, the ability to use summation notation, and basic concepts of the derivative and the integral from calculus. Prepare a 10-minute summary of the module for a classroom presentation.

7.8 Data Analysis and Probability Lesson Plans

Lesson 7: Simulation (Middle School Level)[1]

Objectives: Students will focus on developing methods of approximating the probability of events for which calculating exact probabilities would be difficult. In the pro-

[1] Based on lesson from M. Gnanadesikan, R. L. Scheaffer, & J. Swift (1987). *The art and techniques of simulation*. Palo Alto, CA: Dale Seymour.

cess of working through the lesson, students will get the opportunity to consider the probability of an event both through a simulation and through a theoretical analysis.

In the process, students will

1. Make discoveries about how to use simulations for estimating probabilities.
2. Develop ways to represent events so that random numbers or a random process can be used to simulate the events.
3. Compare and contrast the advantages of simulations and theoretical analyses in determining the "probability" of an event.

This lesson follows up a lesson in which students learned to represent events by flipping coins or spinning spinners to estimate the probability of events with known probabilities associated with some outcomes.

Warm-Up Activity (10 minutes): Have students consider the following warm-up problem posted on an overhead:

> Work with your partner. What is the probability that a family with three children has two female children and one male child?
>
> Use your graphing calculators, as well as coins, dice, and spinners.

Activities:

1. (10 minutes) Opening What question does the problem ask us to consider? How did you approach the problem? Listen to the students' thoughts and ask other students for comments on approaches.
2. (10 minutes) Discussion Discuss using three coins in a box lid and flipping to model the problem. Letting H = male and T = female (alternatively use a random number generator [even—male, odd—female] or spinner with 2 equal parts [1 part male, 1 part female]). Determine the outcome of TTH as a success for this simulation. Compare the results of individual group doing 16 flips of coins in a box lid with the sum of the flips of each group in the class.
3. (5 minutes) Presentation Compare the outcomes with the theoretical probability of .375 derived from an analysis of the sample space. Discuss the steps in a simulation, focusing on the use of random numbers or random number generators:
   a. What assumptions are you making?
   b. What model are you using to generate your outcomes?
   c. What defines a trial?
   d. How can you collect data by repeating a trial?
   e. What probability does your approach suggest?

4. (20 minutes) Group Work Distribute the following exercise and let students work with their partners. Encourage student pairs to share solutions or strategies with other pairs sitting nearby.

   The Chicago Cubs play the St. Louis Cardinals in a five-game series. Assume, although it may be debatable, the teams are evenly matched and the outcome of any game is independent of the games before it. Use a simulation to answer the following questions:

   a. Find the probability that the Cubs win three or more games and thereby win the series.

   b. Find the probability that either the Cubs or the Cardinals win four or more games.

   c. Find the probability that neither team wins two or more games in a row.

   d. Estimate the number of games you would expect the Cubs to win in such a five-game series.

   e. What number of wins for the Cubs has the highest probability of occurring?

   Make sure students also answer the five questions associated with a simulation (activity 3).

5. (5 minutes) Closing Bring together your observations of the students' work as they tackled this problem. Review the five questions associated with monitoring one's work with a simulation and again discuss the role of long-term observations. Assign writing up the findings individually for the next day.

Assessment:

1. Warm-Up Observe student work and make suggestions for those having a hard time getting started.

2. Discussion Watch student participation and ask questions to draw out data from their work in the warm-up. Watch for alternative but equivalent approaches.

3. Presentation Sharing of techniques for simulations with a focus on the selection of the model and defining of the trial. Introduce the group work problem, relating it to the birth problem.

4. Group Work Circulate among groups watching student progress. Take notes on specific students who you need to observe for comments this week. Watch for students having difficulty and get more successful students to talk with them.

5. Homework Display sample of how you want results presented in written form for submission in the next class.

Lesson 8: Line of Best Fit (Secondary Level)[2]

Objectives: Students work with a spreadsheet and its related graphics to explore, make conjectures, and interpret a line of best fit (trend line) for data in a scatterplot. More specifically, students will

- Work in pairs to develop a line of best fit for a set of data presented in a scatterplot.
- Build and test interpretations for the meaning of the line.
- Write a paragraph describing their line and interpretation of its meaning in the context of the data.

Activities:

1. (3–5 minutes) Warm-Up Ask students to discuss how to draw a trend line on data from a local newspaper article. Ask what interpretation might be developed.
2. (5–10 minutes) Class Discussion Discuss their methods for developing a trend line. Get students to conjecture methods and move to consider finding a "mid-point" to the data in a scatterplot. Note that they might draw the "best fitting" line through this data or draw two lines that tightly bound the data set and then draw a line that falls halfway between these two bounding lines. The purpose of this lesson is to use previously learned knowledge to develop new techniques describing data and interpreting its meaning.
3. (20–25 minutes) Distribute worksheet (on page 313) and give students time to explore, develop trend lines, and write interpretations for their lines. Visit groups as they work in order to check understanding and to assist them in getting trend lines developed and interpreted. Sample probing questions:
   - What is the general shape of the data? What kind of change do you see in the y-values as the x-values increase?
   - Can you use some of the points and the slope of what you see to write the equation of your trend line? How?
   - How can you get the software to draw your line of best fit?
   - What story is this line telling about how the values of the factors represented by the two axes are related? Is the pattern correlation or may it be causal? How? Why?
   - How could you use part of the data to test the model developed by other parts of the data?
   - How valid is your interpretation? How could you get other data to test it out?

---

[2] Based on a lesson from COMAP's *Mathematics: Modeling our world*, South-Western Educational Publishing, and J. Landwehr & A. Watkins, *Exploring data*, Palo Alto, CA: Dale Seymour.

4. (10 minutes) Whole-Class Discussion Ask students to share their various trend lines and interpretations with the class. Keep track of the various lines developed and then average their slopes and y-intercepts to develop a class prediction. How does it fit the data? Brainstorm with students some initial ideas for validating their interpretations. Compare and contrast the two methods used to find trend lines. Which did the students find more interesting and helpful? Did they find one worked better in some instances and the other better in the rest?

5. (Remaining time) Foreshadowing Ask pairs to read the material on the median line fit in the handout for the next class, as well as the material in their calculator instruction manual related to it. Discuss how this method moves to consider the process of drawing a line of best fit, or trend line, using a more algorithm-driven procedure.

Assessment:

1. Warm-Up The process of developing a trend line is very interpretive. Selected volunteers will be asked to share their construction techniques.

2. Class Discussion (informal assessment) Watch partner pairs as they work to see who is contributing ideas and how their partners are responding to and working with these ideas. Question students on occasion to get a better sense of their understanding of the problems, the techniques being used, their confidence in their lines and interpretations, and their ideas about how they might validate their lines and their interpretations.

3. Worksheet Group Work Again observe the individual student pairs as they work. Make sure that all are contributing to the work and that students have a good feel for the fact the line should pass through the point with the grand average for x-coordinates and grand average for y-coordinates.

4. Whole-Class Discussion Focus on student work with emphasis on both the lines developed and the interpretation of relationship/change represented by their lines.

5. Foreshadowing Ask students to provide some reasons why we would want to make this process more algorithmic.

Worksheet:

Using the data presented below about the amounts of protein and fat found in common fast-food restaurant products, develop a scatterplot of the data, a line of best fit on the scatterplot, and write an interpretation of the story told by the data according to your line.

Food	Protein (grams)	Fat (grams)
Big Mac—McDonald's	26	33
Cheeseburger—Hardee's	17	17
Double Cheeseburger—Burger Chef	23	22
Cheeseburger Supreme—Jack in the Box	33	54
Single—Wendy's	26	26
Double—Wendy's	44	40
Hamburger—McDonald's	12	10
Quarter Pounder—McDonald's	24	22
Whopper—Burger King	26	36
Roast Beef—Arby's	22	15
Beef-and-Cheese—Arby's	27	22
Roast Beef—Hardee's	21	17
Big Fish—Hardee's	20	26
Ham and Cheese—Hardee's	23	15
Thick Crust Pizza—Pizza Hut	24	10
Super Supreme Thin—Pizza Hut	30	26
Idiot's Delight—Shakey's	14	10
Cheese Pizza—Shakey's	16	12
Chicken McNuggets—McDonald's	20	19
Chili—Wendy's	19	8
French Fries—McDonald's	3	12
Onion Rings—Burger King	3	16
Chocolate Chip Cookies—McDonald's	4	16
Apple Turnover—Jack in the Box	4	24

Directions:

1. Enter the data into the spreadsheet.
2. Graph the data using the spreadsheet. Make sure to choose good scales for the graph's axes and label both the axes and title the graph.
3. Find a line of best fit for the data.
4. Write a paragraph explaining what the data and graph of the line tell us about the relationship of protein and fat in fast-food restaurant items.
5. Why do you think this relationship holds?
6. Does this relationship hold no matter what food items you might select?
7. How might you verify this relationship?

References

Anscombe, F. (1973). Graphs in statistical analysis. *The American Statistician.* **27**:17–21.

Beaton, A. E., Mullis, I. V. S., Martin, M. O., Gonzalez, E. J., Kelly, D. L., & Smith, T. A. (1996). *Mathematics achievement in the middle school years: IEA's Third International Mathematics and Science Study.* Chestnut Hill, MA: TIMSS International Study Center, Boston College.

Burrill, G., et al. (1992). *Data analysis and statistics: Addenda series, grades 9–12.* Reston, VA: NCTM.

Burrill, G., Scheaffer, R., & Rowe, K. B. (1991). *Guidelines for the teaching of statistics.* Alexandria, VA: American Statistical Association.

Fusaro, B. A. (Winter, 1989). Results of the 1989 Mathematical Contest in Modeling. *The UMAP Journal* **10**(4):287–299.

Landwehr, J. M., & Watkins, A. E. (1987). *Exploring data.* Palo Alto, CA: Seymour.

Landwehr, J. M., & Watkins, A. E. (1987). *Exploring surveys and information from samples.* Palo Alto, CA: Seymour.

Moore, D. S. (1990). Uncertainty. *In* L. A. Steen (ed.), *On the shoulders of giants: New approaches to numeracy* (95–137). Washington, D.C., National Academy Press.

National Council of Teachers of Mathematics (1989). *Curriculum and evaluation standards for school mathematics.* Reston, VA: NCTM.

National Council of Teachers of Mathematics (2000). *Principles and standards for school mathematics.* Reston, VA: NCTM.

New York Times (2000). *The New York Times 2000 Almanac.* New York: Author.

Reese, C. M., Miller, K. E., Mazzeo, J., & Dossey, J. A. (1997). NAEP 1996 *mathematics report card for the nation and states.* Washington, D.C.: National Center for Education Statistics.

Rousseeuw, P. J., Ruts, I., & Tukey, J. W. (November, 1999). The Bagplot: A bivariate boxplot. *The American Statistician* **53**(4):382–387.

Tukey, J. W. (1977). *Exploratory data analysis.* Reading, MA: Addison-Wesley.

World Almanac Books (2000). *The world almanac and book of facts 2000.* New York: Author.

Zawojewski, J. S., et al. (1991). *Dealing with data and chance: Addenda series, grades 5–8.* Reston, VA: NCTM.

8 Advanced Topics

Mathematical reasoning penetrates scientific problems in numerous and significant ways. If the secret of technology, as C. P. Snow said, is that it is possible, then the secret of mathematical modelling is that it works. However, the process of developing and employing a mathematical model is both more subtle and more complex than is the traditional solution of mathematics textbook problems. Real models frequently have to be constructed in the presence of more data than can be taken into account; their conclusions are often drawn from calculations in which good approximations play a greater role than do exact solutions; very often there are conflicting standards by which solutions can be judged, so whatever answers emerge can only rarely be labelled as right or wrong.

Ross L. Finney (1981)

Overview

In this chapter we consider opportunities for students at the advanced levels of secondary school mathematics, as well as mathematics organizations and competitions that exist for students at this level. While the previous chapters provide an overview of the content and methods appropriate for the major areas outlined in the NCTM *Principles and Standards*, here we focus on special courses offered for students taking electives in mathematics or focusing on one particular area as preparation for college or career. The individual sections deal with trigonometry, analytic geometry, discrete mathematics, statistics, calculus, and special mathematics activities for students.

Focus on the NCTM *Standards*

While the 1989 NCTM *Standards* provided standards for geometry from an algebraic viewpoint (analytic geometry), trigonometry, discrete mathematics, and the conceptual underpinnings of calculus, the *2000 Principles and Standards for School Mathematics* integrated these recommendations into the other standards for geometry, statistics, and algebra and functions. In order that these topics get some coverage, we have drawn upon the 1989 recommendations and current programs to develop the materials found here. Unlike the previous chapters, this chapter focuses a bit more on the content appropriate for such courses and touches on some of the important topics within these areas.

Focus on the Classroom

Precalculus, trigonometry, discrete mathematics, statistics, calculus, and other elective courses play important roles in the curriculum. They link the first three years of study to university and technical courses in mathematics and related fields. More importantly, they signal an even greater role for modeling as part of the secondary school curriculum. Historically, students rarely saw applications in school math programs until they reached this level of courses. Fortunately, that is no longer the case. However, it is important that these courses continue to serve the transitional roles that help in the articulation between systems and levels of mathematics education.

Depending on where students started their study of the content equivalent to Algebra I, the sequence and year in which a student might encounter these courses will vary. The degree to which they are available also varies. NAEP data suggests that about 15% of the nation's students take a course equivalent to precalculus and about 8% take some course equivalent to, at least, the first semester of university calculus while still in secondary school. A smaller percentage of students have the opportunity to study more advanced coursework while still in high school.

The courses, trigonometry, precalculus, discrete mathematics, statistics, and calculus, can be arranged in a number of ways within a school curriculum. When trigonometry is taught as a separate course, it usually parallels precalculus and is intended for students who are entering technical fields, but who would normally not take precalcu-

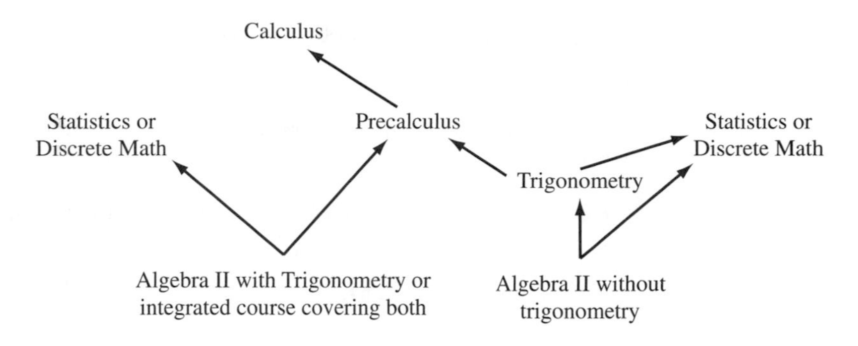

Figure 8.1 Alternative sequencing for advanced mathematics courses.

lus. Statistics and discrete mathematics are often offered as electives for students at the senior level with prerequisites of Algebra II. The availability of statistics has grown substantially in the past decade; there is now an Advanced Placement examination in statistics. Calculus has, and continues to be, the capstone course in many secondary school math programs. Figure 8.1 shows how these courses can be sequenced in a curriculum.

Students completing a standard college preparatory mathematics sequence in high school and who started with Algebra I in eighth grade may take Algebra II and Trigonometry, and then continue to Precalculus and then Calculus, with perhaps an elective course in Statistics or Discrete Mathematics in the senior year. Students taking Algebra II without any trigonometry in their sophomore or junior year might continue to take a semester course in trigonometry and then one in either statistics or discrete mathematics. Some of these students could decide, should they have an additional year remaining to then take precalculus. If the secondary program is composed of courses built around a four-year program of integrated mathematics, like the NSF-based programs, then the Algebra II courses above might be replaced with the Course III level and the Precalculus/Discrete Mathematics course with Course IV in the sequences. It is possible that the trigonometry, statistics, and calculus courses will still be offered as standard courses alongside the Course I–Course IV sequences of these NSF-stimulated sequences.

8.1 Trigonometry

Trigonometry remains as an integrated sequence within the middle and high school curricula and as a free-standing course in many high schools. As an integrated sequence, it is focused on developing both indirect measurement topics and the notion of circular functions. As a free-standing course, trigonometry provides an in-depth look at a number of topics that play central roles in technical vocations and as prerequisite knowledge for many engineering and science programs at the college level. The 1989 NCTM *Standards* calls for the development of a number of concepts and skills related to trigonometry. Among these, the most important direct students to be able to

- Apply trigonometry to problem situations involving triangles.
- Explore periodic real-world phenomena using sin/cos/tan.
- Understand the connections between right triangle trigonometry and circular functions.
- Apply general graphing techniques to circular functions.
- Understand the connections between trigonometric functions, complex numbers, and series. (NCTM, 1989, p. 163)

Trigonometry, whose name historically meant "measurement of triangles," has become somewhat of a stepchild in the secondary school mathematics curriculum with its concepts and applications taught in a variety of locations with little emphasis on pulling together its major ideas. Middle school students see the basic sine, cosine, and tangent ratios as they apply to right triangles in eighth grade, but do little more than calculate a ratio or two and get the general idea of what the ratios can do with the special angles having measures of 0, 30, 45, 60, and 90 degrees. This sequence is usually repeated again in the study of geometric relationships associated with right triangles near the end of the study of geometry. In Algebra II with Trigonometry, students see the ratios again, but extend the relationships beyond the basic right triangles to consider the law of cosines, law of sines, and a number of basic identities, usually involving the trigonometric values of the sum and difference of angles, the basic Pythagorean identities of $\sin^2 \theta + \cos^2 \theta = 1$ and its relatives, and the identities known as the double-angle and half-angle formulas. These laws and identities are then used to solve a number of indirect measurement items and calculate some values of the various trigonometric relationships. In precalculus, students are usually introduced to the unit circle model for the trigonometric functions as circular functions and the graphic representations of these functions. This approach highlights the concepts of amplitude, period, and phase shift in some detail. Unfortunately, few curricula carefully tie together these different views of trigonometry and the relationships and functions comprising them.

At the middle school level, students should receive the opportunity to participate in indirect measurement activities that tie geometry and similar triangles to trigonometry. While students often see the shadow and height of flagpole items in examples and exercise sets, seldom do they get the opportunity to make actual measurements and compare their theoretical results with reality. Students should have some experience in measuring the height of a pole of known height using measurement tools by eighth grade. This can be done in conjunction with the study of similar triangles and the data saved for later work with trigonometry.

The first approach might be to use shadows, assuming a sunlit day. Using the method of the ancient Egyptians, students might use a *gnomon* or shadow-measuring device like the one shown in Figure 8.2. Here they see the origin of the carpenter's square as the shadow of the short arm falls on the measuring scale of the long arm. Using the shadow of a flagpole, measured from the center of the pole, they can set up the appropriate ratio and compute an estimate for the height of the flagpole. It is informative to take a moment and average several student measurements and discuss

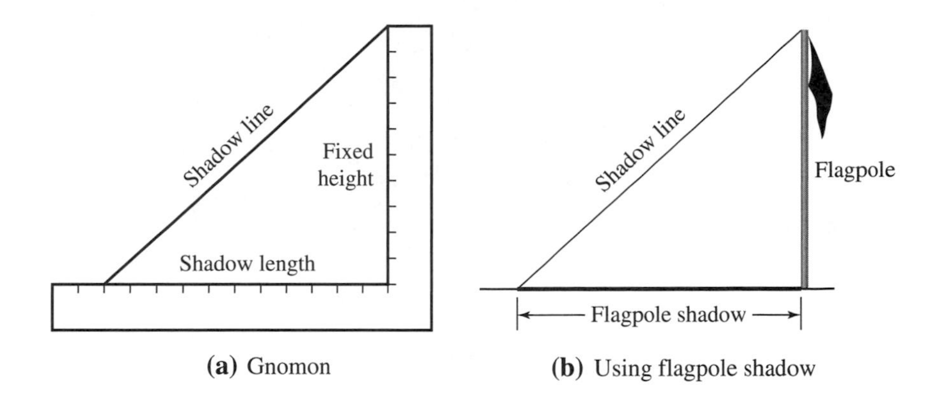

Figure 8.2 Using shadow length to measure flagpole.

experimental error due to measurement error. This is also a good time to discuss the role of significant digits in the calculations.

In a backup method for finding the height using similar triangles, students place a mirror on a stretch of flat ground between the pole and a viewer. When the viewer can see the top of the flagpole reflected in the mirror, the measurements indicated in Figure 8.3 give rise to a second method of calculating the height of the flagpole.

At the next stage, students can approach the problem using a clinometer, an easily made measuring instrument. This tool, shown in Figure 8.4, is made from a wooden staff, a straw, a protractor, a string, and a small fishing weight. Holding the staff vertical, students can sight the top of the flagpole through the straw, which is attached across the 0–180 degree line of the protractor. At the same time, another student reads the angle of elevation of the top of the flagpole from the string that is hanging vertically. This gives rise to the first use of trigonometry, because they can now measure the distance to the flagpole from the base of the wooden staff, and set up a ratio using the tangent of the angle of elevation. Remind students to add in the height of the protractor pivot point as part of the process.

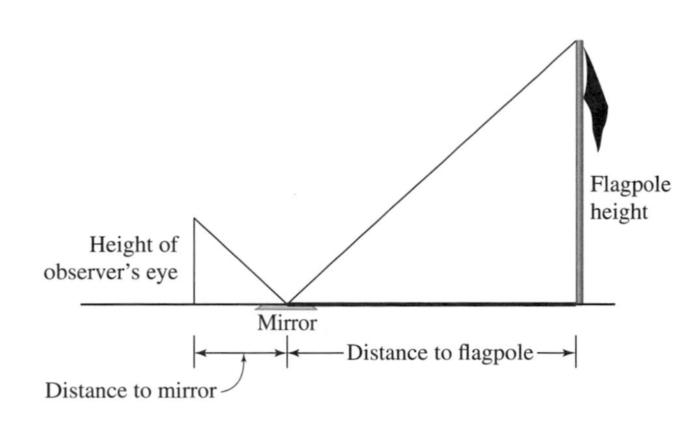

Figure 8.3 Using a mirror to measure flagpole.

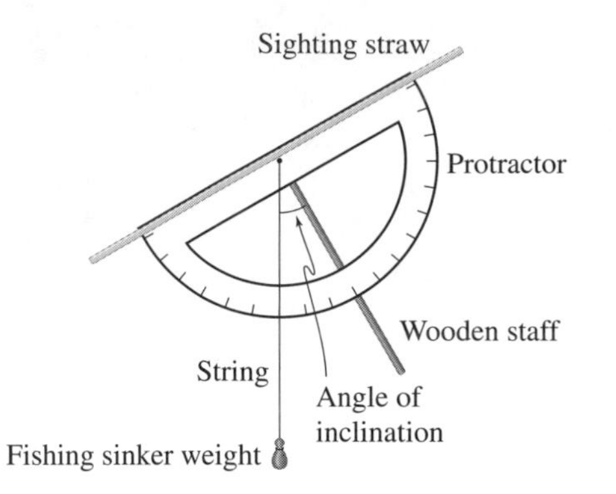

Figure 8.4 Using a clinometer to measure flagpole.

Students both enjoy and need hands-on work with measurement to comprehend the ideas of the trigonometric ratios and their use. At the same time, they need to see that corresponding angle measures in similar triangles remain constant as the sides vary proportionally. These same activities can be repeated as part of the study of trigonometry in geometry, but should involve more work with the sine and cosine relations at this point. Students can measure regions enclosed by sidewalks in the school yard, making use of the *plane table*, shown in Figure 8.5, which measures angles in a horizontal plane, using the same ideas that were central to the use of the clinometer. As the study of trigonometry extends to general triangles and the law of cosines and law of sines are developed, students should again have the opportunity to put these laws into effect and check their work against the direct measurement of a target length or target angle.

Most mathematics students remember the development and verification of trigonometric identities they studied in Algebra II, Trigonometry, or Precalculus. The major identities are listed in Table 8.1. Historically, these identities were developed to calculate the values of the trigonometric functions using values of other known angles. Today, these identities are not needed for this reason, as the values can be calculated directly from series developed in calculus. However, these identities still serve to sim-

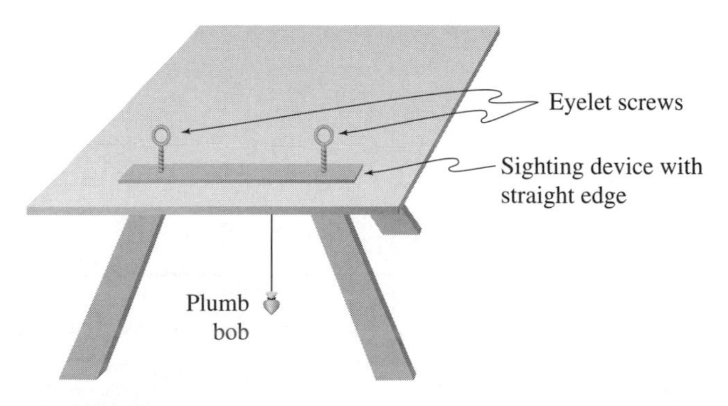

Figure 8.5 Measuring angles using a horizontal plane.

Table 8.1 The trigonometric identities.

Basis Identities

$$\sin(-x) \equiv -\sin x$$
$$\cos(-x) \equiv \cos x$$
$$\tan(-x) \equiv -\tan x$$

Pythagoran Identities

$$\sin^2 x + \cos^2 x \equiv 1$$
$$1 + \tan^2 x \equiv \sec^2 x$$
$$1 + \cot^2 x \equiv \csc^2 x$$

Cofunction Identities

$$\sin\left(x \pm \frac{\pi}{2}\right) \equiv \pm\cos x$$
$$\cos\left(x \pm \frac{\pi}{2}\right) \equiv \mp\sin x$$

Sum and Difference Identities

$$\sin(\alpha \pm \beta) \equiv \sin\alpha\cos\beta \pm \cos\alpha\sin\beta$$
$$\cos(\alpha \pm \beta) \equiv \cos\alpha\cos\beta \mp \sin\alpha\sin\beta$$
$$\tan(\alpha \pm \beta) \equiv \frac{\tan\alpha \pm \tan\beta}{1 \mp \tan\alpha\tan\beta}$$

Double-Angle Identities

$$\sin 2x \equiv 2\sin x\cos x$$
$$\cos 2x \equiv \cos^2 x - \sin^2 x$$
$$\equiv 1 - 2\sin^2 x \equiv 2\cos^2 x - 1$$
$$\tan 2x \equiv \frac{2\tan x}{1 - \tan^2 x}$$
$$\sin^2 x \equiv \frac{1 - \cos 2x}{2}$$
$$\cos^2 x \equiv \frac{1 + \cos 2x}{2}$$

Half-Angle Identities

$$\sin\frac{x}{2} \equiv \pm\sqrt{\frac{1 - \cos x}{2}}$$
$$\cos\frac{x}{2} \equiv \pm\sqrt{\frac{1 + \cos x}{2}}$$
$$\tan\frac{x}{2} \equiv \pm\sqrt{\frac{1 - \cos x}{1 + \cos x}}$$
$$\equiv \frac{\sin x}{1 + \cos x} \equiv \frac{1 - \cos x}{\sin x}$$

plify complex expressions involving trigonometric relationships. They also help set up and solve application problems, such as the one shown here:

> An analyst in examining data surmises that $\cos\theta + \sin\theta = \sqrt{2}\sin(\theta + \frac{\pi}{4})$. The graphs of the functions appear identical. Does the identity conjectured hold?

One issue that comes up in the development of the identities is the form of proof used. Students often want to work simultaneously with both sides of the statement of equality in proving an identity, as shown in Figure 8.6a. However, they need to develop a coherent picture by establishing that the right side of the identity is equivalent to the left side through a series of transformations that results in an equivalent expression, as shown in Figure 8.6b. While the thought processes are somewhat similar, the format shown in Figure 8.6b is to be desired.

With the introduction of the unit circle, shown in Figure 8.7, to study trigonometric functions as circular functions, care must be taken to tie the representations developed for the functions with the unit circle to the ratios used to define right triangle

To prove: $2\sec^2\theta = \dfrac{1}{1-\sin\theta} + \dfrac{1}{1+\sin\theta}$

$$2\sec^2\theta = \frac{1}{1-\sin\theta} + \frac{1}{1+\sin\theta}$$
$$\frac{2}{\cos^2\theta} = \frac{2}{1-\sin^2\theta}$$
$$\frac{2}{\cos^2\theta} = \frac{2}{\cos^2\theta}$$

(a) Wrong way

$$\begin{aligned} 2\sec^2\theta &= \frac{2}{\cos^2\theta} \\ &= \frac{2}{1-\sin^2\theta} \\ &= \frac{2}{(1-\sin\theta)(1+\sin\theta)} \\ &= \frac{1}{1-\sin\theta} + \frac{1}{1+\sin\theta} \end{aligned}$$

(b) Right way

Figure 8.6 A wrong way and a right way.

relationships. For example, knowing that the x-coordinate of the unit-circle point associated with the intersection of the terminal ray of a 45° angle is the cosine of the angle and the y-coordinate is the sine of that angle, students connect the idea that these result from the definitions of adjacent side and opposite side measures in a right triangle with hypotenuse 1.

However, students should see another larger right triangle with the same angle measures inserted into the model, as shown in Figure 8.8, in order to note their similarity and the fact that the unit-circle model captures all of the information using proportion

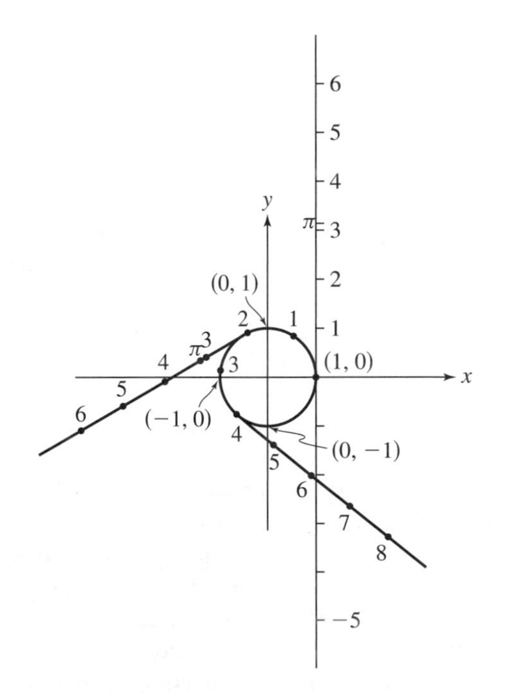

Figure 8.7 The unit circle.

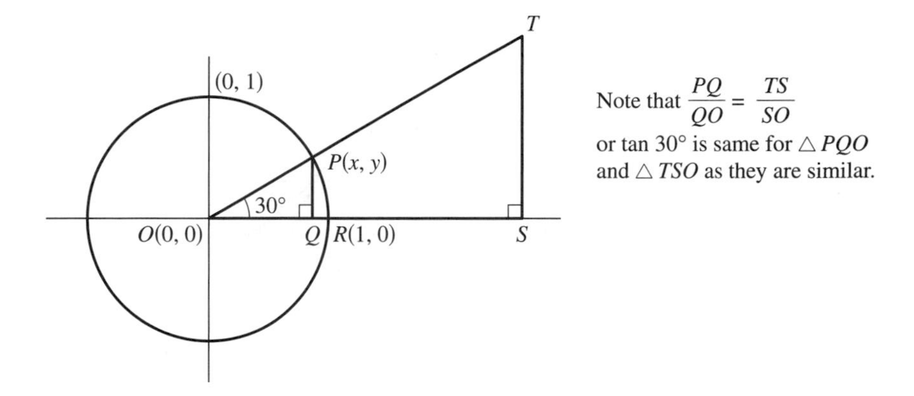

Figure 8.8 Comparison at a right triangle with a similar right triangle inserted in a unit circle.

as a frame of reference. This completes the tie to the earlier work, and new work can go forward on developing the graphs and properties of the trigonometric functions.

Another shift that has to be dealt with in making the transition to the unit-circle model is the shift to radians as the measure of angles. Because the unit-circle model looks at the circular functions, the domain of the functions needs to be the real numbers. Hence, the concept of radian measure is developed in terms of the circle in order to give a real number definition to the measure of angles. Be careful to note at this point as students use technology that students have their calculators, or other forms of technology, properly set for accepting the correct form of angle measurements.

Central to the idea of the unit circle is the wrapping of the real number line around the circumference of the circle to associate the values of the line with points on the circle. This brings the radian measure ideas home. As students see in Figure 8.9, under continued wrappings in both directions, that π is associated with the point $(-1, 0)$ and

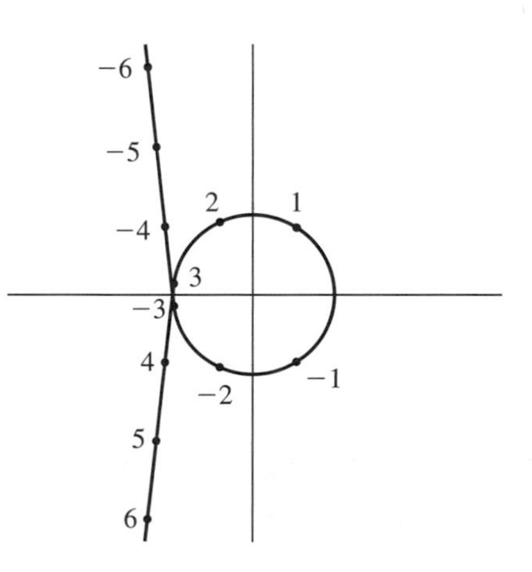

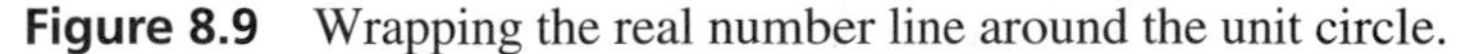
Figure 8.9 Wrapping the real number line around the unit circle.

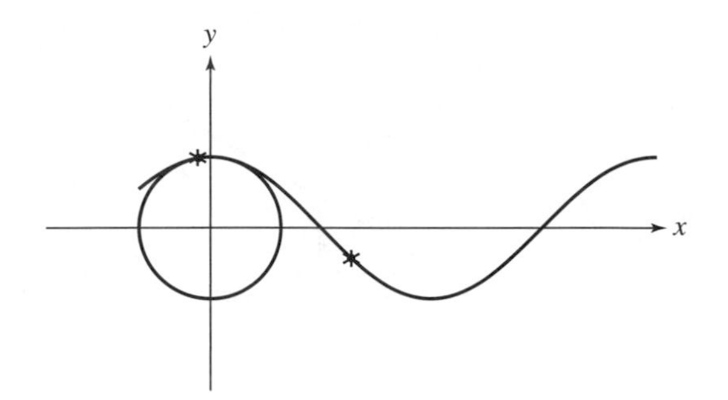

Figure 8.10 Tying the unit-circle and function models together.

2π is associated with the point $(1, 0)$ after one full wrapping, several ideas begin to fall into place. First, the assignment of distances around the circle of unit radius gives rise to the radian measure connection. Second, it makes sense to talk about periodicity as students note that $\cos(\pi/2) = \cos(\pi/2 + 2n\pi)$ for various integer values of n.

When students have several features of the unit circle under control, the next step is to consider the sine, cosine, and tangent functions. You may wish to have the students graph, on their graphing calculators, the following parametric equations, with the graphing done simultaneously:

$$x_1(t) = \cos t$$
$$y_1(t) = \sin t$$
$$x_2(t) = t$$
$$y_2(t) = \cos t$$

where Δt steps by 0.1 radian across the interval $[0, 2\pi]$. Then use the trace function and the up and down arrow keys to move along one of the functions and then the left and right arrow keys to jump back and forth between corresponding points on the two models. The resulting graph, shown in Figure 8.10, ties the unit-circle and functional models together.

This brief overview of trigonometry touches on important issues in teaching this subject. The following exercises will further that discussion.

Exercises 8.1

1. What is the law of cosines and how is it related to the Pythagorean theorem?
2. Given the unit circle shown with $BC = \sin\theta$ and $OB = \cos\theta$, prove that

$$AF = \tan\theta, \qquad OE = \csc\theta, \qquad OF = \sec\theta, \qquad \text{and} \qquad DE = \cot\theta$$

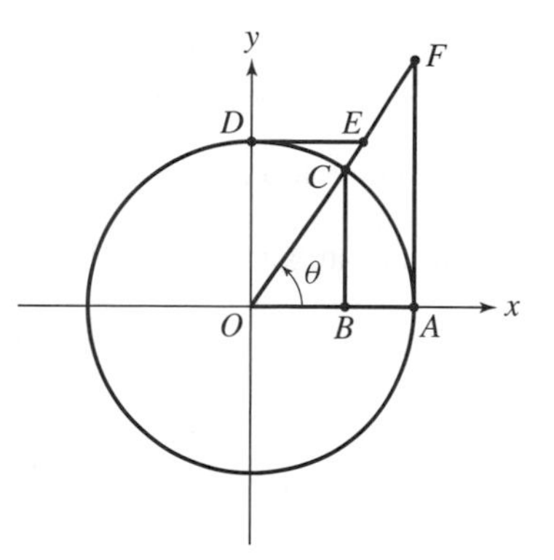

3. Describe how evaluating the series $\sin x = x - x^3/3! + x^5/5! - x^7/7! \cdots$ by its partial sums approximates the value of $\sin x$ for some real number x. Illustrate the process with a graph.

PROJECTS — Find Out for Yourself

1. Interview a high school student taking Algebra II with Trigonometry or equivalent course. What does he or she know about the trigonometric ratios, their relation to triangles, applications in non-right triangles, or the graphs of circular functions?
2. Examine an Algebra II with Trigonometry textbook and a textbook from an integrated series for the junior or senior year. Compare and contrast their coverage of trigonometry.

8.2 Discrete Mathematics

Discrete mathematics is an area of study that entered the secondary mathematics curriculum in the early 1990s. While topics from discrete mathematics had appeared as isolated exercises throughout algebra and geometry curricula prior to that time, the 1989 NCTM *Standards* and the 1991 Yearbook of the National Council of Teachers of Mathematics, *Discrete Mathematics Across the Curriculum, K–12*, legitimized such study in school mathematics.

Central to the introduction of discrete mathematics topics in the curriculum is an understanding of what such topics might be. "Discrete mathematics" can be thought of as the mathematics that is concerned with structure and systems that can be enumerated using the positive integers. The topics most commonly associated with the field are recursion or recurrence relations; finite graphs and networks and their applications; counting, including permutations and combinations; and the study of algorithms. Be sure to distinguish between the use of the word "graph," meaning a network, and "graph," meaning a diagram for a function's values as it is used in discrete mathematics. The word "graph" is more widely used than "network" in discrete mathematics. The 1989 NCTM *Standards* called for the development of student competence in

- Representing problem situations using discrete structures such as finite graphs, matrices, sequences, and recurrence relations.
- Representing and analyzing finite graphs using matrices.
- Developing and analyzing algorithms.
- Solving enumeration and finite probability problems.
- Representing and solving problems using linear programming and difference equations. (NCTM, 1989, p. 176)

While some mention was made of discrete mathematics in Chapter 6 and of recursion in Chapter 5, we have been relatively quiet on discrete mathematics topics. These topics are quite important and serve, in many cases, as models for important contemporary applications of mathematics in business and industry settings (Dossey et al., 2001).

While these topics can be included in the curriculum as part of an overall integrated program, they can also be addressed in a semester-long course at the senior level that focuses on discrete mathematics exclusively. When such a path is taken, the course might include some of the topics listed in Table 8.2.

The topic of graph, or network, theory can be approached in many different ways with secondary students, but the historical approach is an interesting one. Recall the five regular polyhedra that you studied in geometry (see Table 8.3 on page 328). The ancient Greeks proved that these are the only polyhedra with congruent faces, the same number of edges meeting at each vertex, and the same number of faces meeting at each vertex. However, they missed an important relationship that connects the number of edges, faces, and vertices. It wasn't discovered until 1640 by the French philosopher and mathematician René Descartes and rediscovered again in 1752 by the Swiss mathematician Leonhard Euler.

Giving students a worksheet like the one shown in Table 8.4 and models of the polyhedra, they can rediscover the work of Descartes and Euler for themselves.

Ask students to consider various relationships between sums of numbers in two of the columns and the numbers in the remaining third column. Can they see any pattern that might be a general rule? Have them compare their findings with two other classmates and try to write a statement of their findings as a theorem, starting with "For all regular polyhedra, ..."

Euler and Descartes recognized that this theorem extends to a broader set of polyhedra than those you considered. In fact, if the polyhedron, not necessarily regular, is simply convex, the formula holds. A convincing argument for the relationship can be constructed as follows. Without too much effort, students will conjecture that the number of vertices, v, the number of faces, f, and the number of edges, e, are related by one of the equivalent forms of $v + f - e = 2$.

The proof of the formula they discover can be verified through an activity that involves shining a bright light down on the top of a wire model of one of the polyhedrons, say the tetrahedron. This will cast a shadow, like that shown in Figure 8.11, of the edges and vertices in the plane, enclosing regions representing all of the faces except one, the

Table 8.2 Possible topics for a course in discrete mathematics.

I. Graph Theory (4–5 weeks)

- Program evaluation and review techniques (PERT charts) as done in Chapter 6
- Minimal spanning trees (Prim's and Kruskal's algorithms)
- Structure of a graph (basic concepts, representations such as diagrams, adjacency matrix, and adjacency lists)
- Circuits and paths (Euler circuit algorithm, Hamiltonian circuits and paths, traveling salesman problem, Dijkstra's algorithm for shortest paths)
- Network coloring construction
- Structure of trees

II. Counting Techniques (4–5 weeks)

- Logic, sets, and Venn diagrams (disjunction and union, conjunction and intersection, negation and complement, principle of inclusion-exclusion)
- Addition and multiplication principles
- Permutations and combinations
- Pascal's triangle and binomial coefficients
- Discrete probability and applications (mutually exclusive events and addition rule, independent events and the multiplication rule, conditional probabilities, and expected value in finite settings)

III. Mathematics of Recursion/Iteration (difference equations) (4–5 weeks)

- Iterating first-order recursion relations
- Applicaton of iteration (arithmetic and geometric sequences, exponential growth, mathematics of finance, and population dynamics)
- Finding the closed form of a first-order linear recurrence relation
- Iterating second-order difference equations (Fibonacci sequence)

IV. Matrix Models (1–2 weeks)

- Markov chains
- Leslie models for population distribution
- Leontief input-output model of an economy (Dossey, 1991, pp. 7–8)

one the light first shines through. It can be considered as the region surrounding the shadow in the plane.

This shadow in the plane is an example of a network or graph, not to be confused with the graphs of functions we studied earlier. A network or a graph is a nonempty set

Table 8.3 Polyhedra for use in the study of graph theory.

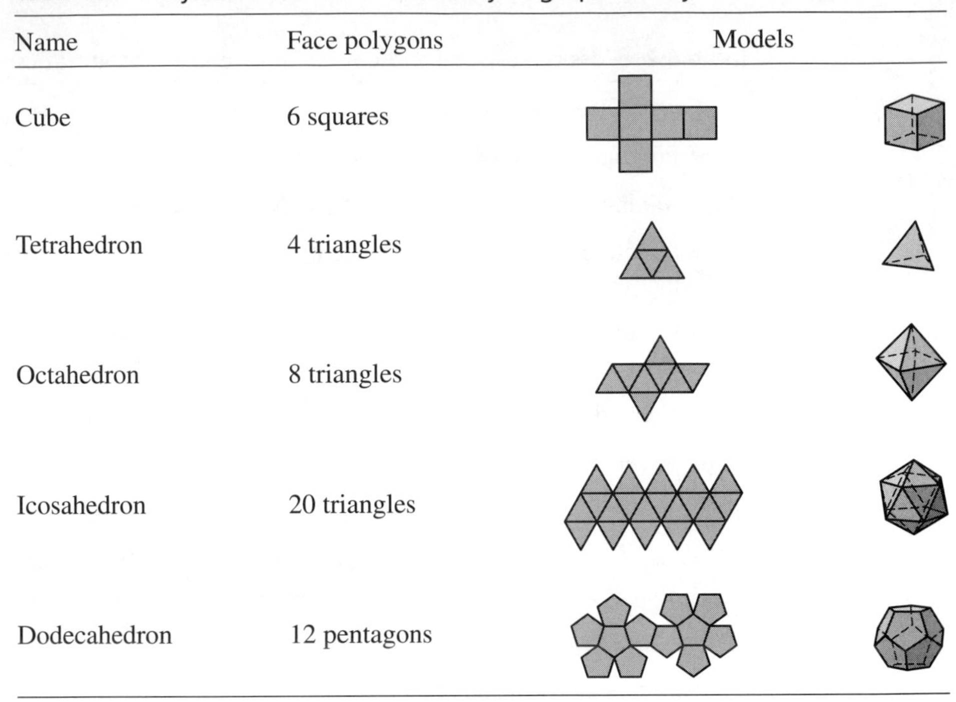

Name	Face polygons	Models
Cube	6 squares	
Tetrahedron	4 triangles	
Octahedron	8 triangles	
Icosahedron	20 triangles	
Dodecahedron	12 pentagons	

Table 8.4 Worksheet for Euler's formula.

Polyhedron	Vertices	Faces	Edges
Tetrahedron			
Cube			
Octahedron			
Dodecahedron			
Icosahedron			

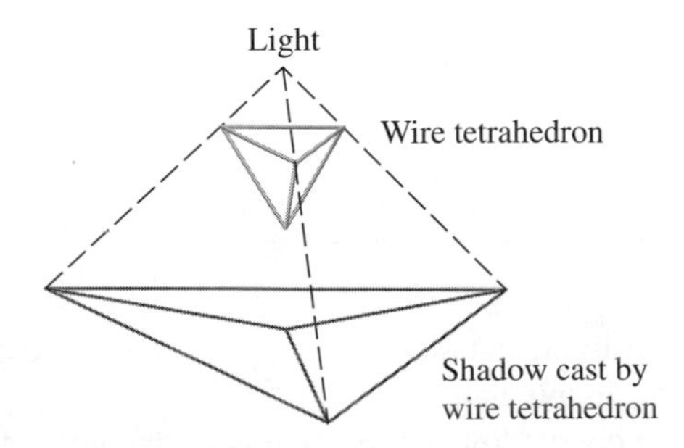

Figure 8.11 Justifying Descartes and Euler's formula.

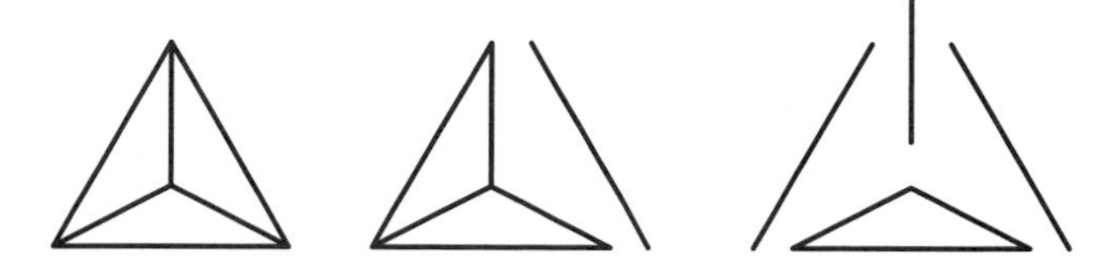

Figure 8.12 Decomposing graph while keeping $v + f - e$ unchanged.

of points, called vertices, and subset of the possible line segments or arcs, called edges, connecting them. Working with the network defined by the tetrahedron, we remove an edge, which also removes a face, leaving the value of the expression $v + f - e$ unchanged, as shown in Figure 8.12.

Continuing to disassemble the figure, we next remove a vertex and two edges at the top left of the figure. This, in effect, also removes a face. Removing 1 vertex and 1 face and 2 edges leaves the value of $v + f - e$ still equal to 1. Then 1 is added to each side of the equation, to account for the uncounted face that is the top of the tetrahedron, represented by the area surrounding the network in the plane. So, $v + f - e = 2$ for the tetrahedron. Using networks of any convex polyhedron, similar analyses can be made to show that the theorem holds in general.

Euler is better known for his work with transversability. This work originated in the Prussian city of Königsberg, located on both banks and two islands in the Pregel River. The city was linked by seven bridges, as shown in Figure 8.13. The legend is that couples liked to stroll around town on weekend afternoons, trying to cross each bridge only once before returning to their starting point.

After a great number of tries, the townspeople asked Euler to study their problem. His paper on the problem, published in 1736, is considered the beginning of the branch of mathematics known as graph theory. In considering this problem, Euler discovered the following theorem: *A graph has an Euler circuit if and only if every vertex in the graph is even. Furthermore, a graph has an Euler path if and only if every vertex has even degree except for two distinct vertices, which have odd degree. In this case, the Euler path starts at one and ends at the other of these two vertices of odd degree.*

To determine whether the system of bridges had an Euler circuit, all Euler had to do was analyze the graph in Figure 8.14, whose vertices represented the regions

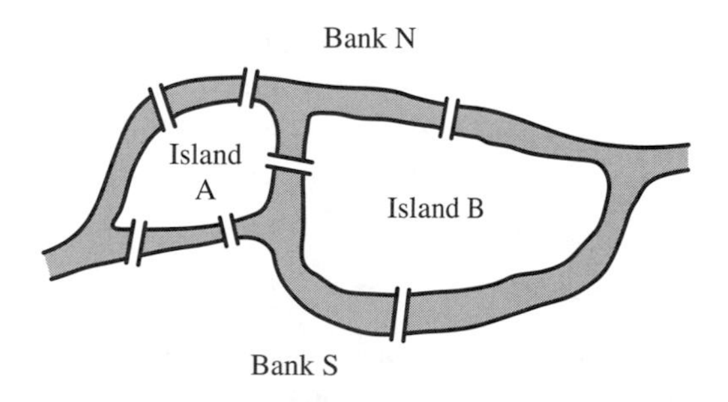

Figure 8.13 The bridges of Königsberg.

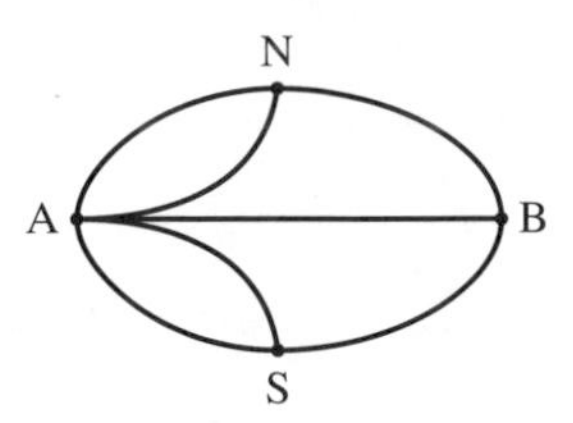

Figure 8.14 Euler's analysis of the Königsberg bridge problem.

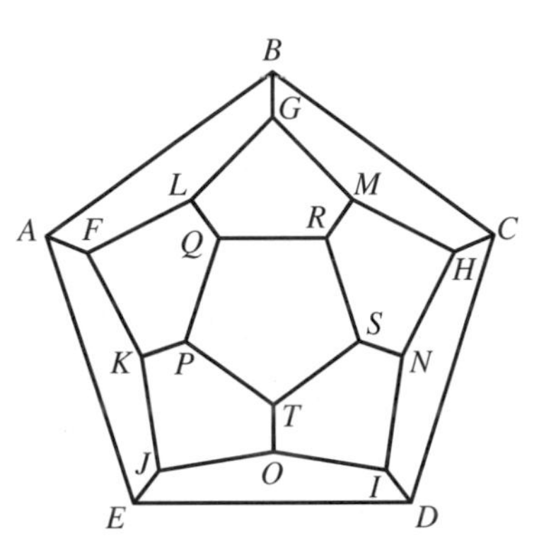

Figure 8.15 Hamilton's puzzle.

connected and whose edges represented the bridges. A quick analysis shows that more than two vertices have an odd number of bridges coming into them (that is, have odd degrees); hence no Euler path exists for the bridges of Königsberg.

Another famous mathematician who worked in graph theory was the Irishman, Sir Rowan Hamilton. In 1859, he invented a puzzle that consisted of the graph, or network, resulting from projecting a dodecahedron into the plane, like we did with the shadows in the previous lesson. The shadow is shown in Figure 8.15. The puzzle involved finding a route that traveled along the edges of the graph; the route must visit each vertex exactly once and return to the original starting point. Hamilton gave individuals pegs to insert at each city, and they wound a string around these pegs to show the route taken. Such a route, if it exists, is called a Hamiltonian cycle. This problem is a version of the traveling salesman problem, in which a salesman wants to leave home, visit each city once, and then return home. It isn't necessary in this case to cover each edge of the graph. Can you find a Hamiltonian cycle for Figure 8.15?

A Question

Can you think of instances where Euler paths, Euler circuits, or Hamiltonian cycles may exist for streets and roads, wiring diagrams, or other configurations that might be represented by a graph?

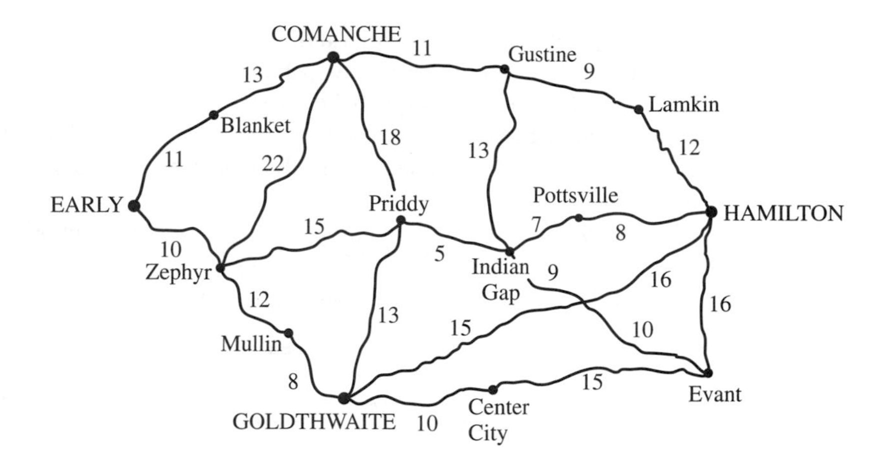

Figure 8.16 Map for cable network installation.

While we can nicely resolve the question of transversability for Euler paths and circuits, no general result is known for the existence of Hamiltonian cycles. It remains one of the great unsolved problems in mathematics.

Consider the mileage between the towns on the map in Figure 8.16. The numbers indicate mileages between the towns (along the roads). Suppose we are installing (if this is possible), a cable communications network that links the cities shown. We want to do this in such a way that every city can connect with every other city using the edges of the network. Further, redundancy is unacceptable because redundancy (unnecessary connections) adds cost.

A graph in which a path connects all edges, but in which no two vertices can be connected in more than one way is called a tree. Figure 8.17 shows one possible way we can install the cable network. Is this graph a tree? Does it satisfy our criteria?

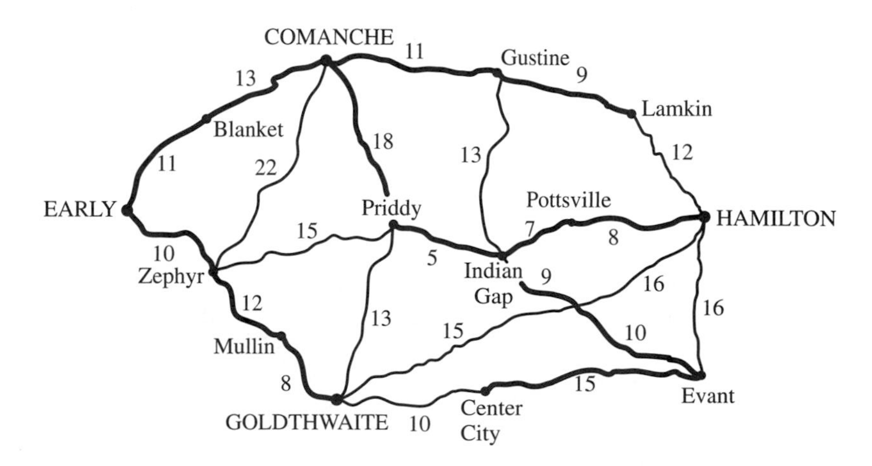

Figure 8.17 One possible cable network.

A tree that satisfies the criteria that the route passes through each town (vertex) in a graph at least once is called a spanning tree, in that it spans the graph to reach each vertex. The companies that build communications networks are most interested in creating spanning trees that use the least amount of wire, as this minimizes their costs. A spanning tree with minimal sum in the length of its edges is a minimum spanning tree.

One way to find a minimum spanning tree uses an algorithm developed by the American mathematician, R. C. Prim. The algorithm, shown here, can be applied to any graph G to construct a minimum spanning tree S.

PRIM'S ALGORITHM FOR FINDING MINIMUM SPANNING TREES

Step 1. Select any vertex A of the graph G as a vertex of the new spanning tree.

Step 2. Consider all edges of the original graph G that have vertex A as an endpoint. Select an edge from this group with as short a length as possible. Place this edge and its other endpoint in the new spanning tree S. (If there are two edges that have the same length, choose one.)

Step 3. Repeat step 2 until all vertices of the original graph G are vertices of the new graph S. (Note, however, that some edges of G may not be included in edges of S.)

See if you can use Prim's algorithm on the city-road graph in Figure 8.16 to develop a minimal spanning tree for the cable communications network.

A final application of graphs is their use in simplifying conflicts. Vertices in the graph shown in Figure 8.18 represent eight chemicals. If two chemicals are connected by an edge, that indicates that they cannot be stored in the same room because an explosion might occur if they mix accidentally. Your task is to decide how many rooms are needed to store these chemicals.

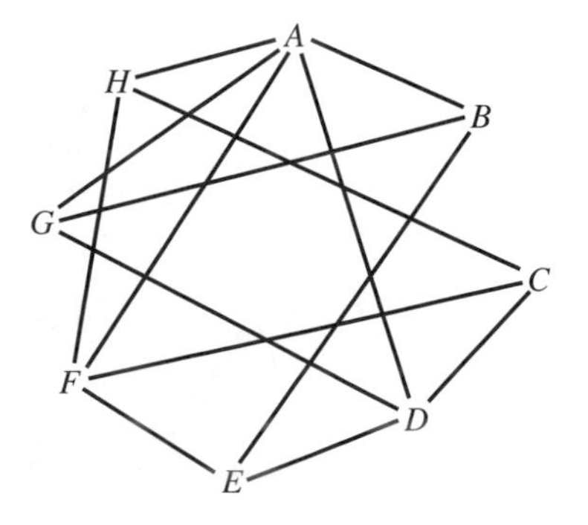

Figure 8.18 The chemical mix conflict problem.

You might begin by asking which chemical makes the largest number of dangerous mixes with the other chemicals. You place it in a room by itself; you then determine the chemical with the next largest number of dangerous mixes. How can you use this approach to minimize the number of storage rooms? What is your estimate of the number of rooms needed? How would you allocate the chemicals to those rooms?

Consider the algorithm given by American mathematicians D. J. Welsh and M. B. Powell. Their algorithm, shown below, "colors" vertices as a means of reducing conflict. The degree of a vertex, $\deg(V)$, in the algorithm refers to the number of edges ending in vertex V. Use this algorithm with the chemical conflict problem to determine the minimal number of rooms needed to store the chemicals.

WELSH AND POWELL'S COLORING ALGORITHM FOR MINIMIZING CONFLICTS

Step 1. Count the number of edges ending at each vertex. Then rename the vertices of the graph $V_1, V_2, V_3, \ldots, V_n$ according to the pattern that the $\deg(V_1) \geq \deg(V_2) \geq \deg(V_3) \geq \cdots \geq \det(V_n)$. Any ties between vertices in their degrees can be broken arbitrarily.

Step 2. Assign a color to the first uncolored vertex in the list of vertices ordered by degree. Then go through the vertices in numerical order, assigning this color to any vertex *not* adjacent, or sharing an edge, with any other vertex having this color.

Step 3. Check to see whether all vertices have color assignments. If not, return to step 2, choose a new color and assign it to the first uncolored vertex in the ordered list. Repeat step 2.

Step 4. Note the number of colors employed. This is an estimation of the minimum number of groups needed to eliminate conflicts. All vertices colored alike are assigned to the same group.

Now apply Welsh and Powell's algorithm to the graph in Figure 8.19. How many colors are needed to minimize "conflicts" here? Can you do it with less? If so, how?

Reducing conflict in class schedules, minimizing the number of colors required to print a map in a textbook, minimizing congestion at traffic lights, and optimizing numer-

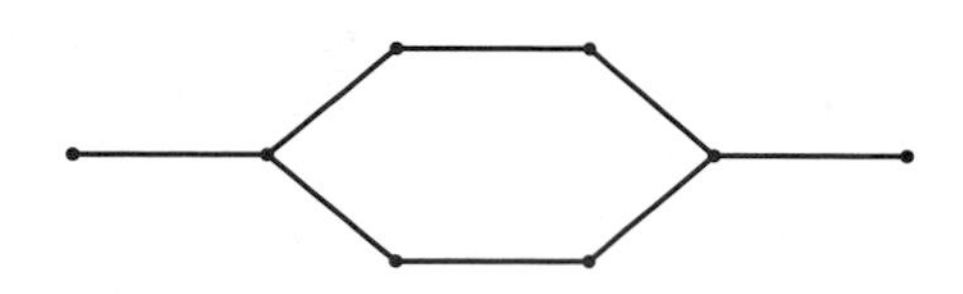

Figure 8.19 Another conflict problem.

ous other network routing problems are important activities. The results of our activities in Figure 8.19 demonstrate that no known algorithm guarantees a minimum assignment in every situation. This is a good open problem for students to ponder. Welsh and Powell's algorithm does, however, work well for most situations. Additional applications of graph theory abound, but these provide a good introduction to the field.

Another major area of study in discrete mathematics is the study of discrete dynamical systems. While we often call this activity recursion or iteration, the end goal is the same. Discrete dynamical systems first enter current school curricula almost "invisibly" as students study arithmetic and geometric sequences in the middle school and secondary school curriculum. These important concepts capture the essence of two of the most powerful forms of change that we can describe mathematically.

Arithmetic growth is described by the difference equation

$$A_n = A_{n-1} + d \qquad \text{for } n = 1, 2, 3, 4, \ldots$$
$$A_0 = a$$

From the initial value of a, each successive term in an arithmetic sequence is found by adding the common difference of d: $a, a + d, a + 2d, a + 3d, \ldots$. Many examples of arithmetic growth are found in elementary mathematics, ranging from the multiplication facts for 3 $(3, 6, 9, 12, \ldots)$ to the effects of simple interest at 5% on a \$100 investment over years (\$100, \$105, \$110, ...) to the growth in values of a linear function $f(x) = 3x + 5$ evaluated at nonnegative integer values $(5, 8, 11, 14, \ldots)$. The central property signaling the presence of the underlying difference equation model in these cases is the fact that the *subtractive* comparison between successive terms $A_n - A_{n-1}$ is a constant difference, d. In this case, the constant difference d reappears in a comparison involving the inverse operation to the nature of the growth.

Geometric growth is described by the difference equation

$$A_n = rA_{n-1} \qquad \text{for } n = 1, 2, 3, 4, \ldots$$
$$A_0 = a$$

From the initial value of a, each successive term in a geometric sequence is found by multiplying by the common factor of r: $a, ar, ar^2, ar^3, \ldots$. Many examples of geometric growth are found in elementary mathematics, from the study of the powers of 5 $(5, 25, 125, 625, \ldots)$ to the effects of compound interest at 5% on a \$100 investment over years (\$105.00, \$110.25, \$115.76, ...) to the growth in values of the exponential $g(x) = e^x$ evaluated at nonnegative integer values $(1, e, e^2, e^3, \ldots)$. The central property signaling the presence of the underlying difference equation model in these cases is the fact that the *divisive* comparison between successive terms A_n/A_{n-1} is the constant ratio, r. In this case, the constant r reappears in a comparison involving the inverse operation to the nature of the growth.

The study of difference equations also exposes students to rich applications. The following example uses difference equations to explain the long-term effect of quantitative decisions in a real-world setting.

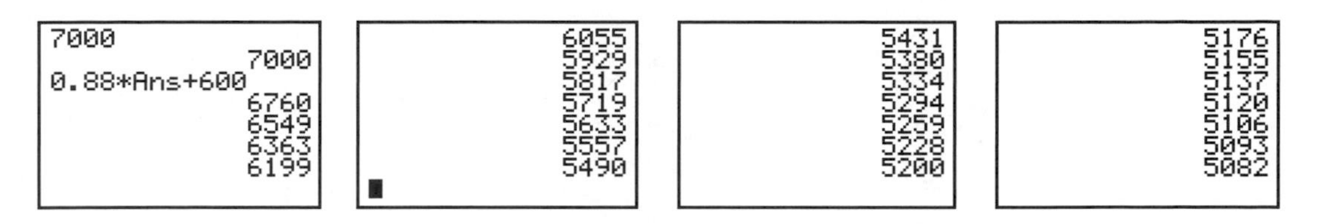

Figure 8.20 Calculator values for p_i in the timber stand model.

Example ***Modeling a Timber Stand***

The Clear Lake Pine Company owns a timber stand with 7000 pine trees (Dossey et al., 2001). Each year the company harvests 12% of its trees and plants 600 seedlings. They are particularly interested in the pine tree population in this timber stand in 10 years and in the long-range future. Examining this situation to determine the nature of change from one year to the next, we can build a first-order linear difference equation model for the pine population p_n in year n as follows:

$$p_n = 0.88p_{n-1} + 600 \qquad p_0 = 7000$$

We can iterate this model on a TI calculator with the following keystrokes: 7000, ENTER, 0.88 *2nd ANS + 600, and then pressing ENTER to get the value for p_1 and for successive values of p_i, as i increases. The first 26 values for the sequence are shown in Figure 8.20.

Iterating the difference equation forward, we see that the pine tree population tends to stabilize at 5000 over the long haul. An examination of our data shows that after ten years the timber stand will contain 5557 trees. Using the solution format for first-order, linear difference equations, we arrive at the general value for the nth year following the initiation of our model:

$$p_n = 2000(0.88)^n + 5000$$

Examining this as a function describing the pine tree population at each year following the first count of 7000 trees, it's easy to see that this model predicts an eventual steady-state population of 5000 trees in the stand. Viewed graphically, allowing a continuous graph for trend, we see that this happens over the first 100 years, as shown in Figure 8.21.

This movement to a limit can be viewed in a stepwise fashion using a cobweb graph. This approach helps us investigate relationships between the difference equation, the initial value, and potential limiting values. Consider Figure 8.22, which graphs the line $y = 0.88x + 600$ as a condition that establishes successive values, y, of the population, given the immediately preceding values, denoted by x. The graph of this line, considered with the auxiliary line $y = x$, allows us to establish geometrically the long-term behavior of the process.

Entering the y-value of 7000 for an initial population on the line $y = x$, we drop vertically to the line defining the difference equation and enter this as an x-value (which determines the next y-value). Then we move horizontally to the $y = x$ line to get the next x-value, then vertically to the difference equation line for the next y-value,

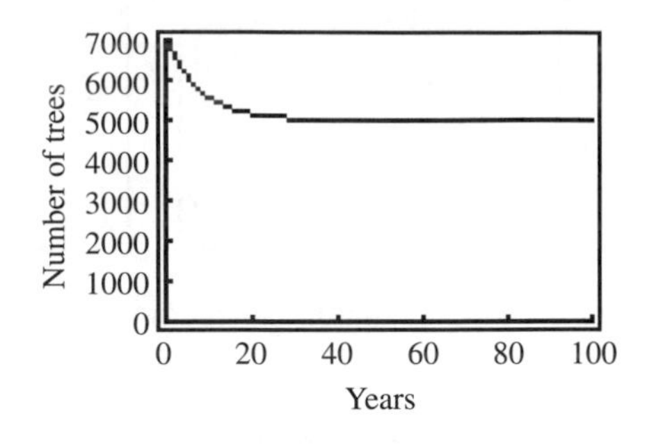

Figure 8.21 The first 100 years of the timber stand model.

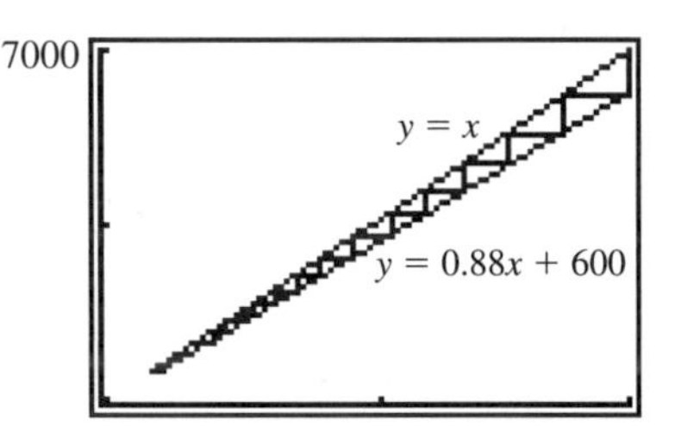

Figure 8.22 Cobweb graph for $p_n = 2000(0.88)^n + 5000$.

transforming back and forth from present value to next value as the $y = x$ line transfers the output value at one stage to input value at the next stage. The resulting pattern helps us visualize the convergence of tree population to the limiting value of 5000 pine trees.

Comparing the various values of the "slope" of the first-order linear difference equations shows that such equations having "slopes" with absolute values of greater than 1 have values that converge to a limit. What relationship might this have to summing geometric series? ■

The study of discrete dynamical systems gives the curriculum a strong warp, allowing many diverse topics to be woven together in a cohesive whole. The nature of arithmetic and geometric patterns, viewed in sequences, can be analyzed in terms of the type of change taking place in each. This allows students to look at the different types of proportionality that exist between successive terms in growth situations. Early examination of the rate of change introduces the concepts that later will be generalized into slope in linear equations in algebra and to the derivative in the study of curves in calculus.

Exercises 8.2

1. Graph (a) in the figure is related to the cube. In (b), the graph has been *triangulated*. The triangulation allows us to verify Euler's formula as we did with the tetrahedron.

The addition of the edge in a quadrilateral-shaped region adds 1 edge and 1 face, leaving $v + f - e$ still equal to 1.

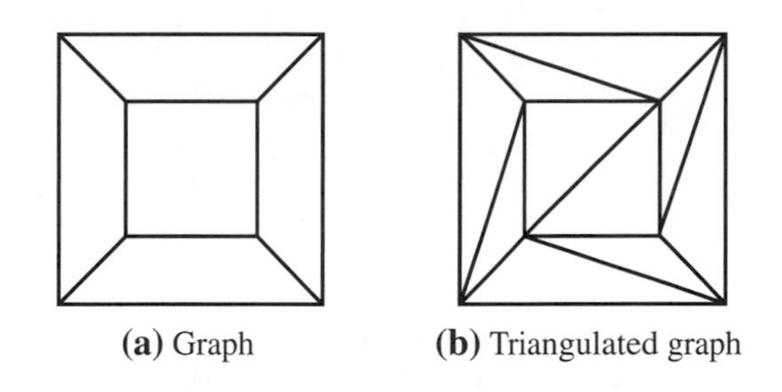

(a) Graph **(b)** Triangulated graph

Verify that Euler's formula holds for the triangulated graph (b).

For Exercises 2 and 3, refer to the following graph:

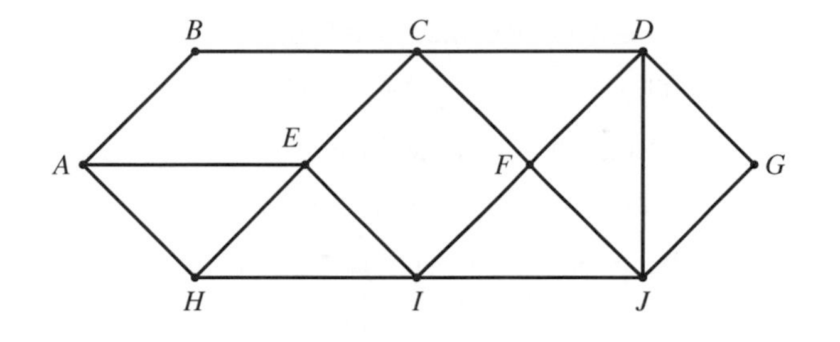

2. Describe an Euler circuit for this graph, if one exists. If not, tell why not.

3. Describe a Hamiltonian cycle for this graph, if one exists.

4. House Tracking. Consider the graph below, in which the edges connect rooms in a house. Is it possible to visit each room in the house without visiting any room twice? What does this say about cycles, circuits, or paths?

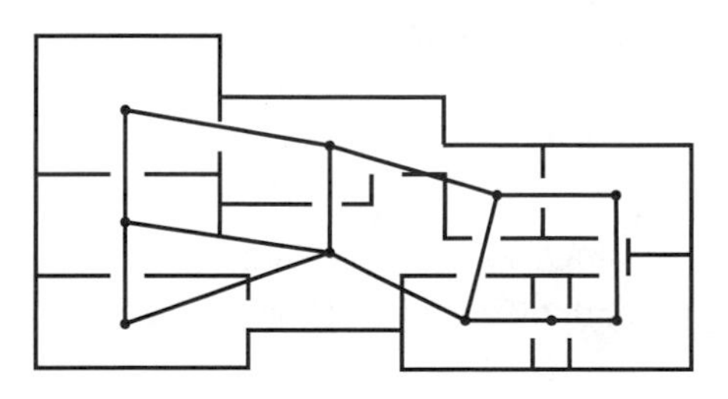

5. Cleaning Streets. You are the sanitary engineer for Somewheresville and you must determine a street sweeper route for the streets shown in the map. Find a path that covers each street at least once, has as short a length as possible, and returns to the starting garage. Your sweeper can clean both sides of the streets in one pass down a block and all blocks in Somewheresville are the same length.

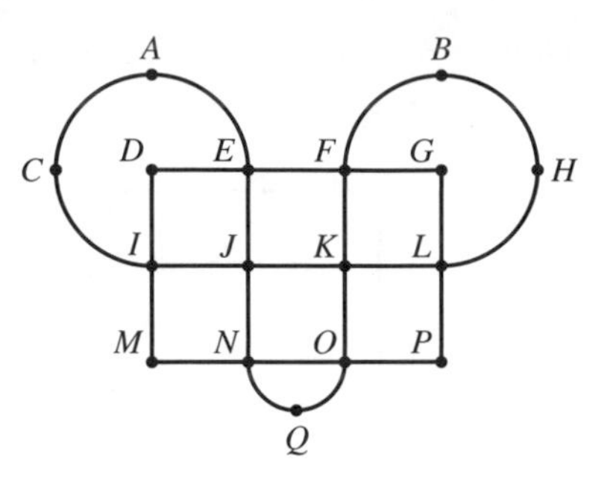

6. The high school student council issues a list of clubs they plan to sponsor next year. They ask students to sign up for clubs in the spring so the council can schedule meetings such that conflicts are minimized. If the clubs are listed as A–G, and students make the following selections, how many different club days will be needed to eliminate conflicts between these club members?

   A: Dossey, McCrone, and Day

   B: Giordano, Presmeg, and Crosswhite

   C: Giordano, Mundy, Wier, and Cooney

   D: Plantholt, McCrone, House, and Henderson

   E: Wier, Cooney, Mundy, and Sennott

   F: Day, Presmeg, McCrone, and Bishop

   G: Bishop, Plantholt, Giordano, and Dossey

7. John needs 25 mg of Vioxx every day to alleviate pain and swelling associated with his arthritis. Further, on average, his body eliminates 25% of the drug through body functions every 24 hours. How many units of Vioxx will be in John's body after the fourth dose? After the tenth dose? If a level of 100 mg is considered dangerous for the stomach lining, does it appear that John will ever be in danger?

8. In an attempt to reduce the spread of a pesticide-resistant fruit fly, sterilized male flies are released to mate with fertile females to cut the growth of the pest population. If the effect of this effort in a controlled environment is to reduce the overall population by 3% per month, what population reduction can be expected in the controlled environment in 6 months? In 1 year?

9. Which is a better investment scheme to employ over a 30-year period: Invest $500 per month over the entire period at 5% interest compounded monthly *or* invest $1000 per month over the last 15 years at 6% interest compounded monthly?

PROJECTS Find Out for Yourself

1. Examine a discrete mathematics textbook designed for secondary school classes. What topics does it cover? How are those topics related to the topics covered in an undergraduate college course in discrete mathematics? Do they have more of an applications focus or not?

2. Develop a lesson plan for constructing a minimal spanning tree for a graph. Your students are seniors in a high school discrete mathematics course.
3. Interview an experienced secondary school math teacher. Find out what discrete mathematics topics they include in their curriculum. In particular, what role do graph theory and discrete dynamical systems (or recursion) play in his or her program?

8.3 Advanced Placement Programs: Statistics and Calculus

The College Board offers a portfolio of examinations that students can take to demonstrate their competence in a particular area of mathematics. At present, the Advanced Placement (AP) program in mathematics has three examinations, one in statistics and two in calculus. The College Board, in cooperation with the Educational Testing Service, presents these examinations for students seeking either proficiency credit or advanced placement as they matriculate to the university level. The College Board also provides sample syllabi for schools to consider as they design curricula and prepare students for the AP examinations.

The AP examination in statistics is meant to be equivalent to a one-semester, introductory, noncalculus-based college course in statistics. Graphing calculators with statistical capabilities are required on the examination, but the College Board stresses that they are not equivalent to computers in the learning and teaching of statistics in the classroom. The curricular outline for the course and suggested classroom activities for the AP statistics course suggests that students have ample opportunities to use and interpret output from statistical software packages. Table 8.5 outlines the topics covered by the AP Statistics examination.

The study of calculus in the school curriculum spans a broad range of years. Students are exposed to the calculus when they consider rates of change in discussing patterns in the study of sequences or when they estimate the areas of irregular regions using smaller and smaller sets of regions. The original NCTM *Standards* called for all students to get some intuitive ideas of the calculus. In doing so, the standards called for students to be able to

- Discuss the notion of change and its representation/determine and interpret maximum and minimum points on a graph.
- Investigate limiting processes and discuss nature of arbitrarily small and large changes in a process.
- Consider limits in terms of sequences, series, and areas.
- Examine rates of change and accumulation in developing derivatives and integrals. (NCTM, 1989, p. 180)

Students should have experience in examining calculus topics as part of their regular mathematics curriculum. The Advanced Placement examination program provides

Table 8.5 Topics covered in the AP Statistics examination.

Topics
Exploring Data: Observing Patterns and Departures from Patterns
Interpreting graphical displays of distributions of univariate data (dot plot, stemplot, histogram, cumulative frequency plot)
Summarizing distributions of univariate data
Comparing distributions of univariate data (dot plots, back-to-back stem-and-leaf diagram, parallel box plots)
Exploring bivariate data
Exploring categorical data: frequency tables
Planning a Study: Deciding What and How to Measure
Overview of methods of data collection
Planning and conducting surveys
Planning and conducting experiments
Generalizability of results from observational studies, experimental studies, and surveys
Anticipating Patterns: Producing Models Using Probability Theory and Simulation
Probability as relative frequency
Combining independent random variables
The normal distribution
Sampling distributions
Statistical Inference: Confirming Models
Confidence intervals
Tests of significance
Special case of normally distributed data

an opportunity for those students whose schools provide a full-year course in calculus to certify their competence. The following paragraphs provide some ideas for developing the intuitive understandings the *Standards* call for and then discuss the AP program in calculus.

Students see the notion of the derivative emerge in the study of linear functions when they consider the slope of a line as representing the rate of change of the value of the function $f(x) = x^2$ for each change of 1 unit in the input. That is, as x increases by 1 unit, the value of the function $f(x) = mx + b$ increases by m. Using coordinate geometry and a family of secant lines all running through a point $(a, f(a))$, students determine a value that approximates the rate of change associated with the function f

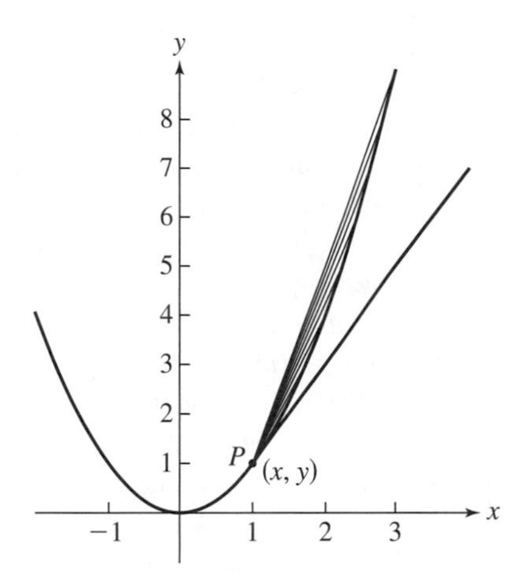

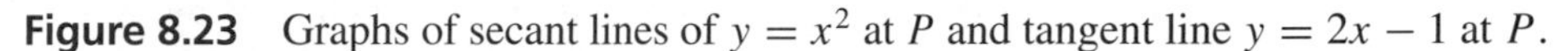
Figure 8.23 Graphs of secant lines of $y = x^2$ at P and tangent line $y = 2x - 1$ at P.

as the values of x approach a. The calculation of a few numerical examples suggest that the slope, or instantaneous rate of change associated with each point on the curve $(x, f(x))$, is given by $2x$ (see Figure 8.23).

In a like manner, students use the notion of limits to intuitively justify the formula for the area of a circle. Starting with a circle of unit radius r, having circumference $2\pi r$, they can repeatedly cut and rearrange the portions of the circles as shown in Figure 8.24.

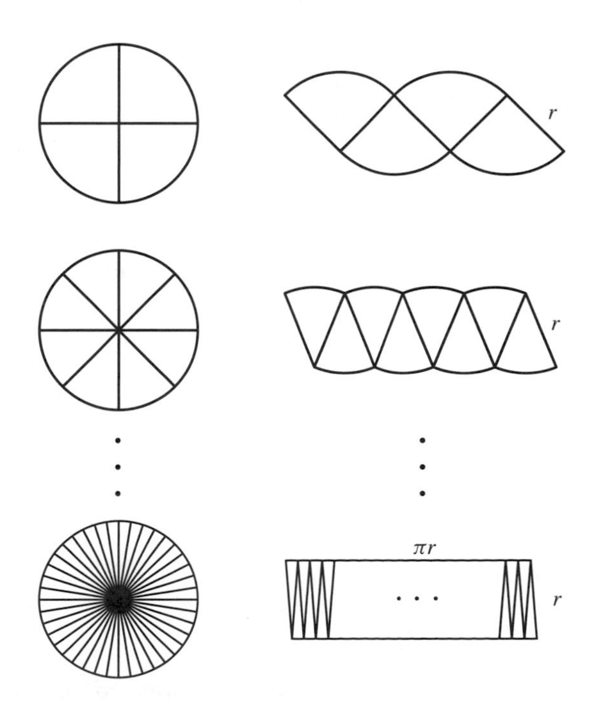

Figure 8.24 An intuitive justification for the formula for the area of a circle.

Each rearrangement has half the circumference, πr, for the entire length of each of the two curve segments, and a length of r for each side that represents the radii of the cut circles. Further, as the subdivisions continue, students see that the shape of the rearrangement approaches that of a rectangle. Hence, by analogy, the area of the circle, as rearranged, must be approximated by the area of the rectangle. But, the product of the "length" and "width" is given by $\pi r \cdot r$, or πr^2. Later, students develop formal proofs, but this informal example provides students with the nugget of the idea.

Key to developing student knowledge is bringing their understanding of both the conceptual and procedural aspects of calculus along. Past assessments of student knowledge of calculus show that students retain much more of the procedural knowledge than they do of the concepts. The three items in Figure 8.25 illustrate three levels of questioning about the concept of an integral and its properties. The first deals with students' knowledge of how to apply the procedure to evaluate a definite integral. The second deals with students' conceptual understanding of the integral as a defining area under a curve. The third deals with students' understanding of the integral as an accumulator and the interaction of that understanding with the concept of a positive function. The performance results of calculus students on these three items in the Second International Mathematics Study showed correct answers of, respectively, 71%, 59%, and 28%. Unfortunately, student understanding drops precipitously as they consider novel and deeper aspects of examination questions.

The AP program in calculus has two different examinations, each accompanied by a syllabus that outlines a solid preparation for its associated examination. Students are scored on a scale of 1–5; scores of 3–5 may earn proficiency credit and/or advanced placement at colleges and universities.

The two different AP examination options for students are known as Calculus AB and Calculus BC. Calculus BC is an extension of Calculus AB. Common topics require a similar degree of understanding. Table 8.6 on page 344 outlines the content covered in the two examinations. Topics marked with asterisks are found only in the Calculus BC exam.

The examinations consist of two multiple-choice sections for 105 minutes and two free-response sections for 90 minutes. Calculators are not allowed in one multiple-choice section and one free-response section, but are allowed in the remaining sections.

In addition to developing the course outline and examinations for students, the College Board also provides teaching ideas, workshops, and practice materials for teachers and students involved in AP courses. If you teach the calculus at the secondary school level, you will find these materials useful.

Exercises 8.3

1. Consider the three calculus problems in Figure 8.25. Work out your own solution to each problem. What errors did students most likely make if they selected wrong answers to these problems?

2. From the elementary standpoint, what concept related to a curve is the child in the cartoon on page 345 talking about?

$\int_1^2 (x^3 - x)\,dx$ is equal to

A $1\frac{3}{4}$ D 3
B 2 E 6
C $2\frac{1}{4}$

I. Assessing only the procedure

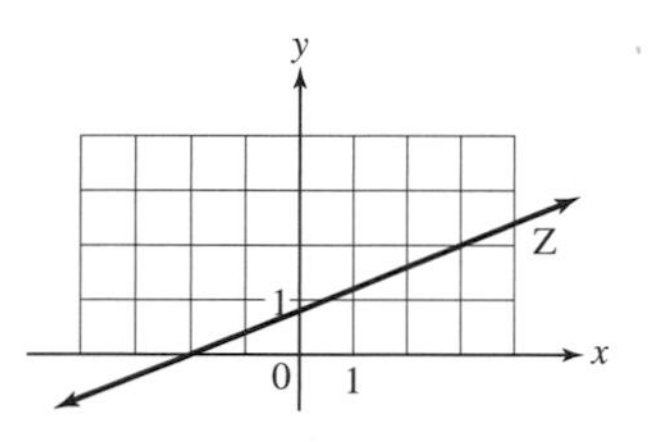

The line Z in the figure is the graph of $y = f(x)$.
$\int_{-2}^3 f(x)\,dx$ is equal to

A 3 D 5
B 4 E 5.5
C 4.5

II. Assessing the concept of integral as area under a curve

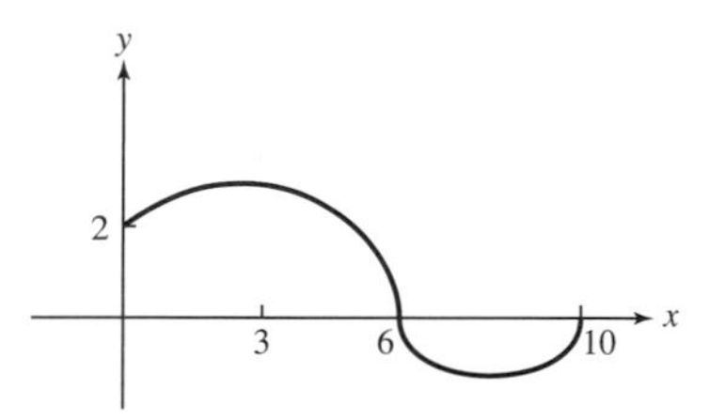

The graph of the function f is shown above for $0 \leq x \leq 10$.
$\int_0^a f(x)\,dx$ attains its greatest value when a is equal to

A 0 D 6
B 2 E 10
C 3

III. Assessing students' real understanding of the integral as an accumulator

Figure 8.25 Three examples of assessing student knowledge.

Table 8.6 Content covered in the AP Calculus AB and BC examinations.

I. Functions, Graphs, and Limits

Analysis of graphs

Limits of functions (including one-sided limits)

Asymptotic and unbounded behavior

Continuity as a property of functions

*Parametric, polar, and vector functions

II. Derivatives

Concept of the derivative

Derivative at a point

Derivative as a function

Second derivatives

Applications of derivatives

Computation of derivatives

III. Integrals

Interpretations and properties of definite integrals

*Applications of integrals

Fundamental theorem of calculus

Techniques of antidifferentiation

Applications of antidifferentiation

Numerical approximations to definite integrals

IV. Polynomial Approximations and Series

*Concept of series

*Series of constants

*Taylor series

*BC topics only.

THE FAMILY CIRCUS®
By Bil Keane
10-22
© 1990 Bil Keane, Inc.
Dist. by Cowles Synd., Inc.
BIL KEANE
"What makes hills in the road flatten out when we get there?"

3. Is infinity a number or a concept? How would you discuss infinity with geometry students, Algebra II students, and Calculus students? Would you make distinctions for each group in how you approach this topic?

8.4 Organizations and Competitions

One feature often associated with advanced mathematics classes is the availability of competitions and special clubs or school activities for students who really enjoy mathematics. Such activities give students with a keen interest in math opportunities to explore complex subjects in depth, hone their skills, meet professionals in the field, and interact with students with a passion for mathematics.

Mu Alpha Theta ($\mu\alpha\theta$), the national honor society for secondary and community college math students, is named for the Greek word for mathematics. Mu Alpha Theta is a national organization with chapters in over 1500 high schools nationwide, and as many as 50,000 student members. Mu Alpha Theta, like the National Honor Society, has a few basic rules for schools to form their own chapters and link to the national society: Student members must be enrolled in, or have completed, the equivalent of Algebra II; have a B average in mathematics; and the school must have a curriculum with courses through at least the precalculus level.

Mu Alpha Theta sponsors regional meetings for students during the year and an annual national meeting. These meetings have competitions and presentations on various topics. In addition, membership provides the school club with a quarterly newspaper on topics of interest to secondary math students. The society, as well as several school chapters, has a web site at

www.matheta.ou.edu

The Mathematical Association of America, through its American Mathematical Competitions Program, runs the largest set of competitions for middle and secondary school students. The AMC portfolio of competitions consists of a middle school contest, the AMC 8. This annual contest involves nearly a quarter of a million middle school students. The contest is designed to increase student interest in mathematics and to develop their problem-solving competencies through competition. The competition is a 25-question, 40-minute multiple-choice examination administered in the students' own schools. Calculators are allowed on the test, but no problem *requires* the use of a calculator. The questions range in difficulty from easy to very difficult in order to appeal to a broad range of students. The AMC 8 covers material normally associated with the middle school mathematics curriculum. Coverage includes, but is not limited to, such topics as

- Estimation
- Reading and interpreting graphs
- Percent
- Spatial visualization
- Graph theory
- Everyday applications

While algebraic thinking may lead to the solutions of some problems, no problem directly requires the use of algebra.

At the high school level, the AMC sponsors the AMC 10 and the AMC 12. These examinations are designed, like the AMC 8, to both interest and challenge students in grades 10–12, excluding calculus. Over 400,000 students take these two examinations in February of each year. Both tests are 25-question, 75-minute examinations.

The highest scorers on the AMC 12 are invited to participate in the American Invitational Mathematics Examination, whose highest performers get invited to participate in the U.S. Mathematical Olympiad. This six-question, six-hour examination is used to select the U.S. team members for the International Mathematical Olympiad, an international secondary school mathematics competition among teams representing nations.

Further information on the AMC's mathematics examinations can be found at their web site:

http://www.unl.edu/amc/

There are additional high school competitions of some national importance. Perhaps the most familiar of these is the American Regions Mathematics League (ARML). This competition is an annual national-level mathematics competition. High school students form teams of 15 to represent their city, state, county, or school and compete against the best from the United States and Canada. The competition takes place the first Saturday after Memorial Day and is held simultaneously at three sites: Penn State, The University of Iowa, and The University of Nevada—Las Vegas. The competition usually involves over 100 teams of students.

The 25th annual meet that took place Saturday, June 3, 2000 at Penn State and Iowa and on Friday-Saturday June 2–3, 2000 at UNLV saw 124 teams participate. Samples of ARML items and other information on this competition can be found at

http://www.armlmath.org/

At the middle school level, The MATHCOUNTS competition is the largest, and most well-known, competition. Each year, more than 500,000 students participate in MATHCOUNTS. The program promotes seventh and eighth grade mathematics achievement through local efforts in every state. MATHCOUNTS promotes student interest in mathematics by making mathematics achievement as challenging, exciting, and prestigious as school sports. Teachers and other volunteers work with groups of students, "mathletes," either as part of in-class instruction or as an extracurricular activity. After a lot of activities at the school and local level, each school selects a team to represent it, both individually and as a team in written and oral competitions. Through a series of competitions working from local to state to national levels, teams vie to show their best. Additional information on the competition can be found at

http://206.152.229.6

Two other programs should be noted as part of the special math activities for secondary school students. The first is the High School Competition in Mathematical Modeling (HiMCM). It is open to teams of students from secondary schools nationwide. The competition takes place in late winter and involves teams of at most four students. These students, working under the sponsorship of a teacher, have 24 hours to craft a solution to one of two mathematical modeling problems posed for that year's competition. Papers are assessed at a regional level with a number of papers declared national finalists from each region. These papers are then evaluated for national awards. HiMCM is hosted by the Consortium for Mathematics and Its Applications (COMAP) with support from the Mathematical Association of America, the National Council of Teachers of Mathematics, and the Institute for Management Science and Operations Research. Additional information on this competition can be found at

http://www.comap.com

The final activity of this type that we consider is the USA Mathematical Talent Search (USAMTS). The USAMTS is designed to encourage and assist the development of problem-solving skills of talented high school students. While most secondary mathematics competitions are timed tests, the USAMTS provides for more reflection on the part of students. Its aim is to foster insight, ingenuity, quick thinking, and creativity, along with writing skills and perseverance in communicating about and attacking significant problems. The USAMTS consists of four rounds, with each round featuring five problems. Students have about 30 days from the date the problem set appears on the USAMTS web site to the date they must mail their solutions. Round 1 is an exception as students have most of the summer to prepare their solutions for this round. The proposed solutions are submitted by mail. They are evaluated by professional mathematicians at the National Security Agency, and each student receives written comments on their work. A newsletter informs students of solutions to previous problems and an overall record is provided about commended participants via the newsletter and web site at

http://www.nsa.gov/programs/mepp/usamts.html

8.5 Lesson Plan for Discrete Mathematics

Lesson 9: Examining Change with Recursive Relations[1]

Objectives: Students will work with graphing calculators to examine the nature of change in the values related to a recursive relationship over time. In the process of examining the situation and sharing their observations, students will

1. Make discoveries about the rate of change in the values of the relationship.
2. Examine the rate of change through a graph of the recursive relationship.
3. Make connections between the parameters and initial conditions in the relationship and the nature of change.
4. Develop an understanding of the stability of a relationship and how that stability is related to initial conditions and parameters.
5. See the usefulness of recursive formulations to relationships in examining real-world situations.

Warm-Up Activity: The timber stand example from Section 8.2. Have the problem on a transparency for students to begin to consider as they enter the classroom.

> Use your calculators to investigate the following situation. What would you tell the company owners about the tree population in 10 years and in the long run?

[1] Based on a lesson in Everyday Learning's *Contemporary Mathematics in Context.*

> The Clear Lake Pine Company owns a timber stand with 7000 pine trees. Each year the company harvests 12% of its trees and plants 600 seedlings. They are particularly interested in the pine tree population in this timber stand in 10 years and in the future.

Activities:

1. (5–10 minutes) Warm-Up and Opening After discussing student responses to the warm-up, explain to students that they may want to develop the graph and table of values for this relationship on their calculators. Be sure students are comfortable with how to generate these displays for the relationship.

2. (5 minutes) Transition to Group Work Organize students into groups (prearranged) and distribute the worksheet shown on page 350.

3. (20 minutes) Group Work Allow students about 20 minutes to work together in their groups. Ask one member of each group to record the findings for each group.

4. (10 minutes) Group Discussion Bring students back together and have groups share their findings and solutions to the problem. Be sure that various approaches to solving the problem are shared. If all of the approaches are similar, work with the students to see other representations for the problem and their relationship to other approaches as part of a whole-class discussion.

5. (5–10 minutes) Follow-up Questions Ask some or all of the following questions:

   - What is the role of the coefficient of a_{n-1} in determining the rate of change?

   - How do you find the value (equilibrium value) associated with a fixed point for the recursive relationship? Does a relationship always have one?

   - What is the relationship between the initial value and the value of the fixed point?

   - Will all recursive relationships approach the value of a fixed point no matter what their initial condition is?

   - How can you see the above results in a sequence graph, in a cobweb graph?

6. (Remaining Time) Allow time for students to rejoin groups and begin discussion of the recursion problem assigned for homework. Inform the students assigned to place the problems on the board at the beginning of the next class period that they will make presentations to the class.

Worksheet:

Consider the following problem. Then answer the questions following it.

Aunt Em feels a scratchy throat and develops a cough and some drainage late one Friday evening. She starts taking a cough/nasal medication immediately. The medication calls for 5-mL dosages of the medication every 4 hours. It is well known that the body loses 20% of the medication during the 4-hour period after taking it. Assuming that the initial dosage is 5 mL and that Aunt Em takes the medicine every 4 hours throughout each 24-hour period, answer the following questions:

1. What recursion relation represents this situation?
2. What is the initial value for the relation?
3. What represents the rate of change in the relationship?
4. How much medicine will gradually accumulate in Aunt Em's body over the week that she continues to take the cold medicine as described?
5. How did you find the long-term (7 days) value? Could you find it another way (tabularly, graphically, symbolically)?
6. What happens if the initial value indicates that Aunt Em took a "swig" of the cough medicine (16 mL) and then settled in to following the dosages of 4 mL per period?
7. What happens if the first dosage period does not take place (initial value is 0 mL), and Aunt Em started the dosage 4 hours later after her trip to the drug store?
8. Which representation (graph or table) best shows the rate of change of the level of medicine in Aunt Em's body? Why?
9. If you were going to describe the long-term (7 days) behavior of the medicine levels in Aunt Em's body, how would you present this information graphically in class? Why?

Assessment:

1. Warm-Up To give me a sense of where students are coming from, what they learned from previous lessons, and who might need extra guidance in today's lesson.
2. Group Work Watch students as they work to see who is contributing ideas and how other group members are responding to and working with these ideas. Question students on occasion to get a better sense of their understanding of the problem, the solution strategies being used, their confidence in the solution,

their ideas about what the solution has to do with the coefficient of a_{n-1} and the initial value for the relationship.

3. Whole-Class Discussion Assess the understanding of those who contribute to the discussion. Try to get a sense of what all students are thinking by asking for a show of hands on some questions or responses.

4. End of Class Assign a problem to be solved and written up for handing in during the next class period. Focus on their ability to handle a problem similar to the worksheet, but with a ratio term greater than 1.

References

Dossey, J. A. (1991). Discrete mathematics: The math for our time. *In* M. Kenney and C. Hirsch (eds.). *Discrete mathematics across the curriculum, K–12*. Reston, VA: National Council of Teachers of Mathematics.

Dossey, J. A., Otto, A. D., Spence, L. E., & Vanden Eynden, C. (2001). *Discrete mathematics* (4th ed.). New York: Scott-Foresman.

National Council of Teachers of Mathematics (1989). *Curriculum and evaluation standards for school mathematics*. Reston, VA: Author.

National Council of Teachers of Mathematics. (2000). *Principles and standards for school mathematics*. Reston, VA: Author.

9
Model Fitting and Empirical Model Construction

The merit of a mathematical model viewed as a scientific contribution to the study of some phenomena is determined by the degree to which the predictions of the model agree with observations. Normally model building is a cyclic process and the first attempt does not yield predictions which agree closely with observations. In such a case the model must be modified and a new set of predictions derived. It is common for a model to be refined several times before adequate agreement is obtained, and this refinement may continue over many years.

M. Thompson (1981)

Overview

This chapter, and the following two chapters, provide extended glimpses into the construction of models and the application of classical models to more advanced parts of the curriculum. They are provided to assist you in understanding the work behind some of the more advanced topics in the secondary school curriculum. As such, these chapters are designed to extend and connect the material you may have had in other courses and relate it to the building of models, the application of optimization techniques, and applying calculus in the development of models. While this content supports teaching in the spirit of the NCTM *Standards*, it goes beyond the level of content directly contained in the high school curriculum. As a result, we do not include in this chapter a Focus on the NCTM *Standards* or a Focus on the Classroom section. Here our focus is on how models are developed using what we know about functions and their behaviors and what we know about how to adjust functional representations to deal with the errors we see between our proposed models and observations of the phenomena being modeled.

9.1 Introduction to Model Fitting

In the mathematical modeling process we encounter situations that cause us to analyze data for different purposes. You have already seen how our assumptions can lead to a model of a particular type. For example, in Chapter 5, we conducted an experiment to measure the stretch of a spring as a function of the mass (measured as a weight) placed upon the string. We hypothesized a model

$$e = Cm$$

where e is the distance the spring is stretched by mass m, and C is some arbitrary constant of proportionality. At this point we can collect and analyze sufficient data to determine if the assumptions are reasonable. If they are, we want to determine the constant C that selects the particular member from the family $e = Cm$ corresponding to the particular spring we are calibrating. What value of C is *best*? What do we mean by *best*?

A different case arises when the problem is so complex as to prevent the formulation of a model explaining the situation. For instance, there may be so many significant variables involved that we would not even attempt to construct an explicative model. (Consider estimating the speed at which people walk in cities of different sizes in distinct countries.) In such cases experiments may have to be conducted to investigate the behavior of the independent variable(s) within the range of the data points.

The preceding discussion identifies three possible tasks when analyzing a collection of data points:

1. Fitting a selected model type or types to the data.
2. Choosing the most appropriate model from competing types that have been fitted. For example, we may need to determine whether the best-fitting exponential model is a better model than the best-fitting polynomial model.
3. Making predictions from the collected data.

In the first two tasks a model or competing models exist that seem to *explain* the behavior being observed. We address these two cases in Sections 9.1 through 9.3 under the general heading of *model fitting*. In the third case, however, a model does not exist to explain the behavior being observed. Rather, there exists a collection of data points that can be used to *predict* the behavior within some range of interest. In essence we wish to construct an *empirical model* based on the collected data. In Sections 9.4 and 9.5 we study such empirical model construction under the general heading of *interpolation*. It is important to understand both the philosophical and the mathematical distinctions between model fitting and interpolation.

The Relationship Between Model Fitting and Interpolation

Let's analyze the three tasks identified to determine what must be done in each case. In task 1 the precise meaning of best model must be identified and the resulting mathematical problem resolved. In task 2 a criterion is needed for comparing models of different types. In task 3 a criterion must be established for determining how to make predictions in between the observed data points.

Note the difference in the modeler's attitude in each of these situations. In the two model-fitting tasks, a relationship of a particular type is strongly suspected, and the modeler is willing to accept some deviation between the model and the collected data points in order to have a model that satisfactorily *explains* the situation under investigation. In fact, the modeler expects errors to be present in both the model and the data. On the other hand, when interpolating, the modeler is strongly guided by the data that have been carefully collected and analyzed, and a curve is sought that captures the trend of the data in order to *predict* in between the data points. Thus the modeler generally attaches little explicative significance to the interpolating curves. In all situations the modeler may ultimately want to make predictions from the model. However, the modeler tends to emphasize the proposed *models* over the data when model fitting, whereas she places greater confidence in the *collected data* when interpolating and attaches less significance to the form of the model. In a sense, explicative models are *theory* driven while predictive models are *data* driven.

Let's illustrate the preceding ideas with an example. Suppose we are attempting to relate two variables, y and x, and have gathered the data plotted in Figure 9.1. If the modeler is going to make predictions based solely upon the data in the figure, he or she

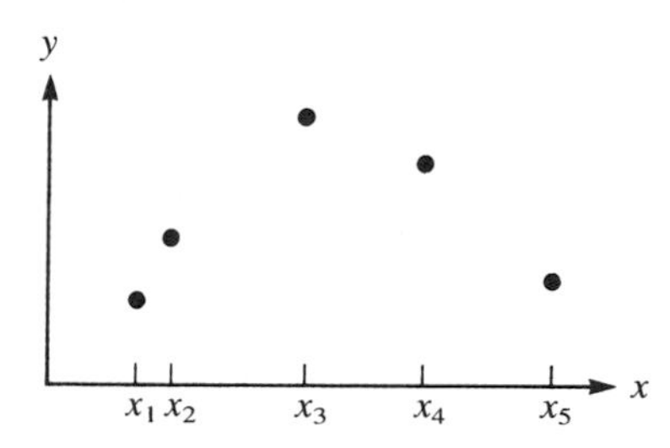

Figure 9.1 Observations relating the variables y and x.

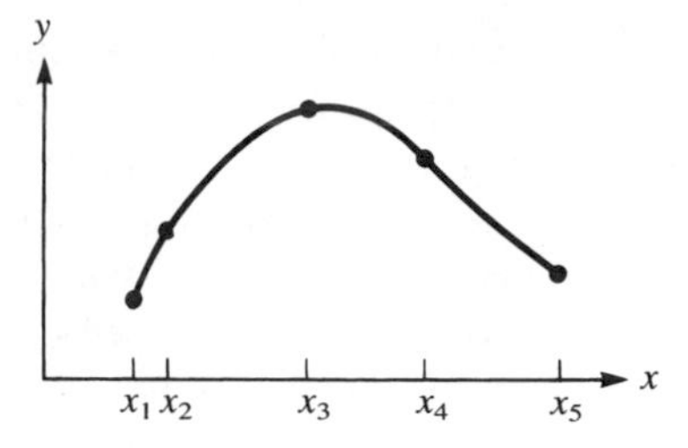

Figure 9.2 Interpolating the data using a smooth polynomial.

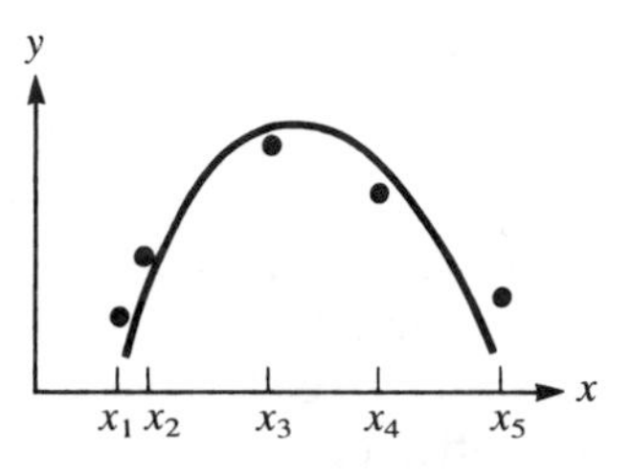

Figure 9.3 Fitting a parabola $y = C_1x^2 + C_2x + C_3$ to the data points.

might pass a polynomial though the points (see Figure 9.2). Note that in Figure 9.2 the interpolating curve passes through the data points and captures the trend of the behavior over the range of observations.

However, suppose that in studying the particular behavior depicted in Figure 9.1 the modeler makes assumptions leading to the expectation of a quadratic model, or parabola, of the form $y = C_1x^2 + C_2x + C_3$. In this case the data of Figure 9.1 would be used to determine the arbitrary constants C_1, C_2, and C_3 to select the best parabola (see Figure 9.3). The fact that the parabola may deviate from some or all of the data points would be of no concern. Note the difference in the values of the predictions made by the curves in Figure 9.2 and Figure 9.3 in the vicinity of the values x_1 and x_5.

Sources of Error in the Modeling Process

Before discussing criteria upon which to base curve-fitting and interpolation decisions, we need to examine the modeling process in order to ascertain where errors can arise. If error considerations are neglected, undue confidence may be placed in intermediate results, causing faulty decisions in subsequent steps. Our goals are to ensure that all parts of the modeling process are computationally compatible and to consider the effects of cumulative errors likely to exist from previous steps.

For purposes of easy reference, we classify errors under the following scheme:

1. Formulation error
2. Truncation error

3. Round-off error
4. Measurement error

Formulation errors result from the assumption that certain variables are negligible or from simplifications arising in describing interrelationships among the variables in the various submodels. Formulation errors are present in even the best models.

Truncation errors are attributable to the numerical method used to solve a mathematical problem. For example, we may find it necessary to approximate $\sin x$ with a polynomial representation obtained from the power series

$$\sin x = x - \frac{x^3}{3!} + \frac{x^5}{5!} - \cdots$$

An error will be introduced when the series is truncated to produce the polynomial.

Round-off errors are caused by using a finite digit machine for computation. Because all numbers cannot be represented exactly using only finite representations, we must always expect round-off errors. For example, consider a calculator or computer that uses eight-digit arithmetic. The number $\frac{1}{3}$ is represented by .33333333 so that 3 times $\frac{1}{3}$ is the number .99999999 rather than the actual value 1. The error 10^{-8} is caused by round-off. The ideal real number $\frac{1}{3}$ is an *infinite* string of decimal digits .3333 . . . , but calculators and computers can do arithmetic only with numbers having finite precision. When many arithmetic operations are performed in succession, each with its own round-off, the accumulated effect of round-off can significantly alter the numbers that are supposed to be the answer. Round-off is just one of the things we have to live with—*and be aware of*—when we use computing machines.

Measurement errors are caused by imprecision in data collection. This imprecision may include such diverse things as human errors in recording or reporting the data, or physical limitations of the laboratory equipment. For example, measurement error would be expected when measuring the elongation of a spring as a function of the weight added.

9.2 Fitting Models to Data Graphically

Assume the modeler has made certain assumptions leading to a model of a particular type. The model generally contains one or more parameters, and sufficient data must be gathered to determine them. Let's consider the problem of data collection.

The determination of how many data points to collect involves a trade-off between the cost of obtaining them and the accuracy required of the model. As a minimum, the modeler needs at least as many data points as there are arbitrary constants in the model curve. Additional points are required to determine any arbitrary constants involved with some technique of "best fit" being used. The *range* over which the model is to be used determines the endpoints of the interval for the independent variable(s). The *spacing* of the data points within that interval is also important because any part of the interval over which the model must fit particularly well can be weighted by using unequal spacing. We may choose to take more data points where maximum use of the model is expected,

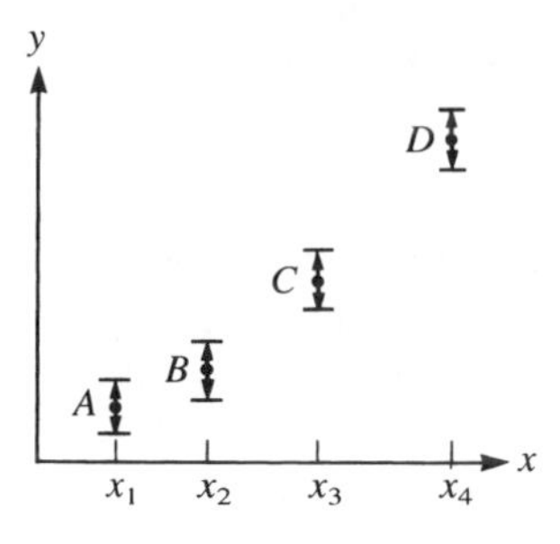

Figure 9.4 Each data point is thought of as an interval of confidence.

or we may collect more data points where we anticipate abrupt changes in the dependent variable(s).

Even if the experiment has been carefully designed and the experiments meticulously conducted, the modeler needs to appraise the accuracy of the data before attempting to fit the model. How were the data collected? What is the accuracy of the measuring devices used in the collection process? Do any points appear suspicious? Following such an appraisal and elimination (or replacement) of spurious data, it is useful to think of each data point as an interval of relative confidence rather than as a single point. This idea is shown in Figure 9.4. The length of each interval should be commensurate with the appraisal of the errors present in the data collection process.

Visual Model Fitting with the Original Data

Suppose it is desired to fit the model $y = ax + b$ to the data shown in Figure 9.4. How might the constants a and b be chosen to determine the line that best fits the data? Generally, when more than two data points exist, all of them cannot be expected to lie exactly along a single straight line, even if such a line accurately models the relationship between the two variables x and y. That is, ordinarily there will be some vertical discrepancy between some of the data points and any particular line being considered. We refer to these vertical discrepancies as **absolute deviations** (see Figure 9.5). For the best-fitting line we might try to minimize the sum of these absolute deviations, leading to the model depicted in Figure 9.5. Although success may be achieved in minimizing the sum of the absolute deviations, the absolute deviation from individual points may be

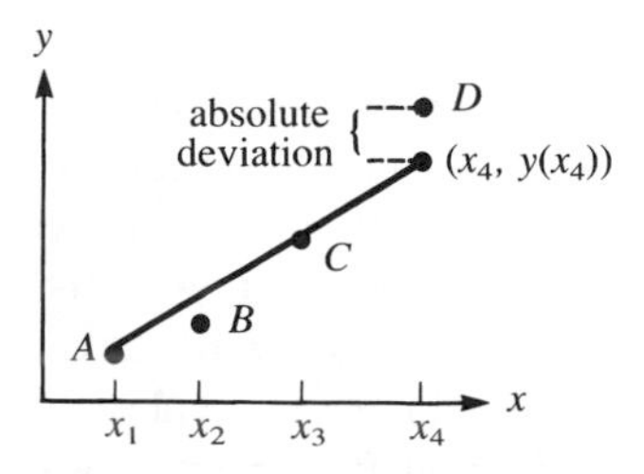

Figure 9.5 Minimizing the sum of the absolute deviations from the fitted line.

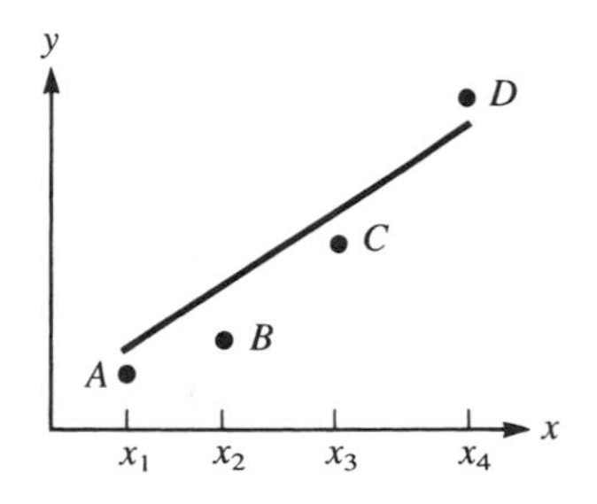

Figure 9.6 Minimizing the largest absolute deviation from the fitted line.

quite large. For example, consider point D in Figure 9.5. If the modeler has confidence in the accuracy of this data point, there would be concern for the predictions made from the fitted line near the point. As an alternative, suppose a line is selected that minimizes the largest deviation from any point. Applying this criterion to the data points might give the line shown in Figure 9.6.

Although these visual methods for fitting a line to data points may appear imprecise, the methods are often quite *compatible* with the accuracy of the modeling process itself. The grossness of the assumptions and the imprecision involved in the data collection may not warrant a more sophisticated analysis. In such situations the blind application of one of the analytic methods to be presented in Section 9.3 may lead to models far less appropriate than one obtained graphically. Furthermore, a visual inspection of the model fitted graphically to the data immediately gives an impression of *how good* the fit is and *where* it appears to fit well. Unfortunately, these important considerations are often overlooked in problems with large amounts of data analytically fitted via computer codes. Because the model-fitting portion of the modeling process seems to be more precise and analytic than some of the other steps, there is a tendency to place undue faith in the numerical computations.

Transforming the Data

Most of us are limited visually to fitting only lines. So how can we graphically fit other curves as models? Suppose, for example, that a relationship of the form $y = Ce^x$ is suspected for some submodel and the data shown in Table 9.1 have been collected.

The model states that y is proportional to e^x. Thus, if we plot y versus e^x, we should obtain approximately a straight line. The situation is depicted in Figure 9.7. Because the plotted data points do lie approximately along a line that projects through the origin, we conclude that the assumed proportionality is reasonable. From the figure,

Table 9.1 Collected data.

x	1	2	3	4
y	8.1	22.1	60.1	165

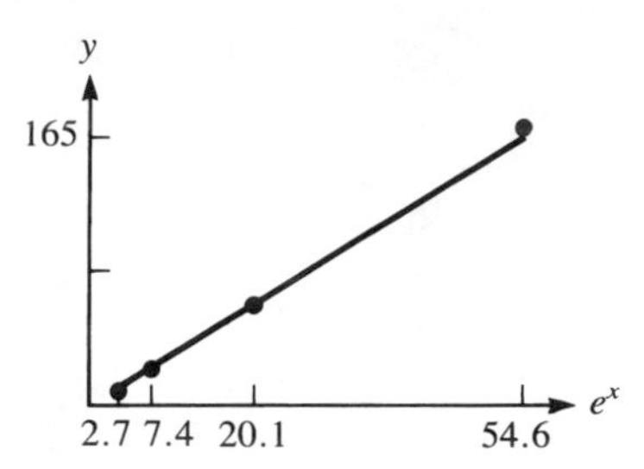

Figure 9.7 Plot of y versus e^x for the data given in Table 9.1.

the slope of the line is approximated as

$$C = \frac{165 - 60.1}{54.6 - 20.1} \approx 3.0$$

Now let's consider an alternate technique that is useful in a variety of problems. Take the logarithm of each side of the equation $y = Ce^x$ to obtain

$$\ln y = \ln C + x$$

Note that this expression is an equation of a line in the variables $\ln y$ and x. The number $\ln C$ is the intercept when $x = 0$. The transformed data are shown in Table 9.2 and plotted in Figure 9.8. Semi-log paper or a computer is useful when plotting large amounts of data.

From Figure 9.8 we can determine that the intercept $\ln C$ is approximately 1.1, giving $C = e^{1.1} \approx 3.0$ as before.

Table 9.2 The transformed data from Table 9.1.

x	1	2	3	4
$\ln y$	2.1	3.1	4.1	5.1

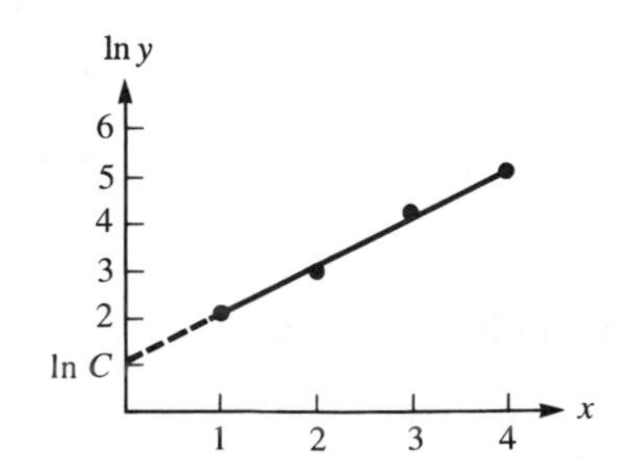

Figure 9.8 Plot of $\ln y$ versus x using Table 9.2.

A similar transformation can be performed on a variety of other curves to produce linear relationships among the resulting transformed variables. For example, if $y = x^a$, then

$$\ln y = a \ln x$$

is a linear relationship in the transformed variables $\ln y$ and $\ln x$. Here log–log paper or a computer is useful when plotting large amounts of data.

Let's pause and make an important observation. Suppose we do invoke a transformation and plot $\ln y$ versus x, as in Figure 9.8, and find the line that successfully minimizes the sum of the absolute deviations of the transformed data points. The line then determines $\ln C$, which in turn produces the proportionality constant C. Although it is not obvious, the resulting model $y = Ce^x$ is not a member of the family of exponential curves of the form ke^x that minimizes the sum of the absolute deviations from the original data points (when we plot y versus x). This important idea will be demonstrated both graphically and analytically in the ensuing discussion. When transformations of the form $y = \ln x$ are made, the distance concept is distorted. While a fit that is compatible with the inherent limitations of a graphical analysis may be obtained, the modeler must be aware of this distortion and *verify the model using the graph from which it is intended to make predictions or conclusions—namely the y versus x graph in the original data rather than the graph of the transformed variables.*

We now present an example illustrating how a transformation may distort distance in the xy-plane. Consider the data plotted in Figure 9.9 and assume the data are expected to fit a model of the form $y = Ce^{1/x}$. Using a logarithmic transformation as before, we find

$$\ln y = \frac{1}{x} + \ln C$$

A plot of the points $\ln y$ versus $\frac{1}{x}$ based on the original data is shown in Figure 9.10. Note from the figure how the transformation distorts the distances between the original data points and squeezes them all together. Consequently, if a straight line is made to fit the transformed data plotted in Figure 9.10, the absolute deviations appear relatively small (that is, small computed on the Figure 9.10 scale rather than on the Figure 9.9 scale). However, if we were to plot the fitted model $y = Ce^{1/x}$ to the data in Figure 9.9, we would see that it fits the data relatively poorly as shown in Figure 9.11.

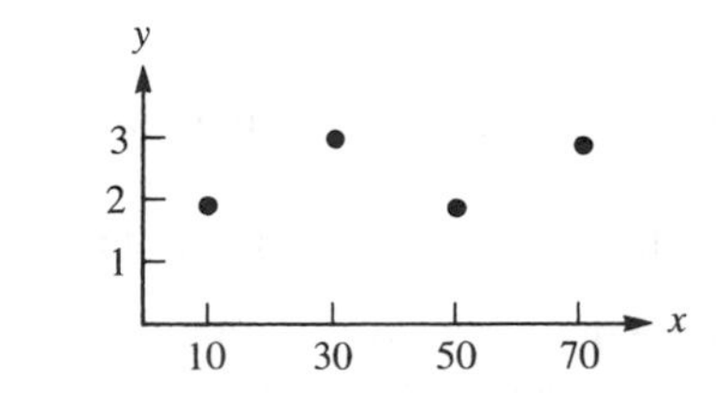

Figure 9.9 A plot of some collected data points.

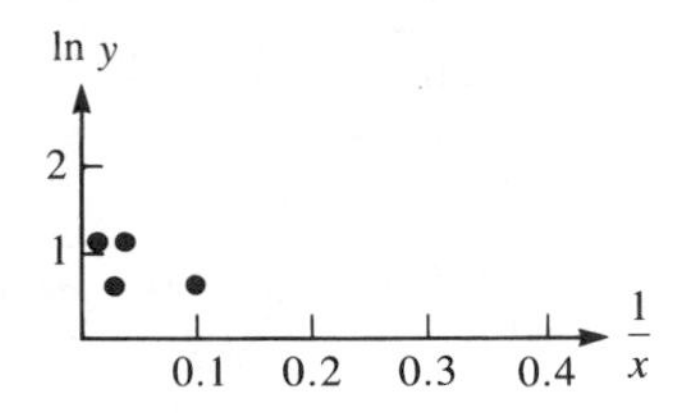

Figure 9.10 A plot of the transformed data points.

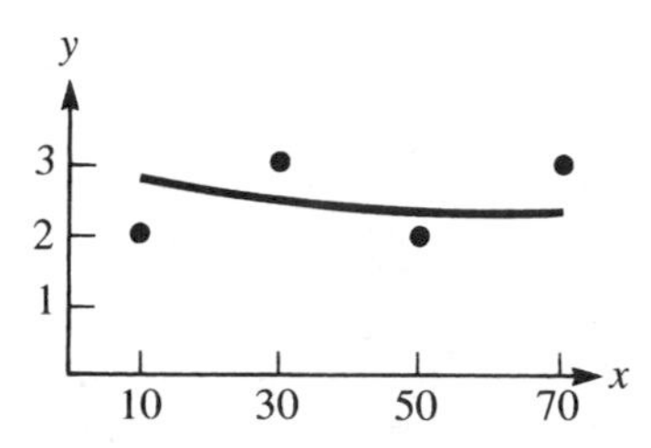

Figure 9.11 A plot of the curve $y = Ce^{1/x}$ based on the value $\ln C = 0.9$ from Figure 9.10.

From the preceding example, we see that if a modeler is not careful when using transformations, he or she can be tricked into selecting a relatively poor model. This realization becomes especially important when comparing alternative models. Serious errors can be introduced when selecting the best model unless all comparisons are made with the original data (plotted in Figure 9.9 in our example). Otherwise, the choice of best model may be determined by a peculiarity of the transformation rather than on the merits of the model and how well it fits the original data. Whereas the danger of making transformations is evident in this graphical illustration, a modeler may be fooled if he or she is not especially observant because many computer codes fit models by first making a transformation. If the modeler intends to use indicators, such as the sum of the absolute deviations, to make decisions about the adequacy of a particular submodel or choose among competing submodels, the modeler must first ascertain how the indicators were computed.

Exercises 9.2

1. The model in Figure 9.2 would normally be used to predict behavior between x_1 and x_5. What would be the danger of using the model to predict y for values of less than x_1 or greater than x_5? Suppose we are modeling the trajectory of a thrown baseball.

2. The following table gives the elongation e in inches per inch (in./in.) for a given stress S on a steel wire measured in pounds per square inch (lb/in^2). Test the model $e = c_1 S$ by plotting the data. Estimate c_1 graphically.

$S(x10^{-3})$	5	10	20	30	40	50	60	70	80	90	100
$e(x10^5)$	0	19	57	94	134	173	216	256	297	343	390

3. In the following data, x is the diameter of a ponderosa pine in inches measured at breast height and y is a measure of volume—number of board feet divided by 10. Test the model $y = ax^b$ by plotting the transformed data. If the model seems reasonable, estimate the parameters a and b of the model graphically.

x	17	19	20	22	23	25	28	31
y	19	25	32	51	57	71	113	141

x	32	33	36	37	38	39	41
y	123	187	192	205	252	259	294

4. In the following data, V represents a mean walking velocity and P represents the population size. We wish to know if we can predict the population size P by observing how fast people walk. Plot the data. What kind of a relationship is suggested? Test the following models by plotting the appropriate transformed data.

a. $P = aV$
b. $P = a \ln V$

V	P	V	P	V	P
2.27	2500	3.85	78,200	4.90	49,375
2.76	365	4.31	70,700	5.05	260,200
3.27	23,700	4.39	138,000	5.21	867,023
3.31	5491	4.42	304,500	5.62	1,340,000
3.70	14,000	4.81	341,948	5.88	1,092,759

5. The following data represent the growth of a population of fruit flies over a 6-week period. Test the following models by plotting an appropriate set of data. Estimate the parameters of the models:

a. $P = c_1 t$
b. $P = ae^{bt}$

t (days)	7	14	21	28	35	42
P (number of observed flies)	8	41	133	250	280	297

6. The following data represent (hypothetical) energy consumption normalized to the year 1900. Plot the data. Test the model $Q = ae^{bx}$ by plotting the transformed data. Estimate the parameters of the model graphically.

x	Year	Consumption Q
0	1900	1.00
10	1910	2.01
20	1920	4.06
30	1930	8.17
40	1940	16.44
50	1950	33.12
60	1960	66.69
70	1970	134.29
80	1980	270.43
90	1990	544.57
100	2000	1096.63

7. In 1601, the German astronomer Johannes Kepler became director of the Prague Observatory. Kepler had been helping Tycho Brahe in collecting 13 years of observations on the relative motion of the planet Mars. By 1609 Kepler had formulated his first two laws:

i. Each planet moves on an ellipse with the sun at one focus.

ii. For each planet, the line from the sun to the planet sweeps out equal areas in equal times.

Kepler spent many years verifying these laws and formulating a third law, which relates the orbital periods and mean distances from the sun.

a. Plot the period time T versus the mean distance r using the updated observational data:

Planet	Period (days)	Mean distance from the sun (km $\times$ 10^{-6})
Mercury	88	57.9
Venus	225	108.2
Earth	365	149.6
Mars	687	227.9
Jupiter	4329	778.3
Saturn	10,753	1427
Uranus	30,660	2870
Neptune	60,150	4497
Pluto	90,670	5907

b. Assuming a relationship of the form

$$T = Cr^a$$

determine the parameters C and a by plotting $\ln T$ versus $\ln r$. Does the model seem reasonable? Try to formulate Kepler's third law.

9.3 Analytical Methods of Model Fitting

In this section we investigate several criteria for fitting curves to a collection of data points. Each criterion gives a way for selecting the best curve from a given family in the sense that the curve most accurately represents the data according to the criterion. We also discuss how the various criteria are related.

Chebyshev Approximation Criterion

In the preceding section we graphically fit lines to a given collection of data points. One of the best-fit criteria used was to minimize the largest distance from the line to any corresponding data point. Let's analyze this geometric construction. Given a collection of m data points (x_i, y_i), $i = 1, 2, \ldots, m$, fit the collection to the line $y = ax + b$, determined by the parameters a and b, that minimizes the distance between any data point (x_i, y_i) and its corresponding data point on the line $(x_i, ax_i + b)$. That is, minimize the largest absolute deviation $|y_i - y(x_i)|$ over the entire collection of data points. Now let's generalize this criterion.

Given some function type $y = f(x)$ and a collection of m data points (x_i, y_i), minimize the largest absolute deviation $|y_i - f(x_i)|$ over the entire collection. That is, determine the parameters of the function type $y = f(x)$ that minimizes the number

$$\text{Maximum } |y_i - f(x_i)| \quad i = 1, 2, \ldots, m \tag{1}$$

This important criterion is often called the **Chebyshev approximation criterion**. The difficulty with the Chebyshev criterion is that it is often complicated to apply in practice, at least using only elementary calculus. The optimization problems that result from applying the criterion may require advanced mathematical procedures or numerical algorithms requiring the use of a computer.

Consider using the Chebyshev criterion to fit the model $y = Cx$ to the following data set:

x	1	2	3
y	2	5	8

If the Chebyshev approximation criterion is applied, a value would be assigned to C in such a way as to minimize the largest of the three numbers $|r_1|$, $|r_2|$, $|r_3|$, where, for example

$$r_2 = y_2 - y(x_2) = 5 - 2C$$

If we call that largest number r, then we want to

$$\begin{aligned} &\text{Minimize } r \\ \text{Subject to:}\quad &|r_1| \le r \quad \text{or} \quad -r \le r_1 \le r \\ &|r_2| \le r \quad \text{or} \quad -r \le r_2 \le r \\ &|r_3| \le r \quad \text{or} \quad -r \le r_3 \le r \end{aligned}$$

Each of these conditions can be replaced by two inequalities. For example, $|r_1| \leq r$ can be replaced by $r - r_1 \geq 0$ and $r + r_1 \geq 0$. If this is done for each condition, the problem can be stated in the form of a classical mathematical problem:

$$\text{Minimize } r$$

$$\begin{array}{llll} \text{Subject to:} & r - (2 - C) \geq 0 & \text{Constraint 1:} & (r - r_1 \geq 0) \\ & r + (2 - C) \geq 0 & \text{Constraint 2:} & (r + r_1 \geq 0) \\ & r - (5 - 2C) \geq 0 & \text{Constraint 3:} & (r - r_2 \geq 0) \\ & r + (5 - 2C) \geq 0 & \text{Constraint 4:} & (r + r_2 \geq 0) \\ & r - (8 - 3C) \geq 0 & \text{Constraint 5:} & (r - r_3 \geq 0) \\ & r + (8 - 3C) \geq 0 & \text{Constraint 6:} & (r + r_3 \geq 0) \end{array}$$

This problem is called a **linear program**. Even large linear programs can be solved by computer implementation of an algorithm known as the simplex method. In the preceding example, the simplex method yields a minimum value of $r = \frac{1}{2}$, and $C = \frac{5}{2}$, as you will see in Section 10.2.

We now generalize this procedure. Given some function type $y = f(x)$, whose parameters are to be determined, and a collection of m data points (x_i, y_i), define the residuals $r_i = y_i - f(x_i)$. If r represents the largest absolute value of these residuals, then the problem is to

$$\text{Minimize } r$$

$$\text{Subject to:} \quad \left. \begin{array}{l} r - r_i \geq 0 \\ r + r_i \geq 0 \end{array} \right\} \text{for } i = 1, 2, \ldots, m$$

We should note here that the model resulting from this procedure is not always a linear program; for example, consider fitting the function $f(x) = \sin kx$. Also note that many computer codes of the simplex algorithm require using variables that are allowed to assume only nonnegative values. This requirement can be accomplished with a simple substitution (see Exercise 4).

As you will see, alternative criteria lead to optimization problems that often can be resolved more conveniently. Primarily for this reason, the Chebyshev criterion is not used often for fitting a curve to a finite collection of data points. However, its application should be considered whenever minimizing the largest absolute deviation is important. We solve linear programs in Chapter 10.

Minimizing the Sum of the Absolute Deviations

When we were graphically fitting lines to the data in Section 9.2, one of our criteria minimized the total sum of the absolute deviations between the data points and their corresponding points on the fitted line. This criterion can be generalized: Given some function type $y = f(x)$ and a collection of m data points (x_i, y_i), minimize the sum of the absolute deviation $|y_i - f(x_i)|$. That is, determine the parameters of the function type $y = f(x)$ to minimize

$$\sum_{i=1}^{m} |y_i - f(x_i)| \tag{2}$$

Figure 9.12 A geometrical interpretation of minimizing the sum of the absolute deviations.

If we let $R_i = |y_i - f(x_i)|$, $1 = 1, 2, \ldots, m$ represent each absolute deviation, then criterion (2) can be interpreted as minimizing the length of the line formed by adding together the numbers R_i. This is illustrated for the case $m = 2$ in Figure 9.12.

Although we geometrically applied this criterion in Section 9.2 when the function type $y = f(x)$ is a line, the general criterion presents severe problems. To solve this optimization problem using the calculus, we need to differentiate the sum (2) with respect to the parameters of $f(x)$ in order to find the critical points. However, the various derivatives of the sum fail to be continuous because of the presence of the absolute values, so we will not pursue the analytic solution to this criterion any further. We return to find numerical solutions to this criterion in Scction 10.5.

Least-Squares Criterion

Currently the most frequently used curve-fitting criterion is the **least-squares criterion.** Using the same notation as shown earlier, the problem is to determine the parameters of the function type $y = f(x)$ in order to minimize the sum

$$\sum_{i=1}^{m} |y_i - f(x_i)|^2 \tag{3}$$

Part of the popularity of this criterion stems from the ease with which the resulting optimization problem can be solved using only the calculus of several variables. The justification for using the least-squares method increases when considering probabilistic arguments that assume the errors are distributed randomly.

We now give a geometric interpretation of the least-squares criterion. Consider the case of three data points and let $R_i = |y_i - f(x_i)|$ denote the absolute deviation between the observed and predicted values for $i = 1, 2, 3$. You can think of the R_i as the scalar components of a deviation vector as depicted in Figure 9.13. Thus the vector

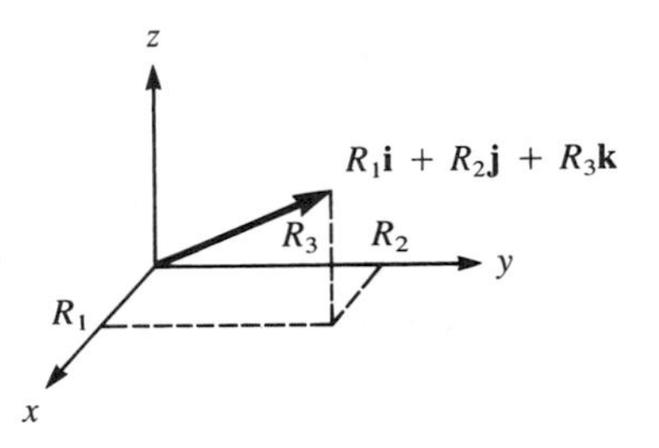

Figure 9.13 A geometrical interpretation of the least-squares criterion.

$\mathbf{R} = R_1\mathbf{i} + R_2\mathbf{j} + R_3\mathbf{k}$ represents the resultant deviation between the observed and predicted values. The magnitude of the deviation vector is given by

$$|\mathbf{R}| = \sqrt{R_1^2 + R_2^2 + R_3^2}$$

To minimize $|\mathbf{R}|$ we can minimize $|\mathbf{R}|^2$ (see Exercise 5 at the end of this section). Thus the least-squares problem is to determine the parameters of the function type $y = f(x)$ such that

$$|\mathbf{R}|^2 = \sum_{i=1}^{3} R_i^2 = \sum_{i=1}^{3} |y_i - f(x_i)|^2$$

is minimized. That is, we may interpret the least-squares criterion as minimizing the magnitude of the vector whose coordinates represent the absolute deviation between the observed and predicted values.

Exercises 9.3

1. For each of the following data sets, formulate the mathematical model that minimizes the largest deviation between the data and the line $y = ax + b$. If a computer is available, solve for the estimates of a and b.

a.

x	1.0	2.3	3.7	4.2	6.1	7.0
y	3.6	3.0	3.2	5.1	5.3	6.8

b.

x	29.1	48.2	72.7	92.0	118	140	165	199
y	0.0493	0.0821	0.123	0.154	0.197	0.234	0.274	0.328

c.

x	2.5	3.0	3.5	4.0	4.5	5.0	5.5
y	4.32	4.83	5.27	5.74	6.26	6.79	7.23

2. For the data given, formulate the mathematical model that minimizes the largest deviation between the data and the model $y = c_1x^2 + c_2x + c_3$. If a computer is available, solve for the estimates of c_1, c_2, and c_3.

x	0.1	0.2	0.3	0.4	0.5
y	0.06	0.12	0.36	0.65	0.95

3. For the following data, formulate the mathematical model that minimizes the largest deviation between the data and the model $P = ae^{bt}$. If a computer is available, solve for the estimates of a and b.

t	7	14	21	28	35	42
P	8	41	133	250	280	297

4. Suppose the variable x_1 can assume any real value. Show that the following substitution using nonnegative variables x_2 and x_3 permits x_1 to assume any real value:

$$x_1 = x_2 - x_3 \qquad \text{where } x_1 \text{ is unconstrained}$$

and

$$x_2 \geq 0 \qquad \text{and} \qquad x_3 \geq 0$$

Thus if a computer allows only nonnegative variables, the substitution allows for solving the linear program in the variables x_2 and x_3, and then recovering the value of the variable x_1.

5. Using elementary calculus, show that the minimum and maximum points for $y = f(x)$ occur among the minimum and maximum points for $y = f^2(x)$. Assuming $f(x) \geq 0$, why can we minimize $f(x)$ by minimizing $f^2(x)$?

9.4 Applying the Least-Squares Criterion

Suppose our assumptions lead us to expect a model of a certain type and that data have been collected and analyzed. In this section the least-squares criterion is applied to estimate the parameters for several types of curves.

Fitting a Straight Line

Suppose a model of the form $y = Ax + B$ is expected and it's been decided to use the m data points (x_i, y_i), $i = 1, 2, \ldots, m$, to estimate A and B. Denote the least-squares estimate of $y = Ax + B$ by $y = ax + b$. Applying the least-squares criterion [equation (3) of Section 9.3] to this situation requires the minimization of

$$\begin{aligned} S &= \sum_{i=1}^{m} (y_i - f(x_i))^2 \\ &= \sum_{i=1}^{m} (y_i - ax_i - b)^2 \end{aligned}$$

A necessary condition for optimality is that the two partial derivatives $\partial S/\partial a$ and $\partial S/\partial b$ equal zero, yielding the equations

$$\frac{\partial S}{\partial a} = -2\sum_{i=1}^{m} (y_i - ax_i - b)x_i = 0$$

$$\frac{\partial S}{\partial b} = -2\sum_{i=1}^{m} (y_i - ax_i - b) = 0$$

These equations can be rewritten to give

$$\left.\begin{aligned} a\sum_{i=1}^{m} x_i^2 + b\sum_{i=1}^{m} x_i &= \sum_{i=1}^{m} x_i y_i \\ a\sum_{i=1}^{m} x_i + mb &= \sum_{i=1}^{m} y_i \end{aligned}\right\} \qquad (1)$$

The preceding equations can be solved for a and b once all the values for x_i and y_i are substituted into them. The solutions for the parameters a and b are easily obtained by elimination and are found to be (see Exercise 1 at the end of this section)

$$a = \frac{m\sum x_i y_i - \sum x_i \sum y_i}{m\sum x_i^2 - \left(\sum x_i\right)^2}, \qquad \text{the slope} \qquad (2)$$

and

$$b = \frac{\sum x_i^2 \sum y_i - \sum x_i y_i \sum x_i}{m\sum x_i^2 - \left(\sum x_i\right)^2}, \qquad \text{the intercept} \qquad (3)$$

Computer algorithms are easily written to compute these values for a and b for any collection of data points. Equations (1) are called the **normal equations**.

Fitting a Power Curve

Now let's use the least-squares criterion to fit a curve of the form $y = Ax^n$, where n is fixed, to a given collection of data points. Call the least-squares estimate of the model $f(x) = ax^n$. Application of the criterion then requires minimization of

$$S = \sum_{i=1}^{m} \left(y_i - f(x_i)\right)^2 = \sum_{i=1}^{m} (y_i - ax_i^n)^2$$

A necessary condition for optimality is that the derivative ds/da equals zero giving the equation

$$\frac{dS}{da} = -2\sum_{i=1}^{m} x_i^n (y_i - ax_i^n) = 0$$

Solving the equation for a yields

$$a = \frac{\sum x_i^n y_i}{\sum x_i^{2n}} \qquad (4)$$

Remember, the number n is *fixed* in equation (4).

The least-squares criterion can be applied to other models as well. The limitation in applying the method lies in calculating the various derivatives required in the optimization process, setting these derivatives to zero, and solving the resulting equations for the parameters in the model type.

Table 9.3 Data collected to fit $y = Ax^2$.

x	0.5	1.0	1.5	2.0	2.5
y	0.7	3.4	7.2	12.4	20.1

For example, let's fit $y = Ax^2$ to the data shown in Table 9.3 and predict the value of y when $x = 2.25$. In this case, the least-squares estimate a is given by

$$a = \frac{\sum x_i^2 y_i}{\sum x_i^4}$$

We compute $\sum x_i^4 = 61.1875$, $\sum x_i^2 y_i = 195.0$ to yield $a = 3.1869$ (to four decimal places). This computation gives the least-squares approximate model

$$y = 3.1869x^2$$

When $x = 2.25$, the predicted value for y is 16.1337.

Transformed Least-Squares Fit

Although the least-squares criterion appears easy to apply in theory, in practice it may be difficult. For example, consider fitting the model $y = Ae^{Bx}$ using the least-squares criterion. Call the least-squares estimate of the model $f(x) = ae^{bx}$. Application of the criterion then requires the minimization of

$$S = \sum_{i=1}^{m} \left(y_i - f(x_i)\right)^2 = \sum_{i=1}^{m} (y_i - ae^{bx_i})^2$$

A necessary condition for optimality is that $\partial S/\partial a = \partial S/\partial b = 0$. Formulate the conditions and convince yourself that solving the resulting system of nonlinear equations would not be easy. Many simple models result in derivatives that are very complex or in systems of equations that are difficult to solve. For this reason, we use transformations that allow us to *approximate* the least-squares model.

In graphically fitting lines to data in Section 9.2, we sometimes found it convenient to transform the data first, and then fit a line to the transformed data. For example, in graphically fitting $y = Ce^x$, we found it convenient to plot $\ln y$ versus x, and then fit a line to the transformed data. The same idea can be used with the least-squares criterion in order to simplify the computational aspects of the process. In particular, if a convenient substitution can be found so that the problem takes the form $Y = AX + B$ in the transformed variables X and Y, then equations (1) can be used to fit a line to the transformed variables. We illustrate the technique with the example that we just worked out.

Suppose we wish to fit the power curve $y = Ax^N$ to a collection of data points. Let's denote the estimate of A by α and the estimate of N by n. Taking the logarithm of both sides of the equation $y = \alpha x^n$ yields

$$\ln y = \ln \alpha + n \ln x \tag{5}$$

Note that in plotting the variables $\ln y$ versus $\ln x$, equation (5) yields a straight line. On that graph, $\ln \alpha$ is the intercept when $\ln x = 0$ and the slope of the line is n. Using equations (2) and (3) to solve for the slope n and intercept $\ln \alpha$ with the transformed variables and $m = 5$ data points, we have

$$n = \frac{5\sum(\ln x_i)(\ln y_i) - (\sum \ln x_i)(\sum \ln y_i)}{5\sum(\ln x_i^2) - (\sum \ln x_i)^2}$$

$$\ln \alpha = \frac{\sum(\ln x_i)^2(\ln y_i) - (\sum \ln x_i)(\ln y_i)\sum \ln x_i}{5\sum(\ln x_i^2) - (\sum \ln x_i)^2}$$

For the data displayed in Table 9.3 we get $\sum \ln x_i = 1.3217558$, $\sum \ln y_i = 8.359597801$, $\sum(\ln x_i)^2 = 1.9648967$, $\sum(\ln x_i)(\ln y_i) = 5.542315175$ yielding $n = 2.062809314$ and $\ln \alpha = 1.126613508$ or $\alpha = 3.085190815$. Thus our least-squares best fit of equation (5) is (rounded to four decimal places)

$$y = 3.0852x^{2.0628}$$

This model predicts $y = 16.4348$ when $x = 2.25$. Note, however, that this model fails to be a quadratic like the one we fit previously.

Suppose we still wish to fit a *quadratic* $= Ax^2$ to the collection of data. Denote the estimate of A by a_1 to distinguish this constant from the constants a and α computed previously. Taking the logarithm of both sides of the equation $y = a_1x^2$ yields

$$\ln y = \ln a_1 + 2 \ln x$$

In this situation the graph of $\ln y$ versus $\ln x$ is a straight line of slope 2 and intercept $\ln a_1$. Using the second equation in (1) to compute the intercept, we have

$$2\sum \ln x_i + 5 \ln a_1 = \sum \ln y_i$$

For the data displayed in Table 9.3, we get $\sum \ln x_i = 1.3217558$ and $\sum \ln y_i = 8.359597801$. Therefore, this last equation gives $\ln a_1 = 1.14321724$ or $a_1 = 3.136844129$, yielding the least-squares best fit (rounded to four decimal places):

$$y = 3.1368x^2$$

The model predicts $y = 15.8801$ when $x = 2.25$, which differs significantly from the value 16.1337 predicted by the first quadratic $y = 3.1869x^2$ obtained as the least-squares best fit of $y = Ax^2$ without transforming the data.

The preceding example illustrates two facts. First, if an equation can be transformed to yield an equation of a straight line in the transformed variables, equations (1) can be used directly to solve for the slope and intercept of the transformed graph. Sec-

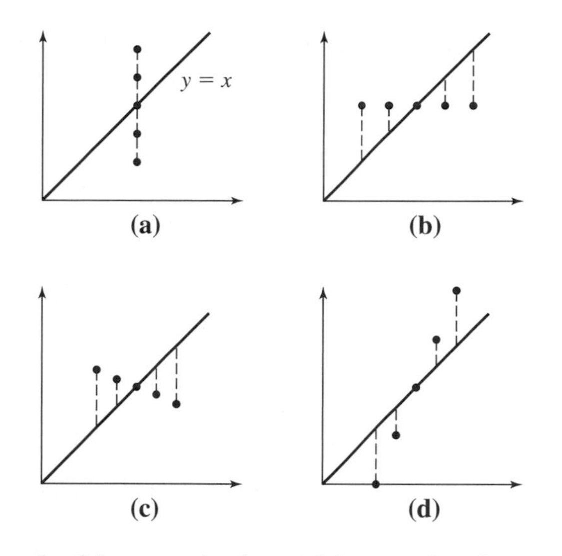

Figure 9.14 In each of these graphs the model $y = x$ has the same sum of squared deviations.

ond, the least-squares best fit to the transformed equations *does not* coincide with the least-squares best fit of the original equations. The reason for this discrepancy is that the resulting optimization problems are different. In the case of the original problem we are finding the curve that minimizes the sum of the squares of the deviations using the original data whereas in the case of the transformed problem we are minimizing the sum of the squares of the deviations using the *transformed* variables.

Choosing a Best Model

When choosing among models or judging the adequacy of a model, we may find it tempting to rely on the value of the best-fit criterion being used. For example, it is tempting to choose the model that has the smallest sum of squared deviations for the given data set or conclude that a sum of squared deviations less than a predetermined value indicates a good fit. However, in isolation these indicators may be very misleading. For example, consider the data displayed in Figure 9.14. In each of the four cases the model $y = x$ results in exactly the same deviations. Without the benefit of the graphs, therefore, we might conclude that in each case the model fits the data about the same. As the graphs show, however, there is a significant variation in each model's ability to capture the trend of the data. A powerful technique for quickly determining where the model is breaking down is to plot the deviations (residuals) as a function of the independent variable(s).

Exercises 9.4

1. Solve the equations given by (1) to obtain the values of the parameters given by (2) and (3), respectively.

2. Use equations (2) and (3) to estimate the coefficients of the line $y = ax + b$ such that the sum of the squared deviations between the line and the following data points is minimized.

a.

x	1.0	2.3	3.7	4.2	6.1	7.0
y	3.6	3.0	3.2	5.1	5.3	6.8

b.

x	29.1	48.2	72.7	92.0	118	140	165	199
y	0.0493	0.0821	0.123	0.154	0.197	0.234	0.274	0.328

c.

x	2.5	3.0	3.5	4.0	4.5	5.0	5.5
y	4.32	4.83	5.27	5.74	6.26	6.79	7.23

3. Derive the equations that minimize the sum of the squared deviations between a set of data points and the quadratic model $y = c_1x^2 + c_2x + c_3$. Use the equations to find estimates of c_1, c_2, and c_3 for the following set of data.

x	0.1	0.2	0.3	0.4	0.5
y	0.06	0.12	0.36	0.65	0.95

4. Make an appropriate transformation to fit the model $P = ae^{bt}$ using equations (4). Estimate a and b.

t	7	14	21	28	35	42
P	8	41	133	250	280	297

5. Examine closely the system of equations that result when you fit the quadratic in Exercise 3. Suppose $c_2 = 0$. What would be the corresponding system of equations? Repeat for the cases $c_1 = 0$ and $c_3 = 0$. Can you suggest a system of equations for a cubic? Check your result. Explain how you would generalize the system of equations (1) to fit any polynomial. Explain what you would do if one or more of the coefficients in the polynomial is zero.

6. A general rule given for computing a person's ideal weight is as follows: For a female, multiply the height in inches by 3.5 and subtract 108; for a male, multiply the height in inches by 4.0 and subtract 128. If the person is small bone-structured, adjust this computation by subtracting 10%; for a large bone-structured person, add 10%. No adjustment is made for an average-size person. Gather some data on the weight versus height of people of differing age, size, and gender. Using equations (1) fit a straight line to your data for males and another straight line to your data for females. What are the slopes and intercepts of those lines? How do the results compare with the general rule?

7. a. In the following data, W represents the weight of a fish (bass) and l represents its length. Fit the model $W = kl^3$ to the data using the least-squares criterion.

Length l (in.)	14.5	12.5	17.25	14.5	12.625	17.75	14.125	12.625
Weight W (oz)	27	17	41	26	17	49	23	16

b. In the following data, g represents the girth of a fish. Fit the model $W = klg^2$ to the data using the least-squares criterion.

Length l (in.)	14.5	12.5	17.25	14.5	12.625	17.75	14.125	12.625
Girth g (in.)	9.75	8.375	11.0	9.75	8.5	12.5	9.0	8.5
Weight W (oz)	27	17	41	26	17	49	23	16

c. Which of the two models fits the data better? Justify fully. Which model do you prefer? Why?

PROJECT Exploring Modeling

Complete the requirements of the module "Curve Fitting via the Criterion of Least Squares," by John W. Alexander, Jr., UMAP 321. This unit provides an easy introduction to correlations, scatter diagrams (polynomial, logarithmic, and exponential scatters), and lines and curves of regression. Students construct scatter diagrams, choose appropriate functions to fit specific data, and use a computer program to fit curves. Recommended for students who wish an introduction to statistical measures of correlation.

Further Reading

Burden, Richard L., & Faires, J. D. (1997). *Numerical analysis*, 6th ed. Pacific Grove, CA: Brooks/Cole.

Hamming, R. W. (1973). *Numerical methods for scientists and engineers*. New York: McGraw-Hill.

Cheney, E. W., & Kincaid, D. (1994). *Numerical mathematics and computing*, 3rd ed. Pacific Grove, CA: Brooks/Cole.

Stanton, Ralph G. (1961). *Numerical methods for science and engineering*. Englewood Cliffs, NJ: Prentice-Hall.

Steifel, E. L. (1963). *An introduction to numerical mathematics*. New York: Academic Press.

9.5 Fitting One-Term Models

Introduction

Often the modeler is unable to construct a model form that satisfactorily explains a behavior under study. If it is necessary to predict the behavior nevertheless, the modeler may conduct experiments (or otherwise gather data) to investigate the behavior of the dependent variable(s) for selected values of the independent variable(s) within some

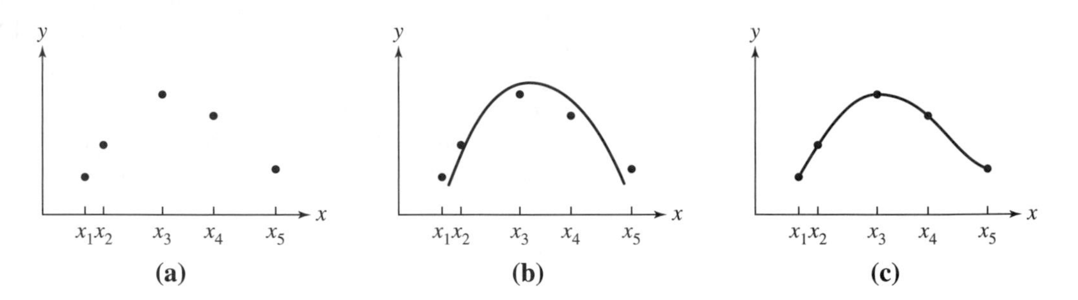

Figure 9.15 If the modeler expects a quadratic relationship, a parabola may be fit to the data, as in b. Otherwise, a smooth curve may be passed through the points, as in c.

range. In essence, the modeler desires to construct an **empirical model** *based on the collected data* rather than select a model based upon certain assumptions. In such cases the modeler is strongly influenced by the data that have been carefully collected and analyzed, so she seeks a curve that captures the trend of the data in order to *predict* in between the data points. For example, consider the data shown in Figure 9.15a. If the modeler's assumptions lead to the expectation of a quadratic model, a parabola would be fit to the data points as illustrated in Figure 9.15b. However, if the modeler has no reason to expect a particular type model, a smooth curve may be passed through the data points instead, as illustrated in Figure 9.15c.

In Sections 9.5 and 9.6, we address the construction of empirical models. In Section 9.5, we study the selection process for simple one-term models that capture the trend of the data. A few scenarios are addressed for which few modelers would attempt to construct an explicative model—namely, predicting the harvest of sea life in the Chesapeake Bay. In Section 9.6, we investigate smoothing of data using low-order polynomials.

Harvesting in the Chesapeake Bay and Other One-Term Models

Let's consider a situation in which a modeler has collected some data but is unable to construct an explicative model. In 1992, the *Daily Press* (a newspaper in Virginia) reported some observations (data) collected over the past 50 years on harvesting sea life in the Chesapeake Bay. We will examine several scenarios using observations from (a) harvesting bluefish and (b) harvesting blue crabs by the commercial industry of the Chesapeake Bay. Table 9.4 shows the data we will use in our one-term models.

A scatterplot of harvesting bluefish versus time is shown in Figure 9.16 and a scatterplot of harvesting blue crabs is shown in Figure 9.17. Figure 9.16 clearly shows a tendency to harvest more bluefish over time, indicating or suggesting the availability of bluefish. A more precise description is not so obvious. In Figure 9.17, the tendency is for the harvesting of the blue crabs to be increasing. Again, a precise model is not so obvious.

In the rest of this section, we suggest how you might begin to predict the availability of bluefish over time. Our strategy will be to transform the data of Table 9.4 in such a way that the resulting graph approximates a line, thus achieving a working model.

Table 9.4 Harvesting the bay 1940–1990.

Year	Bluefish (lb)	Blue crabs (lb)
1940	15,000	100,000
1945	150,000	850,000
1950	250,000	1,330,000
1955	275,000	2,500,000
1960	270,000	3,000,000
1965	280,000	3,700,000
1970	290,000	4,400,000
1975	650,000	4,660,000
1980	1,200,000	4,800,000
1985	1,500,000	4,420,000
1990	2,750,000	5,000,000

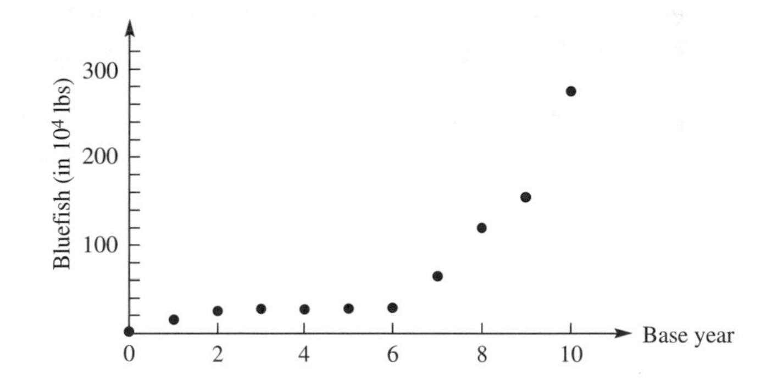

Figure 9.16 Scatterplot of harvesting bluefish versus base year (5-year periods from 1940 to 1990).

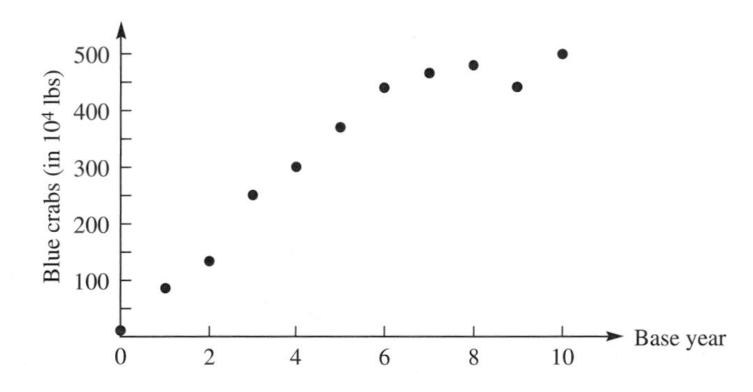

Figure 9.17 Scatterplot of harvesting blue crabs versus base year (5-year periods from 1940 to 1990).

Table 9.5 Ladder of powers.

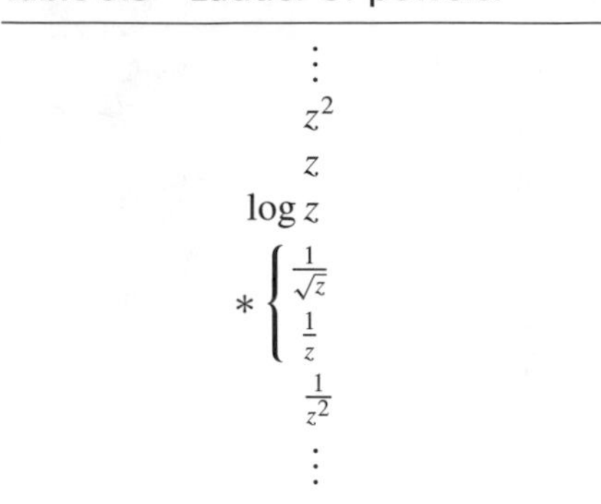

$\vdots$

z^2

z

$\log z$

$* \begin{cases} \frac{1}{\sqrt{z}} \\ \frac{1}{z} \end{cases}$

$\frac{1}{z^2}$

$\vdots$

*Most often used power.

But how do we determine the transformation? We will use the ladder of powers[1] of a variable z to help in the selection of the appropriate linearizing transformation.

Figure 9.18 shows a set of five data points (x, y) together with the line $y = x$, for $x > 1$. Suppose we change the y value of each point to $\sqrt{y}$. This procedure yields a new relation $y = \sqrt{x}$ whose y values are closer together over the domain in question. Note that all the y values are reduced, but the larger values are reduced more than the smaller ones.

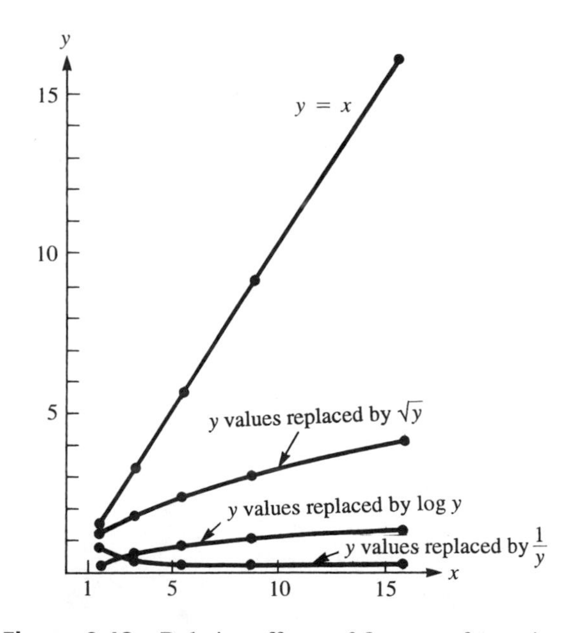

Figure 9.18 Relative effects of five transformations.

[1] See P. F. Velleman & D. C. Hoaglin, (1981), *Applications, basics, and computing of exploratory data analysis* (Boston, Mass.: Duxbury Press), p. 49.

Changing the y value of each point to $\log y$ has a similar, but more pronounced effect, and each additional step down the ladder produces a stronger version of the same effect.

We started in Figure 9.18, in the simplest way—with a linear function. But that was only a convenience. Given any positive-valued function $y = f(x)$, $x > 1$, that is concave up, like this:

then some transformation in the ladder below y—changing the y values to $\sqrt{y}$, or $\log y$, or a more drastic change—squeezes the right-hand tail downward and has a chance of generating a new function more nearly linear than the original. Which transformation should be used is a matter of trial and error and of experience. Another possibility is to stretch the right-hand tail to the right (try changing the x values to x^2, x^3 values, and so on).

Given a positive-valued function $y = f(x)$, $x > 1$, that is increasing and concave down, like this:

We can hope to (more nearly) linearize it by stretching the right-hand tail upward (try changing y values to y^2, y^3 values, and so on). Another possibility is to squeeze the right-hand tail to the left (try changing the x values to $\sqrt{x}$, or $\log x$, or by a more drastic choice from the ladder).

A final comment: Note that, although replacing z by $1/z$ or $1/z^2$, and so on may sometimes have a desirable effect, such replacements also have an undesirable one—an increasing function is converted into a decreasing one. As a result, when using the transformation in the ladder of powers below $\log z$, data analysts usually use a negative sign to keep the transformed data in the same order as the original data. Table 9.6 shows the ladder of transformations as it is usually used.

Table 9.6 Ladder of transformations.

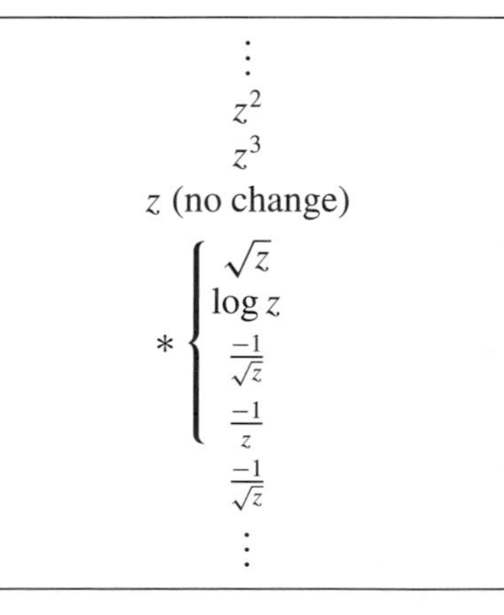

$$\vdots$$
$$z^2$$
$$z^3$$
$$z \text{ (no change)}$$
$$* \begin{cases} \sqrt{z} \\ \log z \\ \frac{-1}{\sqrt{z}} \\ \frac{-1}{z} \end{cases}$$
$$\frac{-1}{\sqrt{z}}$$
$$\vdots$$

*Most often used transformations.

With this information on the ladder of transformations, let's return to the Chesapeake Bay harvesting data.

Example 1 ***Harvesting Bluefish***

Recall from the scatterplot in Figure 9.16, that the trend of the data appears to be increasing and concave up. Using the ladder of powers to squeeze the right-hand tail downward, we can change y values by replacing y with $\log y$ or other transformations down the ladder. Another choice would be to replace x values with x^2 or x^3 values, or other powers up the ladder. We will use the data in Table 9.7 where we make 1940 the base year of $x = 0$ for numerical convenience, with each base year representing a 5-year period.

We begin by squeezing the right-hand tail downward by changing y to values of $\sqrt{y}$ and $\log y$, going down the ladder. Both $\sqrt{y}$ and $\log y$ plots versus x appeared more linear than the transformations to the x variable. (The plot of y versus x^2 and x versus $\sqrt{y}$ were identical in their linearity.) We choose the $\log y$ versus x model. We fit with least-squares criterion the model of the form:

$$\log y = mx + b$$

and obtain the following estimated curve:

$$\log y = 4.7231 + 0.1654x$$

where x is the base year and $\log y$ is to the base 10.

Using the property that $y = \log n$ if and only if $10^y = n$, we can rewrite this equation (with the aid of a calculator) as

$$y = 52{,}857(1.4635)^x \tag{1}$$

Table 9.7 Harvesting the bay: bluefish 1940–1990.

Year	Base year x	Bluefish (lb) y
1940	0	15,000
1945	1	150,000
1950	2	250,000
1955	3	275,000
1960	4	270,000
1965	5	280,000
1970	6	290,000
1975	7	650,000
1980	8	1,200,000
1985	9	1,550,000
1990	10	2,750,000

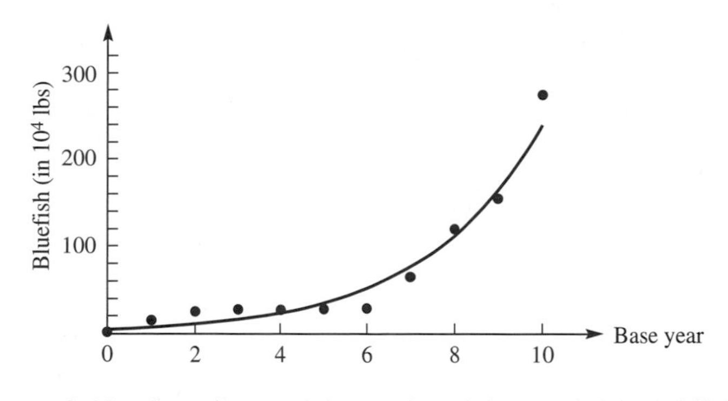

Figure 9.19 Superimposed data and model $y = 52{,}857(1.4635)^x$.

where y is measured in pounds of bluefish and x is the base year. Figure 9.19 shows the graph of this curve superimposed on the scatterplot. The plot of the model appears to fit the data reasonably well. We will accept some error in order to have a simple one-term model. ■

Example 2 *Harvesting Blue Crabs*

Recall from our original scatterplot, Figure 9.17, that the trend of the data is increasing and concave down. With this information, we can utilize the ladder of transformations. We will use the data in Table 9.8, modified by making 1940 (year $x = 0$) as the base year, with each base year representing a 5-year period.

As previously stated, we can attempt to linearize these data by changing x values to $\sqrt{x}$ or $\log x$ values or to others moving down the ladder. After several experiments, we chose to replace the x values with $\sqrt{x}$. This squeezes the right-hand tail to the left.

Table 9.8 Harvesting the bay: blue crabs 1940–1990.

Year	Base year x	Blue crabs (lb) y
1940	0	100,000
1945	1	850,000
1950	2	1,330,000
1955	3	2,500,000
1960	4	3,000,000
1965	5	3,700,000
1970	6	4,400,000
1975	7	4,660,000
1980	8	4,800,000
1985	9	4,420,000
1990	10	5,000,000

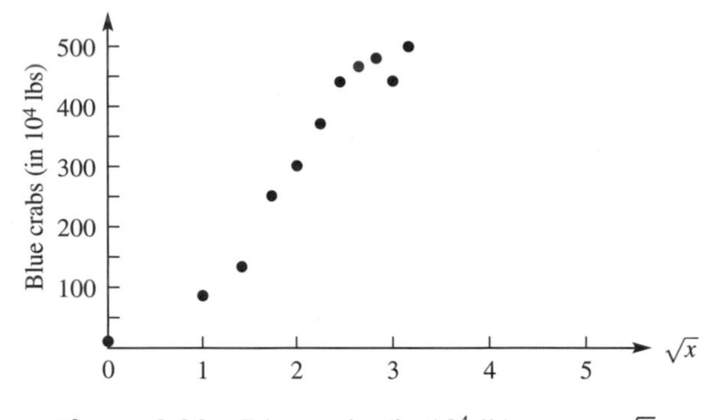

Figure 9.20 Blue crabs (in 10^4 lb) versus $\sqrt{x}$.

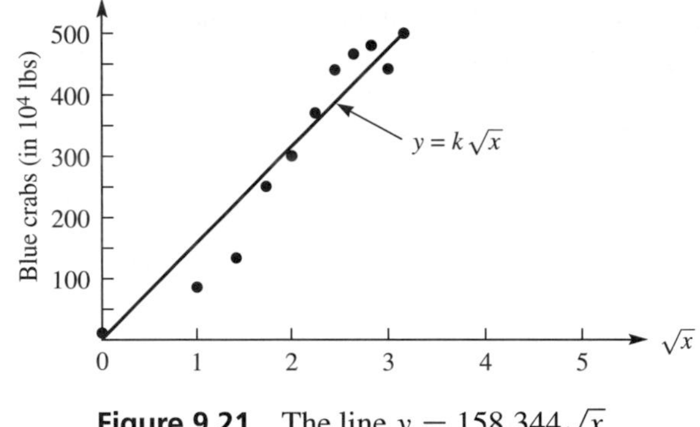

Figure 9.21 The line $y = 158.344\sqrt{x}$.

We provide a plot of y versus $\sqrt{x}$ (Figure 9.20). In Figure 9.21 we superimpose a line $y = k\sqrt{x}$ projected through the origin (no y-intercept). We use least squares from Section 9.3 to find k, yielding:

$$y = 158.344\sqrt{x} \tag{2}$$

Figure 9.22 shows the graph of the preceding model superimposed on the scatterplot. Note that the curve seems to be reasonable because it fits the data about as well as we would expect for a simple one-term model. ■

Verifying the Models

How good are the predictions based upon the models? Part of the answer lies in comparing the observed values with those predicted. We can calculate the residuals and the relative errors for each pair of data. In many cases the question asked of the modeler is to predict or extrapolate to the future. How would these models hold up to predicting the amounts of harvests from the bay in the year 2010?

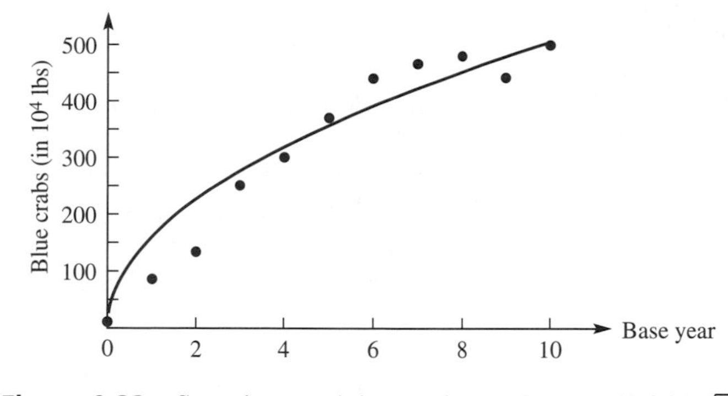

Figure 9.22 Superimposed data and model $y = 158.344\sqrt{x}$.

The following result for the bluefish may be larger than one might predict, whereas the result for the blue crabs may be a little lower than one might predict:

$$\text{Bluefish} \qquad y = 5.2857(1.4635)^{14} = 1092.951(10^4 \text{ lb}) \approx 10.93 \text{ million lb}$$

$$\text{Blue crabs} \qquad y = 158.344\sqrt{14} = 592.469(10^4 \text{ lb}) \approx 5.92 \text{ million lb}$$

These simple one-term models should be used for interpolation and not extrapolation.

Let's summarize the ideas of this section. When we are constructing an empirical model, we always begin with a careful analysis of the collected data. Do the data suggest the existence of a trend? Are there data points that obviously lie outside the trend? If such outliers do exist, it may be desirable to discard them or, if obtained experimentally, to repeat the experiment as a check for a data collection error. When it is clear that a trend exists, we next attempt to find a function that will transform the data into a straight line (approximately). In addition to trying the functions listed in the ladder of transformations presented in this section, we can also attempt the transformations discussed in Sections 9.2 through 9.4. Thus if the model $y = ax^b$ is selected, we would plot $\ln y$ versus $\ln x$ to see if a straight line results. Likewise, when investigating the appropriateness of the model $y = ae^{bx}$, we would plot $\ln y$ versus x to see if a straight line results. Keep in mind our discussion in Section 9.2 about how the use of transformations may be deceiving, especially if the data points are squeezed together. Our judgment here is strictly qualitative; the idea is to determine if a particular model type appears promising. When we are satisfied that a certain model type does seem to capture the trend of the data, we can estimate the parameters of the model graphically or using the analytical techniques discussed in Sections 9.3 and 9.4. Eventually we must analyze the goodness of fit using the indicators discussed in Section 9.3. Remember to graph the proposed model against the original data points, not the transformed data. If we are dissatisfied with the fit, we can investigate other one-term models. Because of their inherent simplicity, however, one-term models cannot fit all data sets. In such situations other techniques can be used, and we will discuss one of these methods in the next section.

Exercises 9.5

In 1976 Marc and Helen Bornstein studied the pace of life.[2] *To see if life becomes more hectic as the size of the city becomes larger, they systematically observed the mean time required for pedestrians to walk 50 feet on the main streets of their cities and towns. In the following table, we present some of the data they collected. The variable P represents the population of the town or city, and the variable V represents the mean velocity for pedestrians walking the 50 feet. Exercises 1–5 are based on the following data:*

Population and mean velocity over a 50-ft course, for 15 locations.*

	Location	Population P	Mean velocity V (ft/s)
1	Brno, Czechoslovakia	341,948	4.81
2	Prague, Czechoslovakia	1,092,759	5.88
3	Corte, Corsica	5,491	3.31
4	Bastia, France	49,375	4.90
5	Munich, Germany	1,340,000	5.62
6	Psychro, Crete	365	2.76
7	Itea, Greece	2,500	2.27
8	Iraklion, Greece	78,200	3.85
9	Athens, Greece	867,023	5.21
10	Safed, Israel	14,000	3.70
11	Dimona, Israel	23,700	3.27
12	Netanya, Israel	70,700	4.31
13	Jerusalem, Israel	304,500	4.42
14	New Haven, U.S.A.	138,000	4.39
15	Brooklyn, U.S.A.	2,602,000	5.05

*Bornstein data.

1. Fit the model $V = CP^a$ to the pace of life data in shown above. Use the transformation: $\log V = a \log P + \log C$. Plot $\log V$ versus $\log P$. Does the relationship seem reasonable?
   a. Make a table of $\log P$ versus $\log V$.
   b. Construct a scatterplot of your log–log data.
   c. Eyeball a line l onto your scatterplot.
   d. Estimate the slope and the intercept.
   e. Find the linear equation that relates $\log V$ and $\log P$.
   f. Find the equation of the form $V = CP^a$ that expresses V in terms of P.
2. Graph the equation you found in Exercise 1f superimposed on the original scatterplot.

[2] M. H. Bornstein and H. G. Bornstein. The pace of life. *Nature* **259** (19 February 1976): 557–559.

3. Using the following data, a calculator, and the model you determined for V (Exercise 1f), complete the following table:

Observed mean velocity for 15 locations.

Location*	Observed velocity V	Predicted velocities
1	4.81	
2	5.88	
3	3.31	
4	4.90	
5	5.62	
6	2.76	
7	2.27	
8	3.85	
9	5.21	
10	3.70	
11	3.27	
12	4.31	
13	4.42	
14	4.39	
15	5.05	

*For location names, see preceding table.

4. From the data shown in Exercise 3, calculate the mean (that is, the average) of the Bornstein errors $|V_{\text{observed}} - V_{\text{predicted}}|$. What do the results suggest about the merit of the model?

5. Do Exercises 1–4 with the model $V = m(\log P) + b$. Compare the errors with those computed in Exercise 4. Compare the two models. Which is better?

6. The following table and figure present data representing the commercial harvesting of oysters in Chesapeake Bay. Fit a simple, one-term model to the data. How well does the best one-term model you find fit the data? What is the largest error? The average error?

Oysters in the bay.

Year	Oysters harvested (10^6 lb)	Year	Oysters harvested (10^6 lb)
1940	3,750,000	1970	1,500,000
1945	3,250,000	1975	1,000,000
1950	2,800,000	1980	1,100,000
1955	2,550,000	1985	750,000
1960	2,650,000	1990	330,000
1965	1,850,000		

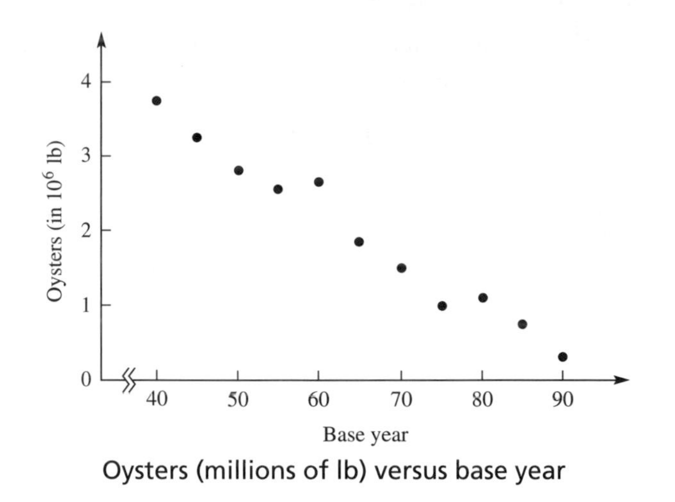

Oysters (millions of lb) versus base year

7. In the table below, X is the Fahrenheit temperature, and Y is the number of times a cricket chirps in 1 min. Fit a model to these data. Analyze how well it fits.

Temperature and chirps per minute for 20 crickets.*

Observation number	X	Y	Observation number	X	Y
1	46	40	11	61	96
2	49	50	12	62	88
3	51	55	13	63	99
4	52	63	14	64	110
5	54	72	15	66	113
6	56	70	16	67	120
7	57	77	17	68	127
8	58	73	18	71	137
9	59	90	19	72	132
10	60	93	20	71	137

*Data inferred from a scatterplot in F. E. Croxton, F. E. Dudley, D. J. Crowden, & S. Klein (1967), *Applied general statistics*, 3rd ed. (Englewood Cliffs, NJ: Prentice-Hall) p. 390.

8. Fit a model to the following data. Do you recognize the data? What relationship can be inferred from them?

Observation number	X	Y
1	35.97	0.241
2	67.21	0.615
3	92.96	1.000
4	141.70	1.881
5	483.70	11.860
6	886.70	29.460
7	1783.00	84.020
8	2794.00	164.800
9	3666.00	248.400

9. The following data measure two characteristics of a ponderosa pine. The variable X is the diameter of the tree, in inches, measured at breast height; Y is a measure of volume—number of board feet divided by 10. Fit a model to the data. Then express Y in terms of X.

Diameter and volume for 20 ponderosa pine trees.*

Observation number	X	Y	Observation number	X	Y
1	36	192	11	31	141
2	28	113	12	20	32
3	28	88	13	25	86
4	41	294	14	19	21
5	19	28	15	39	231
6	32	123	16	33	187
7	22	51	17	17	22
8	38	252	18	37	205
9	25	56	19	23	57
10	17	16	20	39	265

*Data reported in Croxton, Dudley, Crowden, & Klein, *Applied general statistics*, p. 421.

10. The following data set represents fish length and weight (bass). Model weight as a function of the length of the fish.

Length (in.)	12.5	12.625	14.125	14.5	17.25	17.75
Weight (oz)	17	16.5	23	26.5	41	49

11. The following data give the population of the United States from 1800 to 1990. Model the population (in thousands) as a function of the year. How well does your model fit? Is a one-term model appropriate for these data? Why?

Year	Population	Year	Population	Year	Population
1800	5308	1900	75,996	1990	248,710
1820	9638	1920	105,711		
1840	17,069	1940	131,669		
1860	31,443	1960	179,323		
1880	50,156	1980	226,505		

PROJECT Exploring Modeling

Complete the requirements of UMAP 551, "The pace of life, an introduction to empirical model fitting" by Bruce King. Prepare a short summary for classroom discussion.

Further Reading

Bornstein, M. H., & Bornstein, H. G. (1976). The pace of life. *Nature* **259** (19 February): 557–559.

Croxton, F. E., Dudley, F. E., Crowden, J., & Klein, S. (1967). *Applied general statistics*, 3rd ed. Englewood Cliffs, NJ: Prentice-Hall.

Neter, J., & Wasserman, W. (1974). *Applied linear statistical models*. Homewood, IL: Irwin.

Vellman, P. F., & Hoaglin, D. C. (1981). *Applications, basics, and computing of exploratory data analysis*. Boston: Duxbury Press.

Yule, G. U. (1926). Why do we sometimes get nonsense–correlations between time series?—A study in sampling and the nature of time series. *Journal of the Royal Statistical Society* **89**: 1–69.

9.6 Smoothing with Polynomial Models

In the previous section, we investigated the possibility of finding a simple one-term model that captures the trend of the collected data. Because of their inherent simplicity, one-term models facilitate model analysis including sensitivity analysis, optimization, estimation of rates of change and area under the curve. Because of their inherent mathematical simplicity, however, one-term models are limited in their ability to capture the trend of any collection of data. In some cases models with more than one term must be considered. The remainder of this chapter considers one type of multiterm model–namely, the polynomial. Because polynomials are easy to integrate and to differentiate, they are especially popular to use.

Consider the data in Figure 9.23a. Through the two given data points a unique line $y = a_0 + a_1 x$ can be passed. The constants a_0 and a_1 are determined by the conditions that the line pass through the points (x_1, y_1) and (x_2, y_2). Thus,

$$y_1 = a_0 + a_1 x_1$$

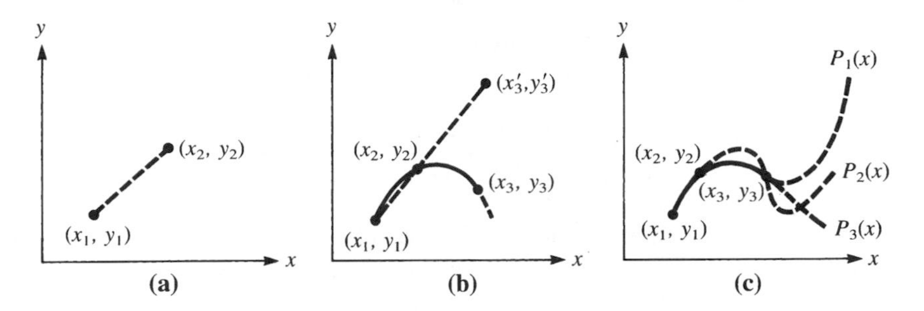

Figure 9.23 A unique polynomial of at most degree 2 can be passed through three data points (b), but an infinite number of polynomials of degree greater than 2 can be passed through three data points (c).

and

$$y_2 = a_0 + a_1 x_2$$

In a similar manner a *unique* polynomial function of (at most) degree 2 $y = a_0 + a_1 x + a_2 x^2$ can be passed through three distinct points, as shown in Figure 9.23b. The constants a_0, a_1, and a_2 are determined by solving the following system of linear equations:

$$y_1 = a_0 + a_1 x_1 + a_2 x_1^2$$
$$y_2 = a_0 + a_1 x_2 + a_2 x_2^2$$
$$y_3 = a_0 + a_1 x_3 + a_2 x_3^2$$

Let's explain why the qualifier "at most" is needed in the next-to-last sentence of the preceding paragraph. Note that if the three points in Figure 9.23b just happen to lie along a straight line, then the unique polynomial function of at most degree 2 passing through the points would necessarily be a straight line (a polynomial of degree 1) rather than a quadratic function, as generally would be expected. The descriptor *unique* is also very important. There are an infinite number of polynomials of degree greater than 2 and passing through the three points depicted in Figure 9.23b. (Convince yourself of this fact before preceding by using Figure 9.23c). There is, however, only one polynomial of degree 2 or less. Although this fact may not be obvious, we later state a theorem in its support. For now, remember from high school geometry that a unique circle, which is also represented by an algebraic equation of degree 2, is determined by three points in a plane. Next, we illustrate these ideas in an applied problem; we then discuss the advantages and disadvantages of the procedure.

Example 1 *Elapsed Time of a Tape Recorder*

We collected data relating the counter on a particular tape recorder with its elapsed playing time. Suppose we are unable to build an explicative model of this system, but are still interested in predicting what may occur. How can this difficulty be resolved? As an example, let's construct an empirical model to predict the amount of elapsed time of a tape recorder as a function of its counter reading.

Thus let c_i represent the counter reading and $t_i(s)$ the corresponding amount of elapsed time. Consider the following data:

c_i	100	200	300	400	500	600	700	800
$t_i(s)$	205	430	677	945	1233	1542	1872	2224

One empirical model is a polynomial that passes through each of the data points. Because we have eight data points, a unique polynomial of at most 7 is expected. Denote the polynomial symbolically by

$$P_7(c) = a_0 + a_1c + a_2c^2 + a_3c^3 + a_4c^4 + a_5c^5 + a_6c^6 + a_7c^7$$

The eight data points require that the constants a_i satisfy the system of linear algebraic equations:

$$205 = a_0 + 1a_1 + 1^2a_2 + 1^3a_3 + 1^4a_4 + 1^5a_5 + 1^6a_6 + 1^7a_7$$

$$430 = a_0 + 2a_1 + 2^2a_2 + 2^3a_3 + 2^4a_4 + 2^5a_5 + 2^6a_6 + 2^7a_7$$

$$\vdots$$

$$2224 = a_0 + 8a_1 + 8^2a_2 + 8^3a_3 + 8^4a_4 + 8^5a_5 + 8^6a_6 + 8^7a_7$$

Large systems of linear equations can be very difficult to solve with great numerical precision. In the preceding illustration we divided each counter reading by 100 to lessen the numerical difficulties. Because the counter data values are being raised to the seventh power, it is easy to generate numbers differing by several orders of magnitude. It is important to have as much accuracy as possible in the coefficients a_i because each is being multiplied by a number raised to a power as high as 7. For instance, a small a_7 may become significant as c becomes large. This observation suggests why there may be dangers using even good polynomial functions that capture the trend of the data when we are beyond the range of the observations. The following solution to this system was obtained with the aid of a handheld calculator program:

$$\begin{aligned} a_0 &= -13.9999923 & a_4 &= -5.354166491 \\ a_1 &= 232.9119031 & a_5 &= 0.8013888621 \\ a_2 &= -29.08333188 & a_6 &= -0.0624999978 \\ a_3 &= 19.78472156 & a_7 &= 0.0019841269 \end{aligned}$$

Let's see how well the empirical model fits the data. Denoting the polynomial predictions by $P_7(c_i)$, we find:

c_i	100	200	300	400	500	600	700	800
t_i	205	430	677	945	1233	1542	1872	2224
$P_7(c_i)$	205	430	677	945	1233	1542	1872	2224

Rounding the predictions for $P_7(c_i)$ to four decimal places gives complete agreement with the observed data (as would be expected) and results in zero absolute deviations. ■

Now we can see the folly of applying any of the criteria of best fit studied in Section 9.3 as the sole judge for the best model. Can we really consider this model to be better than other models we could propose?

The Lagrangian Form of the Polynomial

From the preceding discussion you might expect that given $(n + 1)$ distinct data points, there is a unique polynomial of at most degree n that passes through all the data points. Because there are the same number of coefficients in the polynomial as there are data points, intuitively we would think that only one such polynomial exits. This hypothesis is indeed the case, although we will not prove that fact here.

THEOREM 9.1

If $x_0, x_1, \ldots, x_n$ are $(n + 1)$ distinct points and $y_0, y_1, \ldots, y_n$ are corresponding observations at these points, then there exists a unique polynomial $P(x)$, of at most degree n, with the property that

$$y_k = P(x_k) \text{ for each } k = 0, 1, \ldots, n$$

This polynomial is given by

$$P(x) = y_0 L_o(x) + \cdots + y_n L_n(x) \tag{1}$$

where

$$L_k(x) = \frac{(x - x_0)(x - x_1)\ldots(x - x_{k-1})(x - x_{k+1})\ldots(x - x_n)}{(x_k - x_0)(x_k - x_1)\ldots(x_k - x_{k-1})(x_k - x_{k+1})\ldots(x_k - x_n)}$$

Because the polynomial (1) passes through each of the data points, the resultant sum of absolute deviations is zero. Considering the various criteria of best fit presented in Section 9.3, we are tempted to use high-order polynomials to fit larger sets of data. After all, the fit is precise. Let's examine both the advantages and disadvantages of using high-order polynomials.

Advantages and Disadvantages of High-Order Polynomials

It may be of interest to determine the area under the curve representing our model or its rate of change at a particular point. Polynomial functions have the distinct advantage of being easily integrated and differentiated. If a polynomial can be found that reasonably represents the underlying behavior, it would then be easy to approximate the integral and the derivative of the unknown true model as well. Now consider some of the disadvantages of higher-order polynomials. For the 17 data points presented in Table 9.9, it is clear that the trend of the data is $y = 0$ for all x over the interval $-8 \leq x \leq 8$.

Suppose equation (1) is used to determine a polynomial that passes through the points. Because there are 17 distinct data points, it is possible to pass a unique polynomial of at most degree 16 through the given points. The graph of a polynomial passing through the data points is depicted in Figure 9.24.

Table 9.9

x_i	−8	−7	−6	−5	−4	−3	−2	−1	0	1	2	3	4	5	6	7	8
y_i	0	0	0	0	0	0	0	0	0	0	0	0	0	0	0	0	0

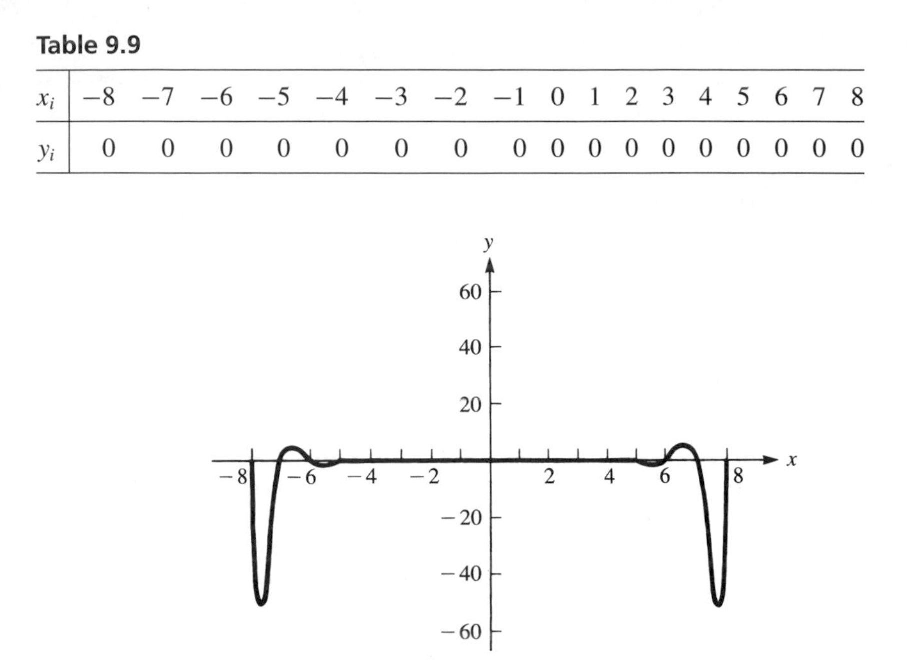

Figure 9.24 Fitting a higher-order polynomial through the data points in Table 9.9.

Note that although the polynomial does pass through the data points (within tolerances of computer round-off error), there is severe oscillation of the polynomial near each end of the interval. Thus there would be gross error in estimating y between the data points near $+8$ or -8. Likewise, consider the error in using the derivative of the polynomial to estimate the rate of change of the data or the area under the polynomial to estimate the area trapped by the data. This tendency of high-order polynomials to oscillate severely near the endpoints of the interval is a serious disadvantage to using them.

Because we do expect measurement error to occur, the tendency of high-order polynomials to oscillate, as well as the sensitivity of their coefficients to small changes in the data, is a disadvantage that restricts their usefulness in modeling. We now consider a technique that addresses the deficiencies noted in this section.

Data Smoothing

We seek methods that retain many of the conveniences found in high-order polynomials without incorporating their disadvantages. One popular technique is to choose a low-order polynomial regardless of the number of data points. This choice normally results in a situation where the number of data points exceed the number of constants necessary to determine the polynomial. Because there are fewer constants to determine than there are data points, the low-order polynomial generally will not pass through all the data points. For example, suppose it is decided to fit a quadratic to a set of 10 data points. Because it is generally impossible to force a quadratic to pass through 10 data points,

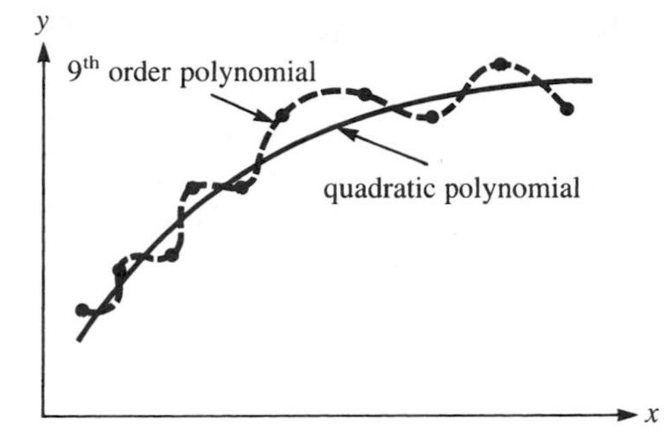

Figure 9.25 The quadratic function smooths the data because it is not required to pass through all the data points.

it must be decided which quadratic best fits the data (according to some criterion, as discussed in Section 9.3). This process, which is called **smoothing**, is illustrated in Figure 9.25. The combination of using low-order polynomials while not requiring that it pass through each data point reduces both the tendency of the polynomial to oscillate and its sensitivity to small changes in the data. This quadratic function smooths the data because it is not required to pass through all the data points.

The process of smoothing requires two decisions. First, the order of the interpolating polynomial must be selected. Second, the coefficients of the polynomial must be determined according to some criterion for the best-fitting polynomial. The problem that results is an optimization problem of the form addressed in Section 9.3. For example, it may be decided to fit a quadratic model to 10 data points using the least-squares best-fitting criterion. We will review the process of fitting a polynomial to a set of data points using the least-squares criterion and later return to a more difficult question of how to best choose the order of the interpolating polynomial.

Example 2 *Elapsed Time of a Tape Recorder Revisited*

Consider again the tape recorder modeled in the previous section. For a particular tape recorder equipped with a counter, relate the counter to the amount of playing time that has elapsed. If we are interested in predicting the elapsed time, but are unable to construct an explicative model, it may be possible to construct an empirical model instead. Let's fit a second-order polynomial of the following form to the data:

$$P_2(c) = a + bc + dc^2$$

where c is the counter reading, $P_2(c)$ is the elapsed time, and a, b, and d are constants to be determined. Consider the collected data for the tape recorder problem in the previous example, shown in Table 9.10.

Our problem is to determine the constants, a, b, and d so that the resultant quadratic model best fits the data. Although other criteria might be used, we will find the quadratic that minimizes the sum of the squared deviations. Mathematically, the problem is to

Table 9.10 Data collected for the tape recorder problem.

c_i	100	200	300	400	500	600	700	800
t_i(s)	205	430	677	945	1233	1542	1872	2224

$$\text{Minimize } S = \sum_{i=1}^{m} [t_i - (a + bc_i + dc_i^2)]^2$$

The necessary conditions for a minimum to exist ($\partial S/\partial a = \partial S/\partial b = \partial S/\partial d = 0$) yield the equations:

$$\begin{aligned} ma + \left(\sum c_i\right) b + \left(\sum c_i^2\right) d &= \sum t_i \\ \left(\sum c_i\right) a + \left(\sum c_i^2\right) b + \left(\sum c_i^3\right) d &= \sum c_i t_i \\ \left(\sum c_i^2\right) a + \left(\sum c_i^3\right) b + \left(\sum c_i^4\right) d &= \sum c_i^2 t_i \end{aligned} \tag{2}$$

For the data given in Table 9.10, (2) becomes

$$\begin{aligned} 8a + 3600b + 2{,}040{,}000d &= 9128 \\ 3600a + 2{,}040{,}000b + 1{,}296{,}000{,}000d &= 5{,}318{,}900 \\ 2{,}040{,}000a + 1{,}296{,}000{,}000b + 8.772 \times 10^{11}d &= 3{,}435{,}390{,}000 \end{aligned} \tag{3}$$

Solution of system (3) yields the values $a = 0.14286$, $b = 1.94226$, and $d = 0.00105$, giving the quadratic:

$$P_2(c) = 0.14286 + 1.94226c + 0.00105c^2$$

We can compute the deviations between the observations and the predictions made by the model $P_2(c)$:

c_i	100	200	300	400	500	600	700	800
t_i	205	430	677	945	1233	1542	1872	2224
$t_i - P_2(c_i)$	0.167	−0.452	0.000	0.524	0.119	−0.214	−0.476	0.333

Note that the deviations are very small compared to the order of magnitude of the times. ■

When we are considering the use of a low-order polynomial for smoothing, two issues come to mind:

1. Should a polynomial be used?
2. If so, what order of polynomial would be appropriate?

The derivative concept can help in answering these two questions.

Divided Differences

Notice that a quadratic function is characterized by the properties that its second derivative is constant and its third derivative is zero. That is, given

$$P(x) = a + bx + cx^2$$

we have

$$P'(x) = b + 2cx$$
$$P''(x) = 2c$$
$$P'''(x) = 0$$

However, the only information available is a set of discrete data points. How can these points be used to estimate the various derivatives? Refer to Figure 9.26 and recall the definition of the derivative:

$$\frac{dy}{dx} = \lim_{\Delta x \to 0} \frac{\Delta y}{\Delta x}$$

Because dy/dx at $x = x_1$ can be interpreted geometrically as the slope of the line tangent to the curve there, we see from Figure 9.26 that, unless Δx is small, the ratio $\Delta y/\Delta x$ is probably not a good estimate of dy/dx. Nevertheless, if dy/dx is to be zero, then Δy must go to zero. Thus, we can compute the *differences* $y_{i+1} - y_i = \Delta y$ between successive function values in our tabled data to gain insight into what the first derivative is doing.

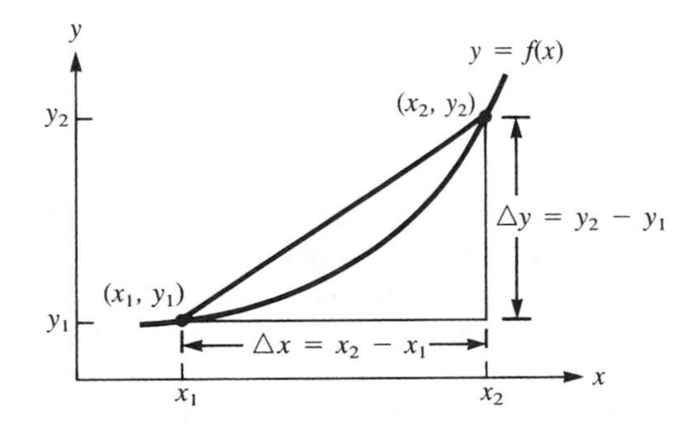

Figure 9.26 The derivative of $y = f(x)$ at $x = x_1$ is the limit of the slopes of the secant lines.

Table 9.11 A hypothetical set of collected data.

x_i	0	2	4	6	8
y_i	0	4	16	36	64

Table 9.12 A difference table for the data of Table 9.11.

Data		Differences			
x_i	y_i	Δ	Δ^2	Δ^3	Δ^4
0	0				
		4			
2	4		8		
		12		0	
4	16		8		0
		20		0	
6	36		8		
		28			
8	64				

Likewise, because the first derivative is itself a function, the process can be repeated to estimate the second derivative. That is, the differences between successive estimates of the first derivative can be computed to approximate the second derivative. Before describing the entire process, we illustrate this idea with a simple example.

You know that the curve $y = x^2$ passes through the points $(0, 0)$, $(2, 4)$, $(4, 16)$, $(6, 36)$, and $(8, 64)$. Suppose the data displayed in Table 9.11 have been collected. Using the data in Table 9.11, we can construct a *difference table*, as shown in Table 9.12.

The first differences, denoted by Δ, are constructed by computing $y_{i+1} - y_i$ for $i = 1, 2, 3, 4$. The second differences, denoted by Δ^2, are computed by finding the difference between successive first differences from the Δ column. The process can be continued, column by column, until Δ^{n-1} is computed for n data points. Note from Table 9.12 that the second differences in our example are constant and the third differences are zero. These results are consistent with the fact that a quadratic function has a constant second derivative and a zero third derivative.

Even if the data are essentially quadratic in nature, we would not expect the differences to go to zero precisely due to the various errors present in the modeling and data collection processes. We might, however, expect them to become small. Our judgment of the significance of small can be improved by computing **divided differences**. Notice that the differences computed in Table 9.12 are estimates of the numerator of each of the various order derivatives. These estimates can be improved by dividing the numerator by the corresponding estimate of the denominator.

Consider the three data points and corresponding estimates of the first and second derivative, called the first and second divided differences, respectively, in Table 9.13. The first divided difference follows immediately from the ratio of $\Delta y / \Delta x$. Because the second derivative represents the rate of change of the first derivative, we can estimate

Table 9.13 The first and second divided differences estimate the first and second derivatives, respectively.

Data		First divided difference	Second divided difference
x_1	y_1		
		$\frac{y_2 - y_1}{x_2 - x_1}$	
x_2	y_2		$\frac{\frac{y_3 - y_2}{x_3 - x_2} - \frac{y_2 - y_1}{x_2 - x_1}}{x_3 - x_1}$
		$\frac{y_3 - y_2}{x_3 - x_2}$	
x_3	y_3		

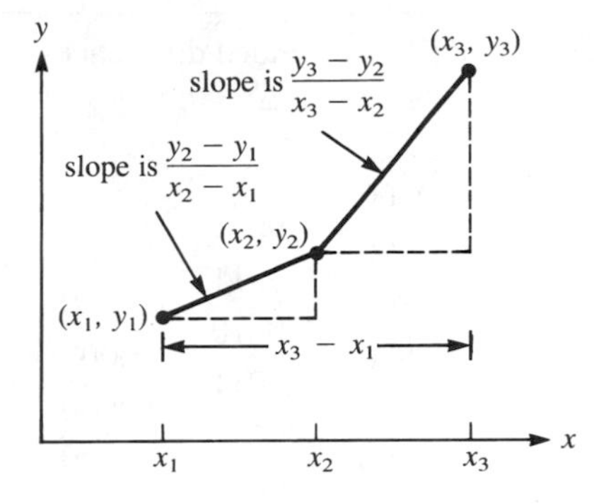

Figure 9.27 The second divided difference may be interpreted as the difference between the adjacent slopes (first divided differences) divided by the length of the interval over which the change has taken place.

how much the first derivative changes between x_1 and x_3. That is, we can compute the differences between the adjacent first divided differences and divide by the length of the interval over which that change takes place ($x_3 - x_1$ in this case). Refer to Figure 9.27 for a geometrical interpretation of the second divided difference.

In practice it is easy to construct a divided difference table. We generate the next higher-order divided difference by taking differences between adjacent current order divided differences and then dividing them by the length of the interval over which the change has taken place. Using Δ^n to denote the nth divided difference, a divided difference table for the data of Table 9.11 is displayed in Table 9.14.

It is easy to remember what the numerator should be in each divided difference of the table. To remember what the denominator should be for a given divided difference, one can construct diagonal lines back to y_i of the original data entries and compute the differences in the corresponding x_i. This is illustrated for a third-order divided difference in Table 9.14. This construction becomes more critical when the x_i's are unequally spaced.

Table 9.14 A divided difference table for the data of Table 9.11.

	Data		Divided differences		
	x_i	y_i	Δ	Δ^2	Δ^3
	0	0			
			$4/2 = 2$		
	2	4		$4/4 = 1$	
			$12/2 = 6$		$0/6 = 0$
	4	16		$4/4 = 1$	
$\Delta x = 6$			$20/2 = 10$		$0/6 = 0$
	6	36		$4/4 = 1$	
			$28/2 = 14$		
	8	64			

Table 9.15 A divided difference table for the tape recorder data in Table 9.10.

Data		Divided differences			
x_i	y_i	Δ	Δ^2	Δ^3	Δ^4
100	205				
		2.2500			
200	430		0.0011		
		2.4700		0.0000	
300	677		0.0011		0.0000
		2.6800		0.0000	
400	945		0.0010		0.0000
		2.8800		0.0000	
500	1233		0.0011		0.0000
		3.0900		0.0000	
600	1542		0.0011		0.0000
		3.3000		0.0000	
700	1872		0.0011		
		3.5200			
800	2224				

Example 3 ***Elapsed Time of a Tape Recorder Revisited***

Returning now to our construction of an empirical model for the elapsed time for a tape recorder, how might the order of the smoothing polynomial be chosen? Let's begin by constructing the divided difference table for the given data from Table 9.10. The divided differences are displayed in Table 9.15.

Note that from Table 9.15 that the second divided differences are essentially constant and that the third divided differences equal zero to four decimal places. The table suggests that the data are essentially quadratic, which supports the use of a quadratic polynomial as an empirical model. ■

Exercises 9.6

1. For the tape recorder problem in this section, give a system of equations determining the coefficients of a polynomial that passes through each of the data points. If

a computer is available, determine and sketch the polynomial. Does it represent the trend of the data?

2. Consider the pace of life data from Exercise 1, Section 9.5. Consider fitting a 14th-order polynomial to the data. Discuss the disadvantages of using the polynomial to make predictions. If a computer is available, determine and graph the polynomial.

For the data sets in Exercises 3–6, construct a divided difference table. What conclusions can you make about the data? Would you use a low-order polynomial as an empirical model? If so, what order?

3.

x	0	1	2	3	4	5	6	7
y	2	8	24	56	110	192	308	464

4.

x	0	1	2	3	4	5	6	7
y	23	48	73	98	123	148	173	198

5.

x	0	1	2	3	4	5	6	7
y	7	15	33	61	99	147	205	273

6.

x	0	1	2	3	4	5	6	7
y	1	4.5	20	90	403	1808	8103	36316

In Exercises 7–12, construct a scatterplot of the given data. Is there a trend in the data? Are any of the data points outliers? Construct a divided difference table. Is smoothing with a low-order polynomial appropriate? If so, choose an appropriate polynomial and fit using the least-squares criterion of best fit. Analyze the goodness of fit by examining appropriate indicators and graphing the model, the data points, and the deviations.

7. In the following data, X is the Fahrenheit temperature and Y is the number of times a cricket chirps in 1 min (see Exercise 7, Section 9.5).

Temperature and chirps per minute for 20 crickets.*

Observation number	X	Y	Observation number	X	Y
1	46	40	6	56	70
2	49	50	7	57	77
3	51	55	8	58	73
4	52	63	9	59	90
5	54	72	10	60	93

*Data inferred from a scatterplot in F. E. Croxton, F. E. Dudley, D. J. Crowden, & S. Klein (1967), *Applied general statistics*, 3rd ed. (Englewood Cliffs, NJ: Prentice-Hall) p. 390.

Temperature and chirps per minute for 20 crickets. *(continued)*

Observation number	X	Y	Observation number	X	Y
11	61	96	16	67	120
12	62	88	17	68	127
13	63	99	18	71	137
14	64	110	19	72	132
15	66	113	20	71	137

8. In the following data, X represents the diameter of a ponderosa pine measure at breast height and Y is a measure of volume—number of board feet divided by 10 (see Exercise 9, Section 9.5).

Observation number	X	Y
1	35.97	0.241
2	67.21	0.615
3	92.96	1.000
4	141.70	1.881
5	483.70	11.860
6	886.70	29.460
7	1783.00	84.020
8	2794.00	164.800
9	3666.00	248.400

9. The following data represents the population of the United States from 1790 to 1990.

Year	Observed population	Year	Observed population	Year	Observed population
1790	3,929,000	1860	31,443,000	1930	122,755,000
1800	5,308,000	1870	38,558,000	1940	131,669,000
1810	7,240,000	1880	50,156,000	1950	150,697,000
1820	9,638,000	1890	62,948,000	1960	179,323,000
1830	12,866,000	1900	75,995,000	1970	203,212,000
1840	17,069,000	1910	91,972,000	1980	226,505,000
1850	23,192,000	1920	105,711,000	1990	248,709,873

10. The following data were obtained for the growth of a sheep population introduced into a new environment on the island of Tasmania. (Adapted from J. Davidson, "On the Growth of the Sheep Population in Tasmania," *Trans. Roy. Soc. S. Australia* **62** (1938): 342–346.)

t (year)	1814	1824	1834	1844	1854	1864
$P(t)$	125	275	830	1200	1750	1650

11. The following data represents the pace of life (see Exercise 1, Section 9.5). P is the population and V is the mean velocity in feet per second over a 50-ft course.

Population and mean velocity over a 50-ft course, for 15 locations.*

	Location	Population P	Mean velocity V (ft/s)
1	Brno, Czechoslovakia	341,948	4.81
2	Prague, Czechoslovakia	1,092,759	5.88
3	Corte, Corsica	5,491	3.31
4	Bastia, France	49,375	4.90
5	Munich, Germany	1,340,000	5.62
6	Psychro, Crete	365	2.76
7	Itea, Greece	2,500	2.27
8	Iraklion, Greece	78,200	3.85
9	Athens, Greece	867,023	5.21
10	Safed, Israel	14,000	3.70
11	Dimona, Israel	23,700	3.27
12	Netanya, Israel	70,700	4.31
13	Jerusalem, Israel	304,500	4.42
14	New Haven, U.S.A.	138,000	4.39
15	Brooklyn, U.S.A.	2,602,000	5.05

*Bornstein data.

12. The following data represent the length of a bass fish and its weight.

Length (in.)	12.5	12.625	14.125	14.5	17.25	17.75
Weight(oz)	17	16.5	23	26.5	41	49

10
Linear Programming and Numerical Search Methods

George B. Dantzig said "The final test of any theory is its capacity to solve the problems which originated it." Linear programming and its offspring ... have come of age and have demonstrably passed this test, and they are fundamentally affecting the economic practice of organizations and management.

SIAM News, November (1994)

Overview

The NCTM *Principles and Standards for School Mathematics* (NCTM, 2000) calls for developing student ability to model mathematical situations. Central to such modeling is the ability to focus on the relevant factors in a problem situation; identify and make assumptions about relationships among those factors; develop a model using algebraic, geometric, probabilistic, or other concepts that represent the situation; and manipulate the model to develop plausible lines of action to resolve the original problem. Nowhere in the secondary curriculum is this process more visible than in the study of linear programming. In fact, linear programming usually constitutes the one and only extended segment of applied mathematics that secondary school students see in their study of mathematics.

This chapter highlights the methods of linear programming and related search methods that resulted from efforts during World War II to allocate scarce materials to operations in a clear and effective manner. Those efforts included using a constrained number of planes and pilots to win the Air Battle for Britain and finding how to elude U-boats and move men and materials safely across the sea to win the Battle of the North Atlantic. The groups that analyzed these problems and generated defensible lines of action were the first operations researchers. Today, individuals working in this field are among the most valuable assets a large corporation has. They examine ways to make the corporation effective, both in service and in cost/benefit fashion, in inventory control and in scheduling.

In the present chapter, we develop the essence of linear programming, a specific operations research technique to optimize plans for action in situations defined by constraints that can be expressed in linear inequalities. In doing so, students make connections between various areas of mathematics they have studied; view such problems from both an algebraic and geometric perspective; and intuitively extend their reasoning about single-variable functions, $f(x)$, to two-variable functions, $f(x, y)$. This material is included in the present text, as individuals preparing for secondary teaching rarely get the opportunity to take courses in linear programming or operations research as part of their state-required courses for certification. We also include it because today linear programming is perhaps the most commonly applied process for decision making in industry.

Focus on the NCTM *Standards*

The NCTM *Principles and Standards* for Patterns, Functions, and Algebra for grades K–12 calls for students to be able to understand and apply various types of functions. This involves recognizing equivalent forms of functions, selecting appropriate representations for these functions, and using such representations to explore and to solve problems based on the relationships identified. In particular, the *Standards* calls for developing student ability to find and interpret intercepts, local extreme values, and interpret their meaning in particular contexts. Linear programming is an excellent way to meet these goals, and material concerning it is contained in almost every secondary school text series. It also provides an avenue for connecting the world of mathematics to the real world of decision making in business, science, and other areas of human affairs. Linear programming is one of the outstanding scientific discoveries of the twentieth century. George Dantzig, of Stanford University, discovered the simplex method for solving linear programming problems in 1947 and was later awarded the National Science Medal for this achievement.

The study of linear programming is important for reasons beyond the issues of motivation and applicability. It is important and meaningful mathematics, it shows the connections between algebra and geometry, and it introduces the concept of optimization in a coherent and important way using situations involving linear constraints.

Focus on the Classroom

The NCTM *Principles and Standards* describes new goals for teachers and for students. The NCTM urges, for example, that teachers and students work together so that students learn to value mathematics, to become confident in their abilities to do mathematics, to become problem solvers, and to communicate and reason mathematically. Teachers and students also create the classroom environment in which this learning takes place. The mathematics discussed in this chapter provides a way in which these goals can be attained, along with the goals outlined in the *Standards* directly related to linear programming content.

The consideration of linear programming problems enables students to progress from attacking the problems with visual and numerical models based on graphs of linear inequalities to more efficient and formal models based on tableau representations of the problems. Students gain the ability to relate mathematical discoveries to the ideas of simplifying processes and more effective decision making. Linear programming provides an excellent opportunity for students to research a local industry, formulate a problem pertinent to the industry, and explore ways to solve it and other social and commerical problems.

10.1 Linear Programming I: Geometric Solutions

Introduction

In Chapter 9, we considered three criteria for fitting a selected model to a collection of data:

1. Minimize the sum of the absolute deviations.
2. Minimize the largest of the absolute deviations (Chebyshev criterion).
3. Minimize the sum of the squared deviations (least-squares criterion).

Also in Chapter 9 we used calculus to solve the optimization problem resulting from the application of the least-squares criterion. Although we formulated several optimization problems using the first criterion to "minimize the sum of the absolute deviations," we were unable to solve the resulting mathematical problem. In Section 10.5, we study several search techniques that allow us to find good solutions to the curve fitting criterion, and examine many other optimization problems as well.

We also interpreted the Chebyshev criterion for several models. For example, given a collection of m data points (x_i, y_i), $i = 1, 2, \ldots, m$, fit the collection to that line $y = ax + b$ (determined by the parameters a and b) that minimizes the greatest distance $r_{\max}$ between any data point (x_i, y_i) and its corresponding point $(x_i, ax_i + b)$ on the line. That is, the largest absolute deviation, $r = \text{Maximum}\{|y_i - y(x_i)|\}$, is minimized over the entire collection of data points. This criterion defines the optimization problem to

$$\text{Minimize } r$$

$$\text{Subject to} \qquad \left.\begin{aligned} r - r_i &\geq 0 \\ r + r_i &\geq 0 \end{aligned}\right\} \text{ for } i = 1, 2, \ldots, m$$

which is a *linear program* for many applications.

You will learn how to solve linear programs geometrically and algebraically in Sections 10.2–10.3, and in Section 10.4, you will learn how to determine the sensitivity of the optimal solution to the coefficients appearing in the linear program. We begin by finding geometrical solutions to curve fitting problems and other linear programs.

Let's use the Chebyshev criterion to fit the model $y = cx$ to the following data set:

x	1	2	3
y	2	5	8

The optimization problem that determines the parameter c to minimize the largest absolute deviation $r_i = |y_i - y(x_i)|$ (residual or error) is the linear program:

$$\text{Minimize } r$$

$$\text{Subject to} \qquad \left.\begin{aligned} r - (2 - c) &\geq 0 \quad \text{(constraint 1)} \\ r + (2 - c) &\geq 0 \quad \text{(constraint 2)} \\ r - (5 - 2c) &\geq 0 \quad \text{(constraint 3)} \\ r + (5 - 2c) &\geq 0 \quad \text{(constraint 4)} \\ r - (8 - 3c) &\geq 0 \quad \text{(constraint 5)} \\ r + (8 - 3c) &\geq 0 \quad \text{(constraint 6)} \end{aligned}\right\} \qquad (1)$$

In this section we solve this problem geometrically. But first, let's use another simpler example to define a linear program for which we describe the geometrical solution procedure.

Defining a Linear Program: The Carpenter's Problem

A carpenter makes tables and bookcases for a net unit profit that he estimates as \$25 and \$30, respectively. He needs to determine how many units of furniture he should make each week. He has up to 690 board feet of lumber to devote to the project weekly and up to 120 hours of labor, but he can use the lumber and labor productively elsewhere if they aren't used in the tables and bookcases. It requires 20 board feet of lumber and 5 hours of labor to complete a table, and 30 board feet of lumber and 4 hours of labor for a bookcase, and he can sell all the tables and bookcases he produces. The carpenter wants to determine a weekly production schedule for tables and bookcases that maximizes his profits.

Let x_1 denote the number of tables to be produced weekly, and x_2 denote the number of bookcases. Then the model becomes

$$\text{Maximize } 25x_1 + 30x_2$$

Subject to

$$\begin{aligned} 20x_1 + 30x_2 &\leq 690 \quad \text{(lumber)} \\ 5x_1 + 4x_2 &\leq 120 \quad \text{(labor)} \\ x_1, x_2 &\geq 0 \quad \text{(nonnegativity)} \end{aligned} \tag{2}$$

This model is an example of a linear program. The function $25x_1 + 30x_2$ is called the **objective function**, and x_1 and x_2 are the **decision variables**. We restrict our attention to linear programs involving two decision variables.

Interpreting a Linear Program Geometrically

Linear programs can include a set of constraints that are linear equations or linear inequalities. In the case of two decision variables, an equality requires that solutions to the linear program lie precisely on the line representing the equality. What about inequalities?

To gain some insight, consider the constraints

$$\begin{aligned} x_1 + 2x_2 &\leq 4 \\ x_1, x_2 &\geq 0 \end{aligned} \tag{3}$$

The **nonnegativity** constraints, $x_1, x_2 \geq 0$, mean that possible solutions lie in the first quadrant. The inequality $x_1 + 2x_2 \leq 4$ divides the first quadrant into two regions. The **feasible region** is the half-space in which the constraint is satisfied. The feasible region can be found by graphing the equation $x_1 + 2x_2 = 4$ and determining which half-plane is feasible, as shown in Figure 10.1.

If the feasible half-plane fails to be obvious, we choose a convenient point (such as the origin) and substitute it into the constraint to determine if it's satisfied. If it is, then all points on the same side of the line as this point will also satisfy the constraint. Now let's find the feasible region and the optimal solution for the carpenter's problem.

Example 1 *The Carpenter's Problem*

A linear program has this important property: The points satisfying the constraints form a **convex set**, which is a set in which any two of its points are joined by a straight-line segment, all of whose points lie within the set. The set depicted in Figure 10.2a fails to be convex, while the set in Figure 10.2b is convex.

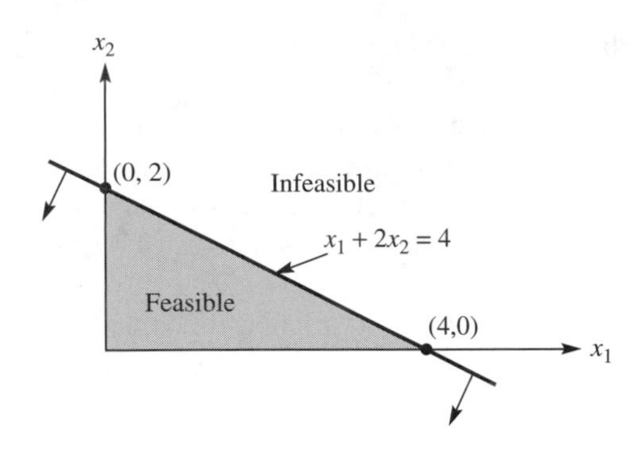

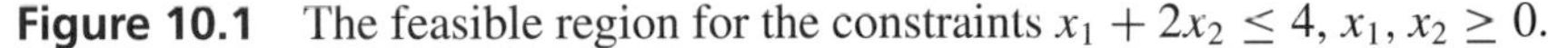

Figure 10.1 The feasible region for the constraints $x_1 + 2x_2 \leq 4$, $x_1, x_2 \geq 0$.

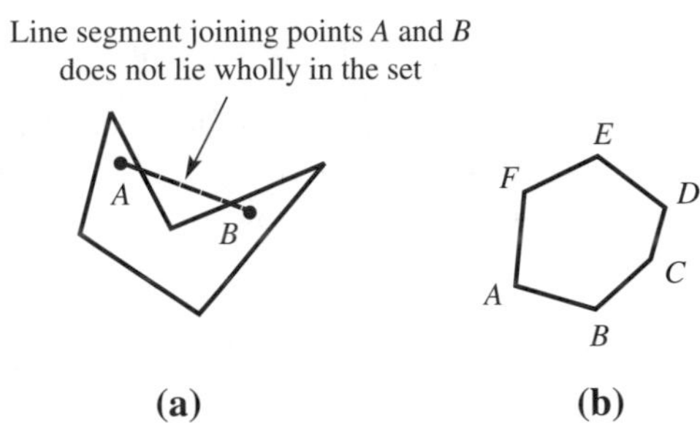

Figure 10.2 The set shown in (a) is not convex, whereas the set shown in (b) is convex.

An **extreme point** (corner point) of a convex set is any boundary point in the convex set that is the unique intersection point of two of the (straight-line) boundary segments. In Figure 10.2b, the points A, B, C, D, E, F are extreme points.

Let's illustrate these ideas with the carpenter's problem. The convex set for the constraints in the carpenter's problem is graphed and given by the polygon region $ABCD$ in Figure 10.3. Note that there are six intersection points of the constraints, but only four of these points (namely, A, B, C, and D) satisfy all of the constraints and hence belong to the convex set. The points A, B, C, D are the extreme points of the polygon. The variables y_1 and y_2 will be explained in a later section.

If an optimal solution to a linear program exists, it turns out that it must occur among the extreme points of the convex set formed by the set of constraints. The value of the objective function (profit for the carpenter's problem) at the extreme points are shown in Table 10.1.

Thus the carpenter should make 12 tables and 15 bookcases each week to earn a maximum weekly profit of \$750. We provide further geometrical evidence later in this section that extreme point C is optimal. ■

Before considering a second example, let's summarize the ideas presented thus far. The constraint set to a linear program is a convex set, which generally contains an infinite number of feasible points to the linear program. If an optimal solution to the

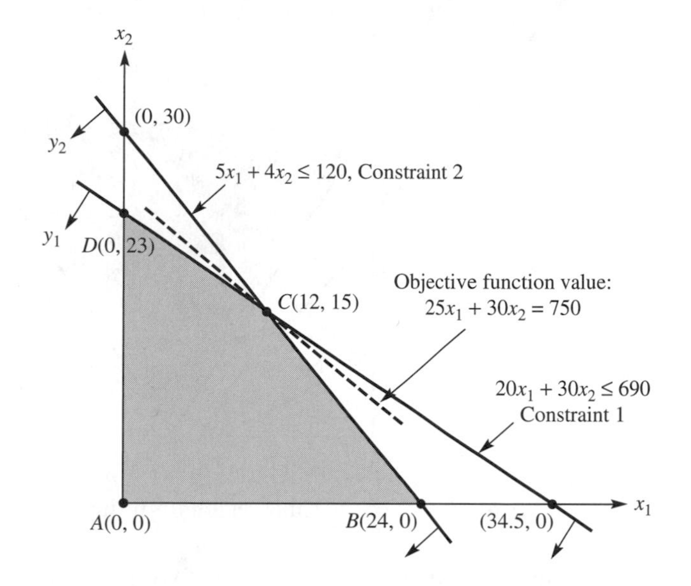

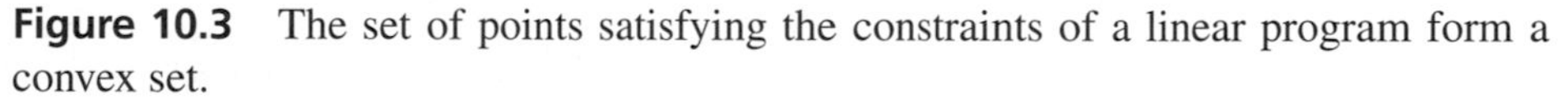

Figure 10.3 The set of points satisfying the constraints of a linear program form a convex set.

Table 10.1 Profit for the carpenter's problem.

Extreme point	Objective function value
A (0, 0)	\$0
B (24, 0)	\$600
C (12, 15)	\$750
D (0, 23)	\$690

linear program exists, it must be taken on at one or more of the extreme points. Thus, to find an optimal solution, we choose from among all the extreme points the one with the best value for the objective function.

Example 2 ***A Data Fitting Problem***

Now let's solve the linear program represented by equation (1). We are given the model $y = cx$ and the data set

x	1	2	3
y	2	5	8

We want to find a value for c such that the resulting largest absolute deviation is as small as possible.

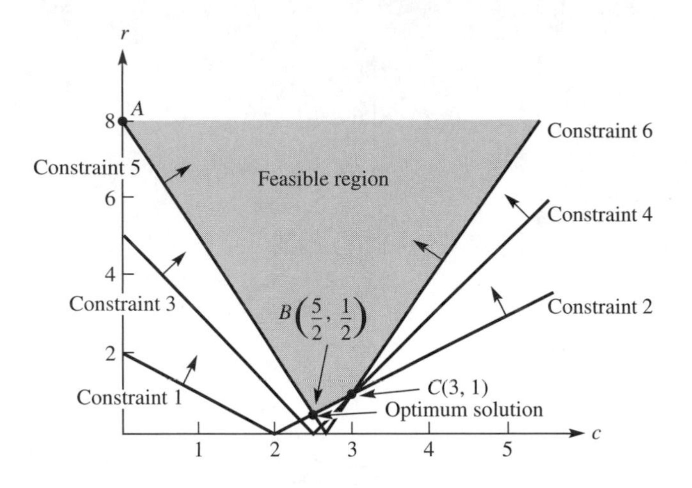

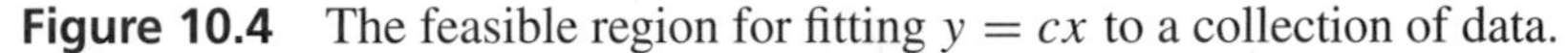

Figure 10.4 The feasible region for fitting $y = cx$ to a collection of data.

In Figure 10.4, we graph the set of six constraints

$$r - (2 - c) \geq 0 \quad \text{(constraint 1)}$$
$$r + (2 - c) \geq 0 \quad \text{(constraint 2)}$$
$$r - (5 - 2c) \geq 0 \quad \text{(constraint 3)}$$
$$r + (5 - 2c) \geq 0 \quad \text{(constraint 4)}$$
$$r - (8 - 3c) \geq 0 \quad \text{(constraint 5)}$$
$$r + (8 - 3c) \geq 0 \quad \text{(constraint 6)}$$

by first graphing the equations

$$r - (2 - c) = 0 \quad \text{(constraint 1 boundary)}$$
$$r + (2 - c) = 0 \quad \text{(constraint 2 boundary)}$$
$$r - (5 - 2c) = 0 \quad \text{(constraint 3 boundary)}$$
$$r + (5 - 2c) = 0 \quad \text{(constraint 4 boundary)}$$
$$r - (8 - 3c) = 0 \quad \text{(constraint 5 boundary)}$$
$$r + (8 - 3c) = 0 \quad \text{(constraint 6 boundary)}$$

Note that constraints 1, 3, and 5 are satisfied above and to the right of the graph of their boundary equations. Similarly, constraints 2, 4, and 6 are satisfied above and to the left of their boundary equations. To convince yourself, pick a point (such as the origin) and determine whether the point satisfies the constraint. If it does, it must be in the feasible region determined by the constraint.

The intersection of all the feasible regions for constraints 1–6 form a convex set in the c, r-plane with extreme points labeled A, B, and C in Figure 10.4. The point A is

Table 10.2 Evaluating the objective function $f(r) = r$.

Extreme point	Objective function value
(c, r)	$f(r) = r$
A	8
B	$\frac{1}{2}$
C	1

the intersection of constraint 5 and the r-axis: $r - (8 - 3c) = 0$ and $c = 0$, or $A = (0, 8)$. Similarly, B is the intersection of constraints 5 and 2:

$$r - (8 - 3c) = 0 \quad \text{or} \quad r + 3c = 8$$
$$r + (2 - c) = 0 \quad \text{or} \quad r - c = -2$$

yielding $c = \frac{5}{2}$ and $r = \frac{1}{2}$, or $B = (\frac{5}{2}, \frac{1}{2})$. Finally, C is the intersection of constraints 2 and 4 yielding $C = (3, 1)$. Note that the set is unbounded. (We discuss unbounded convex sets later.) If an optimal solution to the problem exists, at least one extreme point must take on the optimal solution. We evaluate the objective function $f(r) = r$ at each of the three extreme points in Table 10.2.

The extreme point with the smallest value of r is the extreme point B with coordinates $(\frac{5}{2}, \frac{1}{2})$. Thus $c = \frac{5}{2}$ is the optimal value of c. No other value of c will result in a largest absolute deviation as small as $|r_{\max}| = \frac{1}{2}$. ■

Interpreting the Model

Let's interpret the optimal solution for our data fitting problem. Resolving the linear program, we obtained a value of $c = \frac{5}{2}$, corresponding to the model $y = \frac{5}{2}x$. Further, the objective function value $r = \frac{1}{2}$ should correspond to the largest deviation resulting from the fit. Let's check to see if that's true.

The data points and the model $y = \frac{5}{2}x$ are plotted in Figure 10.5. Note that a largest deviation of $r_i = \frac{1}{2}$ occurs for both the first and third data points. Fix one end of a ruler at the origin. Rotate the ruler to convince yourself geometrically that no other line passing through the origin can yield a smaller largest absolute deviation. Thus the model $y = \frac{5}{2}x$ is optimal by the Chebyshev criterion.

Empty and Unbounded Feasible Regions

We've been careful to say that if an optimal solution to the linear program exists, at least one of the extreme points must take on the optimal value for the objective function. When does an optimal solution fail to exist? Moreover, when does more than one optimal solution exist?

If the feasible region is empty, no feasible solution can exist. For example, given the constraints

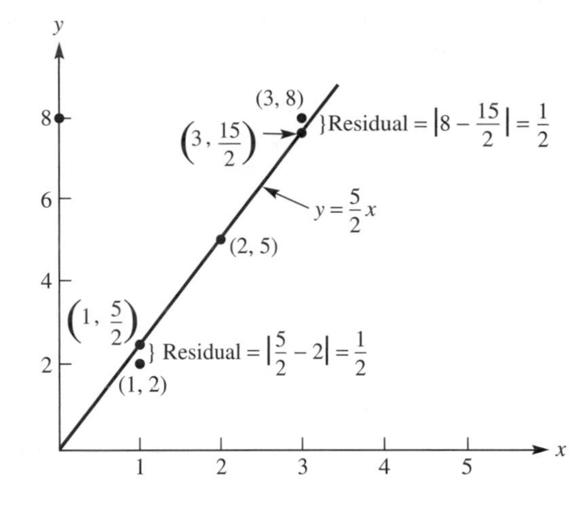

Figure 10.5 The line $y = \frac{5}{2}x$ results in a largest absolute deviation $r_{max} = \frac{1}{2}$, the smallest possible r_{max}.

$$x_1 \leq 3, \qquad \text{and} \qquad x_1 \geq 5$$

no value of x_1 satisfies both of them. We say that such constraint sets are *inconsistent*.

There is another reason an optimal solution may not exist. Consider Figure 10.4 and the constraint set for the data fitting problem in which we noted that the feasible region is *unbounded* (in the sense that either x_1 or x_2 can become arbitrarily large). Then it would be impossible to maximize $x_1 + x_2$ over the feasible region because x_1 and x_2 can take on arbitrarily large values. Note, however, that even though the feasible region is unbounded, an optimal solution *does* exist for the objective function we considered in Example 2. So it is not *necessary* for the feasible region to be bounded for an optimal solution to exist.

Level Curves of the Objective Function

Let's return to the carpenter's problem. The objective function is $25x_1 + 30x_2$, and in Figure 10.6 we plot the lines

$$25x_1 + 30x_2 = 650$$
$$25x_1 + 30x_2 = 750$$
$$25x_1 + 30x_2 = 850$$

in the first quadrant. Note that the objective function has constant values along these line segments. The line segments are called **level curves** of the objective function. As we move in a direction perpendicular to these line segments, the objective function either increases or decreases.

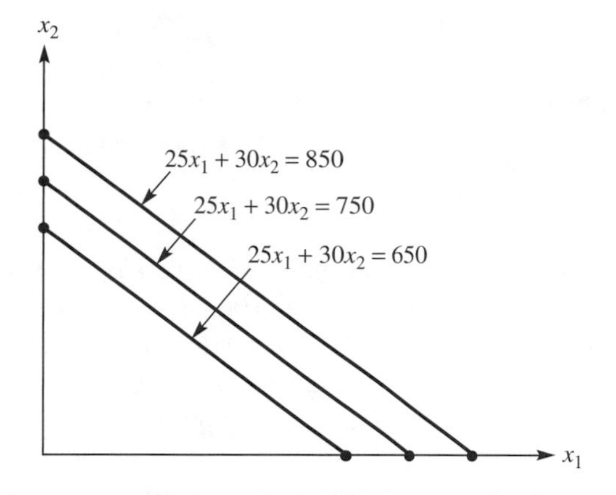

Figure 10.6 The level curves of the objective function f are parallel line segments in the first quadrant. The objective function either increases or decreases as we move in a direction perpendicular to the level curves.

Now let's superimpose the constraint set from the carpenter's problem,

$$\begin{aligned} 20x_1 + 30x_2 &\leq 690 \quad \text{(lumber)} \\ 5x_1 + 4x_2 &\leq 120 \quad \text{(labor)} \\ x_1, x_2 &\geq 0 \quad \text{(nonnegativity)} \end{aligned}$$

onto these level curves (see Figure 10.7). Notice that the level curve with value 750 is the one that intersects the feasible region exactly once at the extreme point $C(12, 15)$.

Can there be more than one optimal solution? Consider the following slight variation of the carpenter's problem in which the labor constraint has been changed:

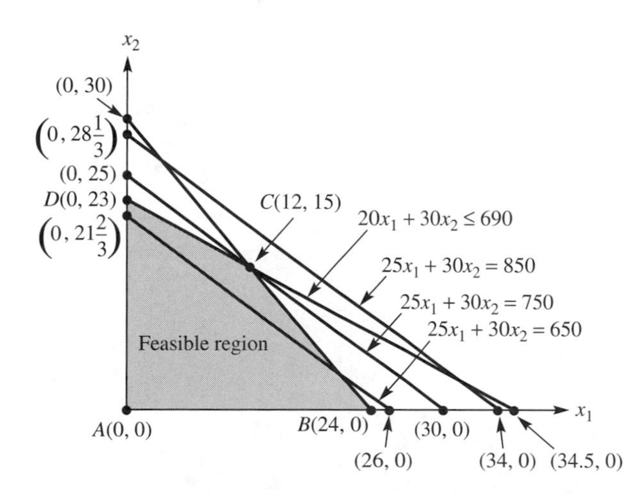

Figure 10.7 The level curve $25x_1 + 30x_2 = 750$ is tangent to the feasible region at extreme point C.

$$\text{Maximize } 25x_1 + 30x_2$$

$$\begin{aligned} \text{Subject to} \qquad 20x_1 + 30x_2 &\le 690 \quad \text{(lumber)} \\ 5x_1 + 6x_2 &\le 150 \quad \text{(labor)} \\ x_1, x_2 &\ge 0 \quad \text{(nonnegativity)} \end{aligned}$$

The constraint set and the level curve $25x_1 + 30x_2 = 750$ are graphed in Figure 10.8. Notice that the level curve and boundary line for the labor constraint coincide. Thus both extreme points B and C have the same objective function value of 750, which is optimal. In fact, the entire line segment BC coincides with the level curve $25x_1 + 30x_2 = 750$. Thus there are infinitely many optimal solutions to the linear program, all along line segment BC.

In Figure 10.9, we summarize the general two-dimensional case for optimizing a linear function on a convex set. The figure shows a typical convex set together with the level curves of a linear objective function. Figure 10.9 provides geometrical intuition for the following fundamental theorem of linear programming.

THEOREM 10.1 Suppose the feasible region of a linear program is a nonempty and bounded convex set, then the objective function must attain both a maximum and minimum value occurring at extreme points of the region. If the feasible region is unbounded, the objective function need not assume its optimal values. If either a maximum or minimum does exist, it must occur at one of the extreme points.

The power of this theorem is that it guarantees an optimal solution to a linear program from among the extreme points of a bounded, nonempty convex set.

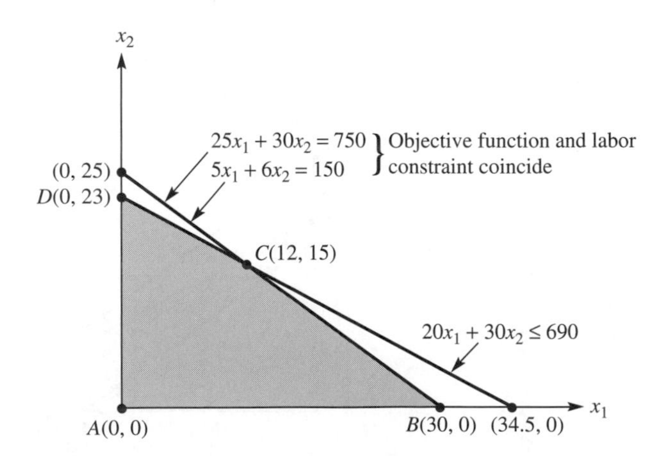

Figure 10.8 The line segment BC coincides with the level curve $25x_1 + 30x_2 = 750$. Every point between extreme points C and B, as well as extreme points C and B, is an optimal solution.

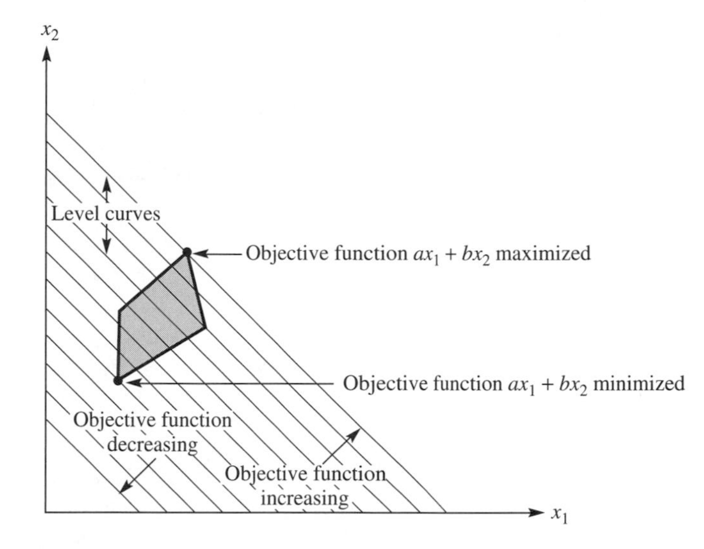

Figure 10.9 A linear function assumes its maximum and minimum value on a nonempty and bounded convex set at an extreme point.

Exercises 10.1

1. A company that carves wooden soldiers specializes in two main types: Confederate and Union soldiers. The profit for each is \$28, and \$30, respectively. A Confederate soldier requires 2 units of lumber, 4 hours of carpentry, and 2 hours of finishing to complete. A Union soldier requires 3 units of lumber, 3.5 hours of carpentry, and 3 hours of finishing to complete. Each week the company has 100 units of lumber delivered. There are 120 hours of carpentry time available and 90 hours of finishing time available. What number of each type of soldier will maximize weekly profits?

2. A local company restores cars and trucks for resale. Each vehicle must be processed both in the paint-refinishing shop and the machine-body shop. On average, each car contributes \$3000 to profit, and each truck contributes \$2000 to profit. The paint-refinishing shop has 2400 work-hours available, and the machine-body shop has 2500 labor-hours available. A car requires 50 labor-hours in the machine-body shop and 40 work-hours in the paint-refinishing shop while a truck requires 50 labor-hours in the machine-body shop and 60 work-hours in the paint-refinishing shop. Use graphical linear programming to determine a daily production schedule that will maximize the company's profits.

3. A Montana farmer owns 45 acres of land. She plans to plant each acre with wheat or corn. Each acre of wheat yields \$200 in profits; each acre of corn yields \$300 in profits. The labor and fertilizer requirements for each follows. The farmer has 100 workers and 120 tons of fertilizer available. Determine how many acres of wheat and corn need to be planted to maximize profits.

	Wheat	Corn
Labor (workers)	3	2
Fertilizer (tons)	2	4

4. Solve the following problems using graphical analysis:

a. Maximize $x + y$

Subject to
$$\begin{aligned} x + y &\le 6 \\ 3x - y &\le 9 \\ x, y &\ge 0 \end{aligned}$$

b. Minimize $x + y$

Subject to
$$\begin{aligned} x + y &\ge 6 \\ 3x - y &\ge 9 \\ x, y &\ge 0 \end{aligned}$$

c. Maximize $10x + 35y$

Subject to
$$\begin{aligned} 8x + 6y &\le 48 && \text{(board-ft lumber)} \\ 4x + y &\le 20 && \text{(hours of carpentry)} \\ y &\ge 5 && \text{(demand)} \\ x, y &\ge 0 && \text{(nonnegativity)} \end{aligned}$$

5. Fit the model to the data using Chebyshev's criterion to minimize the largest deviation.

a. $y = cx$

y	11	25	54	90
x	5	10	20	30

b. $y = cx^2$

y	10	90	250	495
x	1	3	5	7

10.2 Linear Programming II: Algebraic Solutions

The graphical solution to the carpenter's problem suggests a rudimentary procedure for finding an optimal solution to a linear program with a nonempty and bounded feasible region:

1. Find all intersection points of the constraints.
2. Determine which intersection points, if any, are feasible to obtain the extreme points.
3. Evaluate the objective function at each extreme point.

4. Choose the extreme point(s) with the largest (or smallest) value for the objective function.

In order to implement this procedure algebraically, we must characterize the intersection points and the extreme points.

The convex set depicted in Figure 10.10 consists of three linear constraints (plus the two nonnegativity constraints). The nonnegative variables y_1, y_2, and y_3 in the figure measure the degree by which a point satisfies each of the constraints 1, 2, and 3, respectively. The variable y_i is added to the left side of inequality constraint i to convert it to an equality. Thus $y_2 = 0$ characterizes those points that lie precisely on constraint 2, and a negative value for y_2 indicates the violation of constraint 2. Likewise, the decision variables x_1 and x_2 are constrained to nonnegative values. The values of the decision variables x_1 and x_2 thus measure the degree of satisfaction of the nonnegativity constraints, $x_1 \geq 0$ and $x_2 \geq 0$. Note that along the x_1-axis, the decision variable x_2 is 0.

Now consider the values for the entire set of variables $\{x_1, x_2, y_1, y_2, y_3\}$. If two of the variables simultaneously have the value 0, then we have characterized an **intersection point** in the x_1x_2-plane. All (possible) intersection points can be determined systematically by setting all possible distinguishable pairs of the five variables to zero and solving for the remaining three dependent variables. If a solution to the resulting system of equations exists, then it must be an intersection point, which may or may not be a **feasible solution**. A negative value for any of the five variables indicates that a constraint is not satisfied. Such an intersection point would be **infeasible**. For example, the intersection point B where $y_2 = 0$ and $x_1 = 0$ gives a negative value for y_1, and hence is not feasible. Let's illustrate the procedure by solving the carpenter's problem algebraically.

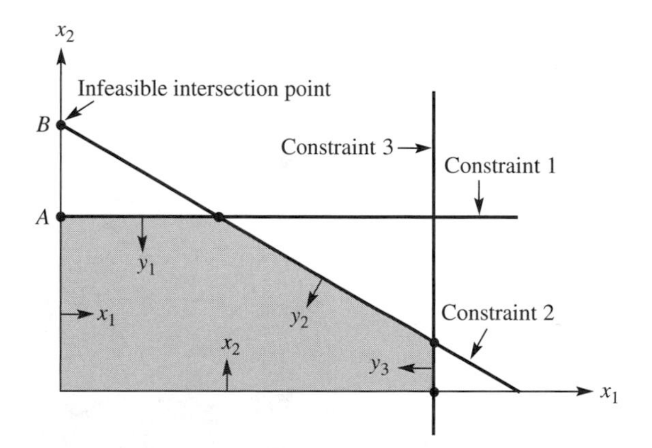

Figure 10.10 The variables x_1, x_2, y_1, y_2, and y_3 measure the satisfaction of each of the constraints. Intersection point A is characterized by $y_1 = x_1 = 0$; intersection point B is not feasible because y_1 is negative. The intersection points (the solid circles) surrounding the shaded region are all feasible because none of the five variables is negative there.

Example 1 ***Solving the Carpenter's Problem Algebraically***

Consider again the carpenter's model:

$$\text{Maximize } 25x_1 + 30x_2$$

Subject to

$$\begin{aligned} 20x_1 + 30x_2 &\le 690 \quad \text{(lumber)} \\ 5x_1 + 4x_2 &\le 120 \quad \text{(labor)} \\ x_1, x_2 &\ge 0 \quad \text{(nonnegativity)} \end{aligned}$$

We convert each of the first two inequalities to equations by adding new nonnegative *slack variables* y_1 and y_2. If either y_1 or y_2 is negative, the constraint is not satisfied. Thus the problem becomes

$$\text{Maximize } 25x_1 + 30x_2$$

Subject to

$$\begin{aligned} 20x_1 + 30x_2 + y_1 &= 690 \\ 5x_1 + 4x_2 + y_2 &= 120 \\ x_1, x_2, y_1, y_2 &\ge 0 \end{aligned}$$

We now consider the entire set of four variables $\{x_1, x_2, y_1, y_2\}$, which are interpreted geometrically in Figure 10.11. To determine a possible intersection point in the x_1x_2-plane, set two of the four variables equal to zero. There are $4!/2!2! = 6$ possible intersection points to consider in this way (four variables taken two at a time). Let's begin by setting the variables x_1 and x_2 equal to zero; this results in the following set of equations:

$$\begin{aligned} y_1 &= 690 \\ y_2 &= 120 \end{aligned}$$

which is a feasible intersection point $A(0, 0)$ since all four variables are nonnegative.

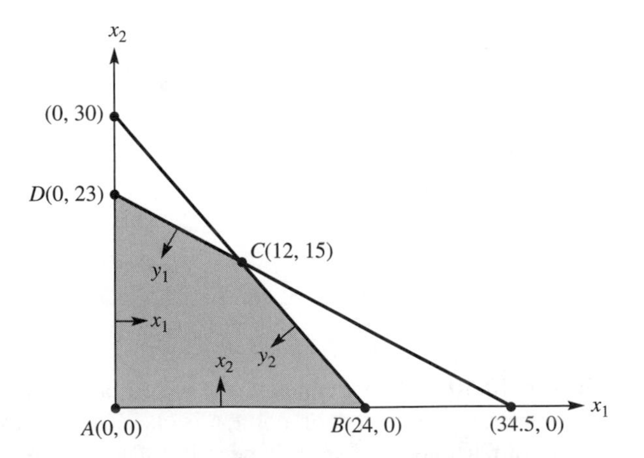

Figure 10.11 The variables $\{x_1, x_2, y_1, y_2\}$ measure the satisfaction of each constraint. An intersection point is characterized by setting two of the variables to zero.

Table 10.3 Finding the objective function by substitution.

Extreme point	Value of objective function
$A(0, 0)$	\$0
$D(0, 23)$	\$690
$C(12, 15)$	\$750
$B(24, 0)$	\$600

For the second intersection point, we choose the variables x_1 and y_1 and set them to zero resulting in the system

$$30x_2 = 690$$

$$4x_2 + y_2 = 120$$

This has solution $x_2 = 23$ and $y_2 = 28$, which is also a feasible intersection point $D(0, 23)$.

For the third intersection point, we choose x_1 and y_2 and set them to zero yielding the system:

$$30x_2 + y_1 = 690$$

$$4x_2 = 120$$

with solution $x_2 = 30$ and $y_1 = -210$. Thus the first constraint is violated by 210 units, indicating that the intersection point $(0, 30)$ is infeasible.

In a similar manner, choosing y_1 and y_2 and setting them to zero gives $x_1 = 12$ and $x_2 = 15$; this corresponds to the intersection point $C(12, 15)$, which is feasible.

Our fifth choice is to choose the variables x_2 and y_1 and set them to zero, giving values of $x_1 = 34.5$ and $y_2 = -52.5$. The second constraint is not satisfied, and the intersection point $(34.5, 0)$ is infeasible.

Finally we determine the sixth intersection point by setting the variables x_2 and y_2 to zero. This gives us $x_1 = 24$ and $y_1 = 210$; so the intersection point $B(24, 0)$ is also feasible.

In summary, of the six possible intersection points in the x_1x_2-plane, four were found to be feasible. For the four, we find the value of the objective function by substitution in Table 10.3.

Our procedure determines that the optimal solution for maximizing the profit is $x_1 = 12$ and $x_2 = 15$. That is, the carpenter should make 12 tables and 15 bookcases for a maximum profit of \$750. ■

Computational Complexity: Intersection Point Enumeration

We now generalize the procedure presented in the carpenter problem. Suppose we have a linear program with m nonnegative decision variables and n constraints, where each constraint is an inequality of the form $\leq$. First, convert each inequality to an equation by adding a nonnegative slack variable y_i to the ith constraint. We now have a total of $m + n$ nonnegative variables. To determine an intersection point, choose m of the variables (because we have m decision variables) and set them to zero. There are $(m+n)!/m!n!$ possible choices to consider. Obviously, as the size of the linear program increases (in numbers of decision variables and constraints), this technique of enumerating all possible intersection points becomes unwieldy, even for powerful computers. How can we improve the procedure?

Note that we enumerated some intersection points in the carpenter problem that turned out to be infeasible. Is there a way to quickly identify that a possible intersection point is infeasible? Moreover, if we have found an extreme point (that is, a feasible intersection point) and know the corresponding value of the objective function, can we quickly determine if another proposed extreme point will improve the value of the objective function? In conclusion, we need a procedure that (1) doesn't enumerate infeasible intersection points and (2) does enumerate only those extreme points that improve the value of the objective function. We study one such procedure in the next section.

Exercises 10.2

1. Using the method of this section, solve Exercises 1–5 in Section 10.1.

2. How many possible intersection points are there in the following cases?

a. 2 decision variables and 5 $\leq$ inequalities
b. 2 decision variables and 10 $\leq$ inequalities
c. 5 decision variables and 12 $\leq$ inequalities
d. 25 decision variables and 50 $\leq$ inequalities
e. 2000 decision variables and 5000 $\leq$ inequalities

10.3 Linear Programming III: The Simplex Method

So far we've found optimal extreme points by searching among all possible intersection points associated with the decision and slack variables. Can we reduce the number of intersection points we actually consider in our search? Certainly, once an initial feasible intersection point is found, we need not consider a potential intersection point that fails to improve the value of the objection function. But can we test the optimality of our current solution against other possible intersection points? Even if an intersection point promises to be more optimal than the current extreme point, it is of no interest if it violates one or more of the constraints. Is there a test to determine if a proposed intersection point is feasible? The **simplex method**, developed by George Dantzig, incorporates both *optimality* and *feasibility* tests to find the optimal solution(s) to a linear program (if one exists).

- An **optimality test** shows whether or not an intersection point corresponds to a value of the objective function better than the best value found so far.

- A **feasibility test** determines whether the proposed intersection point is feasible.

To implement the simplex method, we first separate the decision and slack variables into two nonoverlapping sets, which we call the **independent** and **dependent** sets. For the particular linear programs we consider, the original independent set will consist of the decision variables, and the slack variables will belong to the dependent set. The simplex method consists of the following steps, which we amplify in the discussion to follow.

THE SIMPLEX METHOD

Step 1. **Tableau format:** Place the linear program in tableau format, as explained below.

Step 2. **Initial extreme point:** The simplex method begins with a known extreme point, usually the origin (0, 0).

Step 3. **Optimality test:** Determine if an adjacent intersection point improves the value of the objective function. If not, the current extreme point is optimal. If an improvement is possible, the optimality test determines which variable currently in the independent set (having value zero) should *enter* the dependent set and become nonzero.

Step 4. **Feasibility test:** To find a new intersection point, one variable in the dependent set must *exit* to allow the entering variable from step 3 to become dependent. The feasibility test determines which current dependent variable to choose for exiting, thus ensuring feasibility.

Step 5. **Pivot:** Form a new equivalent system of equations by eliminating the new dependent variable from the equations that don't contain the variable that exited in step 4. Then set the new independent variables to zero in the new system to find the values of the new dependent variables, thereby determining an intersection point.

Step 6. **Repeat steps 3–5** until an optimal extreme point is found.

Before we consider these steps in detail, let's reconsider the carpenter's problem (see Figure 10.12). The origin is an extreme point, so we choose it as our starting point. Thus x_1 and x_2 are the current arbitrary independent variables and are assigned the value zero, while y_1 and y_2 are the current dependent variables with values 690 and 120, respectively. The optimality test determines if a current independent variable assigned the value zero can improve the value of the objective function if it is made dependent and positive. For example, either x_1 or x_2, if made positive, will improve the objective function value. (They have positive coefficients in the objective function we're trying to maximize.) Thus the optimality test determines a promising variable to enter the dependent set. Later we give a rule of thumb for choosing which independent variable

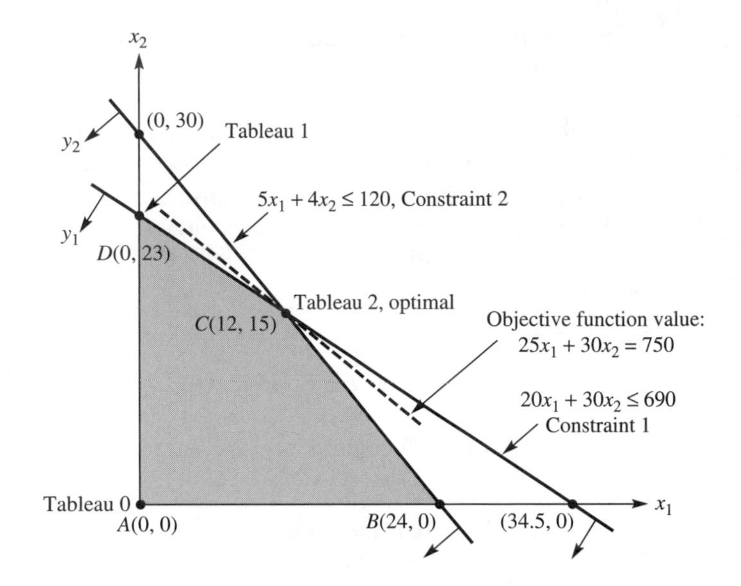

Figure 10.12 The set of points satisfying the constraints of a linear program (the shaded region) form a convex set.

to enter when more than one candidate exists. In the carpenter's problem at hand, we select x_2 as our new dependent variable.

The variable chosen for entry into the dependent set by the optimality condition replaces one of the current dependent variables. The feasibility condition determines which exiting variable this entering variable replaces. Basically, the entering variable replaces whichever current dependent variable can assume a zero value while maintaining nonnegative values for the remaining dependent variables. That is, the feasibility condition ensures that the new intersection point will be feasible and hence an extreme point. In Figure 10.12, the feasibility test leads us to the intersection point (0, 23), which is feasible, and not to (0, 30), which is infeasible. Thus x_2 replaces y_1 as a dependent or nonzero variable; that is, x_2 enters and y_1 exits the set of dependent variables.

Computational Efficiency

The feasibility test does not require that we compute the values of the dependent variables when we're selecting an exiting variable for replacement. Instead, you will see that we select an appropriate exiting variable by quickly determining whether any variable becomes negative if the dependent variable being considered for replacement is assigned the value zero (a ratio test, which will be explained later). If any variable becomes negative, then the dependent variable under consideration cannot be replaced by the entering variable if feasibility is to be maintained.

When a set of dependent variables corresponding to a more optimal extreme point is found from the optimality and feasibility tests, the values of the new dependent variables are determined by pivoting. The pivoting process essentially solves an equivalent system of equations for the new dependent variables after exchanging the entering and exiting dependent variables. The values of the new dependent variables are obtained by setting the independent variables equal to zero. Note that only one dependent variable is replaced at each stage. *Geometrically, the simplex method proceeds from an initial extreme point to an adjacent extreme point until no adjacent extreme point is more optimal.* At that time, the current extreme point is an optimal solution. We now detail the steps of the simplex method.

Step 1 Tableau Format

Many formats exist for implementing the simplex method. The format we use assumes the objective function is to be maximized and that the constraints are less than or equal to inequalities. (If the problem is not expressed initially in this format, it can easily be changed to this format.) For the carpenter's example, our problem is to

$$\text{Maximize } 25x_1 + 30x_2$$

Subject to

$$\begin{aligned} 20x_1 + 30x_2 &\le 690, \\ 5x_1 + 4x_2 &\le 120, \\ x_1, x_2 &\ge 0 \end{aligned}$$

Next we adjoin a new constraint to ensure that any solution improves the best value of the objective function found so far. Take the initial extreme point as the origin, where the value of the objective function is zero. We want to constrain the objective function to be better than its current value, so we require

$$25x_1 + 30x_2 \ge 0$$

Because all of the constraints must be less than or equal to inequalities, multiply the new constraint by -1 and adjoin it to the original constraint set:

$$\begin{aligned} 20x_1 + 30x_2 &\le 690 && \text{(constraint 1, lumber)} \\ 5x_1 + 4x_2 &\le 120 && \text{(constraint 2, labor)} \\ -25x_1 - 30x_2 &\le 0 && \text{(objective function constraint)} \end{aligned}$$

The simplex method implicitly assumes that all variables are nonnegative, so we won't repeat the nonnegativity constraints in the remainder of the presentation.

Next, we convert each inequality to an equality by adding a *nonnegative* new variable y_i (or z), called a *slack variable* since it measures the slack or degree of satisfaction of the constraint. A negative value for y_i indicates the constraint is not satisfied. (We use the variable z for the objective function constraint to avoid confusion with the other constraints.) This process gives the *augmented constraint set*

$$\begin{aligned} 20x_1 + 30x_2 + y_1 &= 690 \\ 5x_1 + \ \ 4x_2 + y_2 &= 120 \\ -25x_1 - 30x_2 + \ \ z &= \ \ 0 \end{aligned}$$

where the variables x_1, x_2, y_1, y_2 are nonnegative. The value of the variable z represents the value of the objective function as we shall see below. (Note from the last equation that $z = 25x_1 + 30x_2$ is the value of the objective function.)

Step 2 Initial Extreme Point

Since there are two decision variables, all possible intersection points lie in the x_1x_2-plane and can be determined by setting two of the variables $\{x_1, x_2, y_1, y_2\}$ to zero. (The variable z is *always* a dependent variable and represents the value of the objective function at the extreme point in question.) The origin is feasible and corresponds to the extreme point characterized by $x_1 = x_2 = 0$, $y_1 = 690$, and $y_2 = 120$. Thus x_1 and x_2 are independent variables assigned the value 0; y_1, y_2, and z are dependent variables whose values are then determined. As we shall see, z conveniently records the current value of the objective function at the extreme points of the convex set in the x_1x_2-plane as we compute them by elimination.

Step 3 The Optimality Test for Choosing an Entering Variable

In the preceding format, a negative coefficient in the last (or objective function) equation indicates that the corresponding variable could improve the current objective function value. Thus the coefficients -25 and -30 indicate that either x_1 or x_2 could enter and improve the current objective function value of $z = 0$. (The current constraint corresponds to $z = 25x_1 + 30x_2 \geq 0$, with x_1 and x_2 currently independent and zero.) When more than one candidate exists for the entering variable, a general rule is to select for the entering variable that variable with the largest (in absolute value) negative coefficient in the objective function row. If no negative coefficients exists, the current solution is optimal. In the case at hand, we choose x_2 as the new entering variable. (The procedure is inexact because at this stage we don't know what values the entering variable can assume.)

Step 4 The Feasibility Test for Choosing an Exiting Variable

The entering variable x_2 (in our example) must replace either y_1 or y_2 as a dependent variable (because z *always* remains the third dependent variable). To determine which of these variables is to exit the dependent set, first divide the right-hand side values 690 and 120 (associated with the original constraint inequalities) by the components for the entering variable in each inequality (30 and 4, respectively) to obtain the ratios $\frac{690}{30} = 23$ and $\frac{120}{4} = 30$. From the subset of ratios that are positive (both in this case), the variable corresponding to the minimum ratio is chosen for replacement (y_1, which corresponds to 23 in this case). *The ratios represent the value the entering variable would obtain if the corresponding exiting variable were assigned the value zero.* Thus only positive values are considered, and the smallest positive value is chosen so that we don't drive any variable negative. For instance, if we choose y_2 as the exiting variable and assign it the value 0, then x_2 assumes a value 30 as the new dependent variable. But then y_1 would be negative, indicating that the intersection point (0, 30) does not satisfy the first constraint. Note that the intersection point (0, 30) is not feasible in Figure 10.12. The

minimum positive ratio rule outlined above obviates enumeration of any infeasible intersection points. In the case at hand, the dependent variable corresponding to the smallest ratio 23 is y_1, so it becomes the exiting variable. Thus x_2, y_2, and z form the new set of dependent variables, and x_1 and y_1 form the new set of independent variables.

Step 5 Pivoting to Solve for the New Dependent Variable Values

Next we derive a new (equivalent) system of equations by eliminating the entering variable x_2 in all equations of the previous system that do not contain the exiting variable y_1. There are numerous ways to execute this step, such as the method of elimination used in Section 10.2. Then we find the values of the dependent variables x_2, y_2, and z when the independent variables x_1 and y_1 are assigned the value zero in the new system of equations. This is called the **pivoting procedure**. The values of x_1 and x_2 give the new extreme point (x_1, x_2), and z is the (improved) value of the objective function at that point.

After performing the pivot, the optimality test is applied again to determine if another entering variable candidate exists. If so, choose an appropriate one and apply the feasibility test for the exiting variable. Then perform the pivoting procedure again. Repeat the process until no variable has a negative coefficient in the objective function row. We now summarize the procedure and use it to solve the carpenter's problem.

SUMMARY OF THE SIMPLEX METHOD

Step 1. *Place the problem in tableau format.* Adjoin slack variables as needed to convert inequality constraints to equalities. Remember that all variables are nonnegative. Include the objective function constraint as the last constraint, including its slack variable z.

Step 2. *Find one initial extreme point.* (For the problems we consider, the origin will be an extreme point.)

Step 3. *Apply the optimality test.* Examine the last equation (which corresponds to the objective function). If all its coefficients are nonnegative, then stop: The current extreme point is optimal. Otherwise, some variables have negative coefficients, so choose the variable with the largest (in absolute value) negative coefficient as the new entering variable.

Step 4. *Apply the feasibility test.* Divide the current right-hand side values by the corresponding coefficient values of the entering variable in each equation. Choose the exiting variable to be the one corresponding to the smallest positive ratio after this division.

Step 5. *Pivot.* Eliminate the entering variable from all the equations that don't contain the exiting variable. (For example, we can use the elimination procedure presented in Section 10.2.) Then assign the value zero to the

(continued)

variables in the new independent set (consisting of the exited variable and the variables remaining after the entering variable has left to become dependent). The resulting values give the new extreme point (x_1, x_2) and objective function value z for that point.

Step 6. *Repeat steps 3–5* until an optimal extreme point is found.

Example 1 ***Solving the Carpenter's Problem Using the Simplex Method***

Step 1. The tableau format gives

$$\begin{aligned} 20x_1 + 30x_2 + y_1 &= 690 \\ 5x_1 + 4x_2 + y_2 &= 120 \\ -25x_1 - 30x_2 + z &= 0 \end{aligned}$$

Step 2. The origin (0, 0) is an initial extreme point for which the independent variables are $x_1 = x_2 = 0$; the dependent variables are $y_1 = 690$, $y_2 = 120$, and $z = 0$.

Step 3. Applying the optimality test, we choose x_2 as the variable entering the dependent set since it has the negative coefficient with the largest absolute value.

Step 4. Applying the feasibility test, we divide the right-hand side values 690 and 120 by the coefficients for the entering variable x_2 in each equation (30 and 4, respectively) and get the ratios $\frac{690}{30} = 23$ and $\frac{120}{4} = 30$. The smallest positive ratio is 23; this corresponds to the first equation, which has the slack variable y_1. Thus we choose y_1 as the exiting dependent variable.

Step 5. We pivot to find the values of the new dependent variables x_2, y_2, and z when the independent variables x_1 and y_1 are set to zero. After eliminating the new dependent variable x_2 from each previous equation that does not contain the exiting variable y_1, we obtain the equivalent system:

$$\begin{aligned} \frac{2}{3}x_1 + x_2 + \frac{1}{30}y_1 \qquad\qquad &= 23 \\ \frac{7}{3}x_1 \qquad - \frac{2}{15}y_1 + y_2 \qquad &= 28 \\ -5x_1 \qquad + \quad y_1 \qquad + z &= 690 \end{aligned}$$

Setting $x_1 = y_1 = 0$, we determine $x_2 = 23$, $y_2 = 28$, and $z = 690$. These results give the extreme point (0, 23) where the value of the objective function is $z = 690$.

Applying the optimality test again, we see that the current extreme point (0, 23) is not optimal (since there is a negative coefficient -5 in the last equation corresponding to the variable x_1). Before continuing, note that we really don't need to write out the entire symbolism of the equations in each step. We merely need to know the coefficient values associated with the variables in each equation together with the right-hand side. A table format, or *tableau*, is commonly used to record these numbers. We complete the carpenter's problem using this format where the headers of each column designate the variables; the abbreviation RHS is the value of the right-hand side. We begin with Tableau 0, which corresponds to the initial extreme point at the origin.

Tableau 0 (original tableau)

x_1	x_2	y_1	y_2	z	RHS
20	30	1	0	0	690 $(= y_1)$
5	4	0	1	0	120 $(= y_2)$
−25	(−30)	0	0	1	0 $(= z)$

Dependent variables: $\{y_1, y_2, z\}$
Independent variables: $x_1 = x_2 = 0$
Extreme point: $(x_1, x_2) = (0, 0)$
Value of objective function: $z = 0$

Optimality Test The entering variable is x_2 (corresponding to -30 in the last row).

Feasibility Test Compute the ratios for the RHS divided by the coefficients in the column labeled x_2 to determine the minimum positive ratio.

x_1	x_2	y_1	y_2	z	RHS	Ratio
20	30	1	0	0	690	(23) (= 690/30) ← Exiting variable
5	4	0	1	0	120	30 (= 120/4)
−25	(−30)	0	0	1	0	*
	↑ Entering variable					

Choose y_1 corresponding to the minimum positive ratio 23 as the exiting variable.

Pivot Divide the row containing the exiting variable (the first row) by the coefficient of the entering variable in that row (the coefficient of x_2), which gives a coefficient of 1 for the entering variable in this row. Then eliminate the entering variable x_2 from the remaining rows (which do not contain the exiting variable y_1 and have a zero coefficient for it). The results are summarized in Tableau 1, which uses five-place decimal approximations to the numerical values.

Tableau 1

x_1	x_2	y_1	y_2	z	RHS
0.66667	1	0.03333	0	0	23 $(= x_2)$
2.33333	0	−0.13333	1	0	28 $(= y_2)$
(−5.00000)	0	1.00000	0	1	690 $(= z)$

Dependent variables: $\{x_2, y_2, z\}$
Independent variables: $x_1 = y_1 = 0$
Extreme point: $(x_1, x_2) = (0, 23)$
Value of objective function: $z = 690$

The pivot determines that the new dependent variables have the values $x_2 = 23$, $y_2 = 28$, and $z = 690$.

Optimality Test The entering variable is x_1 (corresponding to the coefficient -5 in the last row).

Feasibility Test Compute the ratios for the RHS.

x_1	x_2	y_1	y_2	z	RHS	Ratio
0.66667	1	0.03333	0	0	23	34.5 $(= 23/0.66667)$
2.33333	0	-0.13333	1	0	28	12.0 $(= 28/2.33333)$ ← Exiting variable
-5.00000	0	1.00000	0	1	690	*

↑ Entering variable

Choose y_2 as the exiting variable because it corresponds to the minimum positive ratio 12.

Pivot Divide the row containing the exiting variable (the second row) by the coefficient of the entering variable in that row (the coefficient of x_1), which gives a coefficient of 1 for the entering variable in this row. Then eliminate the entering variable x_1 from the remaining rows (which do not contain the exiting variable y_2 and have a zero coefficient for it). The results are summarized in Tableau 2.

Tableau 2

x_1	x_2	y_1	y_2	z	RHS
0	1	0.071429	-0.28571	0	15 $(= x_2)$
1	0	-0.057143	0.42857	0	12 $(= x_1)$
0	0	0.714286	2.14286	1	750 $(= z)$

Dependent variables: $\{x_2, x_1, z\}$
Independent variables: $y_1 = y_2 = 0$
Extreme point: $(x_1, x_2) = (12, 15)$
Value of objective function: $z = 750$

Optimality Test Because there are no negative coefficients in the bottom row, $x_1 = 12$ and $x_2 = 15$ gives the optimal solution $z = \$750$ for the objective function. Note that starting with an initial extreme point, only two of the possible six intersection points had to be enumerated. The power of the simplex method is its reduction of the computations required to find an optimal extreme point. ■

Let's try the following problem.

Example 2 ***Another Maximization Problem***

$$\text{Maximize } 3x_1 + x_2$$

Subject to

$$2x_1 + x_2 \le 6$$
$$x_1 + 3x_2 \le 9$$
$$x_1, x_2 \ge 0$$

The problem in tableau format is

$$2x_1 + x_2 + y_1 = 6$$

$$x_1 + 3x_2 + y_2 = 9$$

$$-3x_1 - x_2 + z = 0$$

where x_1, x_2, y_1, y_2, and $z \ge 0$.

Tableau 0 (original tableau)

x_1	x_2	y_1	y_2	z	RHS
2	1	1	0	0	6 $(= y_1)$
1	3	0	1	0	9 $(= y_2)$
(−3)	−1	0	0	1	0 $(= z)$

Dependent variables: $\{y_1, y_2, z\}$
Independent variables: $x_1 = x_2 = 0$
Extreme point: $(x_1, x_2) = (0, 0)$
Value of objective function: $z = 0$

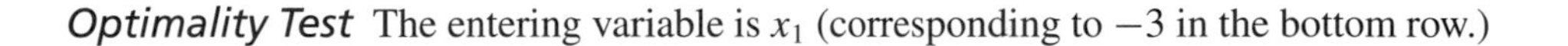

Optimality Test The entering variable is x_1 (corresponding to -3 in the bottom row.)

Feasibility Test Compute the ratios of the RHS divided by the column labeled x_1 to determine the minimum positive ratio.

x_1	x_2	y_1	y_2	z	RHS	Ratio
2	1	1	0	0	6	(3) $(= 6/2)$ ← Exiting variable
1	3	0	1	0	9	9 $(= 9/1)$
(−3) ↑ Entering variable	−1	0	0	1	0	*

Choose y_1 corresponding to the minimum positive ratio 3 as the exiting variable.

Pivot Divide the row containing the exiting variable (the first row) by the coefficient of the entering variable in that row (the coefficient of x_1), which gives a coefficient of 1 for the entering variable in this row. Then eliminate the entering variable x_1 from the remaining rows (which do not contain the exiting variable y_1 and have a zero coefficient for it). The results are summarized in Tableau 1.

Tableau 1

x_1	x_2	y_1	y_2	z	RHS
1	$\frac{1}{2}$	$\frac{1}{2}$	0	0	$3\ (= x_1)$
0	$\frac{5}{2}$	$-\frac{1}{2}$	1	0	$6\ (= y_2)$
0	$\frac{1}{2}$	$\frac{3}{2}$	0	1	$9\ (= z)$

Dependent variables: $\{x_1, y_2, z\}$
Independent variables: $x_2 = y_1 = 0$
Extreme point: $(x_1, x_2) = (3, 0)$
Value of objective function: $z = 9$

The pivot determines that the dependent variables have the values $x_1 = 3$, $y_2 = 6$, and $z = 9$.

Optimality Test There are no negative coefficients in the bottom row. Thus $x_1 = 3$ and $x_2 = 0$ is an extreme point, giving the optimal objective function value $z = 9$.

Remarks We have assumed that the origin is a feasible extreme point. If it is not, then an extreme point must be found before the simplex method, as presented, can be used. We have also assumed that the linear program is not degenerate in the sense that no more than two constraints intersect at the same point. These and other topics are studied in more advanced optimization courses. ■

Exercises 10.3

Use the simplex method to solve Exercises 1–5, Section 10.1.

PROJECT Exploring Modeling

Write a computer code to perform the basic simplex algorithm. Solve Exercise 3 in Section 10.1 using your code.

10.4 Linear Programming IV: Sensitivity Analysis

Mathematical models typically *approximate* a problem under study. For example, the coefficients in the objective function of a linear program may only be estimates. Or the amount of the resources constraining production made available by management may vary, depending on the profit returned per unit of resource invested. (Management may be willing to procure additional resources if the additional profit is high enough.) Thus management would like to know if the additional profit to be realized justifies the cost of another unit of resource. If so, over what range of values for the resources is the analysis valid? Hence, in addition to solving a linear program, we would like to know how sensitive the optimal solution is to changes in the various constants used to formulate the program. In this section we use graphics to analyze the effect on the optimal solution of changes in the coefficients of the objective function and in available resources. Using the carpenter's problem as our example, we answer the following questions:

1. Over what range of values for the profit per table does the current solution remain optimal?

2. What is the value of another unit of the second resource (labor)? That is, how much will the profit increase if another unit of labor is obtained? Over what range of labor values is the analysis valid? What is required to increase profit beyond this limit?

Testing the Optimal Solution for Sensitivity to Change

Let's consider the carpenter's problem again. The objective function is to maximize profits where each table nets \$25 profit and each bookcase \$30. If z represents the amount of profit, then we want to

$$\text{Maximize } z = 25x_1 + 30x_2$$

Note that z is a function of two variables and we can draw the level curves of z in the x_1x_2-plane. In Figure 10.13, we graph the level curves corresponding to the values $z = 650$, $z = 750$, and $z = 850$.

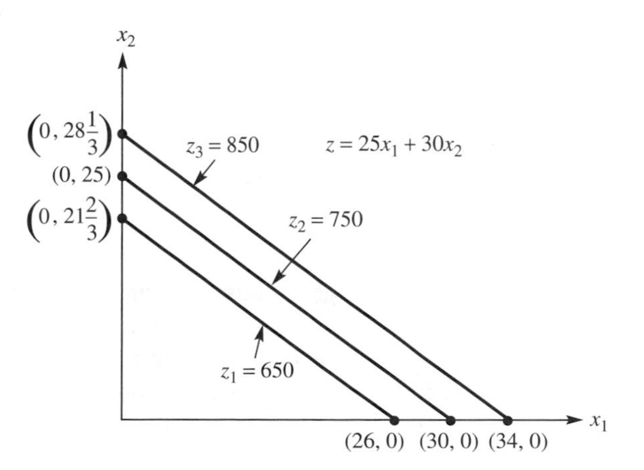

Figure 10.13 The level curves of $z = 25x_1 + 30x_2$ in the x_1x_2-plane have a slope of $-\frac{5}{6}$.

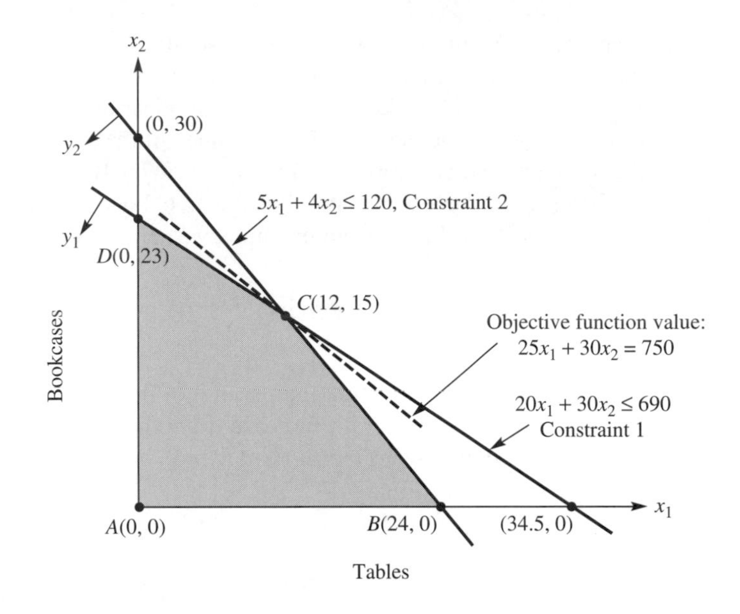

Figure 10.14 The level curve $z = 750$ is tangent to the convex set of feasible solutions at extreme point $C(12, 15)$.

Note that every level curve is a line with slope $-\frac{5}{6}$. In Figure 10.14, we superimpose on the previous graph the constraint set for the carpenter's problem and see that the optimal solution (12, 15) gives an optimal objective function value of $z = 750$.

Now we can ask the question, What is the effect of changing the profit value for each table? Intuitively, if we increase the profit sufficiently, the carpenter eventually makes only tables, (giving the extreme point of 24 tables and 0 bookcases), instead of the current mix of 12 tables and 15 bookcases. Similarly, if we decrease the profit per table sufficiently, he should make only bookcases at the extreme point (0, 23). Note again that the slope of the level curves of the objective function is $-\frac{5}{6}$. If we let c_1 represent the profit per bookcase, then the objective function becomes

$$\text{Maximize } z = c_1x_1 + 30x_2$$

with slope $-c_1/30$ in the x_1x_2-plane. As we vary c_1, the slope of the level curves of the objective function changes. Examine Figure 10.15 to convince yourself that the current extreme point (12, 15) remains optimal as long as the slope of the objective function is between the slopes of the two binding constraints. In this case, the extreme point (12, 15) remains optimal as long as the slope of the objective function is less than $-\frac{2}{3}$ but greater than $-\frac{5}{4}$, the slopes of the lumber and labor constraints, respectively. If we start with the slope for the objective function as $-\frac{2}{3}$, as we increase c_1 we rotate the level curve of the objective function clockwise. If we rotate clockwise, the optimal extreme point changes to (24, 0) if the slope of the objective function is less than $-\frac{5}{4}$. Thus the

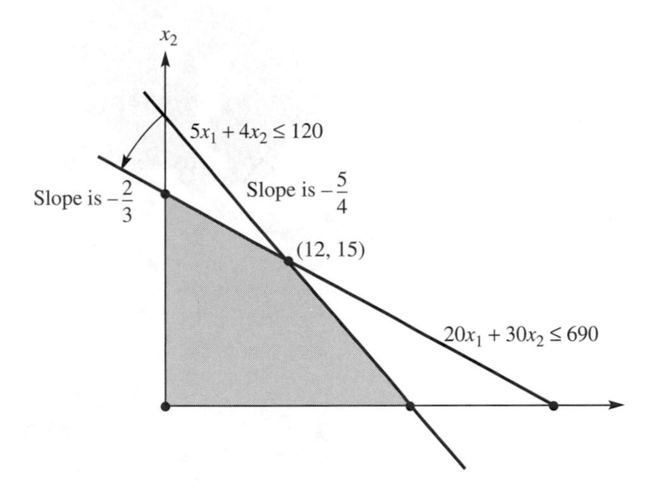

Figure 10.15 The extreme point (12, 15) remains optimal for objective functions with a slope between $-\frac{5}{4}$ and $-\frac{2}{3}$.

range of values for which the current extreme point remains optimal is given by the inequality

$$-\frac{5}{4} \leq -\frac{c_1}{30} \leq -\frac{2}{3}$$

or,

$$20 \leq c_1 \leq 37.5$$

Interpreting this result, if the profit per table exceeds 37.5, the carpenter should produce only tables (that is, 24 tables). If the profit per table is reduced below 20, the carpenter should produce only bookcases (23 bookcases). If c_1 is between 20 and 37.5, he should produce the mix of 12 tables and 15 bookcases. As we change c_1 over the range [20, 37.5], the value of the objective function changes even though the location of the extreme point does not. Because he is making 12 tables, the objective function changes by a factor of 12 times the change in c_1. Note that at the limit $c_1 = 20$, there are *two* extreme points C and B that produce the same value for the objective function. Likewise if $c_1 = 37.5$, the extreme points D and C produce the same value for the objective function. In such cases, we say that there are *alternative optimal solutions*.

Changes in the Amount of Resource Available

Currently, there are 120 units of labor available, all of which is used to produce the 12 tables and 15 bookcases represented by the optimal solution. What is the effect of increasing the available labor? If b_2 represents the units of available labor (the second resource constraint), the constraint can be rewritten as

$$5x_1 + 4x_2 \leq b_2$$

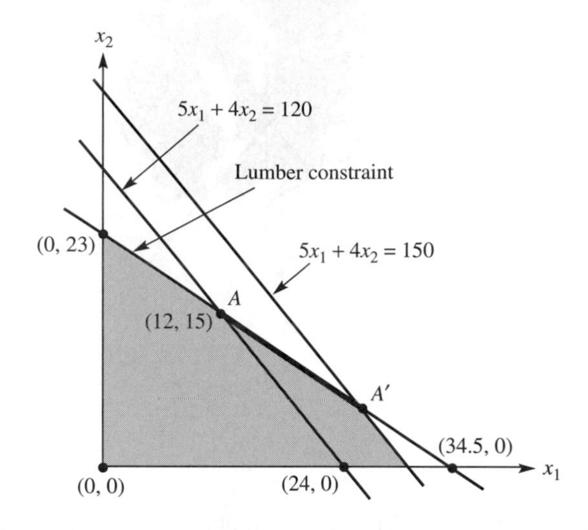

Figure 10.16 As the amount of labor resource b_2 increases from 120 to 150 units, the optimal solution moves from A to A' along the lumber constraint, increasing x_1 and decreasing x_2.

What happens geometrically as we vary b_2? To answer this question, graph the constraint set for the carpenter's problem with the original value of $b_2 = 120$ and a second value, say $b_2 = 150$ (see Figure 10.16). Note that the effect of increasing b_2 is to translate the constraint upward and to the right. As this happens, the optimal value of the objective function moves along the line segment AA', which lies on the lumber constraint. As the optimal solution moves along the line segment from A to A', x_1 increases and x_2 decreases. The net effect of increasing b_2 is to increase the value of the objective function. But by how much? One goal is to determine how much the objective function value changes as *b_2 increases by 1 unit.*

Note that if b_2 increases beyond $5 \cdot 34.5 = 172.5$, the optimal solution remains at the extreme point (34.5, 0). That is, at (34.5, 0) the lumber constraint must also be increased if the objective function is to be increased further. Thus increasing the labor constraint to 200 units results in some excess labor that cannot be used unless the amount of lumber is increased beyond its present value of 690 (see Figure 10.17). Following a similar analysis, if b_2 is decreased, the value of the objective function moves along the lumber constraint until the extreme point (0, 23) is reached. Further reductions in b_2 cause the optimal solution to move from (0, 23) down the y-axis to the origin.

Now let's find the range of values of b_2 for which the optimal solution moves along the lumber constraint as the amount of labor varies. Refer to Figure 10.18 and convince yourself that we want to find the value for b_2 where the labor constraint intersects the lumber constraint on the x_1-axis, or point E (34.5, 0). At point E (34.5, 0), the amount of labor is $5 \cdot 34.5 + 4 \cdot 0 = 172.5$. Similarly, we want to find the value for b_2 at point D (0, 23), which is $5 \cdot 0 + 4 \cdot 23 = 92$. Summarizing, as b_2 changes, the optimal solution moves along the lumber constraint as long as

$$92 \leq b_2 \leq 172.5$$

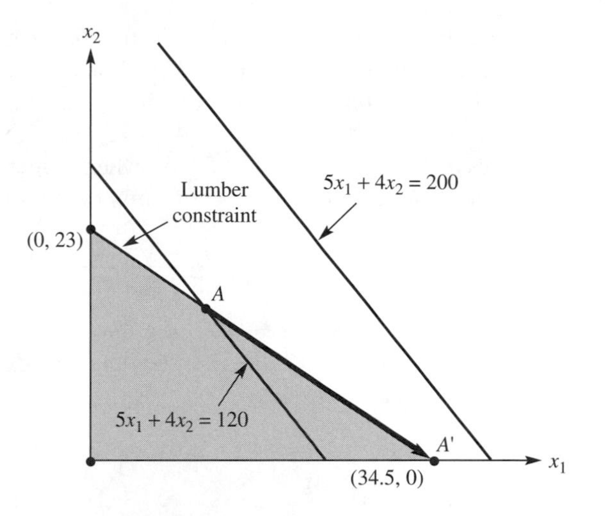

Figure 10.17 As resource b_2 increases from 120 to 172.5, the optimal solution moves from A to A' along the line segment AA'. Increasing b_2 beyond $b_2 = 172.5$ does not increase the value of the objective function unless the lumber constraint is also increased (moving it upward to the right).

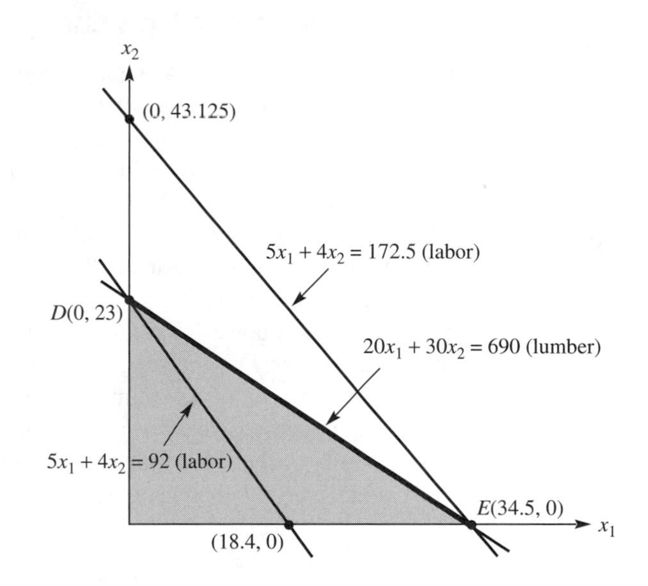

Figure 10.18 As b_2 increases from 92 to 172.5, the optimal solution moves from point $D(0, 23)$ to point $E(34.5, 0)$ along the line segment DE, the lumber constraint.

But by how much does the objective function change as b_2 increases by 1 unit within the range $92 \leq b_2 \leq 172.5$? We'll analyze this in several ways. First, suppose $b_2 = 172.5$. The optimal solution is then the new extreme point E (34.5, 0), and the value of the objective function is $34.5 \cdot 25 = 862.5$ at E. Thus the objective function increases by $862.5 - 750 = 112.5$ units when b_2 increases by $172.5 - 120 = 52.5$ units. Hence the change in the objective function for 1 unit of change in labor is

$$\frac{862.5 - 750}{172.5 - 120} \simeq 2.14$$

Now let's analyze the value of a unit change of labor in another way. If b_2 increases by 1 unit from 120 to 121, then the new extreme point A' represented by the intersection of the constraints

$$20x_1 + 30x_2 = 690$$

$$5x_1 + 4x_2 = 121$$

is the point $A'(12.429, 14.714)$, which has an objective function value of 752.14. Thus the net effect as b_2 increases by 1 labor-hour is to increase the objective function by \$2.14.

Economic Interpretation of a Unit Change in a Resource

In the foregoing analysis, we saw that as one more unit of labor is added, the objective function increases by \$2.14 as long as the total amount of labor does not exceed 172.5 units. Thus in terms of the objective function, an *additional* unit of labor is worth \$2.14. If management can procure a unit of labor for less than \$2.14, it would be profitable to do so. Conversely, if management can sell labor for more than \$2.14 (which is valid until labor is reduced to 92 units), it should also consider this strategy. Note that our analysis gives the value of a unit of resource in terms of the value of the objective function at the optimal extreme point, which is a *marginal value*.

Sensitivity analysis is a powerful method for interpreting linear programs. The information embodied in a carefully accomplished sensitivity analysis is often at least as valuable to the decision maker as the optimal solution to the linear program itself. In advanced courses in optimization, you will perform sensitivity analyses algebraically. Moreover, the coefficients in the constraint set, as well as the right-hand side of the constraints, can be analyzed for their sensitivity.

Exercises 10.4

1. Consider the example problem in this section. Determine the sensitivity of the optimal solution to a change in c_2, using the objective function, $25x_1 + c_2x_2$.

2. Perform a complete sensitivity analysis (objective function coefficients and right hand side values) of the toy soldier problem in Section 10.1 (Exercise 1).

3. Why is sensitivity analysis important in linear programming?

10.5 Numerical Search Methods

If we want to maximize a function $f(x)$ over some interval $[a, b]$, we know from calculus that setting the first derivative equal to zero yields its critical points. The second derivative test may then be employed to characterize the nature of these critical points. We also know we may have to check the endpoints and points where the first derivative fails to exist.

It may be impossible to solve the equation algebraically because we have to set the first derivative equal to zero. For example, consider a machine that earns revenue at the rate of e^{-t} dollars a year for t years and then can be sold for $1/(1+t)$ dollars. If we model maximizing revenues as a function of years, the problem is to maximize

$$f(T) = \int_0^T e^{-t}\, dt + \frac{1}{1+t}$$

Setting the first derivative equal to zero yields

$$f'(T) = e^{-T} + (1+T)^{-2} = 0$$

which cannot be algebraically solved for T in closed form. In such cases, we can use a search procedure to approximate the optimal solution.

Various search methods permit us to approximate solutions to nonlinear optimization problems with a single independent variable. Two search methods commonly used are the dichotomous and golden section methods. Both share several features common to most search methods. We discuss these features, and the methods.

A *unimodal function* on an interval has exactly one point where a maximum or minimum occurs in the interval. If the function is known (or assumed) to be multimodal, then it must be subdivided into separate unimodal functions. (In most practical problems, the optimal solution is known to lie in some restricted range of the independent variable.) More precisely, a function $f(x)$ is a **unimodal function** with an interior local maximum on an interval $[a, b]$, if for some point x^* on $[a, b]$, the function is strictly increasing on $[a, x^*]$ and strictly decreasing on $[x^*, b]$. A similar statement holds for $f(x)$ being unimodal with an interior local minimum. These concepts are illustrated in Figure 10.19.

The unimodal assumption is important for finding the subset of the interval $[a, b]$ that contains the optimal point x to maximize (or minimize) $f(x)$.

The Search Method Paradigm

With most search methods, we divide the region $[a, b]$ into two overlapping intervals $[a, x_2]$ and $[x_1, b]$ after placing two test points x_1 and x_2 in the original interval $[a, b]$ according to some criterion of our chosen search method, as shown in Figure 10.20. We then determine the subinterval where the optimal solution lies and use that subinterval to continue the search based on the function evaluations $f(x_1)$ and $f(x_2)$. Figure 10.21 shows the three cases in the maximization problem (the minimization problem is analogous) with experiments x_1 and x_2 placed between $[a, b]$ according to the chosen search method (fully discussed later):

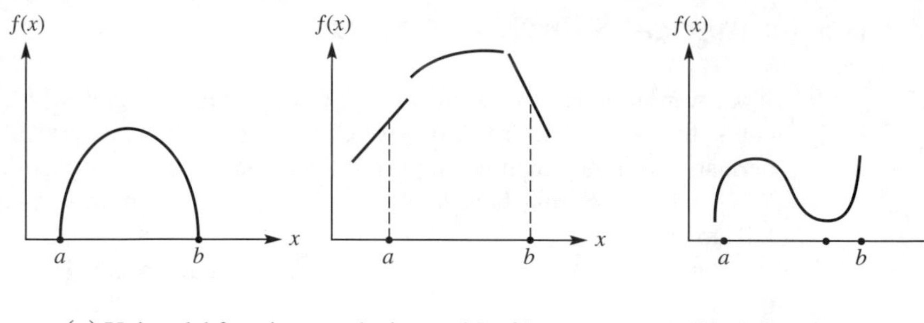

(a) Unimodal functions on the interval $[a, b]$

(b) A function that is not unimodal on the interval $[a, b]$

Figure 10.19 Examples of functional modality.

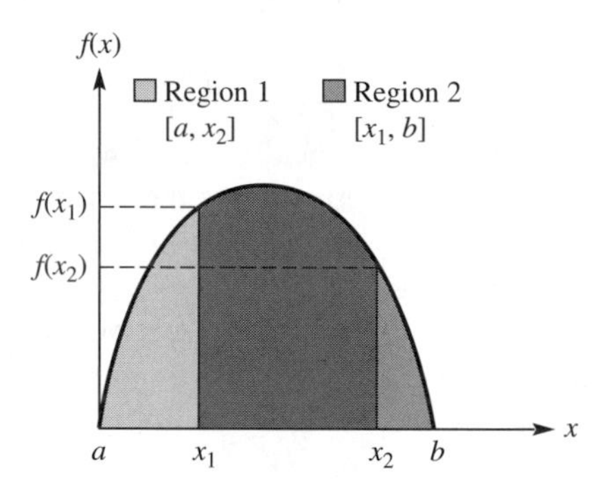

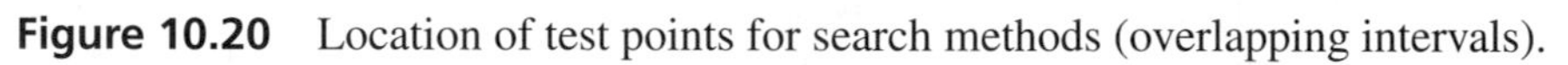

Figure 10.20 Location of test points for search methods (overlapping intervals).

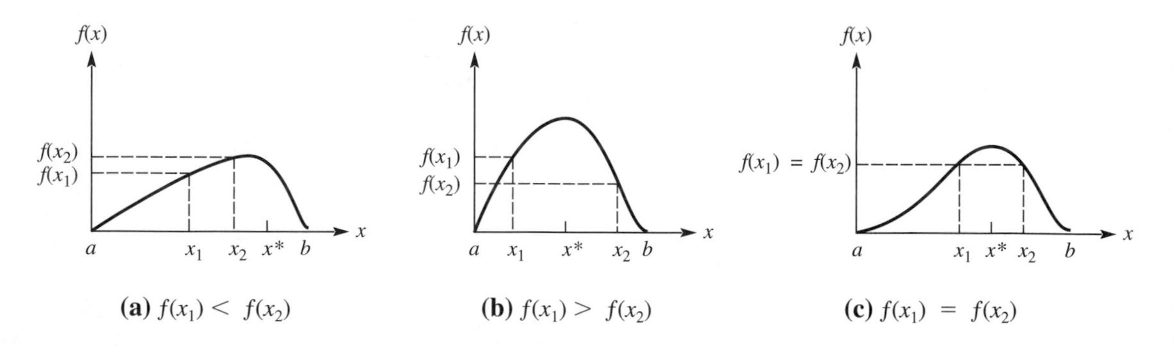

(a) $f(x_1) < f(x_2)$

(b) $f(x_1) > f(x_2)$

(c) $f(x_1) = f(x_2)$

Figure 10.21 The three cases for the maximization problem.

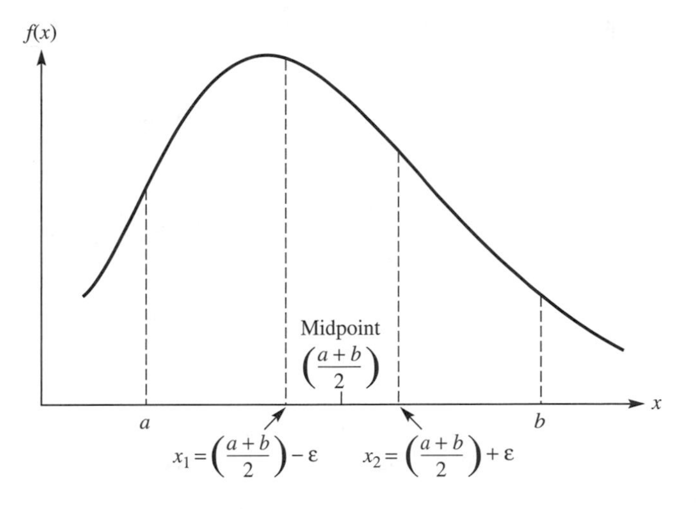

Figure 10.22 Dichotomous search.

Case 1. $f(x_1) < f(x_2)$. Since $f(x)$ is unimodal, the solution cannot occur in the interval $[a, x_1]$. The solution must lie in the interval $(x_1, b]$.

Case 2. $f(x_1) > f(x_2)$. Since $f(x)$ is unimodal, the solution cannot occur in the interval $(x_2, b]$. The solution must lie in the interval $[a, x_2)$.

Case 3. $f(x_1) = f(x_2)$. The solution must lie somewhere in the interval (x_1, x_2).

We next present two commonly used search techniques.

Dichotomous Search Method

Assume we have a function $f(x)$ to maximize over a specified interval $[a, b]$. The dichotomous search method computes the midpoint $(a+b)/2$, and then moves slightly to either side of the midpoint to compute two test points: $(a + b)/2 \pm \varepsilon$, where ε is some very small real number. In practice, the number ε is chosen as small as the accuracy of the computational device will permit, the objective being to place the two experimental points as close together as possible. Figure 10.22 shows this procedure for a maximization problem. The procedure continues until it gets within some small interval containing the optimal solution. Here are the steps in this algorithm.

DICHOTOMOUS SEARCH ALGORITHM TO MAXIMIZE f(x) OVER THE INTERVAL $a \le x \le b$

Step 1. Initialize. Choose a small number $\varepsilon > 0$, like 0.01. Select a small $t > 0$, between $[a, b]$, called the *length of uncertainty* for the search. Calculate the number of iterations n using the formula

(continued)

$$(0.5)^n = \frac{t}{b-a}$$

Step 2. For $k = 1$ to n, do steps 3 and 4.

Step 3.

$$x_1 = \left(\frac{a+b}{2}\right) - \varepsilon$$

and

$$x_2 = \left(\frac{a+b}{2}\right) + \varepsilon$$

Step 4. (For a maximization problem)

(a) If $f(x_1) \geq f(x_2)$, then let

$$a = a$$
$$b = x_2$$
$$k = k + 1$$

Return to step 3.

(b) If $f(x_1) < f(x_2)$, then let

$$b = b$$
$$a = x$$
$$k = k + 1$$

Return to step 3.

Step 5. Let

$$x^* = \frac{a+b}{2}$$

and MAX $= f(x^*)$.

Stop

In this presentation, the number of iterations performed is determined by reductions in the length of uncertainty desired. Alternatively, we may wish to continue to iterate until the change in the dependent variable is less than some predetermined amount, say Δ. That is, continue to iterate until $f(a) - f(b) \leq \Delta$. For example, in an application where $f(x)$ represents the profit realized by producing x items, it might make

Table 10.4 Results of a dichotomous search for Example 1.[1]

a	b	x_1	x_2	$f(x_1)$	$f(x_2)$
−3	6	1.49	1.51	−5.2001	−5.3001
−3	1.51	−0.755	−0.735	0.9400	0.9298
−3	−0.735	−1.8775	−1.8575	0.2230	0.2647
−1.8775	−0.735	−1.3163	−1.2963	0.9000	0.9122
−1.3163	−0.735	−1.0356	−1.0156	0.9987	0.9998
−1.0356	−0.735	−0.8953	−0.8753	0.9890	0.9845
−1.0356	−0.8753				

more sense to stop when the change in profit is less than some acceptable amount. To minimize a function $y = f(x)$, either maximize $-y$ or switch the directions of the signs in steps 4a and 4b.

Example 1 ***Using the Dichotomous Search Algorithm***

Suppose we want to maximize $f(x) = -x^2 - 2x$ over the interval $-3 \leq x \leq 6$, and we want the optimal tolerance to be less than 0.2. We arbitrarily choose ε (the distinquishability constant) to be 0.01. Next we determine the number n of iterations using the relationship $(0.5)^n = 0.2/[6-(-3)]$, or $n \ln(0.5) = \ln(0.2/9)$, which implies that $n = 5.49$ (and we round to the next higher integer, $n = 6$). Table 10.4 gives the results for implementing the search algorithm.

The length of the final interval is less than the 0.2 tolerance initially specified. From step 5, we estimate the location of a maximum at

$$x^* = \frac{-1.0356 - 0.8753}{2} = -0.9555$$

with $f(-0.9555) = 0.9980$. (Examining our table, we see that $f(-1.0156) = 0.9998$, a better estimate.) Note that the number of evaluations $n = 6$ refers to the number of intervals searched. For this example, we can use calculus to find the optimal solution $f(-1) = 1$ at $x = -1$. ■

Golden Section Search Method

The golden section search method uses the **golden ratio**. To better understand the golden ratio, divide the interval [0, 1] into two separate subintervals of lengths r and $1 - r$, as shown in Figure 10.23. These subintervals are said to be divided into the golden ratio if

$$\text{length}_{\text{whole interval}} : \text{length}_{\text{longer segment}} = \text{length}_{\text{longer segment}} : \text{length}_{\text{shorter segment}}$$

[1] The numerical results in this section were computed carrying the 13-place accuracy of the computational device being used. The results were then rounded to four places for presentation.

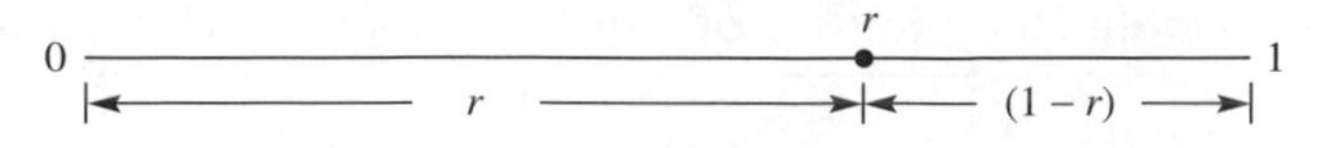

Figure 10.23 Golden ratio using a line segment.

Symbolically, this can be written as

$$\frac{1}{r} = \frac{r}{1-r} \qquad \text{or} \qquad r^2 + r - 1 = 0$$

because $r > 1 - r$ in the figure.

Solving this last equation gives the two roots

$$r_1 = \frac{\sqrt{5}-1}{2} \qquad \text{and} \qquad r_2 = \frac{-\sqrt{5}-1}{2}$$

Only the positive root r_1 lies in the interval [0, 1]. The numerical value of r_1 is approximately 0.618 and is known as the *golden ratio.*

The golden section search method incorporates the following assumptions:

1. The function $f(x)$ must be unimodal over the specified interval, $[a, b]$.
2. The function must have a maximum (or minimum) value over a known interval of uncertainty.
3. The method gives an approximation to the maximum value rather than the exact maximum value.

The method determines a final interval that contains the optimal solution. The length of the final interval can be controlled and made arbitrarily small by the selection of a tolerance value. The length of the final interval will be less than our specified tolerance level.

The search procedure for finding an approximation to the maximum value is iterative. It requires evaluations of $f(x)$ at the test points $x_1 = a + (1 - r)(b - a)$ and $x_2 = a + r(b - a)$ and then determines the new interval of search (see Figure 10.24). If $f(x_1) < f(x_2)$, then the new interval is $[x_1, b]$; if $f(x_1) > f(x_2)$ then the new interval is $[a, x_2]$, as in the dichotomous search method. The iterations continue until the final interval length is less than the tolerance imposed, and the final interval contains the optimal solution point. The length of this final interval determines the accuracy of the approximate optimal solution point. The number of iterations required to achieve the tolerance length can be found as the integer greater than k, where

$$k = \frac{\ln(\text{tolerance}/b - a)}{\ln 0.618}$$

Alternatively, the method can be stopped when an interval $[a, b]$ is less than the required tolerance. The steps of the golden section search method are as follows.

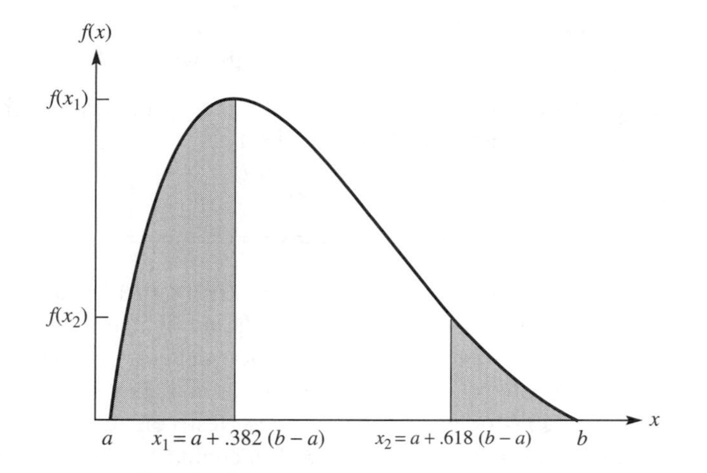

Figure 10.24 Location of x_1 and x_2 for the golden section search.

GOLDEN SECTION SEARCH ALGORITHM FOR MAXIMIZING f(x) OVER THE INTERVAL $a \le x \le b$

Step 1. Initialize. Choose a tolerance $t > 0$.

Step 2. Set $r = 0.618$ and define the test points:

$$x_1 = a + (1 - r)(b - a)$$
$$x_2 = a + r(b - a)$$

Step 3. Calculate $f(x_1)$ and $f(x_2)$.

Step 4. (For a maximization problem) Compare $f(x_1)$ with $f(x_2)$:

(a) If $f(x_1) \le f(x_2)$, then the new interval is $[x_1, b]$ where

a becomes the previous x_1
b does not change
x_1 becomes the previous x_2

Find the new x_2 using the formula in step 2.

(b) If $f(x_1) > f(x_2)$, then the new interval is $[a, x_2]$ where

a remains unchanged
b becomes the previous x_2
x_2 becomes the previous x_1

Find the new x_1 using the formula in step 2.

(continued)

Step 5. If the length of the new interval from step 4 is less than the specified tolerance t, then stop. Otherwise, go back to step 3.

Step 6. Estimate x^* as the midpoint of the final interval $x^* = (a + b)/2$ and compute $\text{MAX} = f(x^*)$.

Stop

To minimize a function $y = f(x)$, either maximize $-y$ or switch the directions of the signs in steps 4a and 4b. The advantage of the golden section search method is that only one new test point (and evaluation of the function at the test point) must be computed at each successive iteration, compared with two new test points (and two evaluations of the function at those test points) for the dichotomous search method. Using the golden section search method, the length of the interval of uncertainty is 61.8% of the length of the previous interval of uncertainty. Thus, for large n, the interval of uncertainty is reduced by approximately $(0.618)^n$ after n test points are computed. [Compare with $(0.5)^{n/2}$ for the dichotomous search method.]

Example 2 *Using the Golden Search Algorithm*

Suppose we want to maximize $f(x) = -3x^2 + 21.6x + 1$ over $0 \leq x \leq 25$, with a tolerance of $t = 0.25$. Determine the first two test points and evaluate $f(x)$ at each test point:

$$x_1 = a + 0.382(b - a) \rightarrow x_1 = 0 + 0.382(25 - 0) = 9.55$$

and

$$x_2 = a + 0.618(b - a) \rightarrow x_2 = 0 + 0.618(25 - 0) = 15.45$$

Then

$$f(x_1) = -66.3275 \qquad \text{and} \qquad f(x_2) = -381.3875$$

Because $f(x_1) > f(x_2)$, we discard all values in $[x_2, b]$ and select the new interval $[a, b] = [0, 15.45]$. Then $x_2 = 9.55$, which is the previous x_1, and $f(x_2) = -66.2972$. We must now find the position of the new test point x_1, and evaluate $f(x_1)$

$$x_1 = 0 + (1 - r)(15.45 - 0) = 5.9017$$

$$f(x_1) = 23.9865$$

Again, $f(x_1) > f(x_2)$, so the new interval is $[a, b]$ is $[0, 9.55]$. Then the new $x_2 = 5.9017$ and $f(x_1) = 23.9865$. We find a new x_1 and $f(x_1)$,

$$x_1 = 0 + (1 - r)(9.55 - 0) = 3.6475$$

$$f(x_1) = 39.8732$$

Because $f(x_1) > f(x_2)$, we discard $[x_2, b]$ and our new search interval is $[a, b] = [0, 5.9017]$. Then $x_2 = 3.6475$ with $f(x_2) = 39.8732$. We find a new x_1 and $f(x_1)$:

$$x_1 = 0 + (1 - r)(5.9017 - 0) = 2.2542$$
$$f(x_1) = 34.4469$$

Because $f(x_2) > f(x_1)$, we discard $[a, x_1)$ and the new interval is $[a, b] = [2.2542, 5.9017]$. The new $x_1 = 3.6475$ with $f(x_1) = 39.8732$. We find a new x_2 and $f(x_2)$:

$$x_2 = 2.2545 + r(5.9017 - 2.2542) = 4.5085$$
$$f(x_2) = 37.4039$$

This process continues until the length of the interval of uncertainty, $b - a$, is less than the tolerance, $t = 0.25$. This requires 10 iterations. The results of the golden section search method for Example 2 are summarized in Table 10.5.

The final interval $[a, b] = [3.4442, 3.6475]$ is the first interval of our $[a, b]$ intervals that is less than our 0.25 tolerance. The value of x that maximizes the given function over the interval must lie within this final interval of uncertainty $[3.4442, 3.6475]$. We estimate

$$x^* = \frac{3.4442 + 3.6475}{2} = 3.5459 \qquad \text{and} \qquad f(x^*) = 39.8712$$

The actual maximum, which in this case can be found by calculus, occurs at $x^* = 3.60$ where $f(3.60) = 39.88$. ■

As illustrated above, we stopped when the interval of uncertainty was less than 0.25. Alternatively, we can compute the number of iterations required to attain the ac-

Table 10.5 Golden section results for Example 2.

k	a	b	x_1	x_2	$f(x_1)$	$f(x_2)$
0	0	25	9.5491	15.4509	−66.2972	−381.4479
1	0	15.4506	5.9017	9.5491	23.9865	−66.2972
2	0	9.5592	3.6475	5.9017	39.8732	23.9865
3	0	5.9017	2.2542	3.6475	34.4469	39.8732
4	2.2542	5.9017	3.6475	4.5085	39.8732	37.4039
5	2.2542	4.5085	3.1153	3.6475	39.1752	39.8732
6	3.1153	4.5085	3.6475	3.9763	39.8732	39.4551
7	3.1153	3.9763	3.4442	3.6475	39.8072	39.8732
8	3.4442	3.9763	3.6475	3.7731	39.8732	39.7901
9	3.4442	3.7731	3.5698	3.6475	39.8773	39.8732
10	3.4442	3.6475				

curacy specified by the tolerance. Because the interval of uncertainty is 61.8% of the interval of uncertainty at each stage, we have

$$\frac{\text{length of final interval (tolerance } t)}{\text{length of initial interval}} = 0.618^k$$

$$\frac{0.25}{25} = 0.618^k$$

$$k = \frac{\ln 0.01}{\ln 0.618} = 9.57, \text{ or 10 iterations}$$

In general, the number of iterations k required is given by

$$k = \frac{\ln(\text{tolerance}/b - a)}{\ln 0.618}$$

Example 3 ***Model Fitting Criterion Revisited***

Recall the curve fitting procedure from Chapter 9 using the criterion

$$\text{Minimize} \sum |y_i - y(x_i)|$$

Let's use the golden section search method to fit the model $y = cx^2$ to the following data for this criterion:

x	1	2	3
y	2	5	8

The function to be minimized is

$$f(c) = |2 - c| + |5 - 4c| + |8 - 9c|$$

and we will search for an optimal value of c in the closed interval [0, 3]. We choose a tolerance $t = 0.2$. We apply the golden section search method until the interval of uncertainty is less than 0.2. The results are summarized in Table 10.6.

Table 10.6 Golden section search method used in model fitting for Example 3.

k	a	b	c_1	c_2	$f(c_1)$	$f(c_2)$
0	0	3	1.1459	1.8541	3.5836	11.2492
1	0	1.8541	0.7082	1.1459	5.0851	3.5836
2	0.7082	1.8541	1.1459	1.4164	3.5836	5.9969
3	0.7082	1.4164	0.9787	1.1459	2.9149	3.5836
4	0.7082	1.1459	0.8754	0.9787	2.7446	2.9149
5	0.7082	0.9787	0.8115	0.8754	3.6386	2.7446
6	0.8115	0.9787				

The length of the final interval is less than 0.2. We can estimate

$$c^* = \frac{0.8115 + 0.9787}{2} = 0.8951$$

with $f(0.8951) = 2.5804$. ■

In the exercises, we ask you to show analytically that the optimal value for c is $c = \frac{8}{9}$.

Example 4 ***Optimizing Industrial Flow***

Figure 10.25 represents a physical system that engineers might consider for an industrial flow process. As shown, let x represent the flow rate of dye into the coloring process of cotton fabric. Based on this rate, the reaction differs with the other substances in the process as evidenced by the function shown in Figure 10.25. The function is defined as

$$f(x) = \begin{cases} 2 + 2x - x^2 & \text{for } 0 < x \le \frac{3}{2} \\ -x + \frac{17}{4} & \text{for } \frac{3}{2} < x \le 4 \end{cases}$$

The function defining the process is unimodal. The company wants to find the flow rate x that maximizes the reaction of the other substances $f(x)$. Through experimentation, the engineers have found that the process is sensitive to within about 0.20 of the actual value of x. They also found that the flow is either *off* ($x = 0$) or *on* ($x > 0$). The process does not allow for turbulent flow, which occurs above $x = 4$ for this process. Thus $x \le 4$ and we use a tolerance of 0.20 to maximize $f(x)$ over $[0, 4]$. Using the

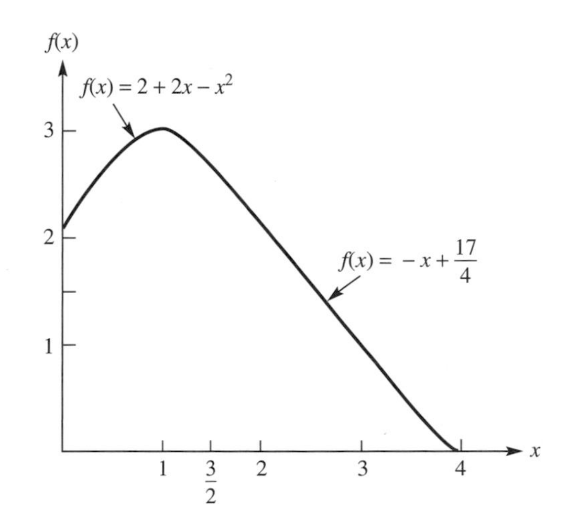

Figure 10.25 Industrial flow process function for Example 4.

Table 10.7 Results of golden section search method for Example 4.

a	b	x_1	x_2	$f(x_1)$	$f(x_2)$
0	4	1.5279	2.4721	2.7221	1.7779
0	2.4721	0.9443	1.5279	2.9969	2.7221
0	1.5279	0.5836	0.9443	2.8266	2.9969
0.5836	1.5279	0.9443	1.1672	2.9969	2.9720
0.5836	1.1672	0.8065	0.9443	2.9626	2.9969
0.8065	1.1672	0.9443	1.0294	2.9969	2.9991
0.9443	1.1672	1.0294	1.0820	2.9991	2.9933
0.9443	1.0820				

golden section search method, we locate the first two test points

$$x_1 = 0 + 4(0.382) = 1.5279$$
$$x_2 = 0 + 4(0.618) = 2.4721$$

and evaluate the function at the test points

$$f(x_1) = 2.7221$$
$$f(x_2) = 1.7779$$

The results of the search are given in Table 10.7.

We stop because $[1.0820 - 0.9443] < 0.20$. The midpoint of the interval is 1.0132 where the value of the function is 2.9998. ■

In the exercises, you are asked to show analytically that a maximum value of $f(x^*) = 3$ occurs at $x = 1$.

Exercises 10.5

1. Use the dichotomous search method with a tolerance of $t = 0.2$ and $\varepsilon = 0.01$.
   a. Minimize $f(x) = x^2 + 2x$, $-3 \leq x \leq 6$.
   b. Maximize $f(x) = -4x^2 + 3.2x + 3$, $[-2 \leq x \leq 2]$.
2. Use the golden section search method with a tolerance of $t = 0.2$.
   a. Minimize $f(x) = x^2 + 2x$, $-3 \leq x \leq 6$.
   b. Maximize $f(x) = -4x^2 + 3.2x + 3$, $[-2 \leq x \leq 2]$.
3. Use the curve fitting criterion that minimizes the sum of the absolute deviations for the following models and data set:
   a. $y = ax$
   b. $y = ax^2$
   c. $y = ax^3$

x	7	14	21	28	35	42
y	8	41	133	250	280	297

4. For Example 2, show that the optimal value of c is $c^* = \frac{8}{9}$. [*Hint:* Apply the definition of the absolute value to obtain a piecewise continuous function. Then find the minimum value of the function over the interval [0, 3].]

5. For Example 3, show that the optimal value of x is $x^* = 1$.

PROJECTS Exploring Modeling

1. *Fibonacci Search* One of the more interesting search techniques uses the Fibonacci sequence. This search method can be employed even if the function is not continuous. The method uses the Fibonacci numbers to place test points for the search. These Fibonacci numbers are defined as follows: $F_0 = F_1 = 1$ and $F_n = F_{n-1} + F_{n-2}$, for $n = 2, 3, 4, \ldots$, yielding the sequence 1, 1, 2, 3, 5, 8, 13, 21, 34, 55, 89, 144, 233, 377, 510, 887, 1397, and so forth.
   a. Find the ratio between successive Fibonacci numbers, using the above sequences. Then, find the numerical limit as n gets large. How is this limiting ratio related to the golden section search method?
   b. Research the Fibonacci search method and present your results to the class.

2. *Methods Using Derivatives: Newton's Method* One of the best-known interpolation methods is Newton's method, which exploits a quadratic approximation to the function $f(x)$ at a given point x_1. The quadratic approximation q is given by

$$q(x) = f(x_1) + f'(x_1)(x - x_1) + \tfrac{1}{2} f''(x_1)(x - x_1)^2.$$

The point x_2 is taken to be the point where q' equals zero. Continuing this procedure yields the sequence

$$x_{k+1} = x_k - \frac{f'(x_k)}{f''(x_k)}$$

for $k = 1, 2, 3, \ldots$. This procedure is terminated when either

$$|x_{k+1} - x_k| < \varepsilon \qquad \text{or} \qquad |f'(x_k)| < \varepsilon$$

where ε is some small number. This procedure can only be applied to twice-differentiable functions if $f''(x)$ never equals zero.
   a. Starting with $x = 4$ and a tolerance of $\varepsilon = 0.01$, use Newton's method to minimize $f(x) = x^2 + 2x$, over $-3 \le x \le 6$.
   b. Use Newton's method to minimize

$$f(x) = \begin{cases} 4x^3 - 3x^4 & \text{for } x > 0 \\ 4x^3 + 3x^4 & \text{for } x < 0 \end{cases}$$

   Let the tolerance be $\varepsilon = 0.01$ and start with $x = 0.4$.

c. Repeat part b starting at $x = 0.6$. Discuss what happens when you apply the method.

Further Reading

Bazarra, M., Sherali, H. D., & Shetty, C. M. (1993). *Nonlinear programming: Theory and algorithms*, 2nd ed. New York: Wiley.

Rao, S. S. (1979). *Optimization: Theory and applications*. New Delhi: Wiley Eastern Limited.

Winston, W. (1995). *Mathematical programming: Applications and algorithms*, 2nd ed. Boston: PSW-Kent.

Winston, W. (1994). *Operations research: Applications and algorithms*, 3rd ed. Belmont: Duxbury Press.

11
Modeling Using Calculus

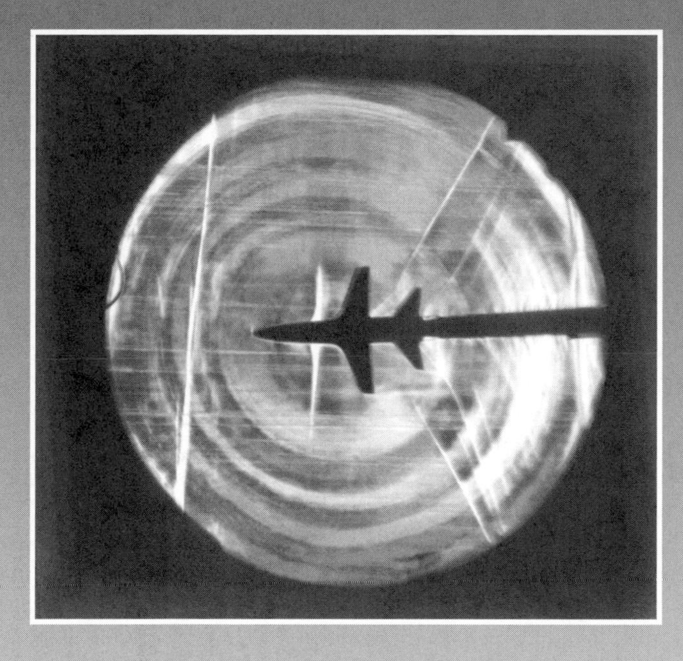

The national spotlight is turning on mathematics as we appreciate its central role in the economic growth of this country. The linkage between mathematics and economic growth needs to be made, and needs to be made stronger than it has been to date. Calculus is a critical way-station for the technical manpower that this country needs. It must become a pump instead of a filter in the pipeline. It is up to you to decide how to do that.

Robert M. White (1988)

Overview

In the earlier chapters of this text we developed models involving functions, difference equations, geometry, and probability. In general, these models are easily tied to the study of patterns and proportion. Their connections to the secondary curriculum are evident, even though their level at times may extend what happens in most secondary school classrooms. One message of our approach is that this type of reasoning and modeling should become a larger factor in the teaching and learning that occurs in secondary school classrooms.

In this chapter modeling is extended through the application of calculus and differential equations. Such an extension is natural and needed to complete the background development that secondary teachers need in order to adequately discuss mathematics and its use in the real-world in their teaching. The models developed in this chapter extend the discussion of rate of change from a discrete setting to a continuous setting, as change is viewed instantaneously. This occurs naturally in the discussion of population dynamics and other settings. The ties are built between the algebraic and geometric interpretations of such change as the nature of the derivative is investigated. The important concepts of the phase line and phase plane are good examples of this.

Difference equations naturally grow into differential equations as the discussion examines relationships between rates of change in a model. The material discussing differential equations provides important, and necessary, background for teachers preparing for teaching positions that might include the teaching of AP calculus. The content in this chapter shows how discrete ideas naturally generalize to related continuous concepts. We see how discrete models of change move to the derivative and how difference equations give way to differential equations. Throughout these changes, we see the general role played by the rate of change and the close ties that exist between the geometric and algebraic interpretation of these concepts. We see that the algebraic power of functions and the accompanying symbolic representation grow with the variety of functions applied. The important logistic function is revisited and extended.

Calculus is often referred to as one of the major intellectual accomplishments of the second millennium. In this chapter we get a glimpse of the important role that it plays in modeling real-world phenomena. Additional examples are included in the exercises and projects. You should look beyond this text to calculus books and professional journals for additional examples of how calculus helps us understand the world about us.

11.1 Modeling Discrete and Continuous Behaviors

Quite often we have information that relates a dependent variable's rate of change to one or more independent variables, and we are interested in discovering the function that relates the variables. For example, if P represents the size of a large population at some time t, then it is reasonable to assume that the population's rate of change with respect to time depends on the current size of P (as well as other factors that we discuss presently). For ecological, economical, and other reasons, we need to determine a relationship between P and t in order to make predictions about P. If the present population size is denoted by $P(t)$ and the population size at time $t + \Delta t$ is $P(t + \Delta t)$, then the change in population ΔP during that time period Δt is given by

$$\Delta P = P(t + \Delta t) - P(t) \tag{1}$$

The factors affecting population growth are developed in detail in Section 11.2. For now, let's assume a simple proportionality: $\Delta P \propto P$. For example, if immigration, emigration, age, and gender are all neglected, we can assume that during a unit time a certain percentage of the population reproduces while a certain percentage dies. Suppose the constant of proportionality k is expressed as a percentage per unit time. Then our proportionality assumption gives

$$\Delta P = P(t + \Delta t) - P(t) = kP\Delta t \tag{2}$$

Equation (2) is a **difference equation**. Note that we are treating a discrete set of times rather than allowing t to vary *continuously* over some interval. Difference equations belong to an important area of mathematics, called **discrete** or **finite** mathematics. In this situation the discrete set of times may give the population in future years at those distinct times (perhaps after the spring spawn in a fish population). Referring to Figure 11.1, observe that the horizontal distance between the points $(t_0, P(t_0))$ and $(t_0 + \Delta t, P(t_0 + \Delta t))$ is Δt, which could represent, say, the time between spawning periods in a fish population growth problem or the length of a fiscal period in a budget growth problem. The time t_0 refers to a particular time. The vertical distance, ΔP in this case, represents the change in the dependent variable.

Assume now that t does vary continuously so that we can take advantage of the calculus. Division of equation (2) by Δt gives

$$\frac{\Delta P}{\Delta t} = \frac{P(t + \Delta t) - P(t)}{\Delta t} = kP \tag{3}$$

We can interpret $\Delta P / \Delta t$ physically as the *average rate of change* in P during the time Δt. For example, $\Delta P / \Delta t$ could represent the average daily growth of the budget. However, in other scenarios it may have no physical interpretation: If fish spawn only in the spring, it is meaningless to talk about the average daily growth in the fish population. Again in Figure 11.1, $\Delta P / \Delta t$ can be interpreted geometrically as the slope of the line

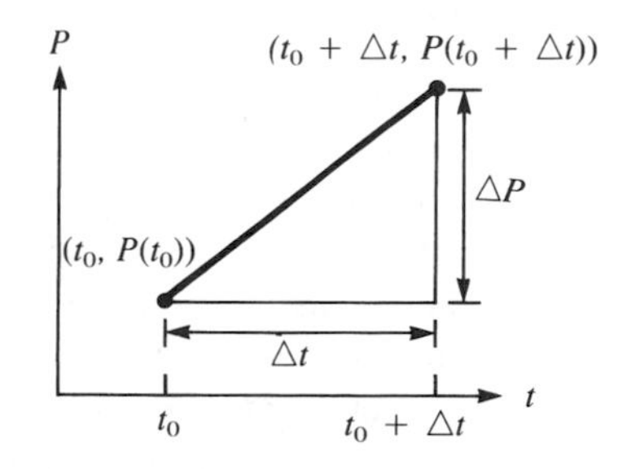

Figure 11.1 The size of the population at future time intervals of length Δt gives a discrete set of points.

segment connecting the point $(t_0, P(t_0))$ and $(t_0 + \Delta t, P(t_0 + \Delta t))$. Next we allow Δt to approach zero. The definition of the derivative gives the *differential equation*

$$\lim_{\Delta t \to 0} \frac{\Delta P}{\Delta t} = \frac{dP}{dt} = kP$$

where dP/dt represents the *instantaneous rate of change*. In many situations the instantaneous rate of change has an identifiable physical interpretation, such as in the flow of heat from a space capsule after entering the ocean or the reading on a car speedometer as the car accelerates. However, in the case of a fish population with a discrete spawning period or the budget process with a discrete fiscal period, the instantaneous change may be meaningless. These latter scenarios are more appropriately modeled using difference equations, but it is occasionally advantageous to approximate a difference equation with a differential equation.

The derivative is used in two distinct roles:

1. To represent the instantaneous rate of change in "continuous" problems
2. To approximate an average rate of change in "discrete" problems

The advantage of approximating an average rate of change by a derivative is that the calculus often helps uncover a functional relationship between the variables under investigation. For instance, the solution to the model (3) is $P = P_0e^{kt}$, where P_0 is the population at time $t = 0$. However, many differential equations cannot be solved so easily with analytic techniques. In such cases, the solutions are approximated using discrete methods. An introduction to numerical techniques is presented in Section 11.4. In case the solution being approximated is a differential equation that is itself an approximation to a difference equation, the modeler should consider using a discrete method with the finite difference equation directly. (See Chapter 1.)

The interpretation of the derivative as an instantaneous rate of change is useful in many modeling applications. The geometrical interpretation of the derivative as the slope of the line tangent to the curve is useful for constructing numerical solutions. Let's review these important concepts from the calculus more carefully.

The Derivative as a Rate of Change

The origins of the derivative lie in humankind's curiosity about motion and our need to develop a deeper understanding of motion. The search for the laws governing planetary motion, the study of the pendulum and its application to clock building, and the laws governing the flight of a cannonball were the kind of problems that captured the imaginations of mathematicians and scientists in the sixteenth and seventeenth centuries. Such problems motivated the development of the calculus.

To remind ourselves of one interpretation of the derivative, consider a particle whose distance s from a fixed position depends on time t. Let the graph in Figure 11.2 represent the distance s as a function of time t, and let (t_1, s_1) and (t_2, s_2) denote two points on the graph.

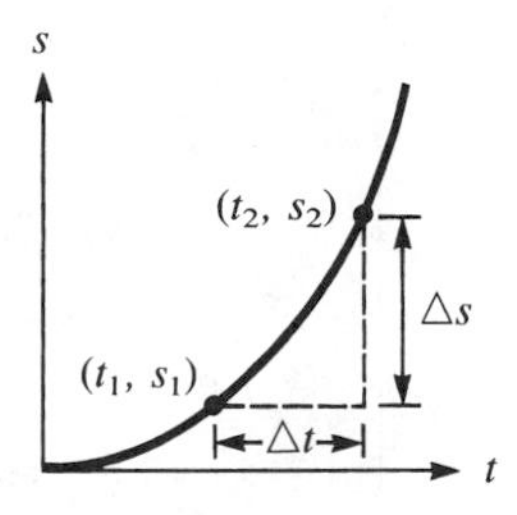

Figure 11.2 Graph of distance s as a function of time t.

Define $\Delta t = t_2 - t_1$ and $\Delta s = s_2 - s_1$, and form the ratio $\Delta s / \Delta t$. Note that this ratio represents a rate: an increment of distance traveled Δs over some increment of time Δt. That is, the ratio $\Delta s / \Delta t$ represents the average velocity during the time period in question. Now remember how the derivative ds/dt evaluated at $t = t_1$ is defined:

$$\left.\frac{ds}{dt}\right|_{t=t_1} = \lim_{\Delta t \to 0} \frac{\Delta s}{\Delta t} \tag{4}$$

Physically, what occurs as $\Delta t \to 0$? Using the interpretation of average velocity, we can see that at each state of a smaller Δt we are computing the average velocity over smaller and smaller intervals with left endpoint at t_1 until, in the limit, we have the instantaneous velocity at $t = t_1$. If we think of the motion of a moving vehicle, this instantaneous velocity would correspond to the exact reading of its (perfect) speedometer at the instant t_1.

More generally, if $y = f(x)$ is a differentiable function, then the derivative dy/dx at any given point can be interpreted as the *instantaneous rate of change* of y with respect to x at that point. Interpreting the derivative as an instantaneous rate of change is useful in many modeling applications.

The Derivative as the Slope of the Tangent Line

Let's consider another interpretation of the derivative. As scholars sought knowledge about the laws of planetary motion, their chief need was to observe and measure the heavenly bodies. However, the construction of lenses for use in telescopes was a difficult task. Grinding a lens to the correct curvature to achieve the desired light refraction requires knowing the tangent to the curve describing the lens surface.

Let's examine the geometrical implications of the limit in equation (4). We consider $s(t)$ now simply as a curve. Let's examine a set of secant lines, each emanating from the point $A = (t_1, s(t_1))$ on the curve. To each secant, there corresponds a pair of increments $(\Delta t_i, \Delta s_i)$, as shown in Figure 11.3. The lines AB, AC, and AD are secant lines. As $\Delta t \to 0$, these secant lines approach the line tangent to the curve at the point A. Because the slope of each secant is $\Delta s / \Delta t$, we may interpret the derivative as *the slope of the line tangent to the curve* $s(t)$ *at the point* A. The interpretation of the derivative evaluated at a point as the slope of the line tangent to the curve at that point is useful in constructing graphical and numerical approximations to solutions of

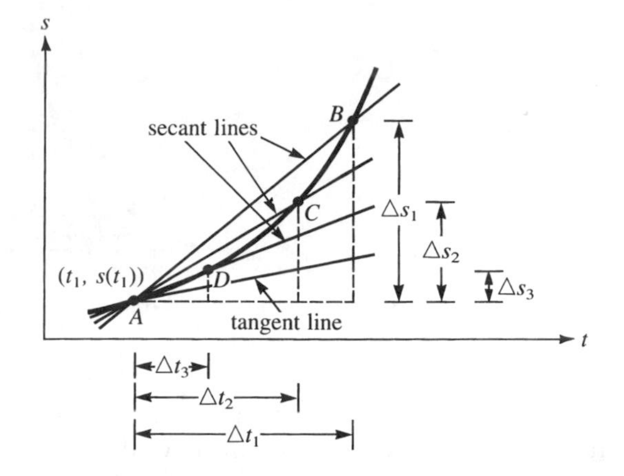

Figure 11.3 The slope of each secant line approximates the slope of the tangent line to the curve at the point A.

differential equations. Graphical solutions are discussed in Section 11.3, and numerical solutions in Section 11.4.

11.2 Modeling Population Growth and Analytic Solutions

Thomas Malthus (1766–1834) sparked widespread interest in how populations grew in the late eighteenth century when he published *An Essay on the Principle of Population as It Affects the Future Improvement of Society*. In his book Malthus proposed an exponential growth model for human population and concluded that eventually the population would exceed the capacity to grow an adequate food supply. Although the Malthusian model leaves out factors important to population growth (the model has thus proved to be inaccurate for technologically developed countries), it is nevertheless instructive to examine this model as a basis for later refinement.

Problem Identification Suppose we know the population at some given time, say P_0 at time $t = t_0$, and we are interested in predicting the population P at some future time $t = t_1$. In other words, we want to find a population function $P(t)$ for $t_0 \leq t \leq t_1$ satisfying $P(t_0) = P_0$.

Assumptions First, we consider some factors that pertain to population growth. Two obvious factors are the *birthrate* and the *death rate*, which are determined by different factors. The birthrate is influenced by infant mortality rate, attitudes toward and availability of contraceptives, attitudes toward abortion, health care during pregnancy, and so forth. The death rate is affected by sanitation and public health, wars, pollution, medicines, diet, psychological stress and anxiety, and so forth. Other factors that influence population growth in a given region are immigration and emigration, living space restrictions, availability of food and water, and epidemics. For our model, let's neglect all these latter factors. (If we are dissatisfied with our results, we can include these factors later in a more refined model, possibly in a simulation model.) Now we'll consider only the birthrate and death rate. Since knowledge and technology have helped humankind diminish the death rate so that it is now lower than the birthrate, human populations have tended to grow.

Let's begin by assuming that during a small unit time period a percentage b (given as a decimal equivalent) of the population is newly born. Similarly, a percentage c of the population dies. In other words, the new population $P(t + \Delta t)$ is the old population $P(t)$ plus the number of births minus the number of deaths during the time period Δt. Symbolically,

$$P(t + \Delta t) = P(t) + bP(t)\Delta t - cP(t)\Delta t$$

or

$$\frac{\Delta P}{\Delta t} = bP - cP = kP$$

Thus we are really assuming that the population's average rate of change over an interval of time is proportional to the size of the population. Using the instantaneous rate of change to approximate the average rate of change, we have the following model:

$$\frac{dP}{dt} = kP, \quad P(t_0) = P_0, \quad t_0 \leq t \leq t_1 \tag{5}$$

where (for growth) k is a positive constant.

Solving the Model We can separate the variables and rewrite equation (5) by moving all terms involving P and dP to one side of the equation and all terms in t and dt to the other. This gives

$$\frac{dP}{P} = k\,dt$$

Integration of both sides of this last equation yields

$$\ln P = kt + C \tag{6}$$

for some constant C. Applying the condition $P(t_0) = P_0$ to equation (6) to find C results in

$$\ln P_0 = kt_0 + C$$

or

$$C = \ln P_0 - kt_0$$

Then substitution for C into (6) gives

$$\ln P = kt + \ln P_0 - kt_0$$

or, simplifying algebraically,

$$\ln \frac{P}{P_0} = k(t - t_0)$$

Finally, by taking the exponential of both sides of the preceding equation and multiplying the result by P_0, we obtain the solution:

$$P(t) = P_0 e^{k(t-t_0)} \tag{7}$$

Equation (7), which is known as the **Malthusian model of population growth**, predicts that population grows exponentially with time.

Verifying the Model Because $\ln(P/P_0) = k(t - t_0)$, our model predicts that if we plot $\ln P/P_0$ versus $t - t_0$, a straight line passing through the origin with slope k should result. However, if we plot the population data for the United States for several years, the model does not fit very well, especially in the later years. In fact, the 1970 census for the U.S. population was 203,211,926, and in 1950, it was 150,697,000. Substituting these values into equation (7) and dividing the first result by the second gives

$$\frac{203{,}211{,}926}{150{,}697{,}000} = e^{k(1970-1950)}$$

Thus

$$k = \frac{1}{20} \ln \frac{203{,}211{,}926}{150{,}697{,}000} \approx 0.015$$

That is, during the 20-year period from 1950 to 1970, the U.S. population increased at the average rate of 1.5% per year. We can use this information together with equation (7) to predict the population for 1980. In this case, $t_0 = 1970$, $P_0 = 203{,}211{,}926$, and $k = 0.015$ yields

$$P(1980) = 203{,}211{,}926 e^{0.015(1980-1970)} = 236{,}098{,}574$$

The 1980 census for the U.S. population was 226,505,000 (rounded to the nearest thousand). Thus our prediction is off by roughly 4%. We can probably live with that magnitude error, but let's look into the distant future. Our model predicts that the population of the United States will be 28,688 billion in the year 2300, a population that exceeds current estimates of the maximum sustainable population of the entire planet! We are forced to conclude that our model is unreasonable over the long run.

Some populations do grow exponentially, provided that the population is not too large. However, in most populations individual members eventually compete with one another for food, living space, and other natural resources, which works to limit growth. Let's refine our Malthusian model of population growth to reflect this competition.

Refining the Model to Reflect Limited Growth Let's consider that the proportionality factor k, measuring the rate of population growth in equation (5), is no longer constant, but a function of the population. As the population increases and gets closer to the maximum population M, the rate k decreases. One simple submodel for k is the linear one

$$k = r(M - P), \quad r > 0$$

where r is a constant. Substitution into equation (5) gives

$$\frac{dP}{dt} = r(M - P)P \tag{8}$$

or

$$\frac{dP}{P(M - P)} = r\,dt \tag{9}$$

Again, we assume the *initial condition* $P(t_0) = P_0$. [Model (8) was first introduced by the Dutch mathematical biologist Pierre-Francois Verhulst (1804–1849) and is referred to as **logistic growth**.] It follows from elementary algebra that

$$\frac{1}{P(M - P)} = \frac{1}{M}\left(\frac{1}{P} + \frac{1}{M - P}\right)$$

Thus equation (9) can be rewritten as

$$\frac{dP}{P} + \frac{dP}{M - P} = rM\,dt$$

which integrates to

$$\ln P - \ln|M - P| = rMt + C \tag{10}$$

for some arbitrary constant C. Using the initial condition, we evaluate C in the case $P < M$:

$$C = \ln\frac{P_0}{M - P_0} - rMt_0$$

Substituting into (10) and simplifying gives

$$\ln\frac{P}{M - P} - \ln\frac{P_0}{M - P_0} = rM(t - t_0)$$

or

$$\ln\frac{P(M - P_0)}{P_0(M - P)} = rM(t - t_0)$$

Exponentiating both sides of this equation gives

$$\frac{P(M - P_0)}{P_0(M - P)} = e^{rM(t - t_0)}$$

or

$$P_0(M - P)e^{rM(t-t_0)} = P(M - P_0)$$

Then,

$$P_0 M e^{rM(t-t_0)} = P(M - P_0) + P_0 P e^{rM(t-t_0)}$$

so that solving for the population P gives

$$P(t) = \frac{P_0 M e^{rM(t-t_0)}}{M - P_0 + P_0 e^{rM(t-t_0)}}$$

To estimate P as $t \to \infty$, we rewrite this last equation as

$$P(t) = \frac{MP_0}{P_0 + (M - P_0)e^{-rM(t-t_0)}} \tag{11}$$

Notice from equation (11) that $P(t)$ approaches M as t tends to infinity. Moreover, from equation (8) we calculate the second derivative

$$P'' = rMP' - 2rPP' = rP'(M - 2P)$$

so that $P'' = 0$ when $P = M/2$. In words, when the population P reaches half the limiting population M, the growth dP/dt is most rapid and then it starts to diminish toward zero. One advantage of recognizing that the maximum rate of growth occurs at $P = M/2$ is that the information can be used to estimate M. In a situation where the analyst is satisfied that the growth involved is essentially logistic, if the point of maximum rate of growth has been reached, then $M/2$ can be estimated. The graph of limited growth equation (11) is depicted in Figure 11.4 for the case $P < M$ (see Exercise 2 for the case $P > M$). Such a curve is called a **logistic curve**.

Verifying the Limited Growth Model Let's test our model (11) against some real-world data. Equation (10) suggests a straight-line relationship of $\ln[P/(M - P)]$ versus t. Let's test this model using the data given in Table 11.1 for the growth of yeast in a

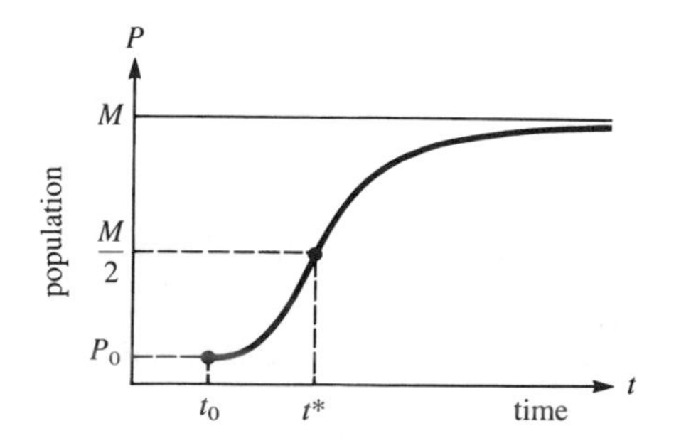

Figure 11.4 Graph of the limited growth model.

Table 11.1 Growth of yeast in a culture.

Time in hours	Observed yeast biomass	Biomass calculated from logistic equation (13)	Percent error
0	9.6	8.9	−7.3
1	18.3	15.3	−16.4
2	29.0	26.0	−10.3
3	47.2	43.8	−7.2
4	71.1	72.5	2.0
5	119.1	116.3	−2.4
6	174.6	178.7	2.3
7	257.3	258.7	0.5
8	350.7	348.9	−0.5
9	441.0	436.7	−1.0
10	513.3	510.9	−4.7
11	559.7	566.4	1.2
12	594.8	604.3	1.6
13	629.4	628.6	−0.1
14	640.8	643.5	0.4
15	651.1	652.4	0.2
16	655.9	657.7	0.3
17	659.6	660.8	0.2
18	661.8	662.5	0.1

Source: Data from R. Pearl (1927). The growth of population. *Quart. Rev. Biol.* **2**:532–548.

culture. To plot $\ln[P/(M-P)]$ versus t, we need an estimate for the limiting population M. From the data in Table 11.1, we see that the population never exceeds 661.8. We estimate $M \approx 665$ and plot $\ln[P/(665 - P)]$ versus t. The graph is shown in Figure 11.5 and approximates a straight line. Thus we accept the assumptions of logistic growth for bacteria. Now equation (10) gives

$$\ln \frac{P}{M-P} = rMt + C$$

and from the graph in Figure 11.5, we can estimate the slope $rM \approx 0.55$, so that $r \approx 0.0008271$ from our estimate for $M \approx 665$.

It is often convenient to express the logistic equation (11) in another form. To this end, let t^* denote the time when the population P reaches half the limiting value; that is, $P(t^*) = M/2$. It follows from equation (11) that

$$t^* = t_0 - \frac{1}{rM} \ln \frac{P_0}{M - P_0}$$

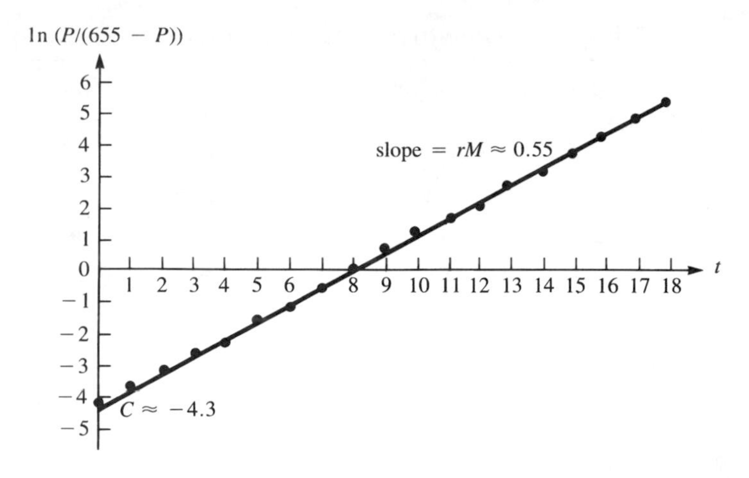

Figure 11.5 Plot of $\ln[P/(665 - P)]$ versus t for the data in Table 11.1.

(see Exercise 1a at the end of this section). Solving this last equation for t_0, substituting the result into equation (11), and simplifying algebraically gives

$$P(t) = \frac{M}{1 + e^{-rM(t-t^*)}} \tag{12}$$

(See Exercise 1b at the end of this section.)

We can estimate t^* for the yeast culture data presented in Table 11.1 using equation (10) and our graph in Figure 11.5:

$$t^* = -\frac{C}{rM} \approx \frac{4.3}{0.55} \approx 7.82$$

This calculation gives the logistic equation

$$P(t) = \frac{665}{1 + 73.8e^{-0.55t}} \tag{13}$$

by substituting $M = 665$, $r = 0.0008271$, and $t^* = 7.82$ in (12).

The logistic model is known to predict quite well for populations of organisms that have very simple life histories, for instance, yeast growing in a culture where space is limited. Table 11.1 shows the calculations for the logistic equation (13), and we can see from the calculated error that there is very good agreement with the original data. A plot of the curve is shown in Figure 11.6.

Next let's consider some data for human populations. A logistic equation for the population growth of the United States was formulated by Pearl and Reed in 1920. One form of their logistic curve is given by

$$P(t) = \frac{197{,}273{,}522}{1 + e^{-0.03134(t-1914.32)}} \tag{14}$$

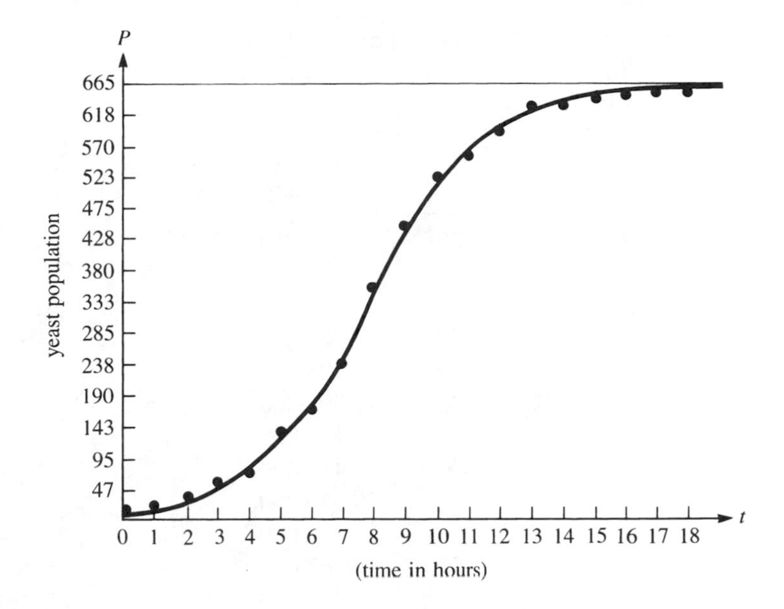

Figure 11.6 Logistic curve showing the growth of yeast in a culture based on the data from Table 11.1 and the model (13). The small circles indicate the observed values.

where $M = 197{,}273{,}522$, $r = 1.5887 \times 10^{-10}$, and $t^* = 1914.32$ were determined using the census figures for the years 1790, 1850, and 1910 (we ask you to estimate M, r, and t^* in the Exercise 4).

Table 11.2 compares the values predicted in 1920 by the logistic equation (14) with the observed values of the U.S. population. The predicted values agree quite well with the observations up to the year 1950, but as we can see, the predicted values are much too small for the years 1970, 1980, and 1990. This should not be surprising because our model fails to take into account such factors as immigration, wars, and advances in medical technology. In populations of higher plants and animals, which have complicated life histories and long periods of individual development, we can expect to observe numerous responses that greatly modify the population growth.

Separation of Variables

The method of direct integration employed in this section works only in those cases in which the dependent and independent variables can be algebraically separated. Any such differential equation can be written in the form:

$$p(y)\,dy = q(x)\,dx$$

For example, given the differential equation:

$$u(x)v(y)\,dx + q(x)p(y)\,dy = 0$$

Table 11.2 Population of the United States from 1790 to 1990, with predictions from equation (14).

Year	Observed population	Predicted population	Percent error
1790	3,929,000	3,929,000	0.0
1800	5,308,000	5,336,000	0.5
1810	7,240,000	7,227,000	−0.2
1820	9,638,000	9,756,000	1.2
1830	12,866,000	13,108,000	1.9
1840	17,069,000	17,505,000	2.6
1850	23,192,000	23,191,000	−0.0
1860	31,443,000	30,410,000	−3.3
1870	38,558,000	39,370,000	2.1
1880	50,156,000	50,175,000	0.0
1890	62,948,000	62,767,000	−0.3
1900	75,995,000	76,867,000	1.1
1910	91,972,000	91,970,000	−0.0
1920	105,711,000	107,393,000	1.6
1930	122,755,000	122,396,000	−0.3
1940	131,669,000	136,317,000	3.5
1950	150,697,000	148,677,000	−1.3
1960	179,323,000	159,230,000	−11.2
1970	203,212,000	167,943,000	−17.4
1980	226,505,000	174,941,000	−22.8
1990	248,710,000	180,440,000	−27.5

we can arrange the equation in the following form:

$$\frac{p(y)}{v(y)}\,dy = \frac{-u(x)}{q(x)}\,dx$$

Because the right-hand side is a function of x and the left-hand side is a function of y only, the solution is obtained by simply integrating both sides directly. Remember from calculus that even if we succeed in separating the variables, it may not be possible to find the integrals in closed form. The method just described is called *separation of variables* and is one of the most elementary methods available for solving differential equations.

Exercises 11.2

1. a. Show that the population P in the logistic equation reaches half the maximum population M at time t^* given by

$$t^* = t_0 - \frac{1}{rM} \ln\left(\frac{P_0}{M - P_0}\right)$$

b. Derive the form given by equation (12) for population growth according to the logistic law.

c. Derive the equation $\ln[P/(M - P)] = rMt - rMt^*$ from equation (12).

2. Consider the solution of equation (8). Evaluate the constant C in (10) in the case that $P > M$ for all t. Sketch the solutions in this case. Also sketch a solution curve for the case $M/2 < P < M$.

3. The following data were obtained for the growth of a sheep population introduced into a new environment on the island of Tasmania. (Adapted from J. Davidson (1938), On the growth of the sheep population in Tasmania, *Trans. Roy. Soc. S. Australia* **62**:342–346.)

t(year)	1814	1824	1834	1844	1854	1864
$P(t)$	125	275	830	1200	1750	1650

a. Estimate M by graphing $P(t)$.

b. Plot $\ln[P/(M - P)]$ against t. If a logistic curve seems reasonable, estimate rM and t^*.

4. Using the data for the U.S. population in Table 11.2, estimate M, r, and t^* using the same technique as in the text. Assume you are making the prediction in 1951 using previous census. Use the data from 1960 to 1990 to check your model.

5. The modern philosopher Jean Jacques Rousseau formulated a simple model of population growth for eighteenth-century England based on the following assumptions:

a. The birthrate in London is less than that in rural England.

b. The death rate in London is greater than that in rural England.

c. As England industrializes, more and more people migrate from the countryside to London.

Rousseau then reasoned that because London's birthrate was lower and its death rate higher and rural people tend to migrate there, the population of England would eventually decline to zero. Critique Rousseau's analysis.

6. Consider a highly communicable disease that is spreading on an isolated island with population size N. A portion of the population travels abroad and returns to the island infected with the disease. You are asked to predict the number of people X who will have been infected by some time t. Consider the following model:

$$\frac{dX}{dt} = kX(N - X)$$

a. List two major assumptions implicit in the preceding model. How reasonable are your assumptions?

b. Graph dX/dt versus X.

c. Graph X versus t if the initial number of infections is $X_1 < N/2$. Graph X versus t if the initial number of infections is $X_2 > N/2$.
d. Solve the model given earlier for X as a function of t.
e. From part d, find the limit of X as t approaches infinity.
f. Consider an island with a population of 5000. At various times during the epidemic, the number of people infected were as follows:

t (days)	2	6	10
X (people infected)	1887	4087	4853
$\ln[X/(N-X)]$	−0.5	1.5	3.5

Do the data collected support the given model?
g. Use the results in part f to estimate the constants in the model, and predict the number of people who will be infected by $t = 12$ days.

7. As a scientist for the Environmental Protection Agency, you are considering the survival of whales. You are assuming if the number of whales falls below a minimum survival level m the species will become extinct. The population is limited by the carrying capacity M of the environment. That is, if the whale population is above M, then it will experience a decline because the environment cannot sustain that large a population level.
a. Discuss the following model for the whale population:

$$\frac{dP}{dt} = k(M - P)(P - m)$$

where $P(t)$ denotes the whale population at time t and k is a positive constant.
b. Graph dP/dt versus P and P versus t. Consider the cases where the initial population $P(0) = P_0$ satisfies $P_0 < m$, $m < P_0 < M$, and $M < P_0$.
c. Solve the model in part a, assuming that $m < P < M$ for all time. Show that the limit of P as t approaches infinity is M.
d. Discuss how you would test the model in part a? How would you determine M and m?
e. Assuming that the model reasonably estimates the whale population, what does this imply for the fishing industry? What controls would you suggest?

8. Sociologists recognize a phenomenon called **social diffusion**, which is the spreading of a piece of information, a technological innovation, or a cultural fad among a population. The members of the population can be divided into two classes: those who have the information and those who do not. In a fixed population whose size is known, it is reasonable to assume that the rate of diffusion is proportional to the number who have the information times the number yet to receive it. If X denotes the number of individuals who have the information in a population of N people, then a mathematical model for social diffusion is given by $dX/dt = kX(N - X)$, where t represents time and k is a positive constant.

a. Solve the model and show that it leads to a logistic curve.
b. At what time is the information spreading fastest?
c. How many people will eventually receive the information?

PROJECTS Exploring Modeling

1. Complete the requirements of the UMAP module, "The Cobb-Douglas Production Function," by Robert Geitz, UMAP 509. A mathematical model relating the output of an economic system to labor and capital is constructed from the assumptions that (a) marginal productivity of labor is proportional to the amount of production per unit of labor, (b) marginal productivity of capital is proportional to the amount of production per unit of capital, and (c) if either labor or capital tends to zero, then so does production.

2. Complete the UMAP module, "The Diffusion of Innovation in Family Planning," by Kathryn N. Harmon, UMAP 303. This module uses finite difference equations to study how public policies are diffused through the population in order to help national governments develop family planning policies.

3. Complete the UMAP module "Difference Equations With Applications," by Donald R. Sherbert, UMAP 322. This module presents a good introduction to solving first- and second-order linear difference equations, including the method of undetermined coefficients for nonhomogeneous equations. Applications to problems in population and economic modeling are included.

4. Write a summary report on the article "Case Studies in Cancer and Its Treatment by Radiotherapy," by J. R. Usher and D. A. Abercrombie (1981), *International Journal of Mathematics Education in Science and Technology 12* **6**:661–682. Present your report to the class.

5. Complete the UMAP module "Selection in Genetics," by Brindell Horelick and Sinan Koont, UMAP 70. This module introduces genetic terminology and basic results about genotype distribution in successive generations. A recurrence relation is obtained from which the nth generation frequency of a recessive gene can be determined. Calculus is used to derive a technique for approximating the number of generations required for this frequency to fall below any given positive value.

6. Complete the UMAP module "Epidemics," by Brindell Horelick and Sinan Koont, UMAP 73. This unit poses two problems: (1) At what rate must infected persons be removed from a population to keep an epidemic under control? (2) What portion of a community will become infected during an epidemic? Threshold removal rate and the extent of an epidemic are discussed when the removal rate is slightly below threshold.

7. Complete the UMAP module "Tracer Methods in Permeability," by Brindell Horelick and Sinan Koont, UMAP 74. This module describes a technique for measuring the permeability of red corpuscle surfaces to K^{42} ions, using radioactive trac-

ers. Students learn how radioactive tracers are used to monitor substances in the body, and learn some of the limitations and strengths of the model described in this unit.

8. Complete the UMAP module "Modeling the Nervous System. Reaction Time and the Central Nervous System," by Brindell Horelick and Sinan Koont, UMAP 67. The process by which the central nervous system reacts to a stimulus is modeled, and the predictions of the model are compared with experimental data. Students learn what conclusions can be drawn from the model about reaction time, and are given an opportunity to discuss the merits of various assumptions about the relation between intensity of excitation and stimulus intensity.

Further Reading

Frauenthal, J. C. (1979). *Introduction to population modeling*. Lexington, Mass.: COMAP.

Hutchinson, G. E. (1978). *An introduction to population ecology*. New Haven, Conn.: Yale University Press.

Levins, R. (1966). The strategy of model building in population biology. *American Scientist* **54**:421–431.

Lotka, A. J. (1956). *Elements of mathematical biology*. New York: Dover.

Odum, E. P. (1971). *Fundamentals of ecology*. Philadelphia: Saunders.

Pearl, R., & Reed, L. J. (1920). On the rate of growth of the population of the United States since 1790. *Proceedings of the National Academy of Science* **6**:275–288.

11.3 Graphical Solutions

Graphical Interpretation of the Derivative

Interpreting the derivative as the slope of the line tangent to the graph of the function is useful for gaining information about the solution to a differential equation. From this information we can sketch qualitatively the solution curves to the equation. These graphs provide considerable insight into the solution, and the graphical techniques are themselves of practical benefit. Since most differential equation models that people construct cannot be solved analytically, other techniques are needed to investigate the behavior of solutions.

Consider again the first-order differential equation

$$\frac{dy}{dx} = g(x, y)$$

that we investigated in Section 11.2. Each time an initial value $y(x_0) = y_0$ is specified, the solution curve must pass through the point (x_0, y_0) and have the slope value $g(x_0, y_0)$ there. Graphically, we can draw a short line segment with the proper slope through each

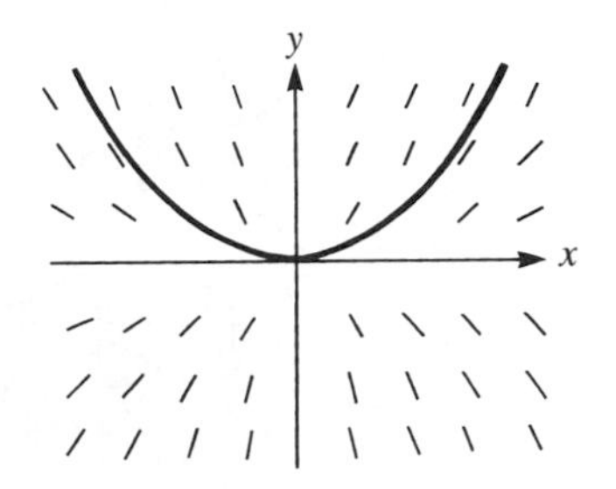

Figure 11.7 The direction field of a first-order differential equation assigns to each point in the plane the slope $y' = g(x, y)$.

point (x, y) in the plane. The resultant configuration is known as the **direction field** or **slope field** of the first-order differential equation and is illustrated in Figure 11.7. A **solution curve** is then tangent to the direction line at each point through which the curve passes. Thus the direction field gives a visual indication of the family of possible solutions to the differential equation. One solution curve is depicted for the family of curves indicated by the direction field in Figure 11.7.

Often the solutions to a first-order differential equation can be expressed in the form $y = f(x)$, where each solution is distinguished by a different constant resulting from integration. For example, if we separate the variables in the differential equation

$$\frac{dy}{dx} = \frac{2y}{x}$$

we obtain

$$\frac{dy}{y} = 2\frac{dx}{x}$$

and integration of both sides then gives

$$\ln|y| = 2\ln|x| + \ln C$$

where the constant of integration is named $\ln C$ for computational convenience. Applying the exponential to both sides of this last equation yields

$$|y| = Cx^2, \quad C > 0 \tag{15}$$

Equation (15) represents a family of parabolas, each parabola being distinguished by a different value of the positive constant C. If $y \geq 0$, the curves $y = Cx^2$ open upward; if $y < 0$, the curves $y = -Cx^2$ open downward. This family is depicted in Figure 11.8. Excluding the origin, exactly one of the parabolas passes through each point in the plane. Thus, by specifying an initial condition $y(x_0) = y_0$, we select a unique solution curve passing through the point (x_0, y_0), $x_0 \neq 0$.

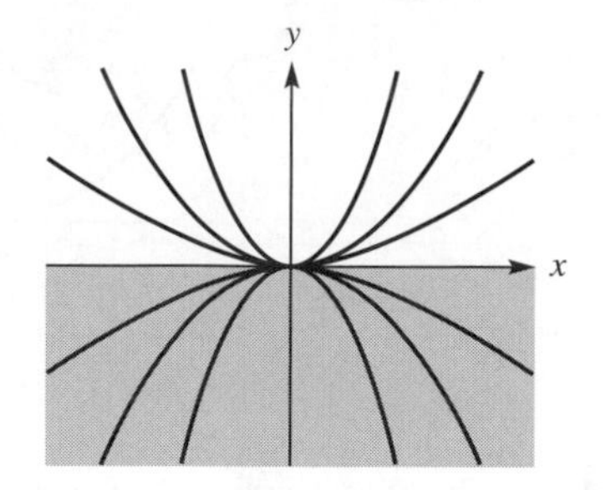

Figure 11.8 Family of parabolas $|y| = Cx^2$, $C > 0$.

Graphing the Derivative

In many modeling applications the instantaneous rate of change dy/dx is assumed to be proportional to some function of the dependent variable y alone. Thus $dy/dx = f(y)$. In such cases, we can obtain qualitative information concerning the nature of a solution curve $y(x)$ by investigating the graph of $f(y)$, that is, by plotting dy/dx versus the independent variable x. We now explore this idea in several examples.

Example 1 *Exponential Growth*

Consider the differential equation describing the Malthusian model of population growth discussed in Section 11.2:

$$\frac{dP}{dt} = kP, \quad k > 0$$

Thus, the Malthusian model assumes a simple proportionality between growth rate and population (see Figure 11.9a). Since $P > 0$ and $k > 0$, you can see that $dP/dt > 0$. It

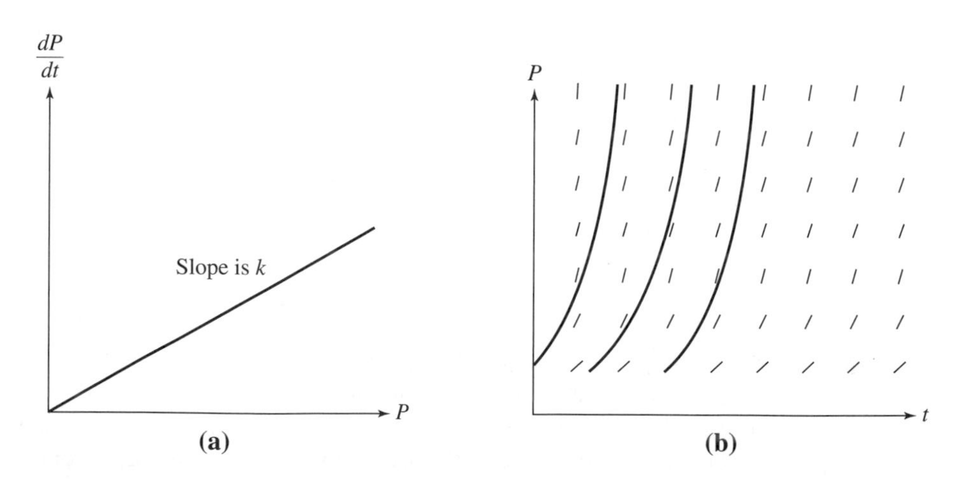

Figure 11.9 Graphs of dP/dt versus P and the solution curves for $dP/dt = kP$, where $k > 0$ is fixed.

follows that the population curve $P(t)$ is everywhere increasing. Moreover, the smaller the value of k, the less rapid is the growth in the population over time. For a fixed value of k, as P increases so does its rate of change dP/dt. Thus the solution curves must appear qualitatively as depicted in Figure 11.9b. Assuming k is constant, at each population level P the derivative dP/dt is constant. Therefore, all the solution curves are horizontal translates of one another. The initial population $P(0) = P_0$ distinguishes these various curves. ■

Example 2 *The Logistic Growth Model*

Recall the logistic population growth model studied in Section 11.2:

$$\frac{dP}{dt} = r(M - P)P \tag{16}$$

The graph of dP/dt versus the dependent variable P is the parabola shown in Figure 11.10. Differentiating equation (16) implicitly with respect to t gives the second derivative

$$P'' = rP'(M - 2P) \tag{17}$$

Since $P'' = 0$ when $P = M/2$, dP/dt is at a maximum at that population level. Note that for $0 < P < M$, the derivative dP/dt is positive and the population $P(t)$ is increasing; for $P > M$, dP/dt is negative and $P(t)$ is decreasing. Moreover, if $P < M/2$, the second derivative P'' is positive and the population curve is concave upward; for $M > P > M/2$, P'' is negative and the population curve is concave downward.

The parabola in Figure 11.10 has zeros at the population levels $P = 0$ and $P = M$. At those levels, $dP/dt = 0$, so no change in the population P can occur. That is, if the population level is at zero, it will remain there for all time; if it is at the level M, it will remain there for all time. Points for which the derivative dP/dt is zero are called **rest points, equilibrium points, stationary points**, or **critical points** of the differential equation. ■

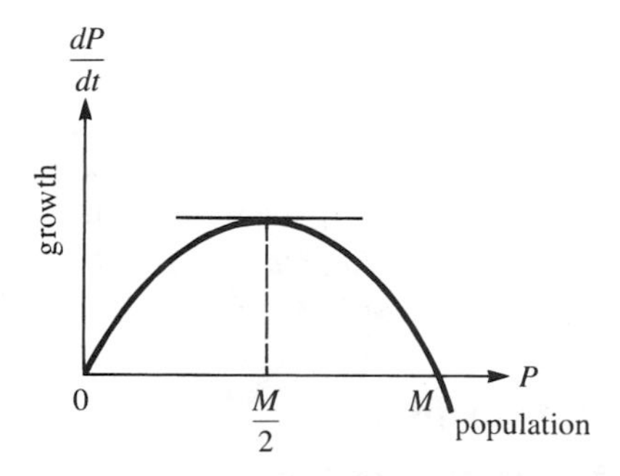

Figure 11.10 Graph of $dP/dt = r(M - P)P$, $r > 0$, $M > 0$.

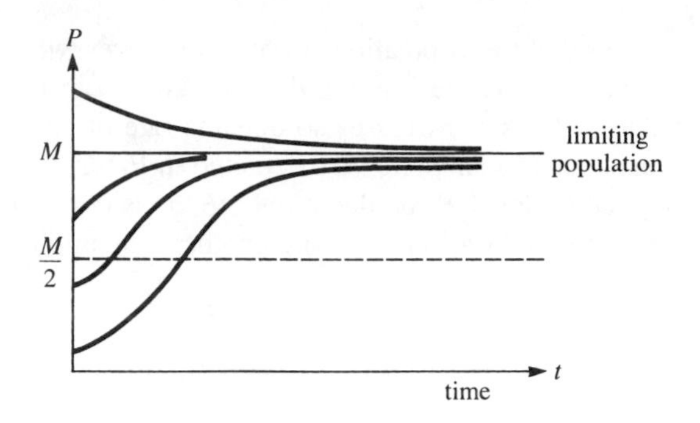

Figure 11.11 Population curves for the logistical model $dP/dt = r(M - P)P$, $r > 0$, $M > 0$.

Stable and Unstable Rest Points

The behavior of solutions near rest points is of significant interest to us. Let's examine what happens to the population when P is near the rest points $P = M$ and $P = 0$.

Suppose the population in model (16) satisfies $P < M$. Then $dP/dt > 0$, and the population increases, getting closer to M. On the other hand, if $P > M$, $dP/dt < 0$, and the population will decrease toward M (see Figure 11.11). Thus, no matter what positive value is assigned to it at the start, the population will tend to the limiting value M as time tends toward infinity. We say that M is an **asymptotically stable** rest point, because whenever the population level is perturbed away from that level it tends to return there again.

Next, consider the rest point $P = 0$ in Figure 11.10. If P is perturbed slightly away from 0 so that $M > P > 0$, then $dP/dt > 0$ and the population increases toward M. In this situation we say that $P = 0$ is an **unstable** rest point, because any population not starting at that level tends to move away from it.

In general, equilibrium solutions to differential equations are classified as stable or unstable according to whether, graphically, nearby solutions stay close to or converge to the equilibrium, or diverge away from the equilibrium, respectively, as the independent variable t tends to infinity.

At $P = M/2$, $P'' = 0$ and the derivative dP/dt is at a maximum (see Figure 11.10). Thus, the population P is increasing most rapidly when $P = M/2$ and a point of inflection occurs in the graph there. These features give each solution curve its characteristic sigmoid ("S") shape. From this information the family of solutions to the limited growth model must appear qualitatively as shown in Figure 11.11.

Notice again that at each population level P, the derivative dP/dt is constant, so that the solution curves are horizontal translates of one another. The curves are distinguished by the initial population level $P(0) = P_0$. In particular, note the solution curve when $M/2 < P_0 < M$ in Figure 11.11.

The Phase Line

Let's consider equation (8) from yet another point of view. In Figure 11.12, we present a phase line depicting the limited growth model. On the phase line the two equilibrium

Figure 11.12 A phase line for logistic growth.

points $P = 0$ and $P = M$ are circled. These are the points where $dP/dt = 0$. Also, on the phase line we identify the sign of dP/dt for all values of P. In this case, dP/dt is positive for $0 < P < M$ and negative for $P > M$. (If $P < 0$ were possible, then we would extend the line to the left of $P = 0$ and dP/dt for $P < 0$ would be negative). From the sign of dP/dt, we then indicate whether P is increasing or decreasing by an appropriate arrow. If the arrow points toward the right, P is increasing; toward the left it indicates P is decreasing. By knowing the sign of the derivative for all values of P, together with the starting values $P_0 = P(0)$, we can determine what happens as $t \to \infty$. A curve approached asymptotically as $t \to \infty$ is called a **steady-state** outcome. The phase line allows us to predict the steady-state outcome of a model by knowing the sign of the derivative and the initial condition.

For the logistic growth model, we note that if $P_0 > 0$, then as $t \to \infty$, the population $P(t)$ approaches M, the maximum sustainable population. However, if $P_0 = 0$, then $P(t) = 0$. (See Figure 11.12.)

Exercises 11.3

1. Construct a direction field and sketch a solution curve for the following differential equations.

a. $dy/dx = y$
b. $dy/dx = x$
c. $dy/dx = x + y$
d. $dy/dx = x - y$
e. $dy/dx = xy$
f. $dy/dx = 1/y$

For Exercises 2–5, sketch a number of solutions to the equations, showing the correct slope, concavity, and any points of inflection.

2. $dy/dx = (y + 2)(y - 3)$

3. $dy/dx = y^2 - 4$

4. $dy/dx = y^3 - 4$

5. $dy/dx = x - 2y$

6. Analyze graphically the equation $dy/dt = ry$, when $r < 0$. What happens to any solution curve as t becomes large?

7. Develop the following models graphically. First graph dP/dt versus P, and then obtain various graphs of P versus t by selecting different initial values $P(0)$ (as in our population example in the text). Identify and discuss the nature of the equilibrium points in each model.

a. $dP/dt = a - bP, \quad a, b > 0$
b. $dP/dt = P(a - bP), \quad a, b > 0$
c. $dP/dt = k(M - P)(P - m), \quad k, M, m > 0$
d. $dP/dt = kP(M - P)(P - m), \quad k, M, m > 0$

8. The Department of Fish and Game in a certain state is planning to issue deer hunting permits. It is known that if the deer population falls below a certain level m, then the deer will become extinct. It is also known that if the deer population goes above the maximum carrying capacity M, the population will decrease to M.

a. Discuss the reasonableness of the following model for the growth rate of the deer population as a function of time

$$\frac{dP}{dt} = kP(m - P)(P - m)$$

where P is the population of the deer and k is a constant of proportionality. Include a graph of dP/dt versus P as part of your discussion.

b. Explain how this growth rate model differs from the logistic model $dP/dt = kP(M - P)$. Is it better or worse than the logistic model? Why?

c. Show that if $P > M$ for all t, then the limit of $P(t)$ as $t \to \infty$ is M.

d. Discuss what happens if $P < m$ for all t.

e. Assuming that $m < P < M$ for all t, explain briefly the steps you would use to solve the differential equation. Do not attempt to solve the differential equation.

f. Discuss the graphical solutions to the differential equation. What are the equilibrium points of the model? Explain the dependence of the equilibrium level of P on the initial conditions. How many deer hunting permits should be issued?

PROJECT Exploring Modeling

Complete the requirements of the UMAP module, "Whales and Krill: A Mathematical Model," by Raymond N. Greenwell, UMAP 610. A predator-prey system involving whales and krill is modeled by a system of differential equations. Although the equations are not solvable, information is extracted using dimensional analysis and the study of equilibrium points. The concept of maximum sustainable yield is introduced and used to draw conclusions about fishing strategies. You will construct a differential equations model, remove dimensions from a set of equations, find equilibrium points of a system of differential equations and learn their significance, and practice manipulative skills in algebra and calculus.

11.4 Numerical Approximation Methods[1]

In the models developed in the preceding sections of this chapter, we found an equation relating a derivative to some function of the independent and dependent variables;

[1] This section is adapted from UMAP Unit 625, which was written by the authors and David H. Cameron. The adaptation is presented with the permission of COMAP, Inc., 57 Bedford St., Lexington, MA, 02173.

that is,

$$\frac{dy}{dx} = g(x, y)$$

where g is some function in which either x or y may not appear explicitly. Moreover, we were given some starting value; that is, $y(x_0) = y_0$. Finally, we were interested in the values of y for a specific set of x values; that is, $x_0 \leq x \leq b$. In summary, we determined models of the general form:

$$\frac{dy}{dx} = g(x, y), \quad y(x_0) = y_0, \quad x_0 \leq x \leq b$$

We call first-order ordinary differential equations with the preceding conditions *first-order initial-value problems*. As seen from our previous models, they constitute an important class of problems. We now discuss the three parts of the model.

First-Order Initial-Value Problems

The Differential Equation $dy/dx = g(x, y)$

As discussed in our models, we are interested in finding a function $y = f(x)$ whose derivative satisfies an equation $dy/dx = g(x, y)$. Although we do not know f, we can compute its derivative given particular values of x and y. As a result, we can find the slope of the tangent line to the solution curve $y = f(x)$ at specificed points (x, y).

The Initial Value $y(x_0) = y_0$

The initial-value equation states that at the initial point x_0, we know the y value is $f(x_0) = y_0$. Geometrically, this means that the point (x_0, y_0) lies on the solution curve (Figure 11.13). Thus, we know where our solution curve begins. Moreover, from the differential equation $dy/dx = g(x, y)$ we know that the slope of the solution curve at (x_0, y_0) is the number $g(x_0, y_0)$. This is also depicted in Figure 11.13.

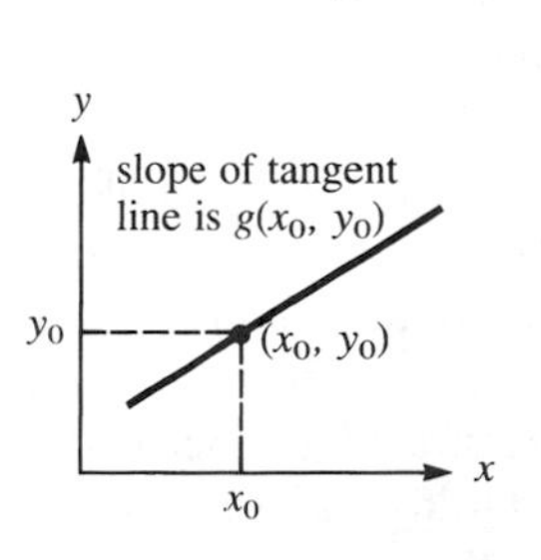

Figure 11.13 The solution curve passes through the point (x_0, y_0) and has slope $g(x_0, y_0)$.

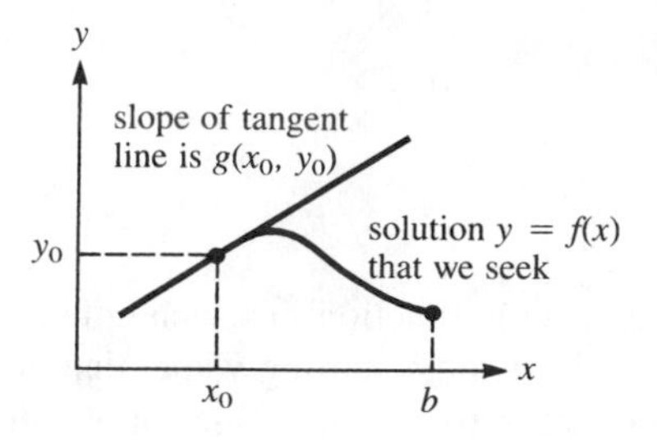

Figure 11.14 The solution $y = f(x)$ to the initial-value problem is a continuous function over the interval from x_0 to b.

The Interval $x_0 \le x \le b$

The condition $x_0 \le x \le b$ gives the particular interval of the x-axis with which we are concerned. Thus, we would like to relate y with x over the interval $x_0 \le x \le b$ by finding the solution function $y = f(x)$ passing through the point (x_0, y_0) with slope $g(x_0, y_0)$ (Figure 11.14). Note that the function $y = f(x)$ is continuous over $x_0 \le x \le b$ because its derivative exists there.

Approximating Solutions to Initial-Value Problems

We shall now study a method that utilizes the three parts of the initial-value problem together with the geometrical interpretation of the derivative to construct a sequence of discrete points in the plane. The sequence numerically approximates the points on the actual solution curve $y = f(x)$. Let's begin with an example.

Example *Interest Compounded Continuously*

Suppose at the end of year 1 we have \$1000 invested at 7% annual interest compounded continuously. We want to know how much money we will have at the end of year 20. Letting $Q(t)$ represent the amount of money at any time t, we have

$$\frac{dQ}{dt} = 0.07Q, \ Q(1) = 1000, \quad 1 \le t \le 20$$

We want to find the function $Q(t)$ solving this initial-value problem. Using a dashed curve to represent the unknown function Q, we sketch the preceding information in Figure 11.15.

We know the derivative of $Q(t)$ for known values of Q and t. In particular, we know that at the point (1, 1000) the derivative is $dQ/dt = 0.07(1000) = 70$. Because the derivative can be interpreted as the slope of the line tangent to the curve, we sketch a tangent line to our unknown function Q at $t = 1$ with slope 70. This situation is depicted in Figure 11.16. Because we don't know $Q(t)$, we cannot determine exactly the value of $Q(20)$. However, we can approximate it by the value on the tangent line when $t = 20$. Now the equation of the tangent line T in point-slope form is given by

$$T - 1000 = 70(t - 1)$$

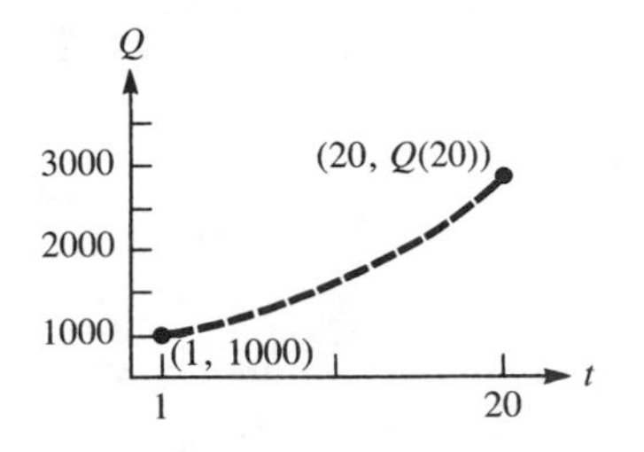

Figure 11.15 The curve satisfying $dQ/dt = 0.07Q$, $Q(1) = 1000$, $1 \leq t \leq 20$.

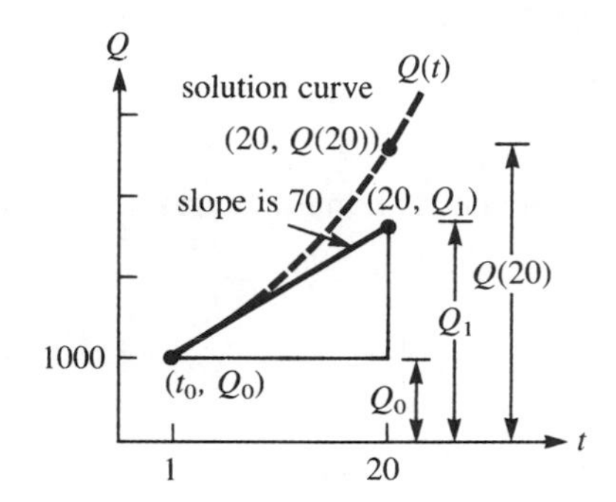

Figure 11.16 The point $(20, Q_1)$ on the tangent line approximates the actual solution point $(20, Q(20))$.

In other words,

$$T = Q_0 + \left.\frac{dQ}{dt}\right|_{t=t_0} \Delta t$$

where $Q_0 = 1000$, $dQ/dt_{t=t_0} = 70$, and $\Delta t = t - 1$. When $t = 20$, we see that

$$T(20) = 1000 + 70(20 - 1) = 2330$$

Then we make the approximation

$$Q(20) \approx 2330 = Q_1$$

to the value of the unknown function at $t = 20$. Thus, starting with \$1000 at the end of year 1, we estimate that we'll have \$2330 at the end of year 20 if interest is compounded continuously at an annual rate of 7%.

Estimating with Two Steps Note that we have used the known starting value $Q(1) = 1000$ to calculate the estimate $Q(20) \approx 2330$. How can we improve the approximation and get a more accurate picture of the solution curve? We have assumed that the derivative Q' is the constant 70 over the interval $1 \leq t \leq 20$, but we know it actually changes as Q and t change. Perhaps our estimate of $Q(20)$ will be more accurate if we make

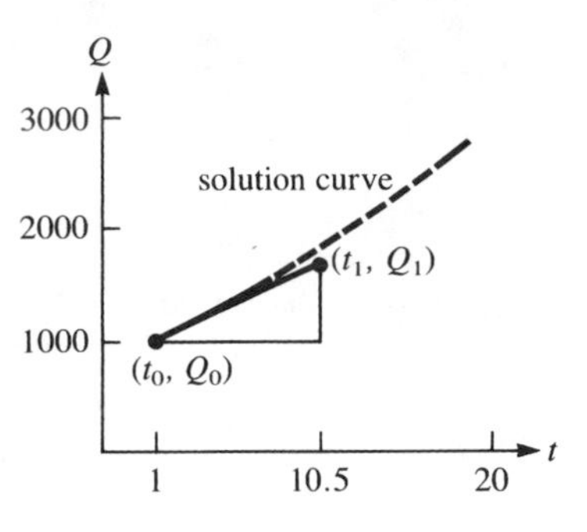

Figure 11.17 The point (t_1, Q_1) estimates the solution curve at the halfway value $t = 10.5$.

another estimate at an intermediate point. Setting $\Delta t = (20 - 1)/2 = 9.5$ in that same problem, we obtain

$$Q(10.5) \approx Q_1 = Q_0 + \left.\frac{dQ}{dt}\right|_{t=1} \Delta t = 1000 + 70(9.5) = 1665$$

This is depicted in Figure 11.17.

Next we use the estimate $Q(10.5) \approx 1665$ to approximate the derivative at $t = 10.5$ from the formula:

$$\left.\frac{dQ}{dt}\right|_{t=10.5} = 0.07Q(10.5)$$

Note an important difference from our first calculation for $Q'(1)$. We *know* the value of the derivative at $t = 1$ exactly because we know $Q(1) = 1000$, but we must *estimate* the derivative at $t = 10.5$ because $Q(10.5) \approx 1665$ is only an estimate. Now we calculate our estimate $Q(20)$:

$$\begin{aligned} Q(20) \approx Q_2 &= Q_1 + \left.\frac{dQ}{dt}\right|_{t=10.5} \Delta t \\ &= Q_1 + 0.07Q(10.5)\Delta t \\ &\approx 1665 + (0.07)(1665)(9.5) = 2772.23 \end{aligned}$$

This two-step process is shown in Figure 11.18. You will see shortly that the approximation $Q(20) \approx 2772.23$ is closer to the actual value of the solution at $t = 20$.

In this way, a table of approximate values to the solution is built up in a step-by-step fashion. The situation is depicted in Figure 11.19. Notice that an error is introduced at each step. As these errors accumulate with more and more steps, the approximations $y_1, y_2, \ldots, y_n$ get farther and farther from the actual solution curve.

Improving the Estimate of the Graph Euler's algorithm can easily be coded for computer implementation to facilitate reductions in step size. Because there was a significant change in the estimate to $Q(20)$ (from 2330 to 2772.23) when we reduced the

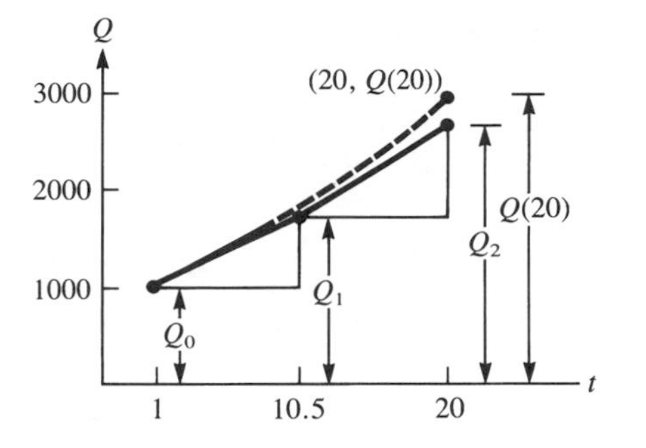

Figure 11.18 The value Q_2 approximates the solution $Q(20)$ in a two-step process.

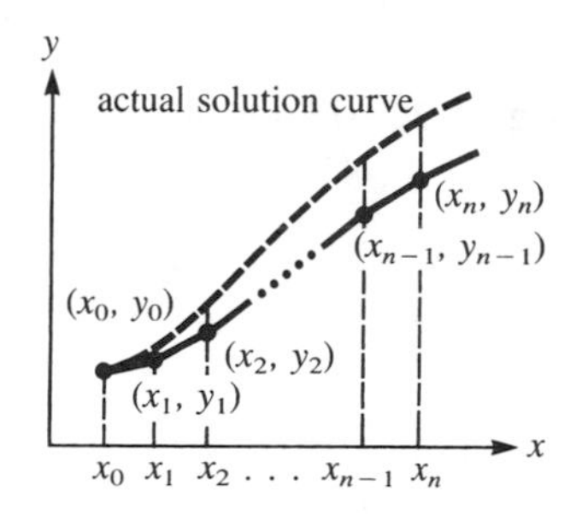

Figure 11.19 The points (x_i, y_i) approximate the solution curve.

step size Δt in the compound interest example, we would not be too confident in our results at this point. Using a calculator program, we applied Euler's algorithm with step sizes $\Delta t = 19$, $\Delta t = 9.5$, and $\Delta t = 1.0$ and obtained approximations for Q every integer value of t between 1 and 20. To get an idea of what the unknown solution function looks like, we plotted the points obtained from the various step sizes on a single graph. The graph is displayed in Figure 11.20.

It might be tempting to reduce the step size even further in order to obtain greater accuracy. However, each additional calculation not only requires additional computer time, but more importantly introduces round-off error. Because these errors accumulate, an ideal method would improve the accuracy of the approximations yet minimize the number of calculations. For this reason, Euler's method may prove unsatisfactory. More refined numerical methods for solving the initial value problem are investigated in courses in the numerical methods of differential equations. We won't pursue those methods here.

Finding the Solution Analytically Just as we did for the population problem in Section 11.2, we can separate the variables and integrate the initial-value problem

$$\frac{dQ}{dt} = 0.07Q, \quad Q(1) = 1000, \quad 1 \le t \le 20$$

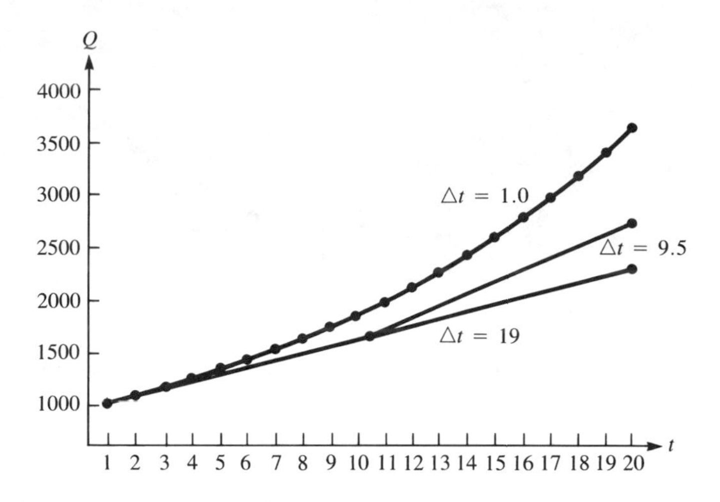

Figure 11.20 Plot of the approximate solution for the various step sizes.

to obtain

$$Q = C_1 e^{0.07t}$$

Finally we apply the initial condition that $Q(1) = 1000$ to evaluate the constant C_1; substituting $t = 1$ into the last equation gives

$$1000 = Q(1) = C_1 e^{0.07}$$

so $C_1 = 1000/e^{0.07} \approx 932.39382$. We now have our desired solution function:

$$Q = \left(\frac{1000}{e^{0.07}}\right) e^{0.07t}$$

If we evaluate this solution at $t = 20$, we obtain $Q(20) = 3781.04$. Thus, our approximations for $Q(20)$ are

$$2330 \quad \text{with } \Delta t = 19$$

$$2772.23 \quad \text{with } \Delta t = 9.5$$

$$3616.53 \quad \text{with } \Delta t = 1$$

They become more accurate as Δt decreases. Note that even with $\Delta t = 1$, an error of 164.51 results. We can further reduce this error by decreasing Δt. With the aid of a calculator, $\Delta t = 0.01$ in Euler's method gives the approximation $Q(20) \approx 3763.56$. ■

Exercises 11.4

1. When interest is compounded, the interest earned is added to the principal amount so that it may also earn interest. For a one-year period, the principal amount P is given

by

$$P = \left(1 + \frac{i}{n}\right)^n P(0)$$

where i is the annual interest rate (given as a decimal) and n is the number of times during the year that the interest is compounded.

To lure depositors, banks offer to compound interest at different intervals: semiannually, quarterly, or daily. A certain bank advertises that it compounds interest continuously. If \$100 is deposited initially, formulate a mathematical model describing the growth of the initial deposit during the first year. Assume an annual interest rate of 10%.

2. Use the differential equation model formulated in the preceding problem to answer the following:
   a. From the derivative evaluated at $t = 0$, determine an equation of the tangent line T passing through the point (0, 100).
   b. Estimate $Q(1)$ by finding $T(1)$, where $Q(t)$ denotes the amount of money in the bank at time t (assuming no withdrawals).
   c. Estimate $Q(1)$ using a step size of $\Delta t = 0.5$.
   d. Estimate $Q(1)$ using a step size of $\Delta t = 0.25$.
   e. Plot the estimates you obtained for $\Delta t = 1.0, 0.5$, and 0.25 to approximate the graph of $Q(t)$.

3. a. For the differential equation model obtained in Exercise 1, find $Q(t)$ by separating the variables and integrating.
   b. Evaluate $Q(1)$.
   c. Compare your previous estimates of $Q(1)$ with its actual value.
   d. Find the effective annual interest rate when an annual rate of 10% is compounded continuously.
   e. Compare the effective annual interest rate computed in part d with interest compounded:

   - Semiannually: $(1 + 0.10/2)^2$
   - Quarterly: $(1 + 0.10/4)^4$
   - Daily: $(1 + 0.10/365)^{365}$

   f. Estimate the limit of $(1 + 0.10/n)^n$ as $n \to \infty$ by evaluating the expression for $n = 1{,}000$; 10,000; 100,000.
   g. What is $\lim_{n\to\infty}(1 + 0.10/n)^n$?

PROJECTS Modeling Explorations

Complete the requirements of the indicated UMAP module.

1. "Feldman's Model," by Brindell Horelick and Sinan Koont, UMAP 75. This unit develops a version of G. A. Feldman's model of growth in a planned economy in

which all means of production are owned by the state. Originally the model was developed by Feldman in connection with planning the economy of the former Soviet Union. Students compute numerical values for rates of output, national income, their rates of change, and the propensity to save, and discuss the effects of changes in the parameters of the model and in the units of measurement.

2. "The Digestive Process of Sheep," by Brindell Horelick and Sinan Koont, UMAP 69. This unit introduces a differential equation model for the digestive processes of sheep. The model is tested and fit using collected data and the least-squares criterion.

Part III

Decisions in Organizing for Mathematics Instruction

12
Getting Ready for Teaching

Ultimately, any attempt to [prepare for teaching] mathematics is an exercise in adaptation from what we are able to do to what we want to do.... But sometimes teacher education programs focus too much on correcting the deficiency in teachers' knowledge (i.e., giving them "more" of something) to the exclusion of considering possibilities of how that knowledge can transform the common, everyday activities of teaching.... Accepting the premise that adaptation should be at the core of teacher education programs does not diminish the importance of acquiring "More," as mentioned earlier. But it requires the realization that "more" can contribute to change only if it is seen as a means for achieving change rather than as an end in itself.

Thomas Cooney (1994)

Overview

Effective planning is key to successful teaching. Even the most experienced teachers are only successful when they have carefully planned and effectively structured a mathematics lesson. The mathematics that students learn stems from what they do and experience in the mathematics classroom. What are the students thinking about? What are they talking about? Are they actively participating in the class? The answers to these questions tell a lot about what the students are learning. In the classroom, it is the teacher who makes the decisions about how to answer these questions. How will you answer them?

This chapter outlines some key components for successful planning, both long-range and daily planning. The chapter also addresses some of the decisions that you will make as a teacher. Among others, these decisions include

1. How to structure and sequence the various topics and concepts to be addressed in a lesson or unit
2. Whether to lecture or allow students to work together in small groups
3. How to motivate and enhance *all* students' mathematical experiences
4. When to incorporate quizzes or other forms of assessment

Focus on the NCTM *Standards*

The NCTM *Principles and Standards for School Mathematics* stresses the need to organize a mathematics curriculum around appropriate goals for the learning of mathematics. In particular, the Curriculum Principle highlights the fact that the decisions made by teachers, such as which topics to teach, when to teach them, and the depth to which to teach them, impact the quality of mathematics programs and the learning of students (NCTM, 2000). In order to make the appropriate choices, the Teaching Principle emphasizes the importance of "knowing and understanding mathematics, students as learners, and pedagogical strategies" (NCTM, 2000, p. 17). All of this comes into play in the planning process.

Through the planning process, teachers reflect on the concepts to be taught, the students in the classroom, and the ways in which the concepts can best be introduced to the students. Teachers should also consider the classroom environment that is needed for providing rich learning opportunities. Planning is the first step toward the development of a quality mathematics program.

Focus on the Classroom

In order to provide appropriate and worthwhile experiences for students, teachers must consider meaningful mathematics for the group of students they teach. They must consider the intended curriculum, the ideal goals of the mathematics instructional program, and make decisions about how to implement this in their classrooms. They must make decisions about what content to emphasize, which pedagogical approaches to use, and when and how to assess student progress. In addition, the math teacher must make decisions about the nature of the learning environment they want to create and how and when to use classroom discourse to support student learning. Teachers need time and resources to make these decisions. Careful planning prior to class can provide the structure to make these important decisions and to achieve desired learning goals.

12.1 The Importance of Planning

Sit in a middle school or high school math class and notice all that is taking place over the course of the lesson. The teacher addresses a chosen set of mathematical ideas. The teacher and students engage in conversations about the ideas. At various points the teacher makes decisions about what questions to ask students or how to answer questions from students. Typically, an assignment is given to help students continue thinking about the concepts addressed in class.

As a student of mathematics, and as a future teacher of mathematics, you may have ideas about the major components of an effective lesson. From observing a lesson, you most likely also have a sense of whether or not the lesson helped students reach the intended goals of the lesson. In the math classroom it is the teacher's responsibility to provide for all students. This includes (1) setting learning *goals* for all students, (2) selecting appropriate *tasks* to address these goals, (3) managing the classroom to provide a safe and stimulating learning *environment* for all students, (4) *assessing* student learning and the instructional decisions made over the course of the lesson (NCTM, 1991). To be successful, a teacher must have a plan.

The lesson plan is like a recipe in that it provides the teacher with a structure or written set of guidelines. The lesson plan can provide guidelines for accomplishing a set of goals, the activities needed to help reach these goals, ideas for creating the appropriate environment, and materials for assessing progress and outcomes. The daily lesson plan may revolve around a set of daily goals, but it is also linked to broader goals that are part of a unit plan (like a single recipe is often part of a larger menu). With broader goals in mind, plans are most effective when the lessons flow smoothly from old topics to new concepts, when the students are engaged in thinking and doing mathematics, and when the lesson revolves around worthwhile mathematical tasks. These and other aspects of the lesson must be considered when planning so that the plan is both effective and useful. That is not to say that every effective teacher spends many hours a day planning lessons for the next school day.

Learning to develop useful and appropriate lessons is similar to learning mathematics. You must understand the concepts involved, the roles played by the various concepts, and how the ideas fit together to create a fuller image. In mathematics this might mean understanding how to find the length and width of a rectangle, understanding that these dimensions are both essential to finding the area of the rectangle, and knowing that these measurements must be multiplied together to obtain the value of the area of the rectangle. In lesson planning, it is often a bit more complicated than this example, but the ideas are the same.

The first step in effective lesson planning is to understand the concepts involved. This pertains to the concept of lesson development as well as to the actual concepts to be taught. That is to say, a teacher must have general pedagogical knowledge (knowledge of teaching), content-specific pedagogical knowledge (knowledge about teaching mathematics), and mathematics content knowledge. Chapters 4–11 provided an overview of the content knowledge necessary for teaching mathematics at the middle school and high school. These same chapters highlighted ideas related to the pedagogy of teaching mathematics by giving a snapshot of the current research on learning and teaching, by providing lesson plan ideas, and by introducing a modeling approach to teaching mathematics. With a good understanding of these three components of pedagogical content knowledge, lesson planning comes together more easily. General pedagogical knowledge is also necessary since the teacher must make decisions about how to create a productive learning environment, how to structure interactions, and how to manage the classroom in ways that promote learning.

Content knowledge is vital to planning. With an understanding of the mathematics content to be taught, the teacher has a sense of what needs to be done and how to sequence the topics and ideas of the lesson. A deep understanding of the concepts is necessary for being able to make connections among concepts and various fields within the mathematical sciences. This is crucial for helping students make these same connections and hence create a bigger picture of the role of mathematics in their lives and in society.

Pedagogical content knowledge also plays an important role in the planning process. To make informed decisions about what to teach and how to teach it, you must have a sense of how your students best learn the concepts involved. For instance, review of previously taught concepts or even the introduction of a somewhat new but straightforward idea may be best approached with some small group work and investigations, while teaching a complex concept might require a more structured or guided lesson. In addition to knowing the important components for effective lesson planning and the roles played by each component, it is essential to know how to put these ideas together to create a worthwhile and effective lesson. You must know how to put together the pieces of a lesson, but you must also have a sense of the bigger picture. An effective lesson is more than the sum of its parts.

How do you put the pieces together to make the greater whole? The most important "greater whole" that needs to be considered is the mathematics. What are the key ideas that you want students to learn? How are these ideas related to other mathematical experiences they have had or will have? The assigned textbook certainly contains many of the key ideas that are important for students in a particular class. However, just knowing what the ideas are is not enough. The math teacher must have a deep understanding of the ideas and how they fit together to build other more complex mathematical ideas. The teacher is responsible for making these ideas come to life for the students, and for helping the students to see how the ideas fit together.

To achieve the goal of students learning these key ideas, the teacher must ensure appropriate mathematical experiences both in the classroom and outside it. In determining appropriate experiences, the teacher must focus on the class as a whole as well as the individual student. How will the students' needs be met? Students need a variety of experiences with the mathematical ideas to get a fuller picture of the important concepts involved. They need opportunities to interact with the mathematics and with one another. Interaction with the teacher or other more knowledgeable students is also important. This may mean developing a long-range plan to ensure that students acquire the tools needed for the next unit or the next course of study. Even when these goals are met, the teacher must also assure that the goals of the state, district, or mathematics department have been met. What big picture do others hold that must be satisfied by your teaching?

Planning helps you develop the greater whole or the big picture. Planning lets you see how to get there. Planning shows others how you will help your students attain an understanding of the big picture. In addition, planning imparts the values of organization and careful consideration to the students.

Exercises 12.1

1. Consider a typical day in class. Name some things that a teacher does that would require
   - **a.** general pedagogical knowledge
   - **b.** content knowledge
   - **c.** pedagogical content knowledge
2. In what ways is a mathematics lesson different from a history lesson? In what ways is a mathematics lesson the same as a lesson in some other discipline?
3. How is a piece of mathematical knowledge different from a piece of knowledge in history or English? What are different kinds of mathematical knowledge?

12.2 Focus on Key Ideas: Setting Goals

Long-Range Planning

As mentioned above, it is pertinent to have a sense of the big mathematical picture that you are hoping to impart to your students. This big picture forms the basis for lesson planning. First, a long-range (semester or year-long) plan is important for clearly stating appropriate goals for student learning across the school year. Once you have a sense of the key ideas (major concepts or units) that will form the big picture, unit plans and daily lesson plans fill in the details. For instance, the teacher of a sophomore math class that focuses on learning mathematics through applications and mathematical modeling must first know which topics are appropriate for the students at this level, which topics build on prior knowledge, and which will be new concepts with which the students will grapple. It is also important to know where these key ideas will lead the students. What courses will they most likely take during their junior and senior years? How will the sophomore curriculum prepare them for these experiences?

Once a picture of the course is established with a sequence of topics or concepts, such as the sequence outlined in Table 12.1, the teacher can begin to block out units that give more detail to the picture and present the mathematics in a coherent way. Finally,

Table 12.1 Sample long-range plan.

Possible Sequence for a Sophomore-Level Course Incorporating Mathematical Modeling[1]

I. EUCLIDEAN GEOMETRY
 A. Two-dimensional
  1. Congruence and similarity
  2. Symmetry
  3. Transformational geometry
 B. Three-dimensional
  1. Regularity
  2. Symmetry
II. COORDINATE GEOMETRY
 A. Taxi-cab Geometry
  1. Distance
  2. Optimization
 B. Graphs of Equations and Inequalities
  1. Linear
  2. Absolute value
 C. Systems of equations
  1. Matrix Representations
  2. Solving systems of equations
  3. Transformations
III. ALGORITHMIC THINKING AND DECISION MAKING
 A. Optimization
  1. Taxi-cab geometry
  2. Graph theory
 B. Simulation
  1. Conjecture building
  2. Probability and statistics
 C. Growth and Decay
  1. Sequences and series
  2. Nonlinear equations
  3. Function inverses

each unit plan provides an outline for weekly and daily lessons. Sample outlines of units and daily lessons for the course described in Table 12.1 can be found in Tables 12.3 and 12.4.

Long-range planning doesn't always imply that the classroom teacher must find and make decisions about the key ideas for an entire school year and then develop a sequence of topics or units. For the most part, a course curriculum is set by the district

[1] Adapted from COMAP (1998). *Mathematics: Modeling our world*, Course 2. Cincinnati: South-Western Educational Publishing.

or by the department, and has been developed and refined over time. At times, the choice of textbook determines the curriculum and the range of key ideas for the school year. However, the mathematics teacher must understand the purpose and the goals of the curriculum that is to be taught. In other words, the teacher must have a sense of the big picture in order to help students construct it for themselves. When the teacher has a sense of that big picture, the teacher is better able to set appropriate shorter-term goals for the students, to develop coherent lessons, and to balance the daily and weekly lessons in ways that keep students motivated and keep the teacher in touch with students' needs.

Knowing how to set appropriate goals for students comes with time and experience. Usually it is best for the first-year teacher to talk with more experienced teachers who have taught the same or similar courses, and study their long-range plans and goals for students. After a teacher has taught a course once or twice, he or she has a sense of the timing to reach desired goals, and even the reasonableness of certain goals. Consider the following example.

Example *Adjusting to Student Needs*

A first-year teacher preparing to teach geometry noticed the emphasis on writing proofs in the book to be used. She got ready to teach the students about writing proofs, two-column proofs as well as other forms. After several weeks of teaching proofs and asking students to write proofs similar to her models, she realized they were continuing to have difficulties. For the remainder of the school year, the teacher required very few written proofs from her students. Even so, the teacher believed that proofs could help students understand some of the geometric relationships and didn't want to leave it out of the curriculum.

Before teaching the same course the following year, the teacher searched for new approaches to teaching proofs. She investigated several sources and came up with a plan that introduced the need for proof through surprising geometry results (Dreyfus & Hadas, 1996). When students were asked to discover why an idea or conjecture was true and to write a complete justification, the teacher realized they were writing proofs. This was a natural way to introduce proofs and it made sense to her students. ■

In this example, the beginning teacher initially used the textbook as a guide for her long-range plan as well as for developing daily lessons for teaching particular topics. She didn't search for alternative resources with which to get a better sense of the sequence to be taught and ways to teach it. Her vision for reaching the long-range goals was restricted to the textbook. Nonetheless, reflections on the long-range goals prompted her to search other resources for reaching these goals the following year.

A first-time long-range set of goals for a particular course and group of students may be based on the textbook being used, the district's curriculum outline, and various notes or lesson plans from colleagues in the school. Other resources that may be helpful include textbooks from a variety of publishers, mathematics and curriculum-based web sites, and your own sense of the course based on personal experiences in high school and college.

Many other factors can influence the long-range plan or goals for a course. The NCTM *Principles and Standards*, for one, suggests goals for students and mathematics curricula at all levels (refer to Chapters 3–8 for outlines of the *Principles and Stan-*

Table 12.2 State of Illinois algebra learning standard for early high school.[2]

STATE GOAL 8: Use algebraic and analytical methods to identify and describe patterns and relationships in data, solve problems, and predict results.	
LEARNING STANDARD	EARLY HIGH SCHOOL
A. Describe numerical relationships using variables and patterns.	a. Use algebraic methods to convert repeating decimals to fractions.
	b. Represent mathematical patterns and describe their properties using variables and mathematical symbols.
B. Interpret and describe numerical relationships using tables, graphs, and symbols.	a. Represent algebraic concepts with physical materials, words, diagrams, tables, graphs, equations, and inequalities and use appropriate technology.
	b. Use the basic functions of absolute value, square root, linear, quadratic, and step to describe numerical relationships.

dards). Teachers who wish to implement these or similar goals may develop a long-range plan that incorporates more of a problem-solving approach to the study of mathematics. State curricular frameworks are another source of recommendations that can influence a teacher's goals. Often, state guidelines for the mathematics curriculum provide specific details on the content to be taught and goals for students at specified grade levels. A sample of the Illinois Learning Standards for Algebra is shown in Table 12.2. It is clear from this sample that a course in first-year algebra is greatly determined by these standards.

A third factor that may influence long-range and unit plans is state-mandated tests. Many states require students at specified levels of school to take standardized tests such as the Iowa Test of Basic Skills (ITBS) or tests developed to assess student achievement in areas outlined in the state curriculum framework. It is well known that students perform better on standardized tests when they are given opportunities to learn about the nature of the tests and to practice test-taking strategies. Hence, many teachers incorporate these experiences into the yearly plan.

It is only fair that the students know what to expect. If teachers prepare a long-range plan for a course, they are better able to convey goals and expectations to the students through discussion of the key ideas, and through the natural flow of mathematics topics. When ideas and concepts are woven together appropriately, students are able to make sense of the larger mathematical structure being built. On a more practical note, a long-range plan or unit plan conveys expectations to the students in terms of the na-

[2] From http://www.isbe.state.il.us/ils

ture of class activities, including plans for assessing learning. This not only helps keep the students motivated to learn, it also helps the teacher ensure that there is a balance of experiences on a daily and weekly basis to enhance student learning. In general, a long-range plan gives the teacher and students a head start on daily lessons.

Unit Planning

When the goals of a course or unit are not predetermined by the choice of textbook, by the department, or by the district, this task falls on the teacher who has the freedom to set the course curriculum. Even when a particular book must be used, the teacher should make decisions about how to organize lessons for a particular unit or topic. One way to begin the process of making these decisions is to create a concept map. A *concept map* is a diagram that illustrates the main concepts in a unit or topic area and the connections between concepts. The process of creating the concept map is often instructive in itself as you begin to realize that mathematics is not as linear as it appears to be in a standard textbook. In fact, many concepts link together in a nonlinear fashion. To create a concept map such as the one shown in Figure 12.1, begin with the central concept around which you will form the teaching unit (such as symmetry). Begin to build the map by adding related concepts. Notice how the concept map allows you to depict concepts that are more closely related to the central idea (for example, reflections and rotations) and those that are subtopics (line of reflection, center of rotation, analytic representations of transformations, tesselations).

A completed concept map is useful for developing the course unit. The map provides a visual aid for planning the sequence of topics to be taught. To add detail to the plan, consider the individual concepts and the experiences that will introduce students to the concepts and their connections within the map. Sometimes it is helpful for students to create concept maps at a few points during the unit to check on the connections they are making.

Once a unit plan has taken shape, consider the following questions:

- Does the unit focus on mathematics content and processes that are important for the students?
- Is the unit organized in a coherent manner that will help students build a deep understanding of the concepts?
- Do the lessons include a variety of experiences that allow for active mathematics learning?
- Do plans focus on the needs of all students as well as the mathematical concepts (not just procedures)?
- Were assignments, long-range projects, tests, and quizzes chosen to reflect the goals of the course and are they appropriately spaced across the unit or semester?
- Have you considered the use of manipulatives, technology, and other supplementary materials to enhance the learning experience?

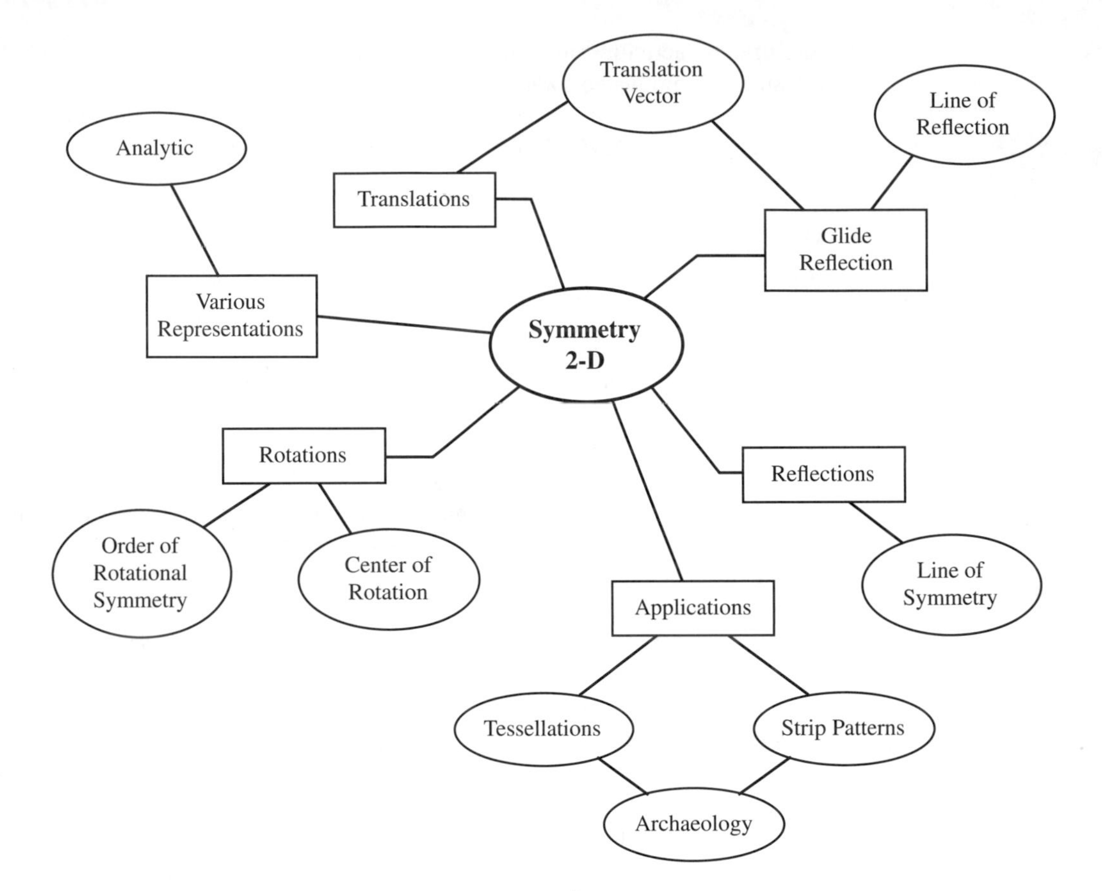

Figure 12.1 Concept map[3] for a unit on symmetry.

Table 12.3 is a sample unit plan that was developed using the symmetry concept map and the long-range plan outlined in Table 12.1.

Daily Lesson Planning

Short-term or daily lesson plans differ from unit and long-range plans in that they include many of the little details that help to ensure a successful lesson. These details may include the nature of questions to ask students to solicit ideas, or the amount of time to be spent with students in small groups. Another feature of a daily lesson plan that makes it unique is its flexibility. The daily lesson plan must be flexible in order to accommodate unforeseen events such as an unannounced assembly or shortened day because of a snow storm. It is also possible that the teacher has prepared a lesson for which the students are not ready and plans must be changed. For instance, at the start of a new

[3] Based on Crisler, N. (1995). *Symmetry and patterns*. Lexington, MA: Consortium for Mathematics and Its Applications.

Table 12.3 Sample unit plan on symmetry.

Day 1
1. Review transformational geometry.
2. Define line symmetry.
3. Explore line symmetry (in various objects found within and outside the classroom).
4. Share Archaeology Problem with students and discuss need for understanding symmetry and patterns to solve problem.

Day 2
1. Explore relations between symmetry and transformations (reflectional symmetry and rotational symmetry).
2. Explore reflectional symmetry (regular polygons, other objects).
3. Explore rotational symmetry (regular polygons, other objects).
4. Informal assessment on symmetry and transformations at end of class (or to take home if no time left).

Days 3–5
1. Use dynamic geometry software to create images with reflectional and rotational symmetry (might need a half class to review use of software).
2. Use ideas of symmetry to create packaging designs—refer to *Mathematics: Modeling our world*,[4] course 2, unit 3, lessons 4 and 5.
3. Presentations of sample designs (discuss use of symmetry to create designs).

Days 6–9
1. Continue presentations of packaging designs if needed, otherwise, begin discussion of other applications of symmetry.
2. Discussion and activities related to strip patterns.
3. Discussion and activities related to tessellations.
4. Assessment on strip patterns and tessellations—analysis of designs.

Days 10–12
1. Return to Archaeology Problem and work with symmetries (use ideas from all previous lessons to solve the mysteries of the patterns).
2. Group presentations on findings/solutions.

---

[4] COMAP (1998). *Mathematics: Modeling our world*, Course 2. Cincinnati: South-Western Educational Publishing.

unit the teacher might notice that many students don't have the background knowledge necessary for understanding the new concepts. An adjustment to the lesson may include a review of topics that will help make sense of the new ideas to be taught. Thus, that day's lesson plan may start differently than planned, or may even be put aside for a day or two. When plans are developed on a daily or weekly basis, such adjustments are simply part of the planning process. In a sense, this allows the teacher to set smaller attainable goals for the students—goals that will help the students reach the larger goals for the course.

Daily lesson planning is essential for the beginning teacher. These plans help the teacher organize all of the details of the classroom and the lesson to be taught. As noted earlier, the lesson plan helps the teacher think about the content to be taught, the mathematics-specific pedagogy to be employed, as well as general pedagogical techniques. Since the teaching of mathematics involves many aspects, including paying attention to student behaviors, it is prudent to make plans for the classroom activities ahead of time when you can think clearly about the best methods for conducting class. Changes to the lesson plan will almost always take place as the lesson is being enacted, but the plan provides a basis for making decisions and links the lesson to the larger unit or semester goals.

Early in a teacher's career, lesson plans should contain all the elements needed to fully implement a quality lesson. Some of the essential elements include

1. *Daily goals or objectives* The daily objectives may describe the main concepts or topics of the lesson as well as the skills or processes to be mastered by the students.

2. *Warm-up or opening activity* An opening activity helps to get the students involved in doing mathematics as soon as they enter the classroom. While students are working on a warm-up, you have the time to take attendance and work with students who were absent the previous day. Thus, class time is not spent taking care of administrative details. A warm-up activity is also a great way to introduce a new topic or provide chances for students to brush up on skills that will be tested on state-mandated or other standardized tests.

3. *Activities for achieving the objectives* This section of the lesson plan should outline what you will do as well as what the students will do to accomplish the stated objectives. These details are the "meat" of the lesson plan and should be carefully spelled out to provide enough detail so that you can glance at the plan to get all the information needed to proceed with an activity. The details might include definitions to be written on the board, a description of specific things you will do, directions for accomplishing a group activity, or examples for demonstrating a procedure. Be sure to consider how each activity contributes to the overall goals of the lesson. It is also important to plan activities that will help students achieve conceptual understanding rather than just how to perform an algorithm for getting the correct answer.

4. *Questions and other means of assessment* It is beneficial to incorporate assessment into each lesson. Assessment provides you with feedback on whether objectives are being met, whether students are grasping key concepts, and whether changes to the lesson must be made. For a beginning teacher, it is often helpful to plan the types

of questions (or even write specific questions) that will elicit information on how the lesson is progressing. Questions that are part of the activities for achieving the lesson objectives should also be written. These may include questions to test students' prior knowledge, questions that stimulate brainstorming among students, or questions that challenge students to investigate specific mathematical ideas. Questions should range from low- to high-order, where low-order questions require a quick answer and high-order questions require more thought.

5. *Summary or closing activity* How will you conclude the lesson? One option is to encourage students to continue a particular activity or investigation, using as much class time as possible for the exploration of mathematical ideas. Another option is to save time for reflection and reporting. When students are asked to think about the lesson in terms of new concepts learned or things that caused confusion, they are afforded the chance to solidify ideas and you are able to see whether the daily objectives have been met. Reflections and summaries may be requested orally or in writing.

Often, math teachers save time (sometimes up to 20 minutes) for students to begin working on assigned homework. There are certainly pros and cons to such a decision. Allowing students time to begin homework while still in class gives you and your students immediate feedback on the lesson and their abilities to grasp the main ideas of the lesson. If needed, you can then take a few moments to work with students' difficulties—one-on-one or as an entire class. If opportunities similar to this are afforded students during the lesson itself, there is often no need to give more time at the conclusion of class. A negative aspect of allotting time for students to work on homework is that time is then lost that could be used for teaching and learning more or deeper concepts. The class period is short. If students are expected to investigate mathematics (rather than sit while mathematics is told to them), time is needed. Students should certainly be expected to work on mathematics until the end of class. But what mathematics will they work on? Will they continue to explore new ideas? Will they reflect on the lesson and report on what they learned? Or will they begin work on homework to test their understanding of new ideas? You may answer the questions differently each day, but these questions must be considered during lesson preparation.

6. *Materials needed or other special notes* This should contain a list of any special materials needed for conducting the lesson. The list might include manipulatives, overhead projector slides, and pens for students to record solutions to be presented to class, or possibly a classroom set of calculators. Notes might remind you (or inform a substitute) that students will meet in the computer lab rather than the regular classroom, or that time will be given for students to go to the library for needed resources.

7. *Approximate time allotments* Although it's often difficult to judge how much time will be needed to complete an activity, it is a good practice to estimate time allotments in order to get a better sense of whether the stated objectives can be met. Time allotments should indicate a range of minutes (or possibly class sessions) needed to complete a stated activity or goal. This will help you decide about how much time might be left over for "extra" activities or by how much will the time run short if activities run at the upper time estimate.

If needed, a daily lesson plan may also include notes about modifications for students with special needs, or possibly notes about how to shorten or lengthen the lesson. Although it is wise to include details to the plan—some advise writing the lesson as if to prepare for a substitute teacher—it is also wise not to include so many details that the lesson plan is cluttered, difficult to read, and hence not useful. With experience, you will learn to keep many of the details in your head. The lesson plan becomes more of an outline or a place to record specific examples or directions. Although not part of the daily lesson, lesson plans are also a good place to record reflections or notes about what changes to make the next day or the next time the lesson is taught. Save space on the plan for recording reflections.

Several sample lessons that contain some or all of these elements appeared in Chapters 4–8. Another sample lesson plan is presented here.

Lesson 10: Sample Daily Lesson from Symmetry Unit

Objectives/Goals: Students will work in groups or pairs to develop an understanding of line symmetry. While working on the activities listed below, students will

- Develop a working definition of line symmetry
- Discover properties of line symmetry
- Explore line symmetry in various objects, geometrical or otherwise
- Consider line symmetry in art, design and other areas
- Come to understand how symmetry can be useful in archeology for discovering patterns and reconstructing ancient objects

Opening Activity: Worksheet review of transformational geometry with emphasis on line reflections and properties of reflections.

Activity Plan:

1. After a quick review and discussion of the opening activity, focus on line reflections and ask students for a definition of symmetry or symmetric. To get started, have students pair up and discuss the following questions. *What does symmetry relate to? What does it mean for something to be symmetric? How is the term* symmetry *used in mathematics? Outside of mathematics?*
2. Have pairs share what they discussed and use these ideas to formulate a definition of **symmetry** that everyone can agree on. Use this definition and some more discussion to define **line symmetry**. Record all definitions and properties on the chalkboard and ask students to place in their notebooks (new section).
3. Ask students to use the definitions to discuss the line symmetry (or lack of symmetry) for various objects on a worksheet, for pictures from art deco magazine, for objects found in the classroom and in the community. *Are there any lines of symmetry in the object? If so, where are they found? How do you know it's a line of symmetry? How do you know you have found all lines of symmetry? Will more lines of symmetry be found if we rotate the object?*

4. Use the discussion to extend their ideas about line symmetry. For example: A square has two different types of lines of symmetry. *One type of line of symmetry passes through opposite vertices. There are two of this type of line. A second type of line of symmetry passes through midpoints of opposite sides. There are two of this type of line.*

5. Have students consider various types of lines of symmetry in two-dimensional figures (have drawings and objects available for students to test). *How many different types of symmetry lines are found in a parallelogram? In a regular polygon with n sides? Does it depend on the number of sides? Explain. What kinds of quadrilaterals have lines of symmetry? How many types? How many of each type? How do you know?*

6. Consider three-dimensional objects: *What kinds of symmetry can we discuss in this case?* This question can be a challenge question that is not answered during class but left for students to think about over a few days.

7. Display several pieces of art (paintings, wallpaper samples, and other designs) and assign small groups of students to each piece of art. Have groups talk about the symmetry of the objects.[5] *How is symmetry of a painting different from the more formal definition used in mathematics? What kinds of symmetries do you see in the wallpaper patterns and other designs? Is the notion of symmetry the same? If not, how would you define symmetry in a nonmathematical way?* Try to limit this discussion to line symmetry although rotational symmetry might need to be briefly addressed.

8. Have a spokesperson from each group share what they found about the line symmetry of their assigned painting, design, or object.

9. Archeology Problem (see notes below)—present to class and ask them to be thinking about how to use their knowledge of symmetry to help solve the problem. We will begin next class with sharing ideas.

Closing/Summary: (5 minutes)

Exit Slips Ask students to write definition of symmetry in their own words. Then have students write about what they discovered about symmetry during today's lesson. Collect this slip of paper as they exit class.

Homework (1) Be thinking about the Archeology Problem; (2) symmetry problems from textbook.

Assessment:

- The opening activity will help assess student understanding of transformational

[5] Refer to Natsoulas, A. (2000). Group symmetries connect art and history with mathematics. *Mathematics Teacher* **93**(5):364–370.

geometry (previous unit) and identify those who may have difficulties with this unit.

- Various questions listed in italics above can be used for informal assessment as class progresses.
- Closing exit slips will allow for assessment of what students are taking away from the lesson.

Materials Needed:

Transformational geometry review sheet (opening activity)
Handouts and other examples of objects with which to investigate symmetry
Reprints of paintings and other design work
Wallpaper samples from local hardware store
Handout with Archeology problem
Large-print copies of all of the above for student with sight disability
3×5 notecards for exit slip

Notes:

1. Hopefully students are somewhat familiar with the terms symmetry or symmetric. If not, provide some basic examples of things that are and are not symmetric (without saying what the definition is) to help spark the conversation.
2. Students with weaker understanding of reflections may have more difficulty with line symmetry. Opening activity will help identify these students.
3. Archeology Problem[6]: A group of archeologists have been digging in three different areas of sub-Sahara Africa and have found many shards of broken pipes and pottery, as well as pieces of woven baskets and cloth dating back thousands of years. Many of the patterns on these artifacts appear to be similar, but the archeologists believe that the three areas where they are digging were separate tribes who may have existed at different times. They want to test this theory. How can they use the shards and pieces of woven materials to test their theory? (More information and sample artifacts will be provided once you have a basic plan for how to go about solving the problem.)

Exercises 12.2

1. Without referring to a textbook or other materials, develop a concept map for the concept of linear functions.
2. Compare your linear functions concept map with others in your class (see Exercise 1). How does it compare with a unit on linear functions in a high school mathe-

---

[6] Adapted from Crisler, N. (1995). *Symmetry and patterns.* Lexington, MA: Consortium for Mathematics and Its Applications.

matics textbook? What modifications would you make to your original concept map? Why?

3. Compare two textbooks and their treatment of the concepts in your linear functions concept map. Are the books helpful for developing a lesson sequence? Do the books agree with each other? If possible, compare a traditional textbook with a nontraditional or "reform" book. How do these suggested lessons compare with your concept map? How would you modify the lessons presented in the books to satisfy your concept map and your sense of how to organize the concepts? Explain your choice of modifications.

4. Choose one topic from your concept map and create or find at least one worthwhile activity that will help students develop an understanding of the concept.

5. Use the activity created for Exercise 4 and develop a daily lesson for the chosen topic. Be sure to consider the list of essential elements listed in this section. Try to keep the lesson to one day.

PROJECTS Find Out for Yourself

1. Observe a middle school or high school math class and take note of the major components of the lesson. How did the teacher begin class? How was class ended? What activities were used to motivate and teach a new topic (or revisit old topics)? Imagine you will teach the class the next day. Use what you learned to develop your lesson plan.

2. Interview a teacher about the planning process. How are long-range plans determined or developed? Do long-range plans accommodate district or school objectives? What details does the teacher include in a daily lesson plan? What details are left out? How does the teacher modify a lesson to accommodate differing student abilities?

12.3 Strategies for Effective Teaching

What makes a lesson or a teacher effective? There are many ways to answer this question, but most will not be satisfactory to a new teacher. There is no one answer, and so achieving effective lessons is not a simple matter. As noted in the previous sections of this chapter, one place to begin is through planning. Motivating and meaningful activities, appropriate resources, and an understanding of one's students all are important contributors to effective teaching. In this section we consider the activities and the in-class experiences that enhance student learning. We also share some resources for finding ideas and developing a meaningful lesson. Keep in mind that the strategies presented here are not effective in and of themselves. Teachers must be able to make decisions about when a small-group learning activity is appropriate, how an interesting application can be woven into a lesson, or when technology may enhance or hinder student learning. We present various ideas about how to structure activities, but many more possibilities exist. A new teacher can likely obtain valuable ideas and useful advice by discussing these and other possibilities with more experienced colleagues.

Cooperative Learning

Cooperative learning, or small-group learning, ideally provides students with a less threatening environment in which to work since they don't feel the pressure to perform in front of their peers. In addition, and possibly more importantly, when students work with others there is the possibility that students will share ideas, build on the ideas of others, justify their ideas to others, and hence create a deeper understanding of concepts being explored. This is, of course, the ideal outcome.

For cooperative learning to work effectively, teachers and students must work together to create a positive learning environment. Students must learn to work together as a team, to communicate with each other. Students must also be willing to take chances, to offer ideas that may not be fully tested but which may move the group forward. Several good books have been written that share ways to effectively make use of cooperative learning (for example, Davidson, 1990; Foster & Theesfeld, 1999). Some ideas presented in this section give the flavor of cooperative group work and what is necessary to ensure learning.

Problem to Investigate:

You may recall playing with a cardboard or plastic tube and using it as a telescope. Although the tube did not actually enlarge what you saw, it did help you focus on a narrower field of vision.

a. Find a tube and collect data to determine a relationship between the distance a viewer's eye is from an object (a vertical wall works well) and the viewer's field of vision on the object (how much of it can be seen?).

b. Use other tubes of various sizes and identify the variables that influence the relationship you determined in part a.

c. Use your solutions to parts a and b to generalize a solution. In other words, if you are given a random viewing tube, how would you calculate the field of vision associated with the tube?[7]

You may notice that the viewing tubes problem given here requires more than one person to solve the problem—one person to look through the tube and a second person to hold a ruler or other object to measure the field of vision. This problem, which is accessible to both algebra students and geometry students, is a great way to incorporate

[7] Thanks to Roger Day for suggesting this problem.

cooperative learning and mathematical modeling in the high school classroom. In planning to use a cooperative activity such as this one, some preparation is needed, not just in terms of getting the ideas for the activity organized, but in terms of organizing the students into groups, setting expectations for working in groups, and much more.

In preparing for cooperative group work, the teacher must make decisions about the size and composition of student groups. Will groups consist of 3, 4, or more students? Ideally groups of 3 or 4 are more productive than larger groups. With a large group, there is a greater need for negotiating roles and a greater possibility for conflicts to arise. The teacher must also decide how to arrange the groups. Will students choose their own groups or will the teacher use predetermined groups based on abilities and strengths of the students? Many advocates of cooperative groups suggest creating diverse groups so that students of low and high ability, different genders, and differing backgrounds are grouped together (Davidson, 1990; Foster & Theesfeld, 1999; Neyland, 1994). This diversity will most likely offer a variety of ideas and opinions that may be valuable for exploring a problem situation. In some cases, students of similar abilities may work well together since they are each as likely to contribute ideas and no one student will feel like he or she is doing all of the work (or that he or she has nothing valuable to contribute). Placing students in groups that work well together is not an easy task and may take several attempts. And a combination that works well for one class of students may not work well for another. It is important, however, to find the best combination for a given class of students so they may all benefit from the cooperative learning experience.

Other important planning elements include

1. Time and experiences to develop *social skills* that will help students learn to work cooperatively with other members of their group;
2. Discussion and negotiation of a *classroom environment* where cooperation among group members is expected;
3. *Clear and accurate directions* for students to follow so they are clear about their responsibilities as group members;
4. A *comfortable physical environment* in which students can work effectively, such as desks grouped in fours, available graphing calculators, and shelves of materials for a variety of mathematical investigations;
5. *Flexibility* to change plans as needed when activities take too long or do not proceed as otherwise planned; and of course
6. *Rich mathematical investigations* that lend themselves nicely to cooperative group work.

Often, the process of getting students settled into new groups or getting them started on a new activity takes more time than was planned, leaving less time for actually working on the activity. It may be more feasible to use cooperative learning if more than one class period is set aside for a given activity. Then, after the groups are assigned and tasks are clearly defined, students can return to class on subsequent days and pick up where they left off, with little time spent on organization and directions. If group work is a regular part of class, and students learn to assemble into groups quickly, cooperative group work can be used for shorter periods of time. Sometimes only a portion of the class period is used for group work, such as when students review homework assignments, or at the very end of class when the teacher wants to review concepts or pull together the ideas discussed during that day's lesson. Examples of long-term and short cooperative group activities are described in Examples 1 and 2.

Example 1 ***Long-Term Group Activities***

Long-term group activities require time to investigate relevant issues, apply appropriate mathematical processes and ideas, and devise a suitable solution. Students must work together to brainstorm idcas and cxplore options. The viewing tubes problem presented earlier in this section is one example of a cooperative activity that may take several days. The modeling problems listed below also set a good tone for cooperative groups.

- Have students collect data on the bounce height of several balls that have a common trait (for example, hollow inside, rubber outside). Have students determine the variables to consider in the problem. Create a model that helps students predict the bounce height of other balls with the same characteristics. Will other types of balls fit this model? Explore student answers.
- Explore the modeling problems found at the end of the following sections of this book: Sections 3.6, 5.7, 6.2, 7.6. ■

Example 2 ***Short-Term Group Activities***

Many short activities also foster group work and mathematics learning. The following two activities are borrowed from Foster and Theesfeld, 1999.

- "Trade-a-problem" is a great way for students to help each other review or practice concepts. In small groups all students create a problem related to the specified concepts that is to be solved by the other students (the student may record the solution on a separate piece of paper). When problems have been created, students trade their problem for one created by someone else (may trade within or between groups). Each student then solves the problem that has been received. When problems are solved, the student may compare solutions with the person who created the problem.
- "Think-Pair-Share" encourages all students to participate in a discussion and gives all students the opportunity to consider a solution before a single person is called on to provide an answer. In this structure, students are paired (or assigned to small groups) at the start of a discussion. When the teacher raises a question or issue to be discussed, the students are asked to first think about the question on their own for a short time. Then, students share their answers or ideas with their partner or team. The teacher may then ask students or groups to share their discussion with the rest of the class. ■

Problems, such as the modeling-based problems, are inherently different in terms of the cooperative group structure needed to complete such projects. These problem-solving activities are discussed in more detail below. The shorter cooperative group activities shared above are highly structured and serve a very specific purpose during class. In addition to reviewing concepts or discussing questions posed by the teacher, these types of activities require students to become active classroom participants. They also help develop the mathematics social environment of the classroom.

Beginning mathematics teachers often approach cooperative learning techniques tentatively, unsure how the students will react and what classroom management problems will arise. A well-structured lesson or project, however, is more likely to be successful and rewarding for the students. They need chances to share ideas with others, and to learn how to work with others in school and in the work place. Most importantly, cooperative learning can be beneficial to all students since group work can require that all members contribute to the best of their knowledge.

Problem Solving and Discovery Activities

One natural choice for cooperative group work in mathematics is a problem-solving task that requires creative thinking, and that is best approached by many minds working together. Mathematics, of course, *is* problem solving. When students are challenged by genuine and interesting problems, they are motivated to learn the mathematics necessary to solve the problem. The *Professional Standards for Teaching Mathematics* (NCTM, 1991) addresses the issue of providing or developing worthwhile tasks that engage students in the doing of significant mathematics. Table 12.4 highlights some of the important features for choosing or developing significant mathematical tasks as described in the *Professional Standards* and elsewhere (NCTM, 1991; Neyland, 1994). The modeling approach to the teaching of mathematics as described in this book, lends itself to problem solving.

Table 12.4 Summary of worthwhile tasks.[8]

A worthwhile mathematical task should include many of these characteristics:

- The problem must be accessible by all students.
- The problem must be challenging yet offer success, at some level, to all students.
- The problem must have natural extensions.
- The problem must involve students in exploration, conjecturing, and testing.
- The problem should promote mathematical communication among students.
- The problem should allow for original thought.
- The problem should be genuine and interesting.

[8] Adapted from D. Neyland, 1994, p. 108.

When using problem solving to enhance student learning of particular concepts, it is crucial that you choose or modify tasks so they are accessible to all students in the class, and so they clearly exemplify the goals to be reached. A problem should offer a challenge to students (probably more challenging for some). It should enhance student learning through the problem-solving process and also lead to new knowledge, or the reorganization of knowledge, when the problem has been solved. Students should not have to be told what they have learned through the problem-solving process.

A *problem-solving* approach and *discovery learning* are often used synonymously. Discovery learning, however, can carry different connotations. Using a true discovery learning approach, a teacher may pose a problem to students and then leave them to devise ways of solving the problem. Alternatively, discovery learning is used to describe a more *guided* approach that requires students to work in groups or independently to either solve a problem or develop conjectures based on their explorations. The process of reaching the intended goal may vary, depending on the level of guidance given to students. It is also true that the students' conclusions may depend on the process they used or a particular aspect of the problem that they chose to explore.

In general, discovery activities, such as the lessons presented at the end of Chapters 4 and 6, often help students learn problem-solving strategies. They may also help students construct an understanding of various new concepts along the way. In addition, discovery activities offer students a sense of accomplishment when they reach a novel conclusion and can justify their solution.

Student Involvement

This section on effective teaching strategies clearly promotes the active doing of mathematics. When students are doing mathematics rather than reading or listening to mathematics, they are more likely to internalize the concepts that they are using. Cooperative learning and a problem-solving focus are not the only ways to promote active student learning. Just about any classroom focus and any mathematical concept can be framed to include students as active learners versus passive learners.

There exist many effective strategies for involving students in the process of doing mathematics whether in small groups, individually, verbally, or in writing. For instance, when a student shares a solution method with the class that offers new insights, the teacher may ask all students to try the method in order to verify it and to help solidify the concepts or methods beyond what they might absorb by passively watching and listening. Even during more procedural parts of a lesson, such as learning a technique for multiplying matrices, students should be engaged in doing mathematics. They can practice an algorithm with paper and pencil or test conjectures on a graphing calculator. Similar to the "think-pair-share" technique described above, teachers can give students a couple of minutes to discuss a question with fellow students or a couple of minutes to verify a solution that was offered by another student. Thus, all students are responsible for being a part of the class, not only those who can perform calculations or recall facts more quickly than others.

Keeping students involved in a lesson is only one step in helping students to learn mathematics. Involved students are more motivated and more likely to ask questions or seek answers on their own. When students are expected to be doing mathematics, they quickly learn that mathematical knowledge is not gained by watching others perform algorithms. Learning mathematics means doing mathematics.

Other Resources for Teachers

Many resources exist that aid the mathematics teacher in motivating and involving students in their own learning. As mentioned, technology is one tool that allows students to test conjectures and visualize mathematics. Calculators with graphing capabilities give students the power to enter equations and modify these equations to discover important properties of functions. They often make problems accessible to a wide range of students. Programming possibilities help students create their own algorithms for performing calculations. With the use of spreadsheet programs and other computer software, students have the tools necessary to partake in problem-solving and modeling projects that can introduce new mathematical ideas. Computers foster critical thinking skills.

Multimedia options are also available for enhancing student involvement. Many terrific videos exist that present mathematical ideas for exploration by students. For instance, the Consortium for Mathematics and its Applications (COMAP) produces videos useful for teaching statistics. *For All Practical Purposes* is another set of well-produced mathematics videos. Teachers may also encourage students to explore media options for presenting projects to their classmates.

Manipulatives, physical objects, or drawings used to represent mathematical situations are valuable tools for getting students to work with mathematics. These objects can represent a particular mathematical phenomenon, such as sectors of circles representing fractional quantities of a whole, and allow students to reason mathematically with the help of a visual or tactile aid. Some manipulatives (commercially sold or hand made) that are appropriate for students at the middle school and high school levels are described below.

1. *Logical blocks* These blocks, also referred to as *attribute blocks*, are found in basic shapes with varying colors, sizes and thickness (for example, large and small circles painted in red, blue, and yellow, some twice as thick as others). Logical blocks can be used to help younger children learn to recognize differences and similarities. With older children and high school students, these blocks help students to understand set inclusion and exclusion as well as conjunction of sets with qualifiers such as *and* and *or*.

2. *Pattern blocks* Pattern blocks are typically found in various shapes and colors. Usually, a particular shape (a triangle) is painted all one color (green). These shapes are often fractional parts of other shapes (three triangles make a trapezoid) and hence can be used to explore fractions. Patterns blocks are also useful when discussing sequences or for exploring symmetry and tessellating shapes.

3. *Geometric peg boards* Peg boards with square, triangular and concentric circle arrays (shown in Figure 12.2) are used in many settings, but particularly geometry. With rubber bands, students can create a wide range of two-dimensional figures. Commercially sold geometry peg boards are available with supporting materials and activity books.

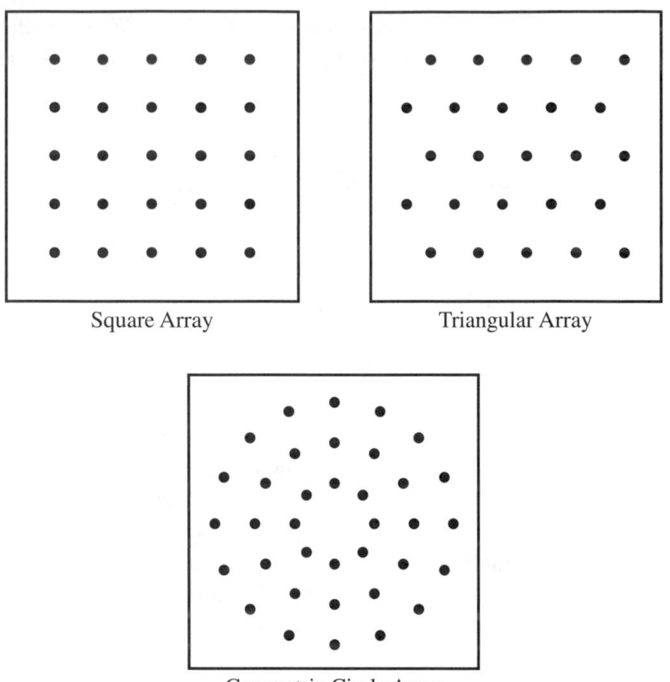

Figure 12.2 Geometry peg boards with various arrays.

4. *Mirrors* A mirror or plexiglass with a beveled edge is great for introducing students to the concepts of reflections and reflectional symmetry. Several commercial brands are available with accompanying activity guides.

5. *Fraction pieces* Whether in the bar or circle form, fraction pieces are useful at many levels of school mathematics. Fraction pieces can help fifth and sixth graders learn algorithms for adding and subtracting fractions. They can also be used to discuss sequences as well as arithmetic and geometric progressions with high school algebra students.

6. *Algebra tiles* As noted in Chapters 4 and 5, algebra tiles are designed to represent variable quantities and to help students learn basic algorithms for performing computations on expressions with variables. Materials that accompany commercially sold tiles provide an array of activities for students.

7. *Three-dimensional solids* Solids of wood or plastic give students a hands-on experience with three-dimensional figures. Some are made so they can be taken apart to demonstrate conic sections or other planar slices. Paper, scissors, and tape can also be used to create three-dimensional shapes. The process of constructing these shapes is often informative. For instance, instead of showing students a dodecagon, you might ask, How many pentagons can be glued together along edges to create a solid?

8. *Snap-together shapes* Several companies that market educational products sell plastic squares and triangles that snap together for creating three-dimensional ob-

jects. As with other manipulatives described above, many of these products are sold with accompanying activity ideas. Rather than purchasing commercially produced shapes, you can also ask students to use gum drops and toothpicks (or any similar combination of tools) to create three-dimensional objects and examine growth patterns for shapes made of cubes and tetrahedrons.

Other sources of ideas for involving students in their own learning can be found in news magazines and journals as well as supplemental materials produced by publishers of the textbooks used in the school. Professional organizations for mathematics teachers (or teachers in general) typically offer periodicals that contain articles of interest, particularly on ideas for the classroom. These are a great resource for ideas, especially if they are accompanied by comments from teachers who have used the activities in their classrooms.

Publishers that produce mathematics textbooks often also produce a variety of supplemental materials to accompany the books such as activity ideas or laboratory manuals. A teachers' set of these materials are sent along with textbooks purchased for the school. These materials may also be purchased separately, directly from the publishers.

No matter where the ideas come from, active learning is one key to a successful lesson. The various components for effective lessons presented in this section are meant to provide ideas for structuring lessons, as well as some of the reasoning behind the ideas.

Exercises 12.3

1. Choose a specific topic from a precalculus course, such as exponential functions, and develop a brief cooperative learning activity that can be done in less than one class period.

2. Using the same topic as chosen in Exercise 1, locate or create a cooperative learning or problem-solving activity to be completed over several class periods. Investigate the world wide web or other resource materials for ideas, being sure to cite all sources.

3. Choose one of the manipulatives discussed in this section. Become familiar with the manipulative, and develop a short activity using it. Be sure to discuss how the manipulative will be used and what mathematical concepts are being explored through the activity.

PROJECTS Find Out for Yourself

1. Visit a high school algebra classroom for 2–5 consecutive days. Describe the level of student involvement in the doing and learning of mathematics. In what ways are the students engaged with the mathematics? In what ways do they interact with the mathematics? What suggestions do you have for getting the students more actively involved in doing mathematics during the lesson?

2. Develop a guided discovery activity suitable for a sixth grade student (you will have to find out what is most appropriate). Interact with the student as he or she works through the activity. Interview a middle school student to determine what mathematical concepts were "discovered" through doing the activity. How did the results meet your expectations? What would you change about the activity before trying it again? Explain.

12.4 Focus on the Students

The main purpose of planning is to establish a lesson or sequence of lessons that promote student learning. One can easily get caught up in the planning process—making sure the mathematics is there and the little details are in place—and forget about the student, the intended beneficiary of all this work. Take a step back from the planning process and consider the students.

- What prior knowledge will the students bring with them that will contribute to the new lesson?
- How will you monitor student progress or how will you know that students are making sense of the new concepts?
- Will the lesson effectively motivate students?
- Will they understand how the ideas fit with prior knowledge?
- Will they see the importance of these ideas and connections to their own lives?
- How will you use the feedback you get from students as the lesson progresses?
- What feedback can you anticipate?

In general, the idea is to keep the students in mind, both during the planning process and during the lesson itself. The preceding questions represent the types of questions to consider before the lesson so that you are better prepared to pay attention to the students during the lesson. As you get to know your students, other questions, more specific to your class or to a particular student, should also be asked during the planning process.

Assessing Prior Knowledge

A sense of the students' prior knowledge is helpful in planning a unit or daily lesson. If a geometry teacher knows that all the students in the class are familiar with a particular dynamic geometry software package, the teacher can plan an activity that utilizes this package without having to plan a couple of days of instruction on using the software. If a middle school algebra teacher knows that most of the students scored poorly on the review of proportional reasoning, she may schedule a follow-up lesson and activities to approach the ideas of proportional reasoning from a different perspective.

How does a teacher find out about the students' prior knowledge? This is typically accomplished in a variety of ways. At the start of a school year or semester when students are beginning a new course or changing teachers, the teacher can assess all student knowledge with a pre-test of the basics needed to succeed in the course. A questionnaire that requires very little mathematical calculations can also provide useful information. Figure 12.3 is an example of a questionnaire that can be given to students at the start of a second-year algebra course. When a teacher begins a new unit, questioning techniques or a carefully chosen problem can be effective ways of gathering information about

Student Information Sheet

Please respond to the questions below as best you can. Any information provided will be kept confidential.

1. Name: ____________________

2. Birthday: ____________________ (to be used for a probability unit in October)

3. For each mathematical software package below, indicate how well you know how to use the software by circling: 1–Have never used it before, 2–Tried it once but don't remember or I'm not sure, 3–I've used it a few times and would like to know more about how to use it in mathematics class, 4–I'm a pro, I could give lessons.

Texas Instruments Scientific Calculators	1	2	3	4
Texas Instruments Graphing Calculators	1	2	3	4
Spreadsheet programs like Excel or Lotus	1	2	3	4
Graphing symbol manipulation programs like Mathematica or MATLAB	1	2	3	4
Dynamic geometry programs like Cabri or Geometer's Sketchpad	1	2	3	4

What mathematics class or classes did you take last year (semester)?

____

What mathematics class or classes do you plan to take next year (semester)?

____

What do you hope to do with the mathematics you learn? (Example: I plan to go to college and major in Chemistry. I know studying mathematics now will help me in my chemistry classes in college.)

____

____

Figure 12.3 Student questionnaire.

students' knowledge base. As an example, consider the unit on symmetry described in Section 12.2. At the start of the unit, the teacher may ask students what they know about symmetry or some of the related concepts. General questions and some brainstorming can reveal a good starting point for the unit. Alternatively, the teacher may have students complete a task (assuming they have already completed a unit on basic transformations) such as this one:

> Find the "center line" of a regular hexagon and then reflect the hexagon over the "center line." What is the image of the reflection? Are there other "center lines" that will produce the same result?

These questions and the task elicit different information, but both methods are important for deciding if the students are ready for the unit as planned or if they can start at a more advanced level (or less advanced level) than was planned. Certainly, you cannot expect all students to be at the same place in their understanding. But when the majority of the students in a class are unprepared for an upcoming unit, you must take responsibility for preparing the students or possibly restructuring the unit activities to allow for a slower beginning. One way to do this is through *scaffolding* questions or tasks. *Scaffolding* questions or tasks are those that allow a student to build an essential framework with some guidance. The tasks or the teacher (sometimes other students) offer just enough information for the students to make sense of the ideas and move ahead to the point where they need to be in order to participate in the planned activities. Although scaffolding does not always guarantee that the background material is understood, it does allow a weaker student to proceed with the rest of the class.

Motivating Student Learning

A mathematical task that reviews necessary background knowledge while connecting to the new ideas to be taught can often motivate the new topic. Motivating student learning does not always require you to be a mathematics cheerleader. You don't have to create games for students to play or show movies to catch their attention. Motivating the lesson is much more about the study of mathematics than about entertaining a classroom of teenagers. Students need to know what the central concepts of the lesson will be. They need to know why this mathematics is considered important or valuable or beautiful, as the case may be. Students need to know how it connects to other ideas, other areas of study, or other facets of their daily lives. Finally, students need to be assured that they can succeed in doing what will be asked of them.

Many modern textbooks do a good job of answering most of the concerns posed here related to the importance of the mathematics and its connections to other disciplines or the physical world. But not all of these concerns are addressed by textbook authors, and even when some motivators are included in a text, many students don't take the opportunity to read the books beyond looking at examples that are helpful for completing the assigned homework.

Teachers should not rely on textbooks to motivate the mathematics for students. Neither should teachers take full responsibility for motivation. In fact, students and the classroom community as a whole play big roles in motivating mathematics learning. Using the ideas presented in textbooks or other resources, and using your knowledge of the students, you have the tools to help students see the importance of the mathematics and to help them recognize how the new concepts fit with those previously discussed. Through the choice of tasks, through discussion, and with a good sense of what is important to students, you can develop a lesson that will keep them engaged.

Monitoring Student Progress

A good lesson plan also contains a means of monitoring student progress. You must be able to ascertain whether students are making sense of new concepts and specifically where students may be having difficulties. This assessment can take the form of questions to individuals or directed to the whole class, a brief problem or two worked out on paper or with a partner, or a visual monitoring as you make your way around the room during a small-group work session. However you gather this kind of information, a plan for information gathering should be included in the lesson plan. With experience, this monitoring will become more of a natural part of what happens in your classroom, and recording these plans in a daily lesson will seem unnecessary.

Through daily monitoring or other assessments, teachers receive feedback about the students and their needs. This feedback should inform subsequent lessons, or even subsequent parts of a class period. Using feedback to reflect on and evaluate a lesson is discussed in the next section. More is said about focusing on the students in Chapter 13, which takes a closer look at special populations of mathematics students and the lesson planning considerations associated with these populations.

Exercises 12.4

1. You are teaching a unit on logarithms in an algebra class. Develop a list of important skills and concepts that you want students to master in order to be successful in learning the new concepts in this unit.

2. Create a pretest that will help you determine if students are ready for the unit on logarithms.

3. Create a short activity (rather than a pretest) that students can work on as they enter class that would help you determine their readiness for a unit on logarithms.

4. Research one or two things you might say or show to students that would help them understand the importance of learning logarithms.

5. Name three different ways that you can monitor student progress during a lesson on logarithms. Be specific in describing what you would do and the nature of the information you would obtain.

6. Suppose, after determining your students' progress, you learn that about 20% of them have not grasped the idea of what a logarithm is, while the other 80% have a good sense of the meaning of a logarithm. What can you do during the next class period to help the 20% who need to review the concepts while not repeating the lesson for the other 80%?

PROJECTS Find Out for Yourself

1. Interview an Algebra 1 student or a student at a similar level to find out if he or she has the background knowledge necessary for learning about quadratic functions.
2. Based on the information you gathered in the interview, how might you begin teaching this student about quadratic functions?

12.5 Reflecting on and Evaluating the Plan

Evaluation is a crucial part of the planning process, even though it takes place after the lesson plan has been executed. Since teaching ideas are recorded in lesson plans (electronically or on paper), the teacher is able to look back on the plans, consider what happened in class, and use these reflections to modify future lessons, whether it is for the following minute, day, week, or year.

Reflecting on your teaching is a good habit to establish early on. Reflection on a regular basis helps keep the flow of lessons consistent and the quality at a high level. This doesn't mean that you have to keep a daily journal, although that may be a way for beginners to get used to the idea of reflecting on their practice. It is often more time efficient to jot notes on lesson plans shortly after the lesson has been taught, noting what went well, where adjustments were made, and what adjustments might need to be made in the future.

The expression "hind-sight is 20/20" can be quite true in the teaching profession. After disappointing results from what might have looked like a well-developed lesson, you can learn to recognize what caused the difficulties and what might have contributed to the students' misunderstandings. Perhaps the concepts were presented too quickly with not enough time for students to explore the ideas and construct an understanding. Maybe many students were off task so that they didn't gain much from the lesson. Or possibly, something about the presentation of the lesson interfered with student learning, such as a lengthy application problem that led them to focus on peripheral ideas. It's best to record these insights as well, so when the lesson is taught a second time, revisions can be made.

At times, hindsight can appear clouded by the lesson that seemed so wonderful on paper. You might wonder what went wrong. Why didn't the students grasp the new ideas of a particular lesson? Sharing your reflections with a colleague or mentor is one way to help evaluate the lesson and tease out the problem areas. A more experienced teacher may quickly notice something about the lesson that you did not, such as the nature of the task students worked on (was it too difficult or could it be completed with skills other than those you anticipated would be required?). An observer in the classroom is even more valuable to obtaining feedback on your lessons. At times, students may be asked to reflect on the lesson. A lesson often looks quite different from the point of view of the student. Their opinions can help you evaluate the lesson from a different perspective.

Although it is not common for beginning teachers to be able to reflect on what is happening while it is happening, this is certainly something to strive for. When you are able to lead a lesson while simultaneously processing the effectiveness of the lesson, you can be more flexible and make decisions about changing parts of the lesson as class progresses. With experience, you will learn to do this kind of reflecting. When this is achieved, you have also learned to focus on the student rather than just on the process of teaching a lesson.

12.6 Creating an Effective Lesson Plan

Ideas for creating, organizing, and enhancing lesson plans were presented in this chapter. Here we highlight some of the important elements of a lesson. As previously noted, there is much more to a lesson than the basic outline of a plan, such as those found in this chapter. The mathematics teacher must take into consideration the many factors that influence student learning in the particular situation of a specific classroom. Varying student populations from class to class demand shifts in thinking and shifts in the way lessons are prepared. An effective lesson must take the students into account.

Elements of a Lesson Plan

1. **Sequence of Topics, Concepts or Flow of Events** This refers to the unit plan or the bigger picture. A mathematics lesson (or a set of lessons) should help students develop a coherent view of mathematical concepts. Mathematical topics should not be taught as isolated pieces of information. A mathematics teacher must have the bigger picture in mind before preparing a daily lesson.

2. **Daily Goals or Objectives** The daily objectives may describe the main concepts or topics of the lesson as well as the skills or processes to be mastered by the students. Goals should reflect a balance of conceptual and procedural aspects of the mathematics. The goals should also address the question, Why is this important for the students to learn?

3. **Warm-Up or Opening Activity** An opening activity helps get the students involved in doing mathematics as soon as they enter the classroom. A warm-up activity is also a great way to introduce a new topic or provide chances for students to brush up on skills that will be tested on state-mandated or other standardized tests.

4. **Activities for Achieving the Objectives** This section of a lesson plan outlines what the teacher will do, as well as what the students will do to accomplish the stated objectives. The activities should follow a logical sequence and should involve students in active learning. Activities should also include connections to prior lessons or prior knowledge, to future lessons, and to the daily lives of the students.

5. **Questions and Other Means of Assessing Student Learning Throughout the Lesson** For a beginning teacher, questions that are part of the activities for achieving the lesson objectives should be written. These may include questions to test students' prior knowledge, questions that stimulate brainstorming among students, or questions that challenge students to investigate specific mathematical ideas. In addition, questions for assessing whether students are grasping key concepts help give the teacher feedback on the lesson as it progresses. This helps the teacher make decisions about whether and how to modify a lesson to meet the needs of all students.

6. **Summary or Closing Activity** Wrapping up a lesson can be as important as the opening. At times it is not important to say or do too much, such as when students will return the next day to continue an activity. At other times, it is valuable to summarize the concepts discussed or give students the opportunity to talk about what they perceive as the lesson outcomes. The summary is not necessarily the same as the closing activity. Often, teachers choose problems for students to work on during the final few minutes of class in order to assess their potential success on homework or other assignments.

7. **Materials Needed or Other Special Notes** This should contain a list of any special materials needed for conducting the lesson. The list might include manipulatives, overhead projector slides, and pens for students to record solutions to be presented to class, or possibly a classroom set of calculators. Notes might include lesson modifications for students with special needs.

8. **Approximate Time Allotments** For the beginning teacher, it is a good practice to estimate time allotments in order to get a better sense of whether the stated objectives can be met.

9. **Possible Extension Problems or Something to Start if Extra Time Is Available**.

References

COMAP. (1998). *Mathematics: Modeling our world.* Cincinnati: South-Western Educational Publishing.

Davidson, N. (ed.) (1990). *Cooperative learning in mathematics*. Menlo Park, CA: Addison-Wesley.

Dreyfus, T., & Hadas, N. (1996). Proof as answer to the question why. *International reviews on mathematical education*, **28**(1):1–5.

Foster, A. G., & Theesfeld, C. A. (1999). *Cooperative learning in the mathematics classroom.* New York: Glencoe McGraw-Hill.

National Council of Teachers of Mathematics (2000). *Principles and standards for school mathematics*. Reston, VA: Author.

——— (1991). *Professional standards for teaching mathematics*. Reston, VA: Author.

Neyland, J. (1994). *Mathematics education: A handbook for teachers*, Volume 1. Wellington, New Zealand: Wellington College of Education.

13

Providing for Individual Differences

One of the most widely accepted ideas within the mathematics education community is the idea that students should understand mathematics.... But achieving this goal has been like searching for the Holy Grail. There is a persistent belief in the merits of the goal, but designing school learning environments that successfully promote understanding has been difficult.

James Hiebert and Thomas P. Carpenter (1992)

Overview

The students in our classrooms are individuals. Although it is certainly easier to teach a class if all students are exactly the same in ability level, it would also make for a boring class. Students' individual differences make teaching them interesting, and it makes teaching more of a challenge. We cannot assume students are similar in the ways they make sense of mathematics, in the ways they do mathematics, or in the ways they communicate about mathematics. Teachers must recognize the differences among students and find ways to work with these differences in order to capitalize on student strengths and move forward with building mathematical knowledge.

Although most lesson planning can be done on a classroom level, teachers must also keep the individual students in mind (refer to Section 12.4, Focus on the Students). This chapter takes a closer look at the individuals that make up the mathematics classroom, including their various learning styles, their cultural backgrounds, their varying senses of what counts as mathematically important, and their interests and abilities. At times the mathematics teacher must also take into account the social structures in the classroom. Providing for the individuals in the classroom doesn't mean that every student should receive individualized instruction. Rather, it implies that teachers should find ways to provide worthwhile mathematical opportunities so that *all* students will benefit.

Focus on the NCTM *Standards*

The original NCTM *Standards* document and subsequent documents emphasize the Council's commitment to providing quality mathematics instruction to *all* students (NCTM, 1989, 1991, 2000). This belief that all students can learn to reason mathematically challenges the more conservative view that some students are not able to succeed in mathematics classes. The belief also challenges all teachers of mathematics to address inequities that exist in the classroom and that persist in society. Certainly, all students are different, and some will achieve higher mathematical goals than others. However, in order to address the needs of all math students, especially those who have traditionally been ignored, teachers must find ways to enhance the teaching and learning process so that all students can partake in the doing and learning of mathematics. The NCTM purports that when teachers expect all students to be successful in mathematics, students will begin to rise to the challenge. Thus, one way to move toward equitable classrooms is to raise expectations of what is possible, and help students discover their strengths and ways to use their strengths to achieve the goals set for them or goals they set for themselves.

Focus on the Classroom

At the middle school and high school levels, students begin to recognize their own individuality and often want others to recognize it, too. Students with learning difficulties, on the other hand, may wish to keep these differences to themselves. Even so, it is important for the math teacher to identify those characteristics that make the students different. Differences will range from the student who seems to understand the mathematics yet performs poorly on tests, to the student who has difficulties with the language of mathematics, to the student who works best when she can discuss the mathematics with her peers. This chapter highlights the range possible in any given mathematics classroom, and provides insights for helping all students achieve success in mathematics learning.

13.1 Differences in How Students Approach Learning

For many years, educational psychologists studied how students differ in their views and representations of concepts, and in their preferred ways of working with and organizing new ideas. From this research has come two bodies of literature—the first on multiple intelligences and the second on learning styles (Gardner, 1993; Butler, 1987).

Multiple Intelligences

Theory of multiple intelligences, first described by Gardner (1983), supports the idea that all individuals possess numerous types of skills or abilities that allow them to acquire knowledge in certain ways for particular purposes. Specific intelligences, but possibly not all, may be more pronounced in some individuals than in others. For instance, an airline pilot has the ability to use and interpret many controls while flying and monitoring a radar screen. Mathematicians are able to use symbols and other notations to describe and make sense of abstract objects while artists are able to visualize and mentally manipulate objects before they create concrete models or pieces of art.

The theory of multiple intelligences views all individuals as intelligent beings because we all possess multiple intelligences. However, theorists recognize that student abilities to perform in one or another specific area differ. More specifically, Gardner (1993) specifies seven intelligences discovered through his research that enable the learner to solve problems and construct understandings. These intelligences include the following:

- *Linguistic intelligence* describes a person's ability to use and understand language, whether native or foreign, such as poets and writers.
- *Logical-mathematical intelligence* describes a person's logical, mathematical, and scientific abilities, such as mathematicians and scientists in many fields.
- *Musical intelligence* refers to the ability to create and understand the creation of music, such as composers and musicians.
- *Spatial intelligence* refers to the ability to develop and manipulate mental models of spatial objects within our world (objects that take up space whether in one, two, or three dimensions), an ability that artists and engineers have.
- *Bodily-kinesthetic intelligence* describes a person's ability to use the body in problem solving or in the development of new products and ideas, like surgeons or dancers.

- *Intrapersonal intelligence* is a form of personal intelligence that allows one to form an accurate portrayal of oneself and use this model effectively.
- *Interpersonal intelligence* is a form of personal intelligence that allows one to understand and portray others and work effectively with others.

Gardner notes that all types of students (not just math students) rely heavily on linguistic and logical-mathematical intelligence, whether they have a propensity toward these types of intelligences or not. However, the other intelligences also play roles in acquiring domain-specific knowledge, including mathematics. In other words, logical-mathematical intelligence is not the only one that facilitates the learning of mathematics. Aspects of other intelligences, such as linguistic and spatial intelligences, also contribute to the acquisition of mathematical knowledge. Gardner claims that, with this view of intelligence, school should be a place to help students develop all of their intelligences, and to reach goals appropriate for their particular "spectrum of intelligences" (1993, p. 9).

Although it may be impossible to achieve an ideal school setting where each individual is helped to nurture their intellectual strengths and develop a full array of their intelligence capacity, theory of multiple intelligences can be useful in making teachers aware of student abilities. In response to multiple intelligence theory, teachers should understand the connections between what they choose to do in the classroom and how this develops student strengths in terms of their intelligences.

Learning Styles

Although multiple intelligence theory is useful for suggesting activities that pertain to various intelligences, the theory does not contribute insights directly to the teaching and learning process. The theory of learning styles, on the other hand, describes an individual's preferred way of using her or his abilities or intelligences. Research on learning styles often centers on determining a person's tendencies, which include the ways in which they see and interpret information as well as the tools they use when problem solving in a given subject area (Butler, 1987, 1988). These tools may be either physical or mental. Research on learning styles also reveals that it is equally important to determine a teacher's learning style, since the teacher's mode of problem solving will most likely be modeled for the students. This model of problem solving has varying effects on students with differing learning styles. For instance, Presmeg[1] comments on a study in which students labeled as visual learners did not perform significantly better when taught by a teacher who emphasized using visual aides in problem solving, versus a teacher who emphasized a more abstract approach. The reason noted is that visual learners, already adept at using visual tools for problem solving, need to see the teacher model other ways of thinking about a situation so that they can develop a more robust understanding of the concepts.

Various theories on learning styles use differing terms to describe learners' aptitudes or ways of seeing and working with ideas. Most theories describe three to four

---

[1] Personal communication with Norma Presmeg, January 2000.

parallel types of learners. Learners are typically described as either visual, tactile, or abstract, although other learning styles, such as auditory, and kinesthetic or a mix of the three already mentioned, are also found in the literature (Butler, 1987; Dunn & Griggs, 1988; Keefe, 1987).

- *Visual learners:* The term *visual learner* most typically describes a student who prefers to use or see drawings or written words to make sense of concepts. The visual learner uses drawings of models to aid in mentally manipulating objects. Visual learners also rely on abstract drawings to represent physical phenomena. Models drawn on paper provide justification for the mathematics to be shown.
- *Tactile learners:* Students who are *tactile learners* learn best with physical models in their hands. These students want to hold a concrete object in order to investigate it. Often, models can be used to represent situations, such as using a small plastic cube to represent a fluid ounce of water. In more abstract situations, such as algebraic expressions with no connections to a specific context, a tactile learner will try to build a model to represent the abstract concepts (the small plastic cube might represent a variable). These models allow the learner to manipulate, explore, and make discoveries that might otherwise seem elusive to them.
- *Abstract learners:* Students who are described as *abstract learners* prefer to think and work in more general terms. They want to see the bigger mathematical picture by making connections between abstract concepts. These students have no difficulties manipulating symbols. In fact, abstract learners often choose to work with symbols and use symbolic notation to represent otherwise physical phenomena.

It is not always obvious which learning style works best for students. And students don't always know either; it is not something we recognize about ourselves without extensive reflection. But what is clear is that students are more successful in learning mathematics when their learning styles are accommodated in the classroom. In other words, when teachers prescribe a certain method of working that interferes with a student's learning style, the student is less likely to develop understanding. When students are allowed to choose a method or problem-solving strategy that makes sense to them, they are more likely to complete the task and understand the underlying concepts. As noted above, the teacher is also a factor in the student's choice of problem-solving method. If a teacher consistently uses the same or similar strategies for demonstrating problem solving, students frequently try to imitate these methods, even when they don't particularly suit the student, in relation to her or his learning style. Thus, teachers will reach a larger range of learners when they teach from multiple perspectives and encourage students to use various representations and solution strategies.

Taken together, theories of multiple intelligences and learning styles inform the mathematics teacher with regard to (1) the nature of activities that will enhance the learning of all students, (2) the importance of addressing individual students, (3) the value of demonstrating various problem-solving strategies, and (4) the importance of encouraging students to develop problem-solving strategies that fit their learning style.

When working with individual students, teachers have the luxury of accommodating the student's learning style. In a classroom, however, individualized instruction isn't practical. But developing tasks that accommodate a variety of learning styles is. Modeling activities are one way to accomplish this. Modeling projects typically require students to work in groups to investigate real-world situations. The investigative work is particularly appealing to visual learners who thrive on reading and seeing. Creating visual graphics to display data is another way to represent the mathematics for a visual learner. Developing models of the situation satisfies the tactile learner who needs to touch and build physical representations of the mathematical ideas. The abstract thinker is well suited to developing the mathematical model of the situation. This individual quickly notices ways to represent models with symbols and mathematical expressions. Since most students are adept at processing information in a variety of ways (through hearing, seeing, touching, and manipulating abstract symbols) it is likely that each small group of students will together have the skills and learning styles needed to construct the model.

Exercises 13.1

1. In what ways do you learn best? Are you a visual or tactile learner? Do you prefer written symbols to words or drawings? Write a short essay to describe your learning style, providing examples to support your claims.
2. Locate and take a Myers-Briggs test of aptitude and discuss how accurately it reflects your responses to Exercise 1.
3. If you believe that the majority of students in your Algebra I class are tactile learners, what is the best way for you to introduce these students to solving equations of one variable? Explain how the tactile learners will benefit from this approach, and also how the nontactile learners will benefit.

13.2 Differences in Student Ability

In any mathematics classroom there is always a range of student abilities. Even when students are grouped by ability, such as in an Honors Geometry class or an eighth grade Algebra I class, there is still great variation among students in terms of achievement and ability. Although it is not always obvious, studies have shown that students in an average ninth grade classroom may perform anywhere from third grade level through beginning college (Kenney & Silver, 1997; Larcombe, 1985). When this is the case, how does the teacher address the needs and abilities of all students? In an ideal setting, teachers would devote equal time to each student, tailoring instruction to each individual's specific needs. This type of individualized instruction is possible on a certain level, but, in general, this scenario is not realistic in the average high school classroom. There are simply too many students and too little time. The teacher, however, still must strive to be aware of the differences in students' abilities, and must work with others in the school or community to find ways of providing for all students.

Addressing Interest and Motivation

At times, teachers mistake student interest for student ability. Just because a student doesn't actively participate in class or scores low on assessments doesn't mean they don't have the capacity to learn mathematics. It may simply mean they need motivation, or need to do the homework, or need to pay attention in class. Alternatively, a student who *is* having difficulties understanding concepts may lose interest in mathematics or become frustrated and give up. Learning difficulties are often mistaken for lack of interest or motivation. Keeping all students interested or motivated to learn is a crucial goal for teachers. Larcombe (1985) describes an affective filter through which students perceive mathematics and mathematics learning. This filter, Larcombe contends, determines the student's motivation level for participating in class (see Figure 13.1).

Just as students vary in their ability to learn mathematics, so do they vary in their interest in studying mathematics. This, too, must be recognized in order to address the learning of all students. It is clear that student interest in mathematics is higher when they feel successful, when mathematics is related to topics of interest to them, and when connections are made that illuminate the usefulness or beauty of mathematics. Yet even when a teacher presents topics in ways that are open to all students for investigation or attempts to make connections to areas of interest to the students, not all students will respond.

How does a teacher choose activities and present concepts that capture the interest of a wide range of students? First, teachers must take an interest in the mathematics themselves, they must demonstrate their enjoyment of mathematics, and they must take an interest in the students. It is unusual to find a math teacher who doesn't enjoy investigating mathematical problems and thinking through mathematical puzzles. And teachers typically choose the teaching profession because they want to work with and help others to grow. But this enthusiasm for doing and teaching mathematics doesn't always manifest itself in a classroom full of students who "have to" be there. Neverthe-

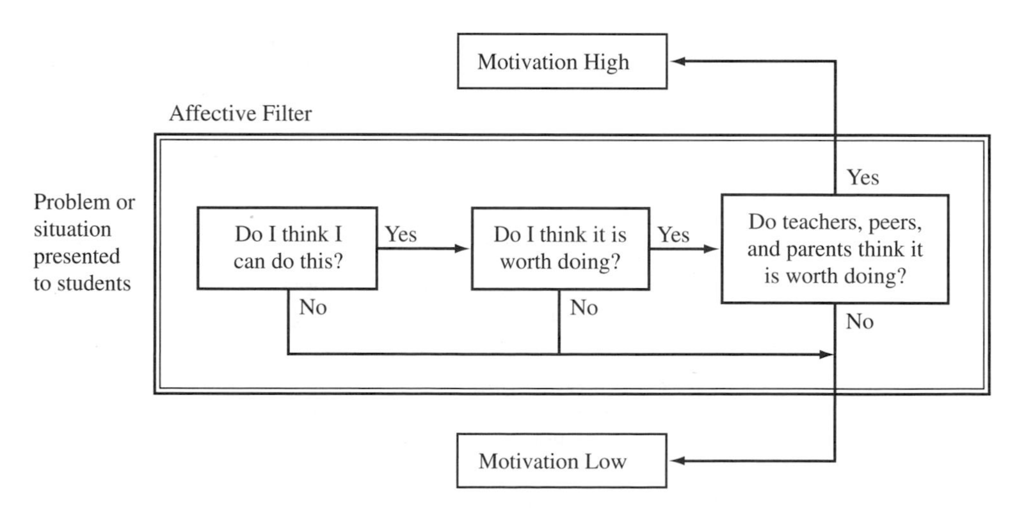

Figure 13.1 The affective filter helps determine motivation level.

less, if you consciously show your enthusiasm for doing mathematics and your interest in the lives of the students, the students will sit up and take notice. David Johnson suggests many wonderful ideas for motivating students in mathematics class in his set of three short books: *Making Every Minute Count, Making Minutes Count Even More*, and *Motivation Counts* (Johnson, 1982, 1986, 1994).

Second, the teacher should investigate and become familiar with topics that demonstrate connections and the usefulness of mathematics. The recent NSF-funded textbook series developed in response to the NCTM *Standards* documents provides a wealth of ideas for motivating the teaching of mathematics by way of practical applications and projects (see Section 2.1 for summaries of the materials). The following two examples have rich possibilities for involving students in using mathematics to determine fairness (Example 1) and to find relationships between variables in data sets (Example 2). Example 1 is adapted from the series entitled *Mathematics: Modeling Our World*, developed under the direction of the Consortium for Mathematics and its Applications (COMAP, 1999). Example 2 comes from *Contemporary Mathematics in Context*, developed by a team of mathematics educators and mathematicians at Western Michigan University (Coxford et al., 1997).

Example 1 *What Is Fair?*

A school district must apportion 90 computers between three secondary schools in the district. School A has 217 students, School B 288 students, and School C 395 students. What is a fair way to allot the computers among the three schools?

1. If 30 computers are given to each school, how many students will there be per computer at each school? For which school is this method the least fair? Explain.
2. You are asked by your school principal to determine fairness according to equal numbers of students per computer at each school.
   a. How should the computers be allotted? Is it possible to make a totally fair allotment? Remember, you can't give a fraction of a computer to a school.
   b. Which school gets the best deal? Which gets the worst deal?
3. If the district superintendent decides to take one computer from the school with the best deal and give it to the school with the worst deal, what happens to the distribution? Would it be more fair? Why?
4. How can you assign a number to determine the "amount of unfairness" in the allotment situations from items 2 and 3 above? ■

Example 2 *Global Warming*

Greenhouse gases such as water vapor, carbon dioxide, and methane keep the earth warm and livable. Recently, there has been concern that the greenhouse effect is being intensified artificially, causing global warming. Scientists want to know if there is a connection between increased amounts of carbon dioxide in the atmosphere and increasing average temperatures over the whole planet. Using the following data, determine what, if any, connection exists.

Year	Carbon dioxide concentration (in parts per million by volume)	Temperature deviation (in °C compared to mean average temperature from 1950–1980)
1960	317.0	0.05
1965	320.4	−0.05
1970	325.5	0.00
1975	331.0	−0.05
1980	338.0	0.15
1985	345.7	0.18
1990	353.8	0.21

- Produce three scatterplots, one each for (year, carbon dioxide concentration), (year, temperature deviation), and (carbon dioxide concentration, temperature deviation). Does there appear to be an association between any pair of variables?
- Do any of the scatterplots reveal a linear association between variables? Why or why not?
- What statistical methods would be useful for helping determine relationships among the variables? Explain.
- Write a brief report that you could submit to a group of concerned citizens. In the report, summarize your analysis of the relationship between carbon dioxide levels and changes in the earth's climate. Include discussion of the overall trend during the period 1960–1990, trends within that period, use of least-squares regression for prediction, and possibilities of a cause-and-effect relationship.

Yet, even with your best efforts to spark interest, some students will remain unaffected. A third bit of advice for motivating student interest: Continue to show enthusiasm and continue to seek ways of connecting mathematics for both your own fulfillment and for those students who will benefit from your efforts.

Addressing Students' Special Needs

When motivation and interest are not the issues underlying a student's poor grades or inability to participate, the teacher must assess and find ways to address the student's needs. Assessment of special needs can be accomplished in a variety of ways. The teacher may administer a pretest at the start of the school year to determine what concepts and skills students lack and need extra help on or time to review. At times, specific learning needs cannot be identified by a pretest, and other methods must be used. If a learning need requires a special education instructor in the room or an aide to help record notes, the classroom teacher will be informed by the Special Education Coordinator in the school. Sometimes the accommodations for students with special needs, such as giving extra time to complete a test or project, can easily be addressed by the classroom teacher. If a student doesn't require special services during class time, even though they are identified as a special needs student, the classroom teacher may never find out what accommodations might enhance the student's performance. It is always best to keep a watchful eye on students who are struggling and talk with them about ways that help them succeed.

Gifted students are another group of students that may require extra attention in order to promote learning and interest in mathematics. Gifted students, like other special needs students, don't always receive the attention they require to succeed and flourish in mathematics. When not appropriately challenged in class, a gifted student may become bored and lazy and lose interest. When students who are particularly gifted in mathematics are identified by a parent, school, or other teacher, the classroom teacher can direct the student to more challenging activities and assignments that hold their interest and increase their knowledge. More suggestions for working with gifted students are discussed below.

The reality is that these groups of special needs students, along with all other students, are together with you for a very short time each day (typically 45–60 minutes a day). Ideally, they are all given the same amount of time with you where they receive the same type of instruction, and all have about the same amount of time to complete their work (more or less). Again, the reality is that even when you are able to spend a few extra minutes with a student having difficulties, this doesn't usually extend to each student every day. Actually, this situation benefits certain students. They learn to operate well or at least adequately in this situation, yet others do not. Addressing the individual needs and interests of your students is a challenging task that should be a part of your daily mode of operating. Individualizing instruction doesn't mean that you work with one student at a time. Individualizing instruction has to do with how you approach the task of teaching to *all* your students. What follows are some details about the various types of students who may be part of or who may make up your entire class. Suggestions for working with these students within and outside the regular classroom are also presented.

Gifted Students

Mathematically gifted students are identified as students who display curiosity and creativity when assessing a problem situation, and possess a high level of task commitment. They also may be able to generalize and transfer mathematical ideas more readily than other students, and they typically (although not always) perform quite well on standardized mathematics achievement tests (House, 1987). Gifted students are usually identified as such before they reach high school, and may participate in gifted programs within the school or district. However, not all gifted students are identified before they enter high school, and even those who are may be placed in regular or "honors track" classes at the high school. How should teachers accommodate students of very high ability?

If a school or district has an established gifted program, students of very high mathematics achievement may be placed in specialized classes as part of that program.

The program may require the student to travel to a central location in the district, or the student may simply participate in a class (sometimes very individualized) within the school building. At the middle school level, the Johns Hopkins program for gifted students provides a mathematically challenging and creative curriculum. Many states and districts provide funding for teachers to attend professional development courses for this program. High school math teachers may also take college classes or attend professional development courses to learn more about working with gifted students. These students will benefit from a program that is academically challenging, that introduces them to worthwhile mathematical tasks, and that nurtures higher-level thinking skills. Gifted programs should encourage and support the development of various talents, particularly creativity and critical thinking.

Another way to address the needs of gifted students is to enroll them in more advanced classes at the school (maybe at the high school for middle school students) or at a local college or university. For instance, a seventh grader with a strong mathematics background and who shows promise for being academically advanced in this area may attend the local high school for Algebra I or Geometry. A gifted high school sophomore may enroll in Pre-calculus or even Advanced Placement Calculus, if offered at the high school, or possibly attend a nearby university for a calculus course. Before placing students with others who are two or more grade levels their senior, it is important for a teacher, counselor or parent to ascertain the student's emotional maturity. Keep in mind that the student will be asked to participate in a different social environment than she or he is accustomed to. The teacher of the higher-level course should also be aware of the younger student's emotional and social maturity levels to appropriately address problems, should they occur.

When gifted students are placed in a regular or possibly an honors track at the middle school or high school, the teacher must be aware of the students' abilities and should encourage them to move beyond the regular content of the course. This can be done several ways. For instance, the teacher can develop or locate various types of enrichment activities for the students to work on once they have completed other assignments or instead of other assignments. This is a good option if the teacher chooses to keep gifted students at the same pace as the rest of the class. Enrichment activities should allow these students to explore concepts more deeply than would otherwise occur. In a modeling context, this might mean introducing new variables into the situation, or simply working to find a more sophisticated model. Another option is to allow gifted students to pursue a mathematical tangent or to explore the uses of mathematics in various fields. For instance, when studying right triangle trigonometry, they may opt to investigate the use of trigonometry in land surveying and in ship navigation. Addressing their needs in this way may mean that teacher and students work together to develop a series of specialized research projects that can be accomplished over time alongside regular class assignments. Yet a third option for serving the needs of gifted students is to allow them to work at a faster pace than the rest of the class, progressing to new topics when ready. In this way, they remain challenged and interested. On the other hand, students who move ahead of the class must work independently and won't have the benefit of working with others in the social environment of the class. Although many gifted students thrive in a self-paced, self-learning environment, they also need to learn to communicate and work with others.

Whether in a special program or classroom, with one-on-one individualized instruction, or in a regular math class, mathematically gifted students have the right to receive an education that gives them the opportunity to realize their abilities. First, these students and their needs must be identified. Then, after consultation with the student and the student's parents, a program of study can be found to fit the student's particular needs. Encouragement and an appropriately challenging environment can spark the student's interest in mathematics for many years to come.

Students with Learning Difficulties

Student mathematical difficulties typically begin as early as third grade and continue through the high school years. As the student progresses through school, the mathematical learning difficulties leave them behind their classmates in terms of content knowledge. Research in this area indicates that for every two years of school, students with mathematical learning disabilities gain one or less than one grade level in mathematics (Rivera, 1998). Thus, by the end of high school, the student with mathematical learning difficulties may have only reached a standard sixth or seventh grade level in mathematics learning.

As noted previously, students' mathematical learning difficulties can result from lack of interest or motivation, fear of poor grades, other affective related problems, or any of a multitude of *learning disabilities*. Conversely, poor grades or bad experiences in math class can result in affective issues (for example, poor motivation, math anxiety, passivity) that continue to limit the student's progress in building mathematical understandings. This is true for students who are identified as learning disabled as well as many other low-achieving students. To compound the problems, students with special needs are not always identified to the classroom teacher. Teachers, however, may be able to identify students based on performance, classroom behavior, attitude, and willingness to accept individualized help. Identifying the cause of the difficulty is not easy, and a professional in the area of learning disabilities should be consulted. Students with severe learning disabilities, such as sight or hearing loss or an inability to record notes, are easily identified, and the school or district often provides for their needs through Special Education programs or full-time assistants in the classroom.

Specific Learning Disabilities Students with mathematics learning disabilities often have difficulties with computation and problem solving, although not always in both areas. These students often require special attention in the form of specialized programs, extra help outside of class, or special accommodations in the regular classroom. Miller and Mercer (1998) identify four major factors that can play a role in students' learning:

- Information-processing factors
- Cognitive and meta-cognitive issues
- Language difficulties
- Affective factors

Information-processing factors have to do with how information is acquired and the form of the information. Various learning disabilities pertain to information processing, such as *attention deficits*, which may mean the student has a difficult time paying attention to instructions, following the steps in an algorithm, or following through with the process of problem solving. Other disabilities related to information-processing include visual-spatial deficits, auditory-processing difficulties, memory problems, and motor disabilities. *Visual-spatial deficits* pertain to factors such as losing one's place on a worksheet, difficulty using and understanding the number line, and difficulties differentiating among various symbols. Students with *auditory-processing difficulties* frequently have trouble following directions that are given orally or responding orally rather than in writing to questions involving computation. *Memory problems* are most often linked to difficulties with recall of facts or new information. This might inhibit a student's performance on multistep tasks or in basic review lessons. Students with *motor disabilities* may write slowly, inaccurately, or illegibly, causing difficulties when they try to process the information at a later time.

Cognitive and meta-cognitive issues have to do with the students' ability to identify and choose appropriate strategies for performing mathematical tasks, to organize information, to monitor or assess the problem-solving process, and to evaluate accuracy. Typically, students use many of these cognitive strategies, but "the strategies they use may not be sufficient" (Miller & Mercer, 1998, p. 88). Hence, students with mathematics disabilities related to cognitive issues need to focus on learning to use strategies appropriately.

Many researchers in the area of mathematics difficulties cite language as a key factor for students with learning disabilities (Larcombe, 1985; Miller & Mercer, 1998; Thornton & Bley, 1994). Mathematics is a precise language. It is our expectation that students interpret this language and imitate the preciseness that can result in learning difficulties. A range of language disabilities exists in students, whether they are identified as mathematics learning disabilities or not. For instance, a student may have a reading disability, and this is the crux of their learning difficulty in math. With the NCTM recommendations and other recent reform efforts in mathematics, there has been an emphasis on real-world applications as well as genuine problem-solving situations. Associated tasks for math students are often accompanied by large amounts of written text that the student must process before they can begin to understand even the nature of the task, let alone start solving it. Word problems have long been the nemesis of students with reading disabilities. More recent textbooks rely even more heavily on the written word for presenting problem situations, and this creates more of a barrier for those students.

Language disabilities in mathematics can be caused by "noise" or distractors. Mathematical concepts, even lower-order concepts, have many properties. When students work with these concepts, they must determine which properties are crucial, which are somewhat relevant to the given situation, and which are irrelevant. Students with language disabilities can have trouble determining which properties are irrelevant. This "noise" compounds when several complex concepts are involved. Further, when students pay attention to the irrelevant distractors, they become frustrated when their efforts don't pay off, or they develop harmful misunderstandings of the concepts themselves (Larcombe, 1985).

Another recent recommendation for enhancing mathematics learning for all students is to increase the emphasis on communication of mathematics through written, graphical, and verbal means. This provides yet another hurdle for students with language disabilities, particularly those whose difficulties are in organizing, recording, and transmitting ideas. What accommodations do these students require?

Affective issues in mathematics learning disabilities include low self-esteem, mathematics anxiety, or a dependence on others for learning, and doing mathematics (learning passivity). We discussed these issues above in terms of low-achieving students who don't have specific mathematics difficulties. Students with special learning disabilities are more prone to repeated academic failure, and hence more likely to confront affective issues in the learning of mathematics. Low self-esteem is frequently the result of years of low grades, failure, or lack of recognition for efforts made. Students with disabilities also learn to wait for others (such as a teacher or special education aide) before they attempt to solve a problem or complete an assigned task. This passivity is a learned reaction because these students often have to start over on assignments ("so I might as well wait for somebody to show me how to do it"). Math anxiety is one outcome of low self-esteem and fear of failure. Students become tense and their self-confidence plummets so that they are unable to think clearly. This causes problems in processing new information as well as in using learned information for problem solving. Such students tend to avoid mathematics whenever possible.

Educational Factors Affecting Learning Difficulties Beyond identifying or labeling specific learning disabilities, many factors in the educational system exacerbate learning difficulties. These factors include recent recommendations for curricula and instructional reform, general student diversity, school and district requirements for graduation, and school or district policies regarding inclusion of learning disabled students in regular classrooms. Figure 13.2 demonstrates how learning disabilities affect student success in mathematics. Students with language disabilities or information-processing

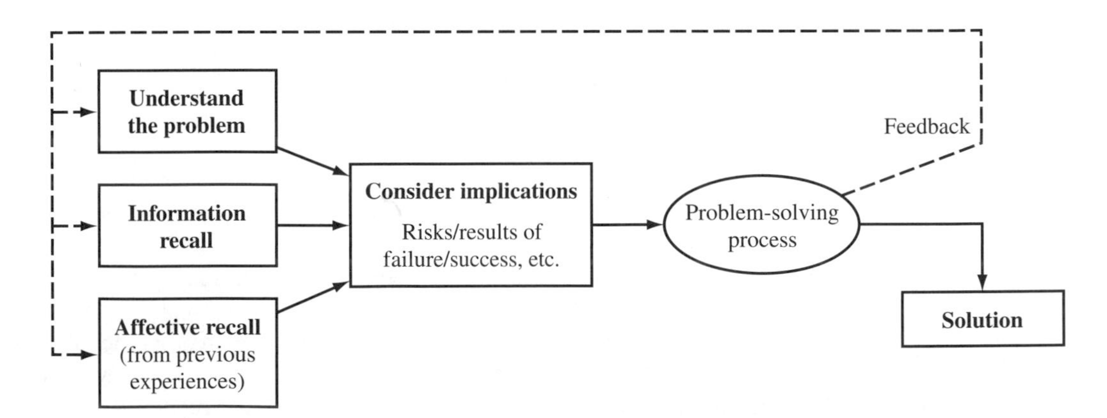

Figure 13.2 The role of various learning strategies in problem solving.[2]

[2] Adapted from Larcombe, 1985, p. 47.

difficulties are easily frustrated as they endeavor to understand what a problem is asking. Students with cognitive or processing difficulties struggle as they try to recall the information they need to solve the problem, while those with affective issues rely on past experiences to judge their ability to proceed with the task, and thus sometimes give up too easily. That is, they focus on their possible failure to solve the problem; this, of course, affects the problem-solving process negatively, thus perpetuating the learning difficulties. This is not to say the problem-solving approach is wrong for students with learning disabilities.

In some respects, the problem-solving approach is beneficial to students with learning disabilities. A more practical, task-oriented curriculum is often helpful to students who need to see and hold objects that represent more abstract concepts in order to make sense of ideas. Figure 13.2, however, does highlight areas that need to be considered in developing such a curriculum for the learning disabled student of mathematics. Curriculum developers and teachers of special needs students must consider whether problems are structured in such a way that all students can engage in the task and the associated mathematics. Teachers must also decide what outcomes are reasonable to expect from students with learning disabilities.

Graduation requirements and the inclusion of special needs students in regular classrooms are two other education factors that affect the mathematics learning of students. Both of these factors point to increased expectations for students with learning disabilities. As states increase the number of mathematics courses or the level of mathematics required for graduation, all students will find it more difficult to reach the required goals for receiving a standard high school diploma. Students with special needs in mathematics will find it particularly difficult to reach these goals and will continue to struggle. What accommodations might be feasible or reasonable for this population of students? This important question must be openly discussed among parents, school personnel, and law makers.

Inclusion or mainstreaming of special needs students, an issue that has gained attention in recent years, has grown out of increased expectations for these students. Advocates of inclusion claim that participation in regular classrooms gives students with learning and physical disabilities a chance to develop socially and to gain confidence in their abilities to participate in wider society. Another hoped-for result was that the attitudes of students without disabilities toward students with disabilities would be positively affected. Research, however, indicates that students with learning and physical disabilities are frequently rejected by nondisabled peers in the classroom (Rivera, 1998). It was also found that special needs students in "inclusion classrooms" showed little to no improvement in mathematics achievement as compared to students who continued to receive individualized instruction through a special education program (Fuchs & Fuchs, 1988). In addition, many "mainstream" teachers who were asked to teach in "inclusion classrooms" were given little to no support for working with special needs students. They tended to judge all students against the same standards (Miller & Mercer, 1998).

Addressing Student Needs Students with a range of learning disabilities related to mathematics, at-risk students, those with physical disabilities, and low achievers all deserve to participate in a mathematics classroom that will help them realize their abilities.

For this to occur, teachers need to be sensitive to the diversity of learners in their classroom. Teachers must work to develop instructional strategies that lead all students to success. It is unrealistic to believe that one curriculum will work for all students. Similarly, it is unrealistic to assume that all students can progress at the same pace and acquire the same mathematical knowledge. Teachers must make accommodations for students with special needs.

Accommodating the needs of a variety of students in one classroom is often impossible. However, the classroom teacher, along with the help of a special education teacher, can modify various aspects of instruction for students whose disabilities leave them behind the rest of the class. This can mean extra time allotted for taking a test or completing a project or creating a special test in very large print for a student with limited sight or other visual-processing difficulties. Often technology enhances the student's functioning in the classroom. If a student's disabilities prevent the usual acquisition of knowledge, the student may be placed in a special education classroom where a teacher can work one-on-one with the student during the entire school day. A regular classroom teacher may be consulted to help the special education teacher develop an instructional plan for the student that fits her or his needs and abilities and that also fits the mathematics curriculum appropriate for the student's attained grade level. No matter where the mathematics is taught, students with special needs will require some sort of specialized instruction to address their specific disabilities.

Larcombe (1985) provides these suggestions for accommodating students with a range of learning difficulties in mathematics:

1. Devise a plan for checking in with students on a regular basis with regard to their work in class and academic progress.
2. Allow for a variety of approaches to solve problems and to go about achieving the goals of the class.
3. Provide opportunities for students to apply their knowledge and skills in practical situations.
4. Ensure that the pace of the class allows slower students to be successful, yet challenges all students to work hard.
5. Provide regular feedback.
6. Help all students to monitor their own progress and help them to find ways to overcome difficulties.

As stated previously, all students have different abilities in learning mathematics. The ideal classroom in which all students learn concepts at the same quick rate cannot be found. Thus, teachers must realize the diversity within a given classroom and work to address the needs of their students. Although it is not always necessary to individualize instruction for all students in a class, addressing learning differences usually means individualization on some level. Using a variety of activities over the course of a unit helps assure that student learning differences (both low and high achievers) are taken into account. Modifications to some activities to accommodate students with special needs (physical, mental, or emotional) helps these students be productive in the mathematics classroom. Such considerations can increase planning time, but can also increase the potential effectiveness of instruction.

Exercises 13.2

1. Write a short essay on why you have chosen to become a mathematics teacher. What is it about mathematics that interests you? What is it about teaching that interests you?
2. How might you use your response to Exercise 1 to bring out student interest in mathematics? You might describe an activity you would like to use in the classroom, or other things you would say or do to draw out student interest.
3. Name at least two ways you can help a student with memory difficulties to be successful in a typical mathematics class.
4. How might you modify a typical or traditional classroom structure to better accommodate students with memory difficulties or other information-processing issues?

PROJECT Find Out for Yourself

Interview a special education teacher at a local high school to discover the following:

1. What is the nature of student mathematics disabilities at the school?
2. What range of accommodations is afforded these students?
3. Is there an inclusion program at the school for mathematics? If so, what support is provided for teachers of these classrooms?

13.3 Cultural Differences and the Role of Culture in Learning

Diversity in the mathematics classroom is not just about differences in student abilities. Cultural diversity, although not an issue in all classrooms, is certainly an issue that requires attention. In the past, many Americans saw mathematics as a filter, a discipline that "weeded out the weak" and produced a strong core of students who would pursue mathematics- and science-related careers.

This seems to contradict another American value, that schooling, and particularly reading and mathematics, is valuable. But calculations or basic computations were often the endpoint for many, particularly those from ethnic minority groups or of lower socioeconomic status. Today, the reforms to mathematics education advocated by the NCTM and others emphasize mathematics as a discipline that should and can be pursued by all, particularly those who have been underrepresented in the past (Cuevas &

Driscoll, 1993; NCTM, 2000; Secada, Fennema, & Adajian, 1995). But why should teachers pay special attention to certain groups of students? Shouldn't all students be treated equally?

Addressing Cultural Backgrounds

The answer to the second question is yes, but in order to do so, teachers must be aware of students' backgrounds and cultural histories. Our own culture underlies how we teach mathematics, and these unacknowledged connections can exclude some groups of students. One example of unnoticed bias in math classes (also true in other subject areas as well) has to do with Native American and Latino populations. Researchers have found that conventions for communicating within these cultures are quite different from the predominant Anglo culture of our schools (Cuevas & Driscoll, 1993; Secada, 1992). Unspoken codes for communicating in these cultures often mean that these students exclude themselves from the discussion and hence are at a disadvantage in the learning process. Similar communication and more general social differences among other cultures may also exclude students in unintended ways. Aspects of various cultures that differ from the dominant American middle-class culture include sense of need and inevitability of competition (for example, Asian societies), eye contact (Native American and Latino cultures), mode of questioning and conventions for working and sharing ideas (African-American culture). When working with students from various cultural backgrounds, teachers need to be sensitive to these differences. The classroom environment established by the teacher and students should accommodate diversity.

Much research indicates that the mathematics performance of students from minority ethnic groups and those of lower socioeconomic status is lower than the performance of their white European-American counterparts (Secada, 1992). Figure 13.3 shows mathematics scores from students participating in the High School and Beyond study.[3] Clearly, both factors (ethnicity and socioeconomic status) appear to be related to student performance. Based on results such as this, students of minority ethnic groups are often grouped with other low-achieving students in the nation's high schools. Unfortunately, "tracking" students based on their ability to perform on standardized mathematics tests does not address issues of cultural diversity and merely propagates social inequities in our schools. If students of African-American and Latino backgrounds typically perform poorly on standardized tests, then we should find other ways of assessing their mathematical potential.

Students from diverse cultures and socioeconomic backgrounds respond in different ways to the instructional strategies used in the mathematics classroom. Often, the traditional lecture is not appropriate for the student whose upbringing emphasizes movement and the open sharing of ideas, such as in many African-American households. An alternative strategy might be to collaborate in small groups so these students will be more engaged in the mathematics. Additional strategies include introducing multicultural mathematics; this helps all students understand and appreciate the different ways in which people do mathematics. Many multicultural resources provide ideas to enhance

[3] The High School and Beyond Study is a longitudinal study of student achievement across racial and ethnic groups conducted in the 1980s.

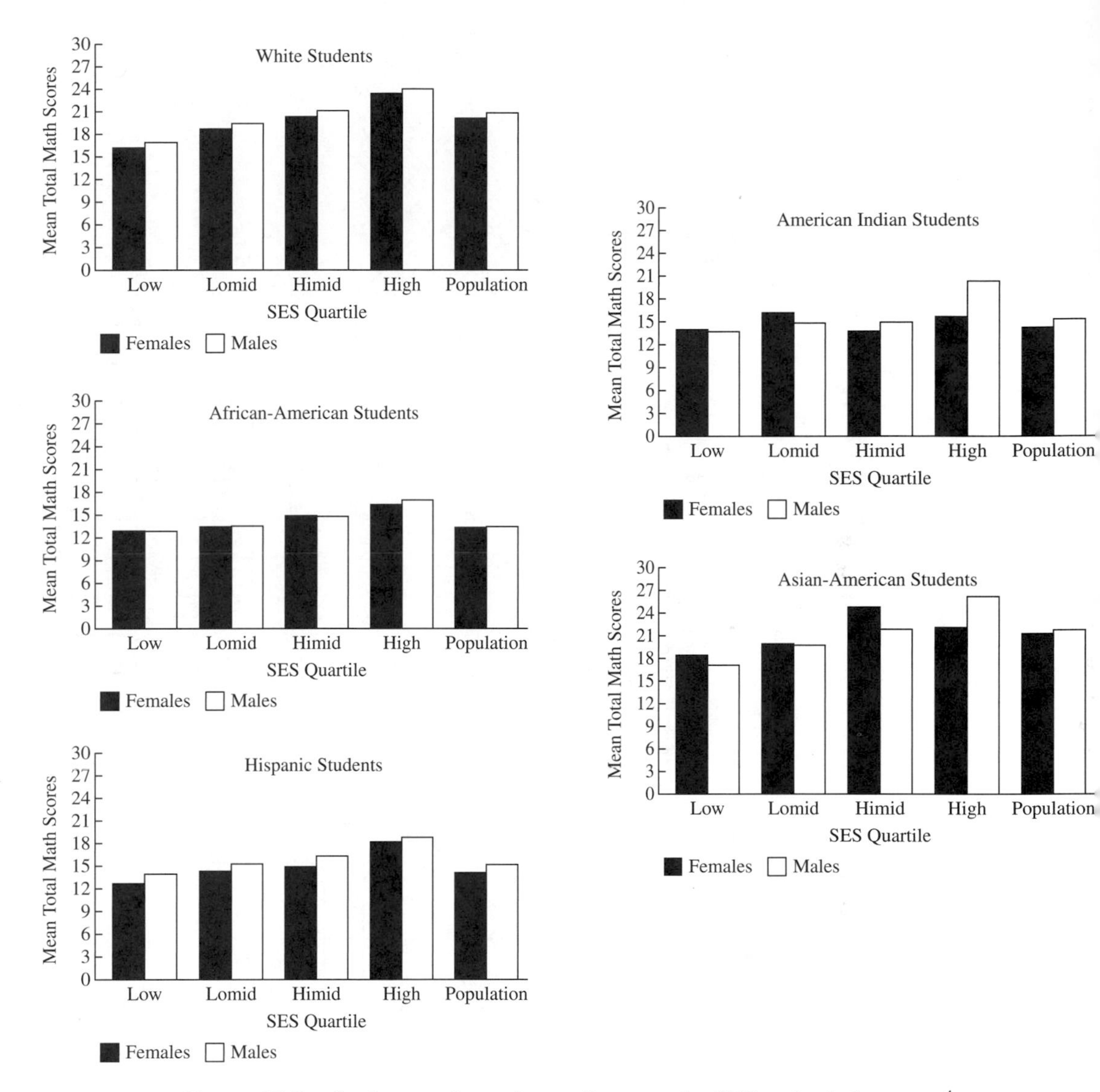

Figure 13.3 Student mathematics performance by SES and ethnic group.[4]

classroom learning for diverse students. One example is the NCTM yearbook from 1997 entitled *Multicultural and Gender Equity in the Mathematics Classroom: The Gift of Diversity.* In this volume, a variety of authors review the research in this area, discuss the major issues, and provide practical suggestions for teachers at all levels of schooling. Chapters address Native American cultures as well as African-American and Latino cultures. Students often find explorations of the mathematics of other cultures such as

[4] Secada, 1992, pp. 636–637.

Figure 13.4 Sand drawings from Angola and Zaire that represent algorithms for finding greatest common divisor of two numbers.[5]

shown in Figure 13.4 intriguing. Valuable resources for the classroom teacher are listed at the end of this section.

Second Language Learners

Spanish and Native American languages have existed in areas of this country longer than the English language. In addition, immigrants from European and Asian countries

[5] Zaslavsky, 1993, pp. 51–52.

enter our society on a daily basis. Even so, we insist that all students in the U.S. education system learn to speak English proficiently. Since most school subjects, including mathematics, are taught in English, students who are proficient English speakers have a better chance of succeeding in mathematics. But many students enter school (at all grade levels) with limited English language skills. Some administrators and teachers falsely believe that these students will perform well in mathematics because the language of mathematics (in terms of numbers and symbols) is universal. This may be true, but mathematics is much more than numbers and symbols. Difficulties arise for minority-language students when they encounter the specialized terms we use to describe mathematical concepts. The following excerpt from a poem by José Franco (1999, pp. 21–22) illustrates this point.

When Mrs. Jones was talking about **addition**,
she used the word **plus**
like 2 **plus** 2 equals 4.
Sounds good to me.
But last week she mentioned the work **combine**,
and she said that meant **addition**, too.

All right. . .

On Monday we were doing some math problems.
(Oh man! I had a hard time reading
TOO MANY WORDS!
I didn't understand them all,
but my buddy Julio helped me out.
Now I owe him one,
but that's another story.)
Anyway. . .
We were doing our math problems.
I read the problem
and it said the herd of elephants was **increased by** three.
Julio and his cousin Julia told me that **increased by**
means **addition** also.
"What Mrs. Jones?
Could you please repeat your question?
What's the sum of all the elephants?
Hmmmmm
What did Julia tell me **sum** meant?
Is that the same as **some**,
like "when **some** of the kids tease me"?
"Sorry, Mrs. Jones.
I don't know what the **sum** is."
(Actually I don't know what that word means.)

Remember, even native English speakers have difficulties learning the language of mathematics.

In addition to the special mathematical terms we use, common terms we use every day have very different meaning in mathematics. Thus, minority-language students must learn English in two different contexts—outside mathematics and inside mathematics.

Typically, these students pick up everyday language quickly and can use this knowledge to apply to mathematical situations. It is still another step in the learning of mathematics (and English) to apply this knowledge to understanding and using formal mathematical language (Flores, 1997; Ron, 1999). One recommendation for helping such students is to allow them to build mathematical language from everyday language. Help them make sense of mathematical situations using informal language, and lead them to use more precise language to talk about mathematical concepts. Ron (1999) also suggests professional development for teachers in bilingual classrooms. In particular, she recommends that teachers "have access to (a) a knowledge base about the linguistic characteristics of the language(s) of the classroom that may facilitate or hinder mathematics understanding; (b) techniques for using the "mathematized" language, which is the building block that links the everyday language to the language of mathematics; and (c) knowledge about cultural practices and issues that may affect the understanding of mathematics" (p. 32).

Other suggestions for working with minority-language students in math classes focus on not only developing their proficiency with English and its relationship to the language of mathematics, but also on developing student confidence in their abilities to understand and communicate in the community of the mathematics classroom. For instance, allowing students to work in small groups provides them with opportunities to both listen to and speak English without the fear of making mistakes in front of the entire class. In small-group settings, students are also more likely to help each other. Others in the group frequently act as interpreters or re-explain the concepts discussed in class.

Using manipulatives or other descriptive objects is another way to help students learn the vocabulary necessary to communicate mathematically. Often when students can see and touch objects, they remember the objects as well as the names or relationships associated with them more easily. For instance, use pattern blocks to discuss relationships such as height, area, vertex, greater than, and congruence. Tasks that lend themselves to multiple points of entry are also helpful. A second-language learner makes sense of some portion of a problem situation, and their work on this portion leads them to solve other aspects of the problem. In general, an important message for teachers in bilingual classrooms is not to focus on second-language learners' mathematical deficiencies. Rather focus on changing the way mathematics is taught in order to help all students achieve their mathematical potential (Flores, 1997).

Gender Issues

Unlike issues of ethnicity, language, and socioeconomic status, the gender gap in mathematics has been well documented over the past several decades (Fennema & Leder, 1990; Peterson & Fennema, 1985). As a result, attention to gender inequity in mathematics has made educators aware of differences in participation and performance between females and males. A closer look at the classroom teacher has revealed the teacher's role in perpetuating these differences (Leder, 1995). For instance, studies found that, overall, teachers tend to pay more attention to male students than to female students. They also ask male students to perform more rigorous tasks and provide more wait time for them to respond to questions. In contrast, female students are typically asked lower-level questions and given less time to respond to a question before moving to another student.

Research results on male-female differences in terms of cognitive abilities has been mixed, but the research on more affective issues clearly shows differences (Leder, 1995; Peterson & Fennema, 1985). Females are less confident, overall, than males in their ability to do mathematics, and females are less likely to persist on difficult tasks. In addition, females are less convinced of the value of learning mathematics for future career choices, and less likely to choose careers in mathematics and science. Many of these beliefs and expectations are most likely based on lagging societal views of the role of women. Although the attention brought to bear on these issues has changed perspectives and attitudes, these attitudes continue to exist and help perpetuate students' beliefs about their mathematical abilities.

Some of the issues raised above, including differential treatment and self-confidence, can be addressed through restructuring the classroom environment. Similar to the suggestions for language-minority students, collaborative small groups allow females the opportunity to express opinions and build confidence in a more supportive environment than the whole-class setting. This also provides all students time to work through more complex or challenging problems, often with the help of others. In addition, carefully crafted tasks and a modeling approach to the learning of mathematics demonstrates the usefulness of mathematics in a wide range of fields. Other suggestions for changing instructional practices discussed in earlier sections are also appropriate for addressing gender equity.

Awareness on the part of mathematics educators, encouragement from teachers, and various special programs for girls in mathematics are addressing many of the gender equity issues. Society is also more attuned to providing equal opportunities to all constituents, which has opened many more doors to females in recent years. The trend toward gender equity in the mathematics classroom (and in mathematics in general) is heartening, but this doesn't mean we can be complacent.

Resources

The following books and book chapters offer much more than we can explore in a few pages. Various perspectives are examined and practical suggestions and pedagogical strategies to meet the needs of all students are given.

- Cuevas, G., & Driscoll, M. (eds.) (1993). *Reaching all students with mathematics*. Reston, VA: NCTM.

The main theme of this book is that all students are capable of learning significant mathematics. Authors share their stories on how they bring mathematics to all students. Diverse student populations are addressed, with a focus on multicultural mathematics, gender equity, and more general heterogeneity of the typical urban classroom.

- Fennema, E., & Leder, G. (eds.) (1990). *Mathematics and gender*. New York: Teachers College Press.

This explores perceived gender differences in the doing and learning of mathematics. These differences include spatial ability, motivation, and discourse. The chapter authors also discuss inequities in the classroom based on these perceived differences, and suggest ways of creating more equitable mathematics learning environments.

- Malloy, C., & Brader-Araje, L. (eds.) (1998). *Challenges in the mathematics education of African American children.* Reston, VA: NCTM.

This book contains the proceedings of a national leadership conference, including papers on the teaching and learning of African-American students, policy issues related to the mathematics education of all students, and professional development efforts for teachers of African-American students.

- Secada, W. G. (1992). Race, ethnicity, social class, language, and achievement in mathematics. *In* D. A. Grouws (ed.), *Handbook of research on mathematics teaching and learning*, (pp. 623–660). New York: Macmillan.

The chapter noted focuses on the research on many areas of equity in the mathematics classroom and considers differences in achievement among various groups. Issues related to language proficiency are also explored. Additionally, this review of the research reports on special programs and their effectiveness in raising achievement levels of typically low-achieving ethnic and social groups.

- Secada, W., Fennema, E., & Adajian, L. B. (eds.) (1995). *New directions for equity in mathematics education.* New York: Cambridge University Press.

The numerous chapter authors take a look at equity in mathematics education from various standpoints. Some report on projects designed to achieve equity while others describe recent trends in our nation's classrooms. Many authors provide practical suggestions for the classroom (elementary through high school).

- Secada, W. (ed.) (1999). *Changing the faces of mathematics.* Reston, VA: NCTM.

This series of anthologies provides multiple perspectives on issues related to the mathematics learning of students from various subgroups of our population. The first two books in the series focus on Latinos and Asian Americans and Pacific Islanders.

- Trentacosta, J. , & Kenney, M. J. (eds.) (1997). *Multicultural and gender equity in the mathematics classroom: The gift of diversity, 1997 Yearbook.* Reston, VA: NCTM.

The first several chapters address overarching issues related to equity in the classroom. Later chapters focus on specific cultural and social groups as well as suggestions for instruction and assessment. A few chapters examine changing perspectives about diversity at various levels including parents and the community.

Exercises 13.3

1. Locate and study results from the National Assessment of Educational Progress (NAEP) and determine trends in the data for various cultural groups. What trends do you notice? How might these trends be accounted for?
2. Choose one of the resources listed above and use it to write a short plan for incorporating multiculturalism in the classroom.
3. Think back to your years in high school. What cultural differences existed among the students in your mathematics classes? How were these cultural differences taken

into account by the school or by your teachers? What suggestions might you have for addressing possible cultural differences?

PROJECTS Find Out for Yourself

1. Develop a questionnaire to be taken by high school students that will help you to understand whether students in today's classrooms feel there are gender inequities in the study of mathematics.

2. Administer the questionnaire and report on the results. In what ways do the results support recent research in this area? In what ways do the results differ from research results? Explain.

13.4 Accommodating Differences

Teachers' Beliefs in Student Ability

The students in our classrooms are different. Although this does not mean that teachers must cater to the needs of each individual student, awareness of these differences is important for developing an instructional plan that meets the needs of all students. The NCTM *Principles and Standards* advocates maintaining high expectations for all students; this helps them achieve goals they once thought impossible. When teachers hold high expectations for all students, they are communicating the belief that all students are capable of learning mathematics. Of course, not all students are capable of learning mathematics to the same academic level. However, failure to hold students to high (but reasonable) standards limits the learning that does take place and hence limits the student's future opportunities.

The math teacher must embody the belief that students can succeed. When you truly believe this and let your expectations be known, students gain confidence. Nonetheless, carrying such a belief is not enough. You must find ways to help this belief come to fruition. As presented in earlier sections of this chapter, recommendations for restructuring traditional formats to accommodate a wide variety of learners include

- Using collaborative small groups.
- Developing tasks that allow for multiple access points into the problem situation.
- Allowing students help to determine how they will demonstrate their understanding (graphically, in written form, verbally, etc.).
- Using manipulatives and other concrete objects to get a feel for mathematical concepts.

- Providing rich learning opportunities that tie mathematics to the lives of the students (through cultural explorations, real-world modeling, and other practical applications).

A strong belief in students' abilities carries over to the students' belief in themselves. Note that getting all students through math class is not the focus here because the practices that facilitate this may be doing a disservice to students' independence and their ownership of their ability "to do it." For example, providing answers to students having difficulty rather than asking scaffolding questions helps them complete the assignment but does not help them learn to accomplish the task on their own. When students learn to expect this kind of enabling help, they often stop thinking for themselves. They may get through mathematics courses, but they won't be successful in learning mathematics. There is a fine difference between accommodating students' needs and enabling students. It is especially crucial to recognize this difference when working with students who are not motivated to learn or those who have always struggled to keep up with the class. It's easy to slip into the mode of giving answers so that all students will "be at the same place" in the plan. The classroom strategies listed above should provide ways for all students to find success on their own.

Students' Self-Confidence

Not only do teachers need to believe that their students are capable of learning significant mathematics, but students also need to hold these beliefs themselves. If they have been unsuccessful in the past with similar activities or skills, they are likely to believe that they won't be successful again. This is a difficult belief to overcome. One way teachers help students bypass these feelings of sure failure is to make the setting new and interesting. For instance, the teacher may present an especially interesting task or group project. In the process of working through a project that has piqued their interest, students will hopefully acquire an understanding of mathematical concepts they thought were unattainable. The interesting task is not the only component of class that will help students develop confidence and motivation to learn. A supportive teacher and learning environment are also key. The teacher and students must respect the diversity of learners in the classroom, and must work together to help each other make sense of the ideas being discussed.

Yes, it sometimes takes more than a safe classroom environment, self-confidence, and a teacher who believes that students can master mathematical concepts and skills. At times, some students will need more accommodations than others. Knowing when and how to accommodate lower-achieving students is not easy. When a student is struggling with an assignment, consider if a different or easier assignment might be warranted. Ask yourself, will the student come to expect easier assignments and, hence, lower her or his work level in general? As you come to know your students, you will come to recognize when they truly need accommodations and when the less-motivated students are simply asking for less work. Accommodating struggling students doesn't always mean giving lesser assignments. You and the student may agree on a different way of doing the assignment, such as changing a written assignment to a graphical or pictorial presentation of the ideas along with short written explanations.

When you have doubts about a student's needs, use the services offered at your school. Perhaps a department chair or more experienced teacher can give advice or ideas for working with a student who needs special services or attention. Perhaps the student should be referred to a special education teacher for this extra attention. Is there a peer tutoring program at your school that would be beneficial? You don't have to make these decisions solo. Most schools encourage conversations, and provide guidelines for dealing with special problems that arise in the classroom.

Easing Math Anxiety

Math anxiety affects many students, and teachers need strategies for dealing with it. For some students, anxiety is *facilitating*. In other words, anxiety about grades or doing well motivates them to work harder and perform better. For most students, however, anxiety is *debilitating*. It causes them to perform poorly, to "choke" on a test, or to stop trying. For these students, math anxiety is often caused by previous failures and fear of poor grades. These feelings build up and manifest themselves negatively as anxiety. For some, this anxiety can be overwhelming—some students intentionally skip classes or don't hand in homework assignments. When mathematics is avoided, the student feels less anxiety.

Math anxiety manifests itself in many other ways as well, according to Tobias (1993). Some students fear that others will think "they're dumb," so they avoid seeking help. This, of course, deepens the lack of understanding and the student continues to fail. For others, a few bad grades may lead them to believe that maybe they really don't understand mathematics, that they have been "faking it" for years. Still others believe mathematical ability is a "talent" ("you either have it or you don't"). These students fear that at some point they will reach the end of their abilities and come up failures.

Only when you recognize that a student is anxious about mathematics, can you begin to help them. A first step is to help the student identify what aspects of mathematics or mathematics class are causing the anxiety. If the student fears making "stupid" mistakes or failing a test, you can help them look at errors in a different light. Errors help both you and the students figure out what misunderstandings are causing the problem and how these can be corrected. Often, students are unwilling to do this kind of reflecting on errors. They just want to know what answers are incorrect and what the correct answer should be. But they must learn to ask themselves, Why did I respond in this way? When they are able to answer this question, you will have an easier time helping them succeed in the future. If the student believes that they don't have an inherent mathematical ability, you can help them discover their abilities. When students who are anxious about mathematics experience success, they are more likely to begin to strive for more success.

Students will always require varying amounts of attention and help. Whether the student has a learning disability, extreme anxiety, or simply a different way of thinking about mathematical concepts, you need to be aware of these differences. You then need to find ways to help all students achieve confidence in their abilities to do mathematics and success in doing mathematics.

13.5 Working with Parents and the Community

One way to begin to address the needs of all students is to get to know the community in which these students are being raised. What is unique about the community and about the way its youths are viewed? Are parents involved in the school and in the broader community? To what degree? What about the parents of your students? Are community agencies, businesses, and social groups invested in the education of their children? How strong is the commitment? In what ways can community groups contribute to the mathematics learning of your students?

Advocates of parent and community involvement in the school recognize that several factors must be addressed. First, the teachers must be willing to open the classroom doors to the community. This means the teacher must be confident about the math program being used and comfortable taking suggestions from others who are not in that classroom on a daily basis. Teachers usually have expectations about what it means to have parents involved in the classroom. These expectations need to be understood by the parents. Teachers may also have strong ideas about what level of parental involvement will be most beneficial for the students. Again, communication with parents in this respect maintains good relationships and provides the best opportunities for the students.

Second, parents' beliefs and expectations must also be made explicit. Are the parents critical and unsupportive of the math program? Do they lack confidence in their own abilities and hence in their ability to contribute to the program in a positive way? Often social and cultural issues inhibit parents from actively engaging in their child's mathematics education. They may believe, for instance, that they have no authority to question the teacher's choices. At the same time, parents need to be made aware of the fact that they are role models for their children in terms of learning mathematics (and learning in general). A parent who claims to not be good at mathematics and hence doesn't help a child with home assignments is most likely instilling similar ideas in the child.

Third, the community and school must support parent involvement. One mathematics teacher in a school system probably cannot have much of an effect on parent involvement. But if several teachers work together to encourage parent involvement, the success of such a program is more likely. Teachers can organize a "back-to-school" night for parents to experience the mathematics that their children are doing in school; this gives parents opportunities to speak with teachers about various aspects of the curriculum. In addition, the school administration can provide direction and resources for the teachers and parents involved.

Community involvement and support are extremely valuable. When businesses, community leaders, and local colleges and universities become involved in the mathematics programs at the middle school and high school levels, this involvement spurs students to pursue mathematics. Businesses can make the value of mathematics very real to students as they show how valuable it is in the workplace. They can also sponsor programs for increased technology use in the schools with financial support or training for teachers and students. Often, businesses organize volunteers from their workforce to participate in tutoring programs at local schools. This is another way of getting students interested in mathematics and getting the community invested in improving the mathematics education of its youths. Local colleges may establish partnerships with schools. Such partnerships can provide special mathematics programs for students with learning disabilities or for gifted students. Often, students who lack motivation to learn mathematics can benefit from enrichment programs such as these. Colleges often offer professional development for mathematics teachers and can help inform the community about new directions in mathematics education.

As noted, parent and community involvement is important and valuable for a multitude of reasons. Such involvement also closes the gap between the student's home environment and the school environment. When this gap is wide, students have difficulty making the transition. This is especially true for immigrant students and those from various cultural backgrounds. Peressini (1997) advocates parent involvement in order to bring the school culture closer to the culture of the community and the students' home lives. He believes parents have much to offer in terms of helping teachers incorporate the local culture in the mathematics classroom. He cites Epstein's (1994) suggestions for how educators can encourage parent involvement on various levels (see Table 13.1).

Table 13.1 Categories of parent involvement.

Parenting
Parents should be supportive of their students' learning. They should provide a positive home environment that helps prepare students for schooling.

Communicating
Teachers and parents should maintain a two-way flow of information. Teachers can do this in parent-teacher conferences, newsletters, parent mathematics nights at school, or written comments on grade reports. Teachers should also encourage parents to ask questions or make suggestions.

Volunteering
This involves parent participation in the mathematics class through career days, helping out when students are working on projects, bringing cultural ideas into the mathematics class (for example, sharing traditional designs or ways of representing numbers).

Learning at Home
Parents should become involved in their students' home assignments. Teachers may need to help parents find ways of assisting a child without providing answers.

Decision Making
Parents should become involved in school governance or school policy decisions. There are often many avenues and levels for becoming involved in school decision making.

Community Collaboration
Mathematics educators can organize community meetings to share their educational ideas and programs, as well as news about current reforms. Various constituencies from the community should voice their needs and concerns. It is also possible to establish partnerships between businesses and schools or between local universities and schools. These partnerships could include career counseling for students, field trips, and tutoring programs.

Rather than waiting for problems to occur, which increases the chances that your relationship with a parent will be confrontational, be proactive. Involve parents early on, rather than later. Your students will thank you.

References

Butler, K. A. (1988). *It's all in your mind: A student's guide to learning style*. Columbia, CT: Learner's Dimension.

Butler, K. A. (1987). *Learning and teaching style: In theory and practice*. Columbia, CT: Learner's Dimension.

Consortium for Mathematics and Its Applications. (1999). *Mathematics: Modeling our world*. Cincinnati, OH: South-Western Educational Publishing.

Coxford, A. F., Fey, J. T., Hirsch, C. R., Schoen, H. L., Burrill, G., Hart, E. W., Messenger, M. J., & Ritsema, B. (1997). *Contemporary mathematics in context: A unified approach*. Chicago: Everyday Learning.

Cuevas, G. & Driscoll, M. (eds.) (1993). *Reaching all students with mathematics*. Reston, VA: NCTM.

Dunn, R. S., & Griggs, S. A. (1988). *Learning styles: A quiet revolution in American Secondary Schools*. Reston, VA: National Association of Secondary School Principals.

Epstein, J. L. (1994). Theory to practice: School and family partnerships lead to school improvement. *In* C. L. Fagnano, & B. Z. Werber (eds.), *School, family and community interaction: A view from the firing lines*, (pp. 39–52). Boulder, CO: Westview Press.

Flores, A. (1997). Si se puede, "it can be done": Quality mathematics in more than one language. *In* J. Trentacosta (ed.), *Multicultural and gender equity in the mathematics classroom: The gift of diversity*, 1997 Yearbook, (pp. 81–91). Reston, VA: NCTM.

Fuchs, D., & Fuchs, L. (1988). An evaluation of the Adaptive Learning Environments Model. *Exceptional Children* **55**:115–127.

Gardner, H. (1983). *Frames of mind*. New York: Basic Books.

Gardner, H. (1993). *Multiple intelligences: The theory in practice*. New York: Basic Books.

House, P. A. (1987). *Providing opportunities for the mathematically gifted*. Reston, VA: National Council of Teachers of Mathematics.

Johnson, D. R. (1982). *Making every minute count*. Palo Alto, CA: Seymour.

——— (1986). *Making minutes count even more*. Palo Alto, CA: Seymour.

——— (1994). *Motivation counts*. Palo Alto, CA: Seymour.

Keefe, J. W. (1987). *Learning Style: Theory and practice*. Reston, VA: National Association of Secondary School Principals.

Kenney, P. A., & Silver, E. A. (1997). *Results from the sixth mathematics assessment of the National Assessment of Educational Progress*. Reston, VA: NCTM.

Larcombe, T. (1985). *Mathematical learning difficulties in the secondary school*. Milton Keynes, England: Open University Press.

Leder, G. (1995). Equity inside the mathematics classroom: Fact or artifact? *In* W. Secada, E. Fennema, & L. B. Adajian (eds.), *New directions for equity in mathematics education*, (pp. 209–224). New York: Cambridge University Press.

Miller, S. P. & Mercer, C. D. (1998). Educational aspects of mathematics disabilities. *In* D. P. Rivera (ed.), *Mathematics education for students with learning disabilities: Theory to practice* (pp. 81–96). Austin, TX: Pro-Ed.

National Council of Teachers of Mathematics. (1989). *Curriculum and evaluation standards for school mathematics*. Reston, VA: Author.

——— (2000). *Principles and standards for school mathematics*. Reston, VA: Author.

——— (1991). *Professional standards for teaching mathematics*. Reston, VA: Author.

Peressini, D. (1997). Building bridges between diverse families and the classroom: Involving parents in school mathematics. *In* J. Trentacosta, & M. J. Kenney (eds.), *Multicultural and gender equity in the mathematics classroom: The gift of diversity, 1997 Yearbook* (pp. 222–229). Reston, VA: NCTM.

Peterson, P. L., & Fennema, E. (1985). Effective teaching, student engagement in classroom activities, and sex-related differences in learning mathematics. *American Educational Research Journal* **22**(3):309–335.

Rivera, D. P. (ed.). (1998). *Mathematics education for students with learning disabilities: Theory to practice*. Austin, TX: Pro-Ed.

Ron, P. (1999). Spanish-English language issues in the mathematics classroom. *In* E. Ortiz-Franco, N. G. Hernandez, & Y. De La Cruz (eds.), *Changing the faces of mathematics: Perspectives on Latinos* (pp. 23–33). Reston, VA: NCTM.

Secada, W. G. (1992). Race, ethnicity, social class, language, and achievement in mathematics. *In* D. A. Grouws (ed.), *Handbook of research on mathematics teaching and learning* (pp. 623–660). New York: Macmillan.

Secada, W., Fennema, E., & Adajian, L. B. (eds.) (1995). *New directions for equity in mathematics education*. New York: Cambridge University Press.

Thornton, C. A., & Bley, N. S. (eds.) (1994). *Windows of opportunity: Mathematics for students with special needs*. Reston, VA: NCTM.

Tobias, S. (1993). *Overcoming math anxiety*. New York: Norton.

Trentacosta, J., & Kenney, M. J. (eds.) (1997). *Multicultural and gender equity in the mathematics classroom: The gift of diversity, 1997 Yearbook*. Reston, VA: NCTM.

Zaslavsky, C. (1993). Multicultural mathematics: One road to the goal of mathematics for all. *In* G. Cuevas & M. Driscoll (eds.), *Reaching all students with mathematics* (pp. 45–55). Reston, VA: NCTM.

14

Assessing and Evaluating Mathematics Achievement

Authentic assessment tasks highlight the usefulness of mathematical thinking and bridge the gap between school and real mathematics. They involve finding patterns, checking generalizations, making models, arguing, simplifying, and extending-processes that resemble the activities of mathematicians or the application of mathematics to everyday life. When we see students planning, modeling, and using mathematics to carry out investigations, we can make valid judgments about their achievement. New forms of assessment are not goals in and of themselves. The major rationale for diversifying mathematics assessment is the value that the diversification has as a tool for the improvement of our teaching and the students' mathematics learning.

Jean Kerr Stenmark (1991)

Overview

The collection of information for describing student progress in mathematics occupies a great deal of teachers' time and effort. The activities range from asking questions in the classroom to developing, scoring, and interpreting final examinations. Some of these activities are directed toward the collection of information to be used in the improvement of teaching and learning as they apply to a student or to a class as a whole. These activities we will refer to as *assessment activities*. Some of the activities related to describing student performance are used to rate or assign value to a student's work. These activities we will refer to as *evaluation activities*. Over time, mathematics teachers have probably spent more time in rating and grading students than in developing skills that improve teaching and learning in the classroom. This chapter focuses on developing the skills associated with both, but with an emphasis on the latter.

Focus on the NCTM *Standards*

The development of the NCTM *Standards* and their sequel, the NCTM *Principles and Standards*, brought with them a shift in focus in the mathematics classroom, which we can view as a shift from teaching to learning or from teacher-centered instruction to student-based instruction. For such a transition to take place, the classroom teacher needs more information about where students are in skill levels, what they know, what is the nature and depth of their understandings, with which representations are they familiar, and a host of other questions. The *Principles* state this as a basic tenet of mathematics education:

> Assessment should support the learning of important mathematics and furnish useful information to both teachers and students (NCTM, 2000, p. 22).

How teachers make use of assessment as an instructional tool relies greatly on the methods they employ in gathering this information. In this chapter we examine the various ways in which teachers and others responsible for student learning of mathematics gather data to describe and assist in student learning.

Focus on the Classroom

We assess students' work to see if they are ready for the coming learning experiences, if they need some review of concepts before proceeding, if they are making progress in solving the problem on which they are working, or for a myriad of other reasons. Assessment provides the teacher with direct input about the nature of the learning taking place and the degree to which students are ready for new or modified learning challenges. The effective use of assessment procedures provides a basis for instructional decision making, for monitoring individual progress, and for evaluating student performance as a whole.

14.1 Assessing and Evaluating

The teaching and learning of mathematics is centered about the development of knowledge and skills. The path to this development is guided by a teacher's knowledge of mathematics and his or her ability to create learning experiences appropriate for students' understandings. The mapping of instruction and learning experiences to students' understandings is based on being able to ascertain what students know and are able to do and targeting activities that build on students' current understandings.

Assessing—Collecting the Information

If "knowing" and "doing mathematics" are related, as the *Principles and Standards* suggests, then teachers must be able to determine where students are performing and understanding in order to create and monitor student learning activities. If the goals are to have students develop into active, reasoning problem solvers who are capable of connecting their mathematics to other subjects, as well as to mathematics, and who can both represent ideas in multiple ways and communicate about these ideas, then assessments must have these same goals. This is what it means to have assessments aligned with instruction.

An assessment system that is aligned with the curriculum will gather information that informs you not only about the content knowledge and skills that the students have, but also about the students' competencies relative to the process standards. Such activities will inform you about the students' grasp of the big ideas, of logical arguments, of the various representations for mathematical ideas and how they communicate mathematical ideas, and of the connections they make between mathematics and other disciplines. Research indicates that classrooms where teachers are constantly collecting information about student progress in an attempt to improve instruction and conditions for student learning are classrooms where conditions for student learning are truly enhanced (Black & Wiliam, 1998).

The assessment of student knowledge of mathematics can best be thought of in terms of types of knowledge and types of processes. Clearly, we want to know what they know about the mathematics they have studied—both the conceptual and process aspects of it. We want to know what they can do in problem-solving and modeling situations.

In assessing students' knowledge of concepts, we examine whether they can

- recognize, label, and generate examples and nonexamples of concepts
- use and interrelate models, diagrams, manipulatives, and varied representations of concepts
- identify and apply principles (that is, valid statements generalizing relationships among concepts in conditional form)
- know and apply facts and definitions
- compare, contrast, and integrate related concepts and principles to extend the nature of concepts and principles

- recognize, interpret, and apply the signs, symbols, and terms used to represent concepts
- interpret the assumptions and relations involving concepts in mathematical settings

In assessing students' knowledge of procedures, we may ask students to perform the skill in at least one way, to create a new procedure for doing a particular task, to explain how a skill works, or to use the skill in different ways to solve a problem. These are captured in seeing whether students can

- connect an algorithmic process with a problem situation
- perform algorithmic processes correctly
- justify the correctness of an algorithmic process
- communicate the meaning of the results of an algorithmic process
- perform noncomputational skills such as estimating and rounding
- read and produce graphs and tables
- execute geometric constructions

In looking at students' problem-solving and modeling skills, we also ask other things of students. For problem solving, we might ask them questions that focus on their understanding of the problem, their ability to select and apply strategies for solving the problem, their capabilities to implement and monitor a given strategy, or to communicate the meaning of a solution and reflect on that solution.

For problem solving and modeling, we need to see whether students can

- recognize and formulate problems and models for situations
- determine the sufficiency and consistency of the data involved
- list assumptions made in working within a given context
- use strategies, data, and relevant mathematics in attempts to understand a mathematical situation
- apply reasoning (inductive, deductive, statistical spatial, or proportional) in correct and productive ways
- judge the reasonableness and correctness of purported solutions or models
- test the robustness or generalizability of a solution or model
- generate, extend, and modify procedures as needed

To carry out such assessments, we rely on a variety of information-gathering methods. These methods include classical ones such as quizzes; newer forms of assessment such as interviews, student journals, projects, and checklists; and activities involving tasks where students have to explain their work on a problem in one form of representation or another. Some of these activities are closed, in that they require a specific answer or the completion of a task using a specified procedure. Others are very open, in that they allow the student considerable flexibility in defining the problem and working toward and through a solution. We shall consider examples of these forms later in the chapter.

Evaluating—Ranking and Valuing Student Work

While assessment activities are associated with gathering information about how students' abilities are developing. Such judgments are called *formative* assessments, as they deal with the growing knowledge of the child. Teachers and educational systems are also called on to provide reports of students' overall performance in which a grade or numerical value is associated with a student's performance. Such activities are summary judgments of the person's work, ranking it against an absolute standard or against the work of others. Such judgments are often called *summative* evaluations. In the case where the judgment is against standards and the call is "meet the standard or not," the evaluation is further called a criterion-referenced evaluation. In the case where the judgment is made relative to the performance of others in the same population of students, the evaluation is called a norm-referenced evaluation.

Some teachers say that they would teach for free. However, they say that they earn their pay when they grade students. Limiting the discussion of a student's work to the issuance of a single grade, A through F, is one of the most difficult things teachers do. Part of the craft of teaching is using one's professional judgment, all of the evidence present, and prior experiences to assign a grade to a student. Grading and evaluating will be discussed further in Section 14.5.

Two Concepts: Validity and Reliability

Validity

Two additional concepts to be borne in mind when we discuss assessment and evaluation are the concepts of validity and reliability. Validity is "... an integrated evaluative judgment of the degree to which empirical evidence and theoretical rationales support the *adequacy* and *appropriateness* of *inferences* and *actions* based on test scores and other modes of assessment" (Messick, 1989, p. 13).

Historically, validity has been divided into content validity, predictive and concurrent validity, and construct validity. These can be described as follows:

> Content validity shows how well test content samples the subject matter of interest.
>
> Criterion-related validity compares the test scores with one or more external variables (called criteria) that directly measure the characteristic or behavior in question.
>
> Predictive validity indicates the extent to which an individual's future performance on the criterion is predicted from prior test performance.
>
> Concurrent validity indicates the extent to which the test scores estimate an individual's present standing on the criterion.
>
> Construct validity investigates what qualities a test measures (Messick, 1989, p. 16).

Considerable discussion is taking place at present about the nature of each of these aspects of validity. Historically, when we only looked at a score and made a judgment about students' performances, these definitions stood us well. They were beacons of whether the content (viewed as facts and generalizations) were correct and representative. Did the student meet or exceed the criterion established? How well did the performance align with that required in a prospective career area? Did the score match well with the student's current standing in class? How did the test reflect the student's problem solving or reasoning?

With the advent of constructivism as a theory of learning, the nature of these questions and their answers take on a different light. Our goals are no longer simply the production of answers and specific facts. New goals for the classroom deal with cognitive processes. Not all evaluations are taken under timed conditions. What influence does the availability of technology have on the ways in which students approach items or how their scores match up with those in a career or field who did not learn in the same fashion?

Today, these questions of validity are still important. However, we must approach them with an eye to including cognitive behaviors (both the students' and the desired cognitive behaviors associated with the *Standards* as well as the career and professional fields students are headed toward) in our decision making about validity.

Reliability

Reliability reports on the variability observed in individual test performances. On any given day, one's performance is likely to be up a bit or down a bit. Such inconsistencies result from numerous factors. From a measurement standpoint, we need to address the concept of one's *true score* for a given test. The obtained score may be a result of error inaccurately skewing the true score. Sources of errors can be context or conditions in which the test was taken, variability in scoring the examination, or a host of other factors. To estimate the reliability of a test, we must quantify the variability: What part of the score is attributable to the learner's actual knowledge and what is attributable to error? Any time we test a student, we want the scores to be reliable, that is, to reflect the true test score as much as possible.

In general, test reliability is given in decimal form between -1 and 1. The reliability coefficient is similar to the correlation coefficient, r_{xx}, which reports the degree to which one administration of the test correlates with other administrations of the same test. Alternatively, it is reported as the proportion of the score that is due to variation in the true score, with $1 - r_{xx}$ giving the variation due to error in the measurement. More will be said about reliability as we move through this chapter.

Exercises 14.1

In Exercises 1–4, work the problem provided and then describe

- *what you think the item is testing*

- *what student work on the problem would tell you about their knowledge of the topic*
- *how you would evaluate student work on that problem*

1. Consider the following two graphs. What do these graphs tell you about the cost of building per foot in the two cities over time?

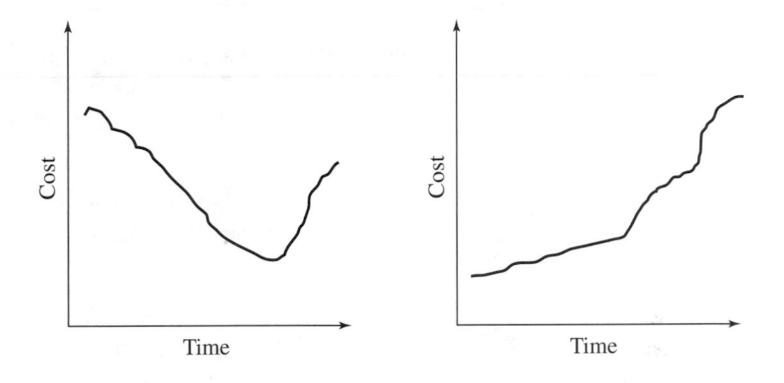

2. The graph of the function f is shown below for $0 \leq x \leq 10$. The integral from 0 to a of $f(x)$ attains its greatest value when a is equal to

a. 0
b. 2
c. 3
d. 6
e. 10

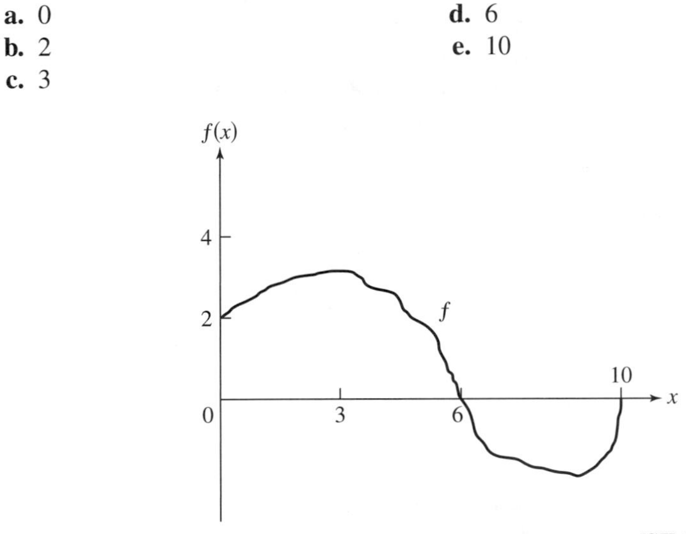

(SIMS, 1985, item 6.13.)

3. What is the greatest number of cuts of pizza you can get if you cut a round pizza using 5 straight cuts?

4. Think about the product of any four consecutive positive whole numbers (such as $3 \times 4 \times 5 \times 6 = 360$ or $10 \times 11 \times 12 \times 13 = 17{,}160$). Shade in the oval, indicating whether the statement is always, sometimes, or never true for the product of any four consecutive positive integers.

	Always true	Sometimes true	Never true
The product is divisible by 4	o	o	o
The product is divisible by 5	o	o	o

What is the *largest* whole number that is *always a factor* of the product of any four consecutive positive whole numbers? Provide a clear and complete justification for your answer.

5. What kinds of knowledge do we want to assess in observing students' growth in mathematical knowledge? Make a list, being as specific as you can about the types of knowledge and why they are important in getting a clear picture of student growth and development.

6. Why is the alignment of assessment with instruction and program goals important? What can happen if the program structure, assessments, and evaluations are not aligned?

7. In a department meeting Sam said, "I don't believe in partial credit for tests in mathematics. The answer is either right or wrong. The sooner students come to learn this, the better off they will be!" Take a position of agreement or disagreement with Sam and provide a rationale for your position.

8. Nancy always allows her students to gain back half of the points they lost on a test by writing out correct solutions to the problems they missed and handing them in within the next two days after they go over the test items in detail. Detail what you believe the pros and cons of this procedure might be. Then, indicate whether you would use this procedure in your own classroom.

14.2 Student-Constructed Work

Using assessments in which students are asked to *produce* their answer rather than to *select* their answer from "forced-choice" lists (that is, multiple-choice and true-false items) is becoming more important and prevalent in mathematics education today. This is due in large part to the influence of the *Principles and Standards* (NCTM, 2000). These recommendations call for the assessment of learning to parallel and to inform the teaching taking place in the classroom. As teachers and those responsible for assessments on a broader scale focus on gaining knowledge of students' grasp of concepts, principles, and skills, they must move to align their work with that seen in the classroom activities.

Thus, the use of student-constructed responses in assessment and evaluation settings has a number of advantages and sends a message or two to students. First, assessments that call for student production of the answers sends the message that such communication skills and ability to represent their thoughts is important and valued. Second, feedback from these activities is most helpful in adjusting instruction and noting student learning deficits.

Central to changes in assessment in the past two decades has been the move away from forced-choice test items to student-constructed response test items. These items require students to provide either a numerical answer they have developed or to write a short, or longer, explanation of their solution. These responses are then compared against criteria to judge a student's level of development relative to the problem features or their overall performance.

Short-Answer or Regular Student-Constructed Response Items

The use of short-answer items and items requiring more extensive student-constructed responses is becoming more common in class assessments, as well as in state, national, and international assessments. In forced-choice test items, students merely pick from the list, and successful guesses can skew the results. However, in student-constructed responses, students have to demonstrate their reasoning and communicate their answers. For large-scale assessments, the grading of student-constructed responses greatly increases assessment costs, both in time and resources. However, such costs are justified, as you will see, in the information garnered from seeing individual student work.

Short-answer questions, those that require students to write a short numerical answer or a short sentence, are referred to as *regular student-constructed response* items. This distinguishes them from *extended student-constructed response* items, which require students to write a paragraph or other extended responses.

Recall the regular student-constructed response item involving the garbage can from Chapter 2. This item, given as part of the NAEP mathematics assessment to Grade 8 students, is shown in Figure 14.1. Recall that the item requires students to

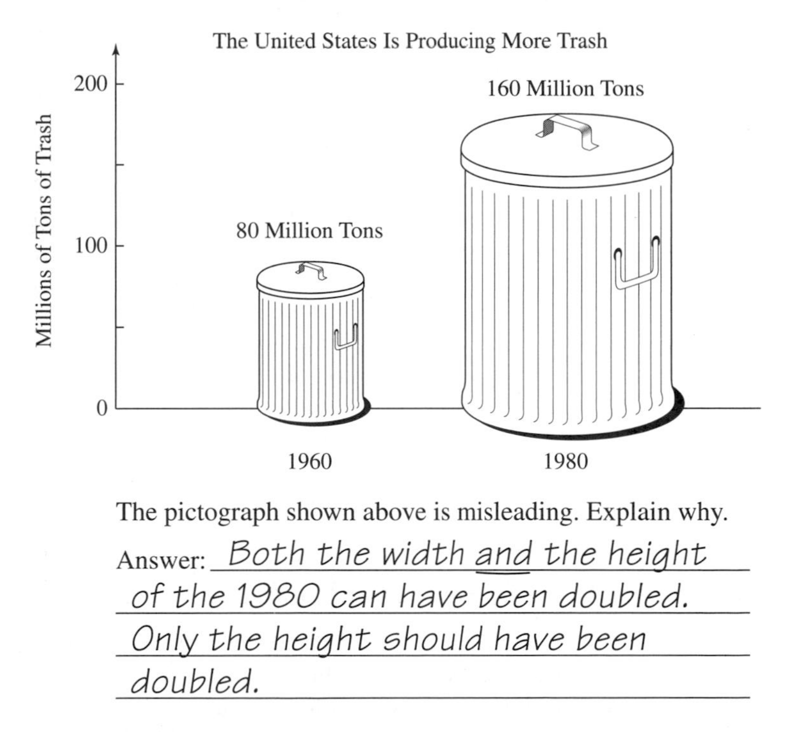
The United States Is Producing More Trash
Millions of Tons of Trash
200
100
0
160 Million Tons
80 Million Tons
1960
1980
The pictograph shown above is misleading. Explain why.
Answer: Both the width and the height of the 1980 can have been doubled. Only the height should have been doubled.

Figure 14.1 Regular student-constructed response item for Grade 8.

explain that the doubling of the can's dimensions more than doubles the volume of the can and it is the volume of garbage that is being discussed.

The work shown is that of one of the 8% of Grade 8 students who received credit for this item in the NAEP assessment. This student explains his solution, but doesn't fully explain why the illustration is misleading. In the NAEP assessment, student responses were accepted that said the can was more than twice as large in volume or that gave a mathematical argument about the mistake in doubling both dimensions.

Scoring guides for student-constructed items are known as *rubrics*. A rubric outlines the various codes that can be given to evaluate an item. The NAEP rubric for the garbage can item is a simple two-point rubric built around the notion of correct-incorrect. A general rubric might be laid out as follows:

X The student leaves a blank paper.

0 The student writes something that is incorrect or irrelevant.

1 The student provides a correct response.

When answer codes are entered into a computer or other recording device, a coding like X for the blank paper code is appropriate. However, if statistical software or a spreadsheet is being used, the X will suffice for managing the data. The blank paper code is important because it signals one of three things: (1) The student ran out of time (struggling with all the answers). (2) The student doesn't understand *this* problem. Or, (3) The student has a motivation problem and hence is dawdling on the test.

A rubric particular to the garbage can item might look something like this:

X The student leaves a blank paper.

0 The student writes something that is incorrect or irrelevant.

1 The student states one of the following:

(a) The volume of the 1980 can is more than double that of the 1960 can.

(b) Both the radius and height have been doubled, only the height should have been doubled.

Such rubrics are known as *holistic* rubrics, because they deal with the whole response at once. That is, the scoring codes in the rubric assess the students' mathematical correctness, their communication of the response, and the strategies employed. Other rubrics, to be discussed later, are known as *analytic* rubrics; they supply individual codes for each area of mathematical correctness, communication, and strategy applied.

Example 1 contains another regular student-constructed response item. This item allows for one level of partial credit between the incorrect and fully correct response codes.

Example 1 *Regular Student-Constructed Response Item*

A cereal company packs its oatmeal into cylindrical containers. The height of each container is 10 inches and the radius of the bottom is 3 inches. What is the volume of the box to the nearest cubic inch? (The formula for the volume of a cylinder is $V = \pi r^2 h$.)

Answer: __________ cubic inches (Mitchell et al., 1999) ■

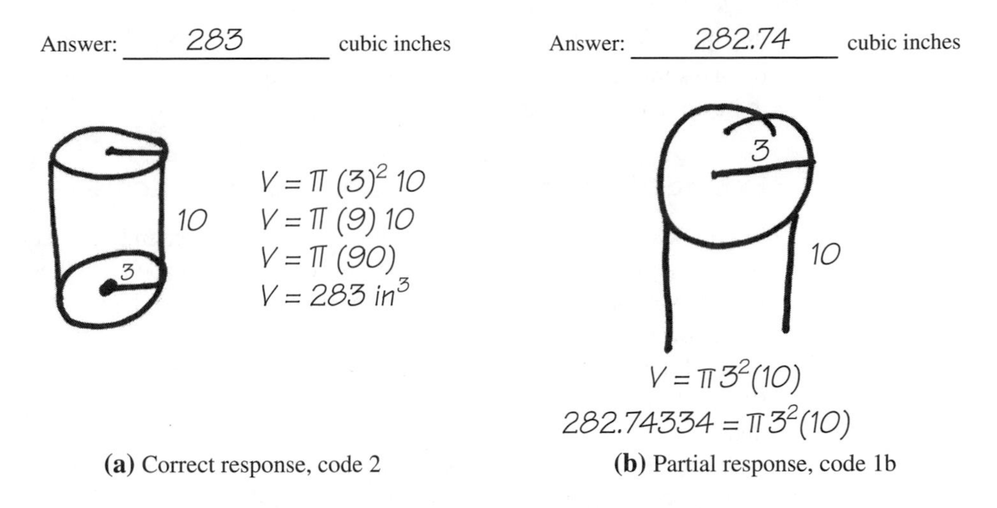

(a) Correct response, code 2 **(b)** Partial response, code 1b

Figure 14.2 Coded answers for Example 1.

The rubric for this item is constructed as follows:

X The student leaves a blank paper.
0 The response was incorrect or irrelevant.
1 The response showed any of the following:
- (a) correct substitutions into the formula, but incorrect rounding
- (b) 282.74334, suggesting multiplication by the π key on the calculator, but no work shown
- (c) 282.6, suggesting multiplication by 3.14, but with no work shown
- (d) 282.8571, suggesting multiplication by 22/7, but with no work shown

2 The correct answer of 283.

Code 2 answers are said to be correct, code 1's are called partial answers, and codes 0 and X are, respectively, incorrect or no responses. Figure 14.2 shows examples of a code 2 and code 1 answer for this item.

Interpreting Rubric Codes

It is important to remember that rubric codes show teachers and researchers involved with large-scale assessments what the student has done on the item and the answer's overall correctness. The code is not a grade. It is not the case that a 2 is an A, a 1 is a C, and anything else is an F. The notion that everything produced in math class is a score evaluating the student must be put aside; rather we want to focus our assessments on what students know and are able to do in a given problem.

Many teachers especially want to turn rubric codes into grades as they move to extended student-constructed responses, because rubrics for these responses often have 4 or 5 levels of answer codes, which look even more like A–F grading scales. Remember that the information on one item is not a grade, it is an indication of student performance on one item. A summative evaluation would require a broader base of the student's work to be both valid and reliable.

Extended Student-Constructed Response Items

In contrast to the regular student-constructed items, extended student-constructed response items require students to produce expanded responses; these test items often leave the problem far less structured. In some cases, the problems are broken into pieces that allow the student to attack the problem in stages. Extended response items are commonly found on European examinations in mathematics. This process of "guiding" the student through the problem by breaking the problem into parts is known as "scaffolding."

Some educators and researchers argue that scaffolded items should be coded as a sequence of regular response items, because the entire item is no longer extended in its demands on the student. Opinions on this controversy vary. However, one thing is clear, if scaffolded items are used, the individual parts have to be *locally independent*. That is, answers to subsequent items cannot depend on earlier answers in the scaffolded problem. If they do, the students' overall performance depends too heavily on their previous answers. Others argue that scaffolding merely reflects a sum of parts, but does not necessarily give a picture of student growth in cognitive activity. That is, each part may be relatively low in cognitive demand.

Consider the item displayed in Example 2. This extended student-constructed response item was developed by teachers in Kentucky.

Example 2 ***Extended Student-Constructed Response Item for Grades 9–12***

Your cousin Fred has built a new swimming pool in his backyard. He has just e-mailed that he desperately needs your help. The water company will be filling the pool three days from now, and Fred still needs to seal it with pool paint. He promised that you'd have use of the pool any time you wish if only you would help him figure how much paint to buy. The paint is very expensive and he doesn't want to buy any more than he needs. The pool is rectangular in shape, 20 ft. wide by 40 ft. long. The shallow end is 3 ft. deep. The bottom of the pool is horizontal for 8 ft. and then slants down to a depth of 7 ft. at the deep end. The directions on the paint indicate that a gallon of paint will cover approximately 300 square feet. (Bush & Greer, 1999). ■

Note that what students are being asked to find—how much paint to buy—is embedded in the middle of the problem, so students must read the problem carefully. This problem requires students to decide whether they should develop a representation of the problem (for example, it doesn't diagram the pool for them). To solve it, they will have to communicate in more detail and apply their knowledge of geometry and measurement to a real-life situation. Students must decide what assumptions they should make: Are the walls of the pool perpendicular to the base? Should they allow for a tile rim around the pool?

The second test item shown in Chapter 2 is an extended student-constructed response item that deals with students' recognition of a pattern and the number of tiles

The first 3 figures in a pattern of tiles are shown below. The pattern of tiles contains 50 figures.

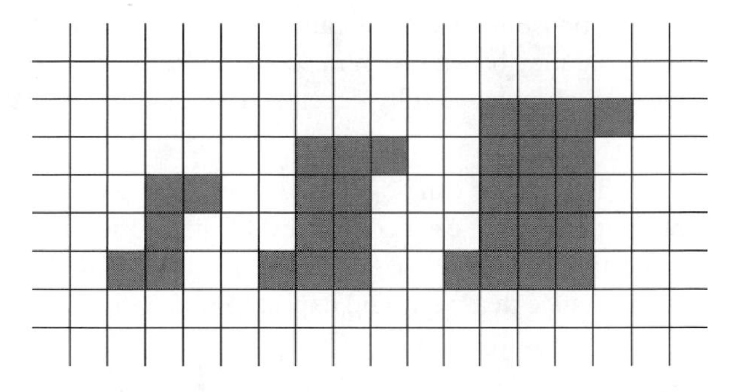

Describe the 20th figure in this pattern, including the total number of tiles it contains and how they are arranged. Then explain the reasoning that you used to determine this information. Write a description that could be used to define any figure in the pattern.

Figure 14.3 NAEP extended student-constructed response for Grade 12 (Mitchell et al., 1999).

needed to represent a particular member in the pattern. The problem is shown in Figure 14.3.

Rubrics for extended student-constructed response items allow for a greater variation in student performance. The general holistic rubric used in the NAEP coding of student performance is shown in Example 3.

Example 3 ***NAEP Extended Student-Constructed Response Item Rubric***

X No Response

0 Incorrect. The response is incorrect or consists of irrelevant remarks.

1 Minimal. The response indicates a minimal understanding of the problem posed but does not suggest a feasible approach to a solution. Although there may or may not be some correct work signifying a mathematical approach, the response is incomplete, contains major errors, or reveals other serious flaws. Examples are absent.

2 Partial. The response contains evidence of an understanding of the problem at a conceptual level evidenced by the mathematical approach taken. However, on the whole, the response is not well developed. Although there may be serious mathematical errors or flaws in the reasoning, the response does contain some correct work. Examples provided are incorrect or inappropriate.

3 Satisfactory. The response demonstrates a clear understanding of the problem and provides an acceptable approach. The response is generally well developed and coherent but contains minor weaknesses in the development. Examples are provided, but not fully developed.

4 Extended. The response demonstrates a complete understanding of the problem, is correct, and the methods of solution are clear, appropriate, and fully developed. The response is mathematically sound, clearly written, and contains no errors beyond ones that may be a result of miscopying from elsewhere in the student's work. Examples are well chosen and fully developed. (Dossey et al., 1993) ■

This rubric can be easily adapted to a particular test rubric by adding particulars of the problem to each coding level. The rubric for the test item in Figure 14.3 is given in Example 4. Here we see the various levels defined in terms of the type of student performance that we would expect to find at the level.

Example 4 ***Rubric for Tile Pattern Test Item***

X No Response.

1 Incorrect. The student shows no attempt to go beyond what was shown in the question.

2 Minimal. The student has attempted to draw or describe the pattern or an additional figure in the pattern or make some attempt to go beyond what is shown in the question.

3 Partial. The student illustrates or describes at least one additional figure in the pattern correctly or states that there are 442 tiles in the 20th figure, but does no more.

4 Satisfactory. The student describes the 20th figure, gives the number of tiles, and provides some evidence of sound reasoning.

5 Extended. The student has included all of the following: a correct count of 442 tiles for the 20th figure, a verbal or graphical explanation of the reasoning used, and an accurate generalization, based on inductive reasoning. (Mitchell et al., 1999) ■

Examples of student work at each of these coding levels is shown in Figure 14.4.

Developing a good extended student-constructed response item is difficult. These test items must allow for student responses at several levels and yet be clear about what the student is supposed to do.

Some states or school districts use what is called an analytic rubric to score their extended response items. Such a rubric scores the same material several different times, each time for a different purpose. One such rubric might look like this:

Understanding the Problem

0 Complete misunderstanding of the problem

1 Part of the problem is understood but another part misinterpreted

2 Complete understanding of the problem

* Each figure increases 1 layer in height and one middle layer in width for every succession, relative to the first.
For example, for the nth section the figure will be n+1 units across at the base, n units wide, n+1 units across at the top, and n+2 units high. This is the pattern
The 20th figure will be 21 units across on the bottom length, 20 units wide in the middle, 22 units high; and 21 units wide at the top. The increase is linear.
Total number of tiles it contains:

21 + (20×20) + 21 = 442

The inner square is always (n×n) units in area (n^2)

(a) Sample extended response.

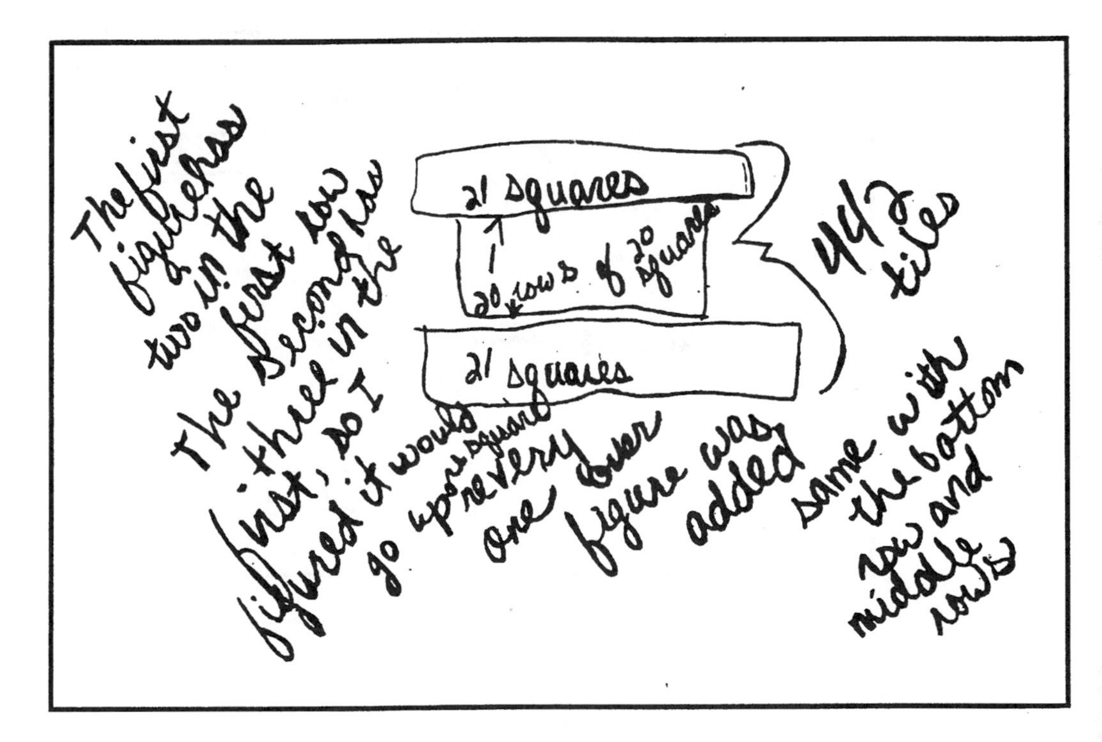

(b) Sample satisfactory response.

Figure 14.4

20 . Rows across
22 Rows up
21 Rows across at just top
21 Rows across at just bottom
21 blocks
22 Rows of blocks
20 rows of blocks
21 blocks

(c) Sample partial response.

increase 1 across w/ each new figure
22
21 spaces →

(d) Sample minimal response.

Figure 14.4 *(continued)*

The 20th figure is on the right hand side in the 3rd pattern and is one up on the corner; like this. The figure could be a box.

(e) Sample incorrect response.

Figure 14.4 *(continued)*

Planning a Solution

0 No attempt, or totally inappropriate plan

1 Partially correct plan based on part of the problem information

2 Plan could lead to a correct solution if fully implemented

Getting an Answer

0 No answer, or wrong answer based on inappropriate attack

1 Copying error, computational error, or partial answer for a problem with multiple parts

2 Correct answer and correct label for the answer (O'Daffer et al., 1987)

Another form of analytical rubric makes use of "getting a solution, selecting a strategy, and communicating the response." Central to using and interpreting analytic rubrics is assessing a student in their overall performance on a scaffolded test item.

Exercises 14.2

In Exercises 1–6, decide if the problem should be a regular student-constructed response item or an extended response item. Then design a rubric to code student responses.

1. Golf balls come in packs of 3. A carton holds 24 packs. Mr. John, the owner of a sporting goods store, ordered 2160 golf balls. How many cartons did Mr. John order?

2. A tennis club held a tournament for its 47 members. If every member played one match against every other member, how many matches were played?

3. You are given a checkerboard and 32 dominoes. Each domino covers exactly two adjacent squares on the board. Thus, the 32 dominoes can cover all 64 squares of the checkerboard. Now suppose the two squares at opposite corners of the board are removed. Is it possible to place 31 dominoes on the board so that all of the 62 remaining squares are covered? If so, how?

4. In an Asian country there exists a mountain with a temple on its summit. On a given day, Naho the monk arose at the base of the mountain and walked, starting at 4 A.M., up to the temple at the top to meditate. Naho arrived at noon. The next morning, he awoke and descended to the base of the mountain, starting at 4 A.M. and arriving back at the base at noon. Is it possible that the monk was at the same elevation on the mountain at the same time on each of the two mornings? If so, would this be a happenstance or a dead certainty? Describe your reasoning.

5. How many squares does a 6×6 square checkerboard design contain if the six units on each side are the same length? How many squares does an $n \times n$ square contain if the n segments on each side are the same length?

6. Suppose two trains travel toward each other on the same track. One train, A, starts at the west end of the 200-mile section of track and travels east at 60 miles an hour. The second train, B, starts at the east end of the track and travels west at 80 miles an hour. A superfly starts on the lamp of engine A, flies east until it turns on the lamp of train B, then flies west until it turns on the lamp of engine A, and then flies east.... The superfly travels at 200 miles per hour and turns on a lamp with no loss of speed. How far does the superfly travel until it is squashed to death between the lamps of the two colliding engines?

7. How can you use the rubric scores for an extended student-constructed response item to ascertain how the class as a whole is doing relative to the objective being assessed by the test item?

8. Develop an analytic rubric to assess student performance on the following item:

> Suppose a cube of modeling clay is sliced carefully in a planar fashion with a straight wire. Describe the different types of polygonal regions that could result from the cut in terms of the number of sides each region might have. For example, if a corner is cut off by moving back a short distance on each of the three edges coming into the corner, a triangle would be the resulting boundary for the piece cut off.

9. Consider the student work shown for the following problem. Use the rubric given to assign codes to the answers.

This question requires you to show your work and explain your reasoning. You may use drawings, words, and numbers in your explanation. Your answer should be clear enough so that another person could read it and understand your thinking. It is important that you show <u>all</u> of your work.

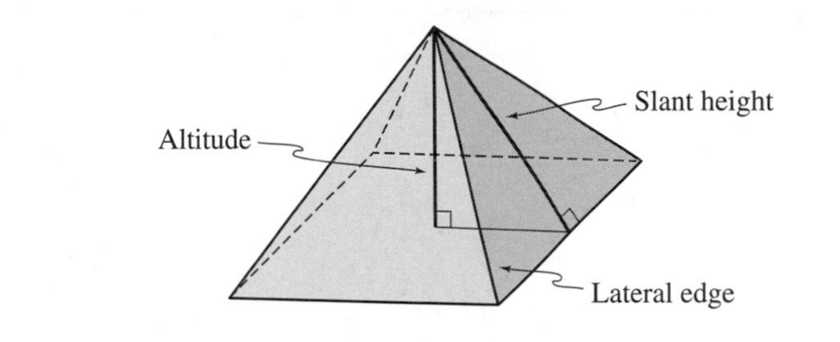

10. The figure above shows the altitude, lateral edge, and slant height of a square pyramid. Each of the three square pyramids P, Q, and R (not shown) has a base with side 10. In each pyramid the lateral edges are of equal length. Pyramids P, Q, and R have the following characteristics.

a. The altitude of pyramid P is 10.

b. The lateral edge of pyramid Q is 10.

c. The slant height of pyramid R is 10.

List the pyramids by order of volume from smallest volume to largest volume.

Support your conclusion with mathematical evidence.

[The formula for the volume of a pyramid with base area B and height (altitude) h is $V = \frac{1}{3}Bh$.] (Mitchell et al., 1999)

The rubric for coding the student work is as follows:

4 Extended. Student gives ordering of Q, R, P and a mathematical argument supporting that ranking.

3 Satisfactory. Student gives correct volumes of two of the three pyramids,
or
finds the altitudes of Q and R,
or
correctly compares the heights of Q, P, and R.

2 Partial. Student finds a correct solution for either the height of pyramid Q
or
the height of pyramid R.

1 Minimal. Student shows or states that the three pyramids have the same base area

or

that $\frac{1}{3}$ the area of the base is the same for all three pyramids.

0 Incorrect or irrelevant writing.

Student 1:

P

$V_p = \frac{1}{3}Bh$

$V_p = \frac{1}{3}(100)(10)$

$V_p = 333.\bar{3}$

Q

$a^2 + 14^2 = 10^2$

$a^2 + 196 = 100$

$V_Q = \frac{1}{3}(100)(9.8)$

$V_Q = 326.\bar{6}$

R

$10^2 + 5^2 = c^2$

$100 + 25 = c^2$

$125 = c^2$

$11 = c$

$14^2 + b^2 = 10^2$

$196 + b^2 = 121$

$b^2 = 75$

$b = 8.6$

$V_R = \frac{1}{3}(100)(8.6)$

$V_R = 286.\bar{6}$

Smallest to largest
Q, P, R

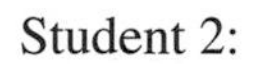
Student 2:

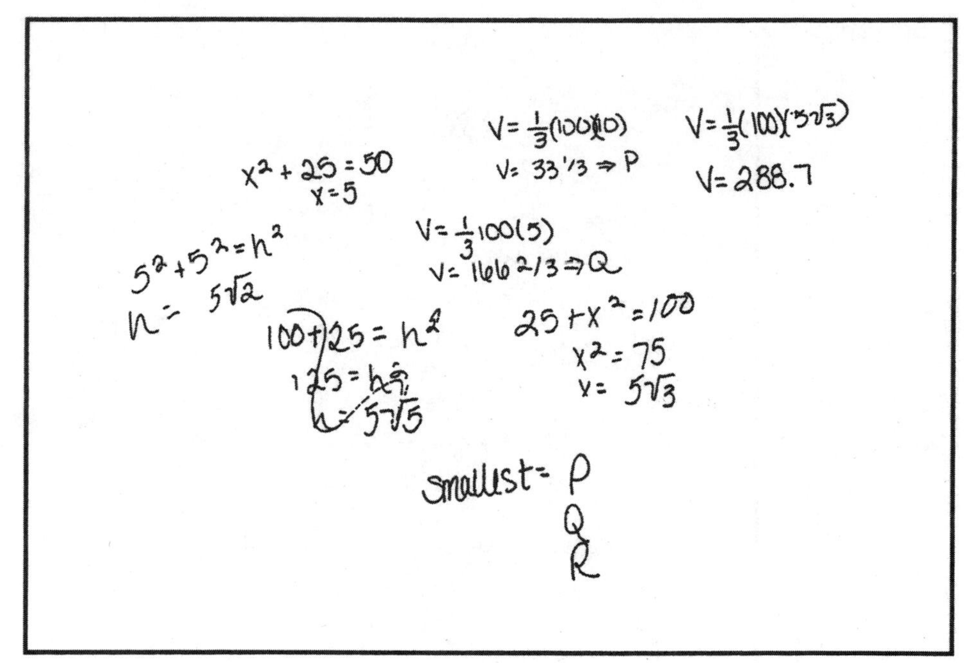
x² + 25 = 50
x = 5
5² + 5² = h²
h = 5√2
V = 1/3(100)(10)
V = 33 1/3 ⇒ P
V = 1/3 100(5)
V = 166 2/3 ⇒ Q
V = 1/3(100)(5√3)
V = 288.7
100 + 25 = h²
125 = h²
h = 5√5
25 + x² = 100
x² = 75
x = 5√3
smallest = P
Q
R

Student 3:

$V = \frac{1}{3}Bh$

$B = 10$

Base of pyramid Q = Base of pyramid P = Base of pyr. R

Lateral edge " " lateral " " " = lateral " " "

Pyramid P: Base = 10

Alt = 10

lateral = 10

$V = \frac{1}{3} \cdot 10 \cdot 10$

$V = 33.\overline{3}$

$V = \frac{1}{3}BH$

Pyramid Q

$V = \frac{1}{3} \cdot 10 \cdot 10$

$= 33.3$

Pyramid R

$V = \frac{1}{3} \cdot 10 \cdot 10$

$= 33.3$

Student 4:

V of P = 1/3 Bh B = 100 h = 10

$V_P = \frac{1000}{3} \doteq 333$

V of Q = 1/3 Bh B = 100 le = 10 so

$\frac{100\sqrt{50}}{3} \doteq 236$ Sh = $\sqrt{75}$, so h = $\sqrt{50}$

V_R = 1/3 Bh B = 100 sh = 10 so h = $\sqrt{}$

$V_R \doteq 289$.

$V_Q < V_R < V_P$

Because the area of the base is equal, it can be cancelled as can the 1/3. So, Therefore, the volume is directly proportional to the height.

14.3 Projects and Portfolios

In order to assess student progress, teachers often have students demonstrate and compile their work and class experiences over time. Two of the most common approaches use projects and portfolios.

Projects

Getting students involved in modeling in the mathematics classroom means having them work on tasks that take longer than a class period to complete. Such experiences call on students to examine a situation with some care, make assumptions, gather background information or research a topic in greater depth, develop a plan to confront the problem, and carry it through by watching the behavior of critical variables. Once a "solution" has been developed, the student needs to see if the solution makes sense in a real-world setting. How dependent does it appear to be on the data used to frame the situation? How easy is it to implement? To what degree does it generalize to related problems?

Consider the project described in Example 1.

Example 1 ***A Mini-Project Investigation***

The state government is considering increasing gasoline taxes 10 cents per gallon in an attempt to build a fund to deal with environmental problems.

a. Estimate the increased cost of gasoline on an average driver over a one-year period.

b. Do you think the increased gasoline tax will result in less driving? Justify or explain your answer.

c. Sketch a possible graph of the increased costs to an average driver over a yearly period for tax increases of 10–50 cents in increments of 10 cents. Justify or explain your reasoning.

d. Write a report and recommendation for the committees that are considering this increase (Stenmark, 1991, p. 19). ■

This project is closed-ended and directive. However, it does give the student a chance to investigate a rather straightforward situation where the creative leaps demanded are small and the outcome is relevant to real-world politics and economics.

Example 2 describes a project that requires more thought and productivity on the part of a student or students. It draws on student knowledge of how highways work and issues that relate to the efficient flow of traffic. Example 2 was the featured problem in the 1999 High School Mathematical Contest in Modeling.

Example 2 ***A Highway Design Project***

Major thoroughfares in big cities are usually highly congested. Traffic lights are used to allow cars to cross the highway or to make turns onto side streets. During commuting hours, when the traffic is much heavier than on any cross street, it is desirable to keep traffic flowing as smoothly as possible. Consider a 2-mile stretch of a major thoroughfare with cross streets every city block. Build a mathematical model that satisfies both the commuters on the thoroughfare as well as those on the cross streets trying to enter the thoroughfare as a function of the traffic lights. Assume that there is a light at every intersection along your 2-mile stretch. First, you may assume the city blocks are of a constant length. You may then wish to generalize to blocks of variable length (HiMCM, 1999). ■

Solutions to project problems such as this require a great deal of thinking and work on the part of students. They will need to make a number of assumptions dealing with the design and flow restricting/enhancing features of the highway. The evaluation of students' work centers on their communication of their answer. First, you may wish to take a look at their assumptions, their follow-through to getting a good representation of the problem, the identification of variables and their use in developing the model. Students will need to show how the model holds up under realistic conditions. Finally, they need to present the sensitivity tests of their original assumptions.

Portfolios

Portfolios are samples of student class work and associated student artifacts. They provide a broad picture of what the student has produced in class, along with a sample of their writing, project work, and other items that help paint a longitudinal picture of who the student is mathematically. For example, a class test, no matter when it is taken only records what a student was able to do that day. It is thus referred to as a cross-sectional view of the student's work; that is, it is a picture that cuts across the student's knowledge, skills, attitudes, and beliefs on that given day. However, we also want to see how a student changes over time, or longitudinally, as a result of their experiences in class.

A portfolio is an effective way to gather, organize, and reflect on such information. Generally, a portfolio is an organized folder or scrapbook of the student's classroom work, quizzes, tests, homework, projects, problems, essays, and other pertinent artifacts related to the student's mathematics experiences in some frame of time.

Methods of Development

The actual development of a portfolio can be done in a variety of ways. The following provides some direction in what might be included and how it might be organized in order to best answer questions a teacher, parent, or student might have relative to the student's progress in mathematics.

In general, we can think of a portfolio as a reflection of a student's growth in mathematics over time. As such, we would want information in the portfolio that reflects, among other things, their conceptual knowledge, their procedural understandings, their attitudes and beliefs, and their ability to apply mathematics in both theoretical and applied settings.

With respect to the student's conceptual knowledge and procedural understandings, an effective portfolio might include

- work that shows the student's conceptual understanding of the same topic at two or more points in time
- work from other subject matter areas that reflects the student's ability to connect the concepts in mathematics with concepts in other disciplines
- work that demonstrates the student's ability to use different representations to examine core concepts, principles, and procedures
- work that shows the student's growth in problem solving and modeling
- the student's corrections of items where they initially had misconceptions
- work on open-ended problems along with their rubric scores

- a copy of a group project addressing a significant problem, along with a student reflection on their part of the project and what they learned as a result of the experience
- a "mathematical autobiography" illustrating their view of themselves as a mathematics student, their view of the worth of mathematics as a discipline, or their view of the difficulty of mathematics as a subject
- your observation notes or a written transcript of your interview with the student
- a written piece illustrating the student's ability to read a mathematical text with meaning and then write a summary of it

Portfolios should thus contain information relative to the student's *growth*. Growth in understanding mathematics as in cognitive development and affective change. Growth in the student's ability to apply cognitive processes to mathematical situations and contexts. And growth in performance as an individual and in groups. The written pieces in the portfolio should include student reflections on their mathematical development and knowledge, as well as what individual experiences in class have contributed to that development.

Methods of Portfolio Management and Review

In order to get good portfolios from students, teachers need to think through the organization and management of the portfolios before embarking on their use. Many teachers find that by starting small and experimenting with a variety of the possible artifacts they find what works best for them and their students. Clearly you want contents that help assess students' growth.

An interesting way to begin is to have students write a "mathematical autobiography" that reflects on a list of topics. Across the year, you can make sure that artifacts are collected and added to the portfolio that relate to these topics. As the year proceeds, papers and class examinations should be added to all portfolios. This allows for cross-student comparisons of progress as well. Regularly ask students to add something to the portfolio that requires them to reflect, in writing, about what has been going on in class or how their view of mathematics is changing, or something along these lines. This helps students begin to monitor their own development and begin to assess their own growth and change. Meaningful assessment of one's own work is a difficult and important skill to develop. Giving students a checklist and copies of interview/observation notes you have compiled may help them frame questions to ask themselves.

Some teachers keep two different portfolios of student artifacts. One is the portfolio that the student compiles. The other the teacher compiles from assessment information and evaluation products. For example, a student's portfolio might contain examples of daily work, written assignments, autobiographies, and journal-like entries. The teacher's assessment portfolio for the student might include examples of quizzes, tests, group projects, observation notes, and interview checklists.

The review of portfolios can take as many forms as the portfolios themselves. Clearly teachers cannot be expected to examine all students' portfolios every day or even at one period in time. You might consider establishing a regular rotation for examining the contents of student portfolios—say four or five per day. These evaluations might be done for the purpose of noting changes in student performance or belief that need support and attention, making sure that students are remaining up to date in getting things into their portfolios, or for tracking both the student's progress and their maintenance of their portfolios.

At the end of a particular segment of study, you may wish to evaluate all of the portfolios in a short period of time as part of your grading activity. As part of this process, it is usually advisable to give students a writing assignment that includes revising the contents of their portfolio, pruning its contents to a specified list of required artifacts, plus other items they wish to save that they feel reflects their best work. At this time it's advisable to ask them to again write some form of reflective work that looks at one or more aspects of their development as a student or "doer" of mathematics.

Portfolio evaluation is best done using a checklist with room for notes and an observation form to leave, with comments, in the student's portfolio. You might look for evidence of the following as you evaluate a student's portfolio:

- what portion of the assigned portfolio work the student has completed;
- evidence of student growth in problem solving, reasoning, and representation;
- evidence of student connections of mathematics to other disciplines and within mathematics itself;
- the quality of the student's communications;
- evidence of growth in areas where misconceptions had been noted earlier;
- the student's ability to carry out extended projects and reflect on the final results; and
- the student's changes in affective matters related to mathematics and his or herself.

Stenmark (1991) suggests using a rubric to evaluate student portfolios. You might make modifications to this rubric as you begin to use it and continually modify it until you find a form that works well for you. Here are the four levels in Stenmark's rubric:

Level 4. The student's portfolio is exciting to look through because of the richness of understanding and uses of mathematics that it portrays. It employs a variety of representations of the mathematical concepts in displaying the student's understanding of mathematical concepts and procedures. The work is presented in an orderly, but very reflective way. It shows organization and growth in the student's work, by reflecting on what was and what now is. It is not neatness, but a structure that supports reflection that is the key ingredient of the organization. The student has included a piece providing reflective information relative to the contents of the portfolio.

Level 3. The student's portfolio is a solid piece of work. A variety of contents includes all but perhaps one of the assigned pieces. The student has written a piece that provides evidence of the organizational structure of the portfolio, but it is not as reflective as that of a Level 4 student. The student is able to make an adequate explanation of their work and the papers reflect some aspects of the student's work over time. Gaps occur in the coverage of the papers included, both in topics and in work over time.

Level 2. The student's portfolio is bound by assigned work, but it mainly reflects responses to textbook exercise sets and required assessments. There is little to provide evidence of the student's work on projects, group work, or other nontext forms of assignments. The reflective pieces do not show evidence of effort or attempts to describe/monitor the student's work. While some attempts at organization are visible, there are gaps in the work.

Level 1. The student's portfolio contains little or no evidence of the student's work outside textbook exercise sets. The assigned reflective pieces are shallow or incomplete. The student's attempts at organization show little thought or understanding of the task at hand.

Exercises 14.3

1. Develop a list of four projects that you might use at the middle school level to assess student understanding in the areas of direct proportion, simple probability involving independent events, finding the volume of a cylinder, and developing a general rule for a geometric sequence.
2. Develop a list of four project ideas for the eleventh grade that reflect student knowledge of combinations, similar triangles, applications of the law of cosines, and use of logarithmic or exponential functions in a modeling situation.
3. Devise a checklist for evaluating a student portfolio based on homework and test items. Explain why you included these items.
4. Write up the directions for a portfolio assignment for middle school or high school students, explaining what you would like them to include in their portfolio of work and how you will be evaluating the portfolio.
5. How would you describe the value of project work to a group of parents who are very concerned that their students have complete mastery of the basic procedural skills and are worried that a focus on project work and applications rather than on procedural skills will hurt their performance on college entrance examinations?

14.4 Multiple Choice and Other Forms of Assessment

There are many other forms of information about how students are progressing in their learning of mathematics. Classical forms of assessment include multiple choice, matching, and true-false items. These types of items are often used in large-scale assessments, like achievement tests, state assessments, and national surveys of student learning. They offer the advantage of gathering a large amount of data in a small amount of time, as students don't have to write their answers, only select an alternative to indicate their choice. They also increase the reliability of tests, that is, the degree to which the test results will be replicated in repeated testings under similar conditions.

On the other hand, these forms of collecting data don't allow the teacher to see what a student would have constructed on their own. The forced-choice nature limits the responses that a student can give. Hence when a student gets an item correct, we only know that the student has marked the response form attached to the correct answer, not that they would have constructed this response themselves under other conditions. Teachers need to be aware of the advantages and disadvantages of these types of items and how to maximize the information these more classical forms of assessment items provide.

Multiple Choice

Multiple-choice items can be used to measure conceptual, procedural, and problem-solving/modeling outcomes. However, they are best suited to measuring students' knowledge of facts, terminology, conventions, classifications, or generalizations. These are aspects of mathematical knowledge that can be stated in a clear and unambiguous way and have a single correct answer.

For example, consider the following item for eighth graders used in the Second International Mathematics Study (Crosswhite et al., 1986):

Example 1 *A 1986 Multiple-Choice Item for Grade 8*

One bell rings every 8 minutes, a second bell rings every 12 minutes.

They both ring at exactly 12 o'clock. After how many minutes will they next ring together?

A. 8

B. 12

C. 20

D. 24

E. 96

■

This item measures student knowledge of problem solving in whole number settings. A student who answers the item correctly probably has some understanding of the least common multiple of 8 and 12.

Another example of such an item is the following item for eighth graders from the 1992 National Assessment of Education Progress (NAEP, 1994):

Example 2 *A 1992 Multiple-Choice Item for Grade 8*

For each figure below, the lengths of 3 sides are given. Which figure could have a perimeter of 28?

A. Figure A

B. Figure B

C. Figure C

D. Figure D

E. Figure E

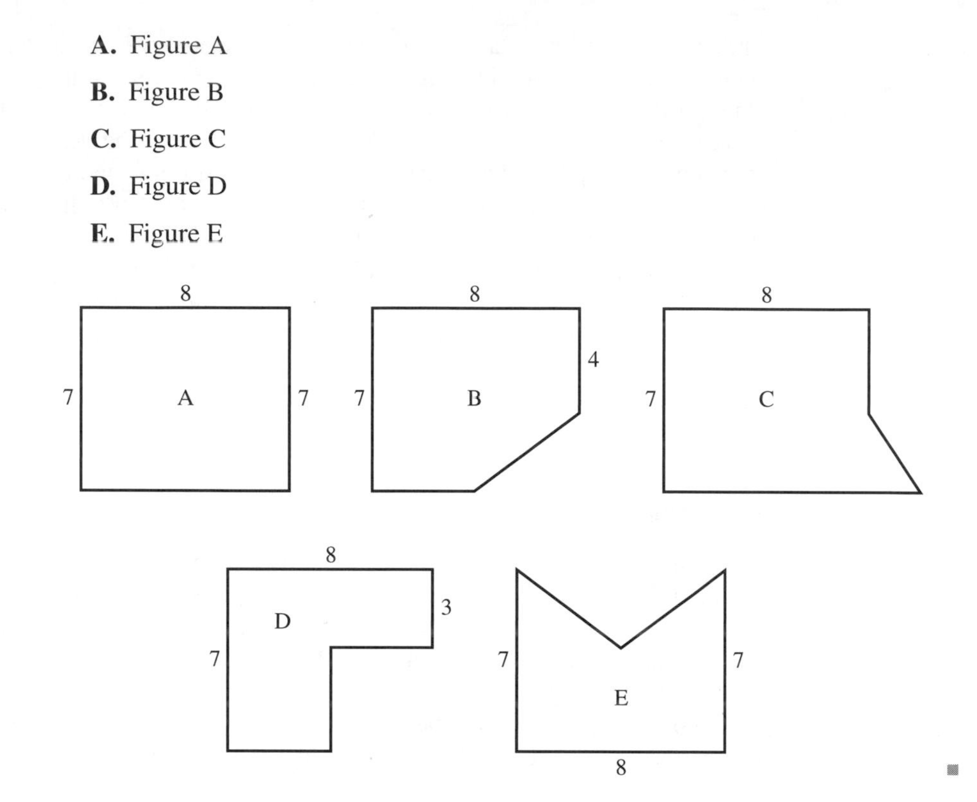

This item assesses a student's knowledge of perimeter and the relationships expressed by the lengths of the sides of the figures shown. An analysis of the figures shows that the perimeter of the shapes A and D are equal, as the sides of the "indent" in D are equal in length to the same sides had they not been "indented." These shapes appear to have perimeters of 30 units, if one assumes that the angles shown are really right angles. The triangle inequality, which states that the sum of the lengths of any two sides of a triangle must be greater in length than the length of the third side, shows that C and E are both greater than the perimeter of A and D. Hence their perimeters are greater than 30 units. Shape B, on the other hand, has a perimeter that is bounded above by 30 units. It is the only possibility here for a shape with perimeter of 28.

In general, multiple-choice items should

- measure a single, well-defined outcome
- be stated in simple, clear language
- contain as much of the wording as possible in the stem of the item
- be stated in a positive form or have negative wording emphasized when used
- have all alternatives grammatically consistent with the stem
- use alternatives that are plausible and attractive to the uninformed
- avoid words such as "all, " "none," "each," or "never"

- vary the position of the correct answer in the listing of alternatives
- be independent of other items on the examination

Constructing a good set of multiple-choice items is a difficult task. To do so, first list outcomes that match the facts, terminology, conventions, classifications, or generalizations related to the area of mathematics to be assessed. Then select from this list outcomes where multiple-choice items might be most appropriate. Develop the stems for the items, choosing matching alternatives that reflect the correct answer and common misconceptions or miscalculations often found in student work. Each alternative, besides the correct answer, should be equally plausible so that they can't be easily eliminated. Avoid using the alternative of "none of the above" as it doesn't show whether the student knows the correct answer—only that they could eliminate those shown. Likewise, don't use it as a nonfunctioning "filler." Students will soon discern this and immediately eliminate it as never being the correct answer on your tests.

Examine students' responses to multiple-choice items at the alternative level. The patterns or percentages of incorrect choices reveal a great deal about students' understanding of the area being tested.

Consider, for example, the answer patterns to Examples 1 and 2. For Example 1, the percentage of eighth grade students selecting each alternative was as follows:

A. 8 (7%) B. 12 (12%) C. 20 (18%) D. 24 (41%) E. 96 (18%)

For Example 2, the response pattern was

A. (15%) B. (32%) C. (15%) D. (20%) E. (15%)

In Example 1, 41% of the students selected the correct answer (D). The pattern of the remaining answers suggests that students tended to either add the two numbers (C) or multiply the two numbers (E) to get their answer. Since this accounts for 36% of the students, a few moments of class time to clear up this misconception would be worthwhile.

Responses to Example 2 indicate that except for those who chose the correct answer, students appeared to be essentially guessing, because the percentages show a pretty even distribution. This may indicate a more in-depth review of this concept is in order.

True-False

True-false items, of course, only have two choices. They are useful in situations where there are only two logical choices possible. Beyond that, they should probably be used sparingly. For example, wrong answers don't help you diagnose misconceptions. No diagnostic information like we saw for multiple-choice items can be drawn from true-false items. Secondly, the student either gets credit or no credit for the item. This limits the range of scores possible and has a negative effect on the reliability of the test, as it decreases the variability.

Here is an example of an appropriate true-false item:

The logical statement $(p \vee \widetilde{p}) \rightarrow p$ is a tautology.

True False

In this case, the statement is either a tautology or it isn't. There is no room for other choices.

Role in Reliability

The reliability of a set of test scores is a measure of the consistency of the scores over time. That is, how stable are the scores for an individual over a set of repeated measurements? Or how stable is the placement of individuals in the same group over a set of administrations of the exam? While we expect a certain amount of variation in scores or positions over different administrations, we hope that the correlation of the scores or positions remains high. When the correlation is computed for two applications of the same measurement instrument, we call the correlation coefficient a reliability coefficient.

This can be done in a number of ways. We can give the same test or measure to a group on two different occasions and correlate the scores. We can create "equivalent" forms of the test and administer each of them and correlate individual's scores. And we can subdivide a test internally into two or more equivalent "subtests" and correlate the results of individual's performances on the subtests. Each approach has advantages and disadvantages, and numerous texts on measurement detail this.

An easier way to calculate the reliability of a test with dichotomously scored items, that is, items scored as right or wrong, is the Kuder-Richardson formula. This approach, which provides a conservative estimate of test reliability, is based on assumptions parallel to those of the split-half test approach to reliability. The formula is

$$r_{xx} = \frac{n}{n-1} \, \frac{s_x^2 - \sum_{i=1}^{n} p_i q_i}{s_x^2}$$

where r_{xx} = reliability of the test

n = number of items on the test

s_x^2 = variance of scores on the test $\left[\sum(x-\overline{x})^2/n\right]$

$p_i q_i$ = product of proportion of passes and fails for item i

Calculating the estimate of reliability results in a coefficient of reliability, r_{xx}, between -1 and 1. Reliability coefficients for classroom tests should range between 0.60 and 0.80. Similar statistics for large-scale assessments, like state assessments or standardized tests, should be between 0.90 and 1. When interpreting the meaning of a reliability statistic, say $r_{xx} = 0.89$, we say that 89% of the variation on the test is due to variation in the true score of individuals and 11% is due to error in the measurement.

Examining the Kuder-Richardson formula, we see the role that test length plays in determining reliability. As the number of test items, n, grows, the finite sampling correction factor, $n/n-1$, approaches 1 in value. However, the second factor approaches 1 also, as the value of s_x^2 gets smaller as n increases. Hence, having a larger number of items on a test increases the reliability significantly (Thorndike & Hagen, 1969). However, at the same time, we want to make sure that what we are measuring is valid and worthwhile. It does no good to measure worthless information with great precision.

Exercises 14.4

Examine the multiple-choice items shown in Exercises 1–4, along with the percentages of students selecting each alternative. What do the results tell you and what actions would you take based on the information if you were these students' teacher?

1. What is the volume of a rectangular box with interior dimensions 10 cm long, 10 cm wide, and 7 cm high? (Grade 8)
a. 27 cm^3 (29%)
b. 70 cm^3 (9%)
c. 140 cm^3 (7%)
d. 280 cm^3 (4%)
e. 700 cm^3 (50%)

2. A painter is to mix green and yellow paint in the ratio of 4 to 7 to obtain the color he wants. If he has 28 liters of green paint, how many liters of yellow paint should be added? (Grade 8)
a. 11 (10%)
b. 16 (27%)
c. 28 (13%)
d. 49 (41%)
e. 196 (6%)

3. If $a_1 = 1$ and $a_{n+1} = a_n + 2n + 1$ for $n \geq 1$, then for all n, (Grade 12)
a. $a_n = 4$ (4%)
b. $a_n = 4n + 2$ (9%)
c. $a_n = 2n - 1$ (24%)
d. $a_n = 2n + 2$ (12%)
e. $a_n = n^2$ (33%)
Students omitting (18%)

4. If $2x^2 - 12x + 9 = 2(x - a)^2 + b$, then (Grade 12)
a. $a = -3$ and $b = -9$ (5%)
b. $a = 3$ and $b = -9$ (67%)
c. $a = -3$ and $b = 9$ (11%)
d. $a = -3$ and $b = 27$ (5%)
e. $a = 3$ and $b = 27$ (6%)
Students omitting (6%)

5. Write a five-item multiple-choice test that examines student knowledge of the existence and nature of solutions to the general quadratic equation.

6. Write a four-item true-false examination to assess student knowledge about the rules for divisibility by 6.
7. What are the strongest arguments that you can make for the use of multiple-choice items to evaluate student knowledge in mathematics?

14.5 Student Evaluation and Grades

Assigning grades to evaluate and report on student work is one of the most difficult tasks that teachers engage in. This act calls on the teacher's professional judgment, just as surgical decisions depend on the surgeon's professional knowledge and experience. A lot of people think that "teachers just have to add the scores and compare the final numbers with some fixed scale for grades A–F. So the only decisions remaining are the borderline cases. I could do that!"

In reality it is not this easy. As you enter the classroom, you will see students whose group work reflects strong understanding, but whose performance on daily work may be considerably lower. You will have students whose daily work is consistently high, but who occasionally misfire on a major examination. And you will have students consistently on the borderline of excellent work on both daily and group assignments. What type of grading system is fair and equitable to all of them?

Keeping Daily Grades

One point made earlier is that the greater the number of scores or items that enter into a value, the more likely the resulting value is reliable. Keeping daily notes or scores provides a basis for determining what is normal behavior and what is not. It also allows you to catch problems early as a student's performance changes. Teachers collect daily work information in a variety of ways:

1. Teachers assign homework grades based on student effort: 3—most work done and correct; 2—majority of work done and partially correct; 1—some work done and signs of understanding; 0—not much more than their name on the paper; and blank box in the grade book for a student doing nothing. This procedure allows teachers to make a quick assessment of the students' individual work and maintain a focus on the learning in the classroom. Students with an excessive number of 0's or blanks will show up quickly in the grade book. Here the focus is not on the grading and punishing of students who are still in a stage of learning.
2. Some teachers use a homework quiz to assign daily grades. This is done after considering homework and dealing with student questions. Students then work out one problem that was on the homework assignment in detail and hand it in for scoring. This can also be evaluated on a short 0, 1, 2, 3 rubric. This gives students a chance to ask questions about problems they had difficulty with before the quiz. It also stimulates note taking on the part of the students. Some teachers allow the use of student notes in such quizzes, others do not.

3. Some teachers have students keep a notebook of their homework and pick it up and grade it from time to time. This grade may reflect individual day's work or be a more summative grade of greater magnitude reflecting the number of assignments during the period the grade reflects.

While a myriad of approaches exist, the ones noted here allow for student motivation and reflect on the efforts that a student has made in getting the work done. Approaches 1 and 2 provide teachers with rapid feedback on situations when problems are developing. Approaches 2 and 3 allow for students to make some corrective action prior to the collection of the evaluative marks. The main feature in such day-to-day collection of information is that

- it provides the teacher with accurate information on the student's continuing work in mathematics;
- it can be done quickly and reliably without detracting from the activities of the classroom; and
- it allows for students to get some corrective feedback prior to any marking that focuses strictly on the correctness of the work.

More Major Summative Assessments

In addition to daily grades, there are quizzes and tests to be considered, along with group work, projects, portfolios, and other sources of information concerning students' performances. Teachers must consider what role each of these plays in their final evaluative mark for students for each grading period and over the course as a whole. It is important that this be communicated to your students in a policy document and that students, and their parents, understand how you will evaluate the students. If you don't get a chance to meet with parents and discuss this as part of an open house at the beginning of the year, it is worthwhile to send a note home. With student grades playing such an important role in college admissions and other important decisions, it's best to have a policy in place that has been discussed prior to any questions being raised about individual grades.

As noted earlier in the discussion on building a test, a major grade should reflect the broad range of objectives covered during the period being evaluated. In a like manner, the test should sample the students' knowledge of concepts, principles, procedures, and the major processes. The balance of these is a personal decision for the teacher and the mathematics department as a whole, but it should be balanced and representative of the learning that the evaluative remark is reporting.

Some teachers grade to an absolute scale for tests. That is, they establish point values of various problems by constructing partial-point rubrics for evaluating items worth more than one point. Then they compare the point totals to a scale like A: 90–100; B: 80–89; and so on. They are careful to establish ways for dealing with rounding scores in the boundary areas, from 89–90, for example.

Other teachers look at the total distribution of scores from an examination and look for natural breaks in the scores. For example, consider the following scores for a 100-point examination that were observed in a classroom:

99	85,	76, 76, 76, 76	62
96	84, 84	75, 75	57
95	83	74	55
93, 93	81	71	43
91	80	68	
89	79	67	
87	78, 78	65	
86, 86			

These 34 scores have a mean of 78.32, a median of 78.5, and a mode of 76. The standard deviation is 12.4, the maximum is 99, and the minimum is 43. Examining the scores, one teacher might see the gap between 91 and 89 and make a break for A's at 91. Then looking further down the list, the gap between 83 and 81 might be selected for the boundary for B's, the gap between 74 and 71 for C's, and the gap at 62 for D's. Thus, this test's scale might be:

A: 91–100

B: 83–90

C: 74–82

D: 62–73

F: 0–61

This scale is built around natural gaps at the 60, 70, 80, 90 cut points often associated with 60–69/D, 70–79/C, 80–89/B, and 90–100/A, but uses the natural breaks in student performance. Further, the gaps were chosen so that, where possible, no one is cut off by one point. Assuming that all tests in a course have the same number of points, these cut points can be summed to get overall cut points for the tests themselves over a grading period.

Yet other teachers write an examination and then look through the examination, writing down the number of points they would expect an "A student" to get on each item. They sum these points to find the A cut point, then do the same for the B–D cut points. The teacher can then examine these points and adjust them somewhat depending on the pattern that results. Frequently, teachers using this approach will also make some adjustments based on the distribution of actual scores looking for a balance between their predicted breaks and actual score distribution.

There are other ways to develop grades, but all involve some manner of looking at the total distribution of students and at how a student rates relative to the entire distribution. This is called a norm-referenced approach. Another approach compares students against an absolute set of standards that meet the objectives set for the students at the outset of the chapter or unit.

Deciding Final Grades

Students and parents should be aware of the percentages of each activity entering into the grade and the weights that each will carry. When rubric-based scores and checklist information is entered into the mix, it is important that students, and parents, see how you will evaluate such activities. Many feel that this is "subjective" and can weigh against a student who is "good with the numbers." Here you have a responsibility to talk about the broad range of objectives that students need to reach as part of their mathematics learning experience. You also need to talk with parents, and students, about the fact that rubric scoring can be quite objective. In fact, that is the reason for the development of a coding guide.

Beyond the development of assurances that you are acting in a fair and equitable manner, you need to make the hard decisions about the relative weights you will assign to different portions of your overall evaluation of the students. This might take the following form:

Daily grades 20%

Quizzes (5 per period) 10%

Tests (3 per period) 55%

Portfolio 10%

Teacher observations 5%

You will have to find a balance that suits you, but once you have made the contract with the students, you need to stick to it, or announce any changes prior to the start of a grading period.

In deciding on weights, you may also want to check with your fellow teachers concerning the weights they assign to topics. The development of a departmental approach to evaluation helps students become used to the procedure, regardless of the individual instructor they have. It also helps establish some uniformity in student expectations across the mathematics curriculum.

Dealing with Questions on Grading

Every teacher will experience questions about their evaluation of students, either from the students themselves or from their parents. When a question comes up about an individual daily grade, it is important to get the question answered rapidly and any corrections made immediately. This is not something to be dealt with after a summative grade has been assigned and then the student or parent wants to go back and check through each assignment. It is important to have a period of time for corrections to be dealt with after a grade is given. Then, with few exceptions, that grade stands.

When a parent, or student, brings up a question about a grade, keep two important maxims in mind:

- A grade is a report of what happened, not what might have happened.
- A grade is earned by the student, not given by the teacher.

With these maxims in mind, you can examine the nature of the student or parent question about the grade. Did the grade report what happened? If so, you can describe what remedial actions the student might take to improve their performance. Was the grading consistent with the policies established? If so, what did the student or parent not understand? Using these maxims, you can sit on the same side of the issue as the parent or student in examining where the student's efforts did not earn the mark desired. Remember that all that you did was report on the student's performances and products.

Dealing with Cheating

One of the most disturbing situations a teacher has to deal with involves potential student dishonesty. Such situations cause a number of difficulties. The first is the potential destruction of relationships between the affected student and teacher. The second is the potential of not being able to accurately evaluate the student at all. The third is the time and effort that such situations consume.

It is best to discuss the problems associated with cheating with students openly. Often, the use of a stern glance at a student during a test, walking over near their desk, or a general remark to "Keep your eyes on your own papers," is sufficient to handle such situations. Comments to students noting the similarity of their paper to that of a neighbor's paper may be a next level of approach should there be some evidence of one or more students having the same incorrect student work and either sitting close to one another or potentially having worked together on an individual out-of-class assignment.

Dealing with a major infraction requires more careful action. The first step that you must take is to collect all of the evidence that exists and make copies of it for further records. The second thing is usually to discuss the matter with the department chair or school dean/principal who deals with such cases. Then there are a number of potential remedies. The first is to give a grade of 0 for the assignment; however, this does not reflect the student's work. It is probably better to retest the student individually with an equivalent examination, perhaps one written by another teacher, but covering the same material. If such an examination is not available, you might write a parallel examination for the student. Bear in mind that the student already knows the balance and content, so they have an unfair advantage over their counterparts that took the examination without signs of misconduct. There is no quick remedy for dealing with this that does not make a presumption of guilt. Hence, you must let the warning and public questioning of dishonesty stand as the warning and punishment. However, a written description of the incident should be maintained in case there is a recurrence.

The use of alternative forms of tests is another possibility for cutting down on the possibilities of dishonesty. However, be prepared to deal with the fact that one test may be slightly harder than the other. Also, be sure to put student's names on the tests to keep students from trading them once they've been passed out.

In any case, questions of dishonesty making significant changes in a student's performance must be dealt with, as they alter the overall performance of that student relative to others. The student not only alters the assessment of their own work, but the change in their position relative to that of their classmates affects the interpretation of the work of everyone in the class.

Exercises 14.5

1. Some schools have different grading scales and expectations for different tracks, or levels, of mathematics courses. What are the advantages and disadvantages of differentiated grading (for example, for honors courses, for special education courses)?

2. Sam argues that grading homework is unfair, in that you are grading students on their first attempts to master material. Susan argues that it is OK to grade homework, as long as you don't make it a major portion of the final period grade. Where do you agree and disagree with Sam or Susan? What do you think your policy will be?

3. What are the pros and cons of take-home examinations? What is their role in middle school mathematics? In secondary school mathematics?

4. Obtain a copy of a standardized achievement test. Examine the items on the test and then write a paragraph detailing what role you think that a student's performance on such a test should play in their evaluation in mathematics class.

References

Black, P., & Wiliam, D. (1998). Inside the black box. *Phi Delta Kappan* (October):139–48.

Bush, W. S., & Greer, A. S. (1999). *Mathematics assessment: A practical handbook for grades 9–12*. Reston, VA: NCTM.

Crosswhite, F. J., Dossey, J. A., Swafford, J. O., McKnight, C. C., Travers, K. J., Cooney, T. J., Downs, F. L., Grouws, D. A., & Weinzweig, A. I. (1986). *Second international mathematics study: Detailed report for the United States*. Champaign, IL: Stipes.

Feldt, L. S., & Brennan, R. L. Reliability. *In* R. L. Linn (ed.). *Educational measurement* (3rd ed.) (pp. 105–146). New York: Macmillan.

High School Mathematical Contest in Modeling (1999). *Traffic flow problem*. Lexington, MA: COMAP, Inc.

Messick, S. (1989). Validity. *In* R. L. Linn (ed.) *Educational measurement* (3rd ed.) (pp. 13–103). New York: Macmillan.

Mitchell, J. H., Hawkins, E. F., Jakwerth, P. M., Stancavage, F. B., & Dossey, J. A. (1999). *Student work and teacher practices in mathematics*. Washington, D.C.: National Center for Education Statistics.

Mitchell, J. H., Hawkins, E. F., Stancavage, F. B., & Dossey, J. A. (1999). Estimation skills, mathematics-in-context, and advanced skills in mathematics. Washington, D.C.: National Center for Education Statistics.

Mullis, I. V. S., Dossey, J. A., Owen, E. H., & Phillips, G. W. (1993). *NAEP 1992 Mathematics report card for the nation and states: Data from the national and trial state assessments*. Washington, D.C.: National Center for Education Statistics.

Second International Mathematics Study (1985). *U.S. Technical Report I; Item-level achievement data*. Urbana, IL: U.S. National Coordinating Center.

Stenmark, J. K. (ed.) (1991). *Mathematics assessment: Myths, models, good questions, and practical suggestions*. Reston, VA: NCTM.

Thorndike, R. L., & Hagen, E. (1969). *Measurement and evaluation in psychology and education*. New York: Wiley.

15

Classroom Management in Mathematics

Communication skills and the promotion of classroom discourse in mathematics should be approached developmentally. We cannot assume that if we give students an unstructured task such as "use data from current newspaper advertisements to examine the economics of buying a car versus leasing" that they will give complete explanations of procedures, solutions, and conclusions. Teachers should guide students in the development of mathematical communication until students achieve the skills and comfort level to communicate mathematical ideas effectively.

Gilbert Cuevas (1998)

Overview

Developing a routine that allows you to be efficient in the classroom and to fruitfully engage your students in the study of worthwhile mathematics, while still having time to individually affect your students, is a goal you will continually work toward across your teaching career. The creation of an inviting environment that allows students to confront mathematical questions in a no-holds-barred fashion without worrying about peer pressure or what grade will result is a context that all teachers strive for, but few achieve. Such classrooms create and support high expectations for students—expectations for their cognitive engagement and for their behavior. In this chapter we examine techniques that teachers can employ to make their classrooms move in these directions, while at the same time making them fun places to be as a teacher as well.

Focus on the NCTM *Standards*

The NCTM *Principles and Standards for School Mathematics* (2000) calls for the development of mathematics programs that hold high expectations for *all* students, while at the same time changing teaching to make the learning of mathematics more of a participatory activity for both students and teachers. Such instruction requires that teachers know their students, their capabilities, and their knowledge of the prerequisites for the topics of study. These expectations require teachers to increase their involvement with students in considering challenging problems that brings the mathematics in them alive. These are easy words to quote and write, but hard words to bring to life in the classroom. Such engagement, both with students and with the content, requires a great deal of planning and classroom management as lessons unfold.

The topic of classroom management is not merely a consideration of maintaining discipline and focus on the part of students, it is the overall administration of the learning environment for your classroom. Hopefully, some of the processes and approaches to doing mathematics that you encourage in the classroom will follow your students out of your classroom to study hall, to their individual work at home, and finally to their approach to doing mathematics in life.

Focus on the Classroom

Central to having an effective learning environment are the expectations that you set for your students at the beginning of the year. The setting of these expectations and consistent follow-through over the year is essential to making your time together with the students a pleasurable and successful experience. As you enter the year, or enter the classroom each day, two important thoughts should be crossing your mind. The first is "What are we here to do today?" (What mathematics is the central objective for this period? What learning goals do I want my students to reach today?) The second thought is "What expectations do I hold for my behavior and my students' behavior?" This question focuses on the consistency that was mentioned above. What are the classroom rules or expectations for behavior and interacting with one another? How are they met on a day-to-day basis? Serving as an adult role model in a consistent fashion is key to leading a successful class to learning in mathematics.

While the questions of discipline and administrative efficiency loom large in the minds of new teachers, they should not forget that they are interwoven with the questions of mathematical engagement of students. Classrooms that are alive with mathematics and both recognize and react to students' interests and styles of learning often assist in getting the mundane tasks accomplished quickly and helping hold down problems that arise when students are not engaged or recognized as individuals. While establishing routines and approaches to accomplish these tasks takes a few years, most teachers adapt quickly to what works well for them. The secret is to continue working with peers and looking for new techniques in professional journals and at professional meetings to enlarge your repertoire of teaching skills in these areas. Growth and change is important. It also adds some variety to the classroom, variety that students appreciate and react positively toward. When students see you working to change your methods to help them, especially when you explain to them what you are attempting to do, they react positively and will give you feedback on what they like and don't like.

Learning to teach is a journey. Along the way you will encounter a number of individuals that will significantly alter your approach to teaching mathematics. Some of them will be peers from your undergraduate years, some will be teachers you have already had as instructors, others will be teaching peers in schools where you teach, and others will be people who write or speak about the teaching of mathematics. Perhaps the ones that you will learn the most from are your students. Watch what they do and say as you work to get them involved in learning and using mathematics. See what works well. Examine why some methods work better than others. Reflect on your teaching and share your reflections with other teachers so that you can learn to grow better together.

15.1 Developing a Classroom That Works

Perhaps the most important facet to one's success as a teacher of mathematics is the development of a classroom that functions in an efficient and learning-oriented way. When teachers stumble, it is more likely that they have been unable to develop a routine for operating their classroom than that they cannot handle the mathematical knowledge required for their position. Part of the routine for the classroom involves the handling of administrative tasks (taking role, handling school administrative tasks that apply to the students in class, dealing with discipline, and so on), and part of the task involves creating and engaging students in quality learning experiences involving important mathematical content.

Before Classes Begin

As soon as you know what you will be teaching is the time to begin your thinking about your courses. From that point forward, you can begin to draw on materials that you have, talk with those who have taught the courses before, and begin to make your plans. Occasionally, factors at the school will not make this possible, but in most cases these decisions are made before the summer break begins.

Getting a Scope of the Year

Foremost in this planning is the school curriculum for the given courses. What written expectations exist for the course and the content coverage associated with the course? If there is no written curriculum, you might wish to consult with a couple of your colleagues who teach at other schools having a similar course and get a copy of their curricular guides for the course. In states with definitive sets of outcomes or state syllabi for courses, you may wish to review the state outcomes.

This is the time for making the big decisions first—about what major topics, or units, you will attempt to cover in a year. Yearly and unit planning were discussed in Chapter 12. It is here that you determine what the overall reach of the course should be. This is the time to make some rough estimates of the time required to accomplish the objectives associated with each of the big pieces of the year's study. In doing so, recognize that events at the school during the year may interrupt your plan and place the material at the end of the curriculum in some jeopardy. This must be kept in mind across the year, necessitating some adjustments as time moves forward.

Within the individual topics making up the year's plan, establish some objectives for what you wish to accomplish. These objectives begin to help form themes that will connect your course together over the year. Note where placing emphases on certain topics earlier when they are first encountered will save time and lead to better learning later in the year. From here you can begin to plan the first couple of units of study that will open the year. It is good to have the first two units in each of your courses laid out at the beginning of the year as other things will be competing for your time as you open the year and get your schedule off and running.

Organizing Your Room

Parallel with considering the courses you will be teaching, you need to begin to think about how you will want your room organized and how that organization will help you. Where do you wish to place your desk and filing cabinet(s)? Where will resources students frequently use be stored? Where are the screen and outlets for the overhead projector located? The answers to these and other questions will frame your decisions about how to make your room function effectively.

Clearly, you want the screen for the overhead projector to be at the "front" of the room where all students can easily see it. Using the screen as a focal point is probably more important than using a chalk- or whiteboard as the focal point. Anything that can be done at the board can also be done at the overhead. By using the overhead you also have visual command of the classroom as you are facing the students.

Many teachers have found that arranging their desks or student tables in a slightly curved pattern, as shown in Figure 15.1, with students in pairs is an efficient way to organize student seating. If you share a classroom with other teachers, you will need to work with them about what seating pattern works well for all of you. Daily rearranging the desks and chairs from one period to another costs valuable instruction time and removes the focus from mathematics immediately at the beginning of the hour—something that you do not want to happen. If desks have to remain in rows, have students slide rows together to form working groups of two students. In any case, have a pattern that allows

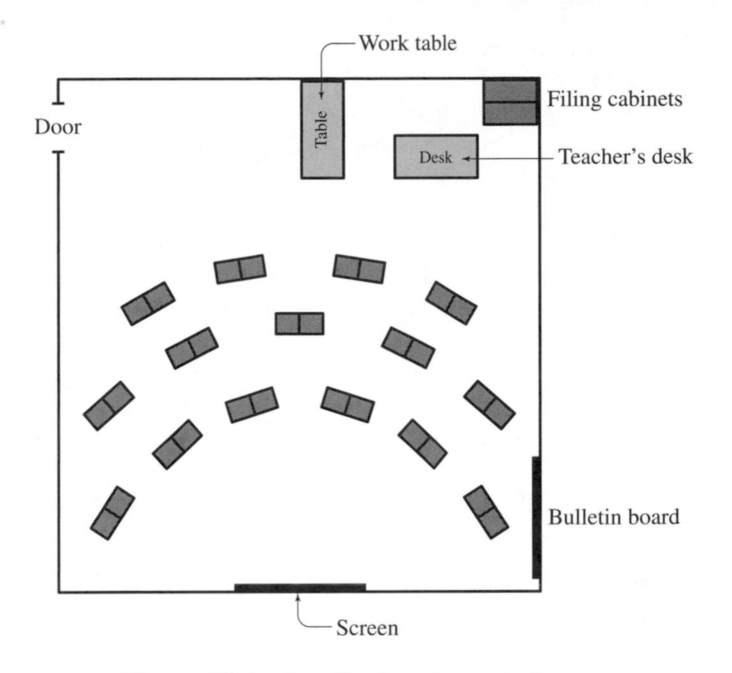

Figure 15.1 An effective classroom layout.

you to move about the classroom easily and to walk between the pairs of desks on all sides. It's nice to keep the view out the doorway out of students' easy view. This keeps passing events in the hall from disturbing the classroom activities.

In choosing to locate your desk, several possibilities exist. When possible, it is nice to put it at the back of the room, as it allows you to visually track student activities when they are working independently without them seeing what you are doing. It also gives you a bit more privacy, at least visually, when working with a student at your desk. Filing cabinets are best located where you can reach the most frequently used drawers without having to get up from your chair, if possible. In a like manner, it is good to have the computer you use for administrative purposes located near your desk, but maybe not on it. Sometimes the location of the computer on your desk takes away valuable planning and instructional space for working with students individually. However, it is nice to have it nearby so that you can move to it easily to work on instructional materials or deal with administrative tasks. Some teachers like the computer on their desk and like to have another table nearby to use for spreading out materials or working with students individually. Either plan works well and is a matter of individual preference.

The location of the overhead screen in the classroom is best when it is centered at the focal front of the classroom. (Remember that you may need to seat some students near the screen because of visual problems. Handle this early in the year by asking which students need such accommodations and deal with them discreetly. Make sure at the same time that some students are not just trying to change their seating.) It is best when there can be a seat at a low table at the front of the classroom where you can sit and write on the overhead. This allows for all students to see the screen without you or the overhead blocking the line of sight. It is nice to have a surface here, perhaps a small table, to put your materials on, hold the calculator or a text, and any other materials you might need. If you can get a stool, it will allow you to sit, but still have total visual command of the classroom, and make you less the focus of attention and thus turn the focus to the mathematics on the screen. In making these decisions, keep in mind the lighting in the classroom. How can you manage the blinds, the overhead lights, and other factors that make the screen easy to view while still giving you and students light to take notes, to monitor other activities in the classroom, and to make use of other instructional resources? When lights need to be dimmed, a specific student in each class can be asked to handle that task.

Another decision to make in organizing the classroom is where to put resources that you will use frequently. If there are calculators that have to be passed out and picked up, measuring materials, geometric manipulatives, or other instructional materials, decide where to locate them. If security is an issue, develop a way of coding the materials with serial numbers for specific students and a way of visually checking for their presence before class starts and before students are dismissed at the end of the hour. Some teachers have used the cloth shoe holders that hang on doors to hold calculators that are used in the classroom (see Figure 15.2). This is an efficient way of storing and visually checking they are all there. It is also important to have a couple of calculator batteries on hand for times when they go down during the hour. Check with your departmental chair or colleagues about how the purchase of such things or the access to school supplies is handled in your school.

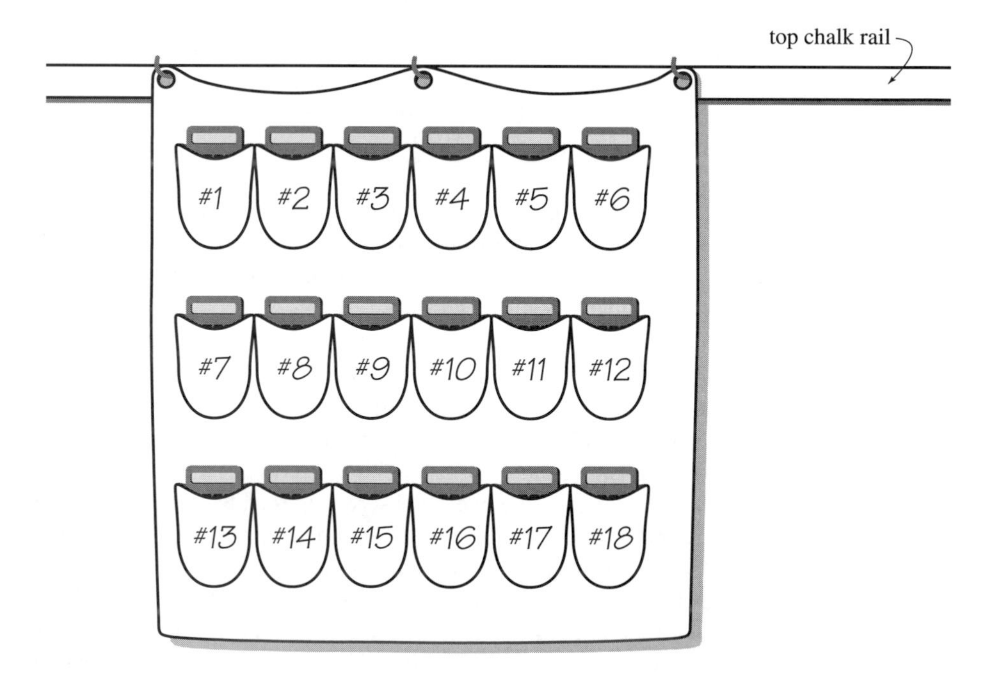

Figure 15.2 One way to account for classroom calculators.

The last organizational issue is making the room look attractive and inviting. Part of this is in having the room exude mathematics, but in an attractive and informative way. Bulletin boards can contain required information, such as fire drill and emergency procedures, classroom rules, as well as an attractive poster or problem of the week. Other posters can be placed in other prominent display areas about the classroom. Work to collect posters that both inform and, at the same time, uplift the classroom environment. It is handy to have a large poster showing the key locations on the calculators most frequently used by the students near the focal front of the classroom. Such posters are helpful in assisting students master new sequences of keystrokes on their calculators.

Developing Your Routine

Along with thinking about organizing your room to maximize your instructional efforts, you need to also consider how to most effectively arrange activities within the time you have been allotted for each class you will teach. How do you intend to deal with opening the class, to handle administrative tasks, to get the students involved in new material while still handling questions over previous material, and to make assignments for work to be completed before the next class period? These questions all deal with developing an operating routine for your classroom. You will bring some ideas from your methods classes, some from your student teaching experiences, and some ideas you will develop on your own. It is important to think through these and have a routine ready to go at the beginning of the year.

Often new teachers, and some experienced teachers, allot too much time to going over homework exercises. This leads to a time trap that allots too little time to the consideration of new material. This in turn causes more questions the next day, and soon the need to teach through answering questions on homework has taken over the time allotted for your class. The secret to avoiding this problem is to make sure that the development of new material is well done and that students are ready for the successful practice of the new material as they move into their homework. This does not mean that every piece of the homework assignment will be easy for the students. There are several ways to handle the homework to assure that questions are answered without eating up the time required to keep the flow moving. Some of these are discussed in this section.

Think of how you will open the class period. Many teachers use a "warm-up" problem, which is projected on the screen for all students to work individually or in pairs. This can be up and ready when the students enter the classroom, immediately focusing their attention on the mathematics of the day. Set a reasonable, but not too long, period of time for them to work on the problem. This time allows for you to quickly take roll and deal with any message or information that must be given to a student, perhaps one who missed the previous period.

Think through how you will signal students that it is time for them to focus their attention on you. It can be something as simple as turning off the overhead or walking to the front of the room and standing by the overhead or asking for their attention in a calm voice. You may simply say "Your Attention" or "Heads Up." Be careful not to phrase this as a request, as it allows for them to decide not to respond to the request. Some teachers ask students to raise one hand until everyone is focused. This involves students in helping get the focus for instruction.

Other decisions about how you will conduct class involve how you arrange for needed materials, papers, calculators, and other resources to be distributed. Think through how a few students can collect materials from a regular place and both distribute and collect the materials at the start and end of classes. This allows for the minimum disruption of the instructional activities. If papers are to be distributed to students, how can it be done as quickly as possible, preserving private information for students should grades or other comments be contained on the papers?

Another issue to consider in getting ready for class is the form of evaluation you will use. How will you gather evaluative information and combine it to determine grades for grading periods and semester/final grades? Such information is of vital interest to students and, once established, should not be changed unless for a very important reason. If such a change is necessary, it should not be to the detriment of any student and should be announced far enough in advance of its change to allow students to adjust to its impact on their personal position. With these decisions made and necessary arrangements carried out in the organization of the classroom, you are almost ready for the start of classes.

Beyond the Classroom

Along with getting things ready in your classroom, it is important to visit with support staff in the school to get to know them and their function in working with students you will share across the year. In particular, this set of individuals will include your department chair (if any), the counselor or a member of the counseling staff, the dean or individual in charge of disciplinary issues, and the librarian. In meeting with each of these individuals, it makes sense to have a list of questions relative to how your classroom interfaces with their duties and how you can work together to effect the best result for the students for whom you both have responsibilities. With your department chair, you may wish to explain your plans and get reactions before finalizing them. With the counselor, you may wish to determine their role—is it counseling for course selection, dealing with problem students, helping with college admissions, or a mixture of all of these? Getting to know them before you need their help in working with a student allows for establishing interpersonal relationships that might later get entangled with the situation involving some student having a problem. Find out the procedures involved in getting help for a student with problems and how they would like you to deal with such referrals if and when they might occur.

The same is true of the meeting with the dean or disciplinary officer for the school. Find out what the overall school policies are for discipline and how they are handled administratively. Are there referral forms? How are tardy students handled? What are daily attendance reporting procedures? Getting answers to these and other questions allows you to plan ahead and get this information factored into your operating style prior to school starting and prior to ever having to make use of these resources. Again, getting to know these support individuals personally before needing their assistance is to your advantage when and if you should need their aid.

A final set of resources to check on are the library holdings in mathematics and the technology support facilities. At the library you may wish to see if the school receives *The Middle School Mathematics Journal, The Mathematics Teacher*, or any other professional journals. Do they have back copies if they do? Also see what books they have on mathematics for students to use. How are audio-visual materials—movies, tapes, and so on—ordered for classroom use? The NCTM and Mathematical Association of America have lists of recommended materials, and every copy of their journals contains reviews of recently released materials (Steen, 1992; Thiessen, Mathias, & Smith, 1998). An updated listing of the MAA listing in the Steen work is available online at the MAA as well and covers the high school level and up, while the Thiessen text covers the elementary and early middle school grades. Between the two lists, you have a start at building a hardbound collection. In a like manner, you need to become acquainted with the audio-visual and technology coordinators for your school to learn about the availability of computer resources, both software and hardware, and how to schedule their use for your classes.

Beginning the School Year

The first session with your students is an important one. This is the time that you both form impressions about each other. These impressions will remain as the initial conditions that other information is added to as the year unfolds. Hence, it is important to get this session off on the right foot. Following are some suggestions for handling this session with your students. While these suggestions are given in a bulleted list, they should flow in a friendly and supportive way in the classroom.

- Greet the students at the door of your classroom, recognizing any that you know by their names. Have a problem up on the overhead for them to consider as they enter the classroom. The problem can serve as a basis for opening the class. It might be an application of the subject area that has played a role in a recent news story or a recreational puzzle whose solution could be covered in the class during the year. Greeting students at the door across the year is a good habit. It provides a break, or transition, from the hall topics and behavior of students and the mathematical focus that will take place in your classroom. It also allows you to make comments to, or recognize accomplishments of, individual students without having to do so in front of the entire class.
- When the bell rings, enter the classroom and greet the class as a group. Introduce yourself, giving a little information about yourself and assuring the students that you will want to get to know them better as well as the next few weeks unfold. As part of this introduction, you may wish to inform the students how they are to address you, how they can contact you at school if necessary, and other information of that type.
- Provide students with an overview of the course, preferably giving them a handout showing the major topics to be discussed. This handout might have your contact information, a rough timeline for the topic coverage, and evaluation information on it. In giving such coverage, indicate the applications of the topics and why they are important. Do not belabor the information, just give students a feel for the course. As part of this introduction, make sure to connect the opening activity from the overhead with the course outline and contents.

- Indicate what materials they will need in addition to their textbook or other curricular materials provided by the school. If a calculator is required, indicate what model is preferable. In so doing, you may wish to contact other teachers to see how they handle this. If students are to keep a notebook or journal, explain what you require here. It is also good to list as much as possible of this on the handout as well.
- This is a good time to turn to the expectations and rules that you have for them as students in the classroom. These comments can contain, but not be limited to, how you expect class to begin and end, homework expectations, how to get your attention, what level of talking is permitted, gum-chewing and hat-wearing, and other topics of importance. This discussion should be matter of fact, but firm. Don't mention the consequences of noncompliance, just state that these are expectations. It is also good to make a short typed list of these that can be distributed to students, but also posted on the bulletin board. This allows you to remind them of the rules by simply pointing at the list when needed.
- Discuss your policies on evaluation and grades. Be clear about how you will determine their evaluations with them. Alert them to keep their papers and other important documents should any questions arise about a point total or a grade. Also indicate to them that you will hold periodic conferences to check with them on their progress during the year. As part of this coverage of evaluation policies, let them know how daily homework, quizzes, tests, and projects or other major assignments are factored into the grading process.
- From here, you might wish to indicate how your daily classes operate. If they are to come in and begin work on a warm-up problem on the overhead, let them know. This is a good time to indicate what you want them to have out and on their desks for mathematics class at the beginning of each period. Make sure that you carry through on this expectation consistently over the first couple of weeks until it becomes routine. If some students are to get calculators or other materials to distribute, indicate how this will work. Here you might mention that there will be a seating chart on the overhead the next class period. Ask if any students having visual problems would let you know so that you can seat them appropriately.
- This would be a good time to collect some information from the students. You might ask them to fill out a notebook information sheet that asks for the students' name, what name they like to be called in class, student identification number (if they are used), home address and telephone number, e-mail contact, class schedule, and then ask them to write a short introductory autobiography for you. You might indicate that they could tell you about their interests, hobbies, school activities, whether they have a job, and such. Make sure that you make use of this information in talking with them over the coming weeks, thus valuing your interest in them as individuals.
- Depending on the amount of time left, begin the study of the syllabus. This should model the class period as being a period that has a businesslike, but expecting focus. If the time left is short, preview the text or curricular materials

showing them its organization and helpful features (glossary, tables of formulas, and so on) and making a reading and review assignment for the next class period. If there is more time, a more involved start can take place with a regular assignment.

The most important feature of the day is the setting of the tone for the remainder of the year. Hopefully the outcome of the session will be that you are concerned about them as individuals and learners, that the class will be focused and businesslike in organization, and that you are approachable and interested in helping them progress in their study of mathematics. At the same time, students should leave the class with knowledge that you have expectations for them as learners and for their deportment while they are in your class.

Homework

Handling the coverage of homework is a difficult task. There are days when almost everyone in the classroom needs some help or clarification about a particular problem. There are other days when essentially no one in the classroom needs any help with the problems that were assigned. As you grow in experience and get to know your students, discerning between these two situations will become easier. Until that time, you need to think through how you will deal with homework and what amount of time you will give to it in your daily lesson planning. As discussed earlier, it can take over your lesson development time and subvert all of your good intentions and planning if you are not careful.

Effective teachers have developed a variety of ways of dealing with homework. Some of these are

- Posting a list of worked examples that students can use to check their work against. In doing so, focus on solutions to the questions with which students most likely had difficulty and then allow students to ask other questions at a point near the end of the hour if they still haven't grasped how to deal with the question.
- Having students list the problems they had difficulty with at a particular point on the board when they enter class. Then look at the list, group the problems by type, and work particular examples to illustrate the main methods. After the new lesson has been finished, deal with individual questions on other homework problems.
- Having students help one another in pairs on the homework with the answers displayed on the overhead. Then take questions for approximately five minutes over a couple of the problems.
- Assigning particular problems to students to put on the board the next day when they enter the classroom. Allocate such coverage to four or five problems a day. Students should be responsible for working the problem or seeing you before school for help to be assured that the problem is up at the beginning of the hour.

There are other ways to handle homework. These give some alternatives other than just standing at the front of the classroom working through exercise after exercise. Remember that in many cases it is only one or two students who are having problems. It is easier to keep the whole class moving forward and deal with these students individually. Don't let students use homework they do not need to see slow down the pace of the class. Learn to discern when students know or don't know how to work particular homework assignments, and deal with these situations accordingly.

Discipline

In your opening remarks, you should make some comments about your expectations of student performance in your class. These might be limited to getting started at the bell and remaining on task until you, not the bell, dismisses them at the end of the period. This conversation should cover raising hands to speak and no calling out or blurting out answers. It should also cover your expectations about no gum-chewing and no hat-wearing or other nonproductive behaviors. Explain your reasons for such rules and move on with your expectations for their working with one another in group sessions and the degree of talking and volume that is expected during such work. Rules should be justified in terms of how such behavior supports a learning atmosphere in your class.

Such discussion should be handled in a straightforward and professional manner. However, it should be clear that you expect compliance. It is to be expected that students will test your consistency relative to these rules in the coming weeks. Make sure that you enforce your rules consistently each time your students probe at the boundaries of your expectations. Maintaining a classroom that keeps the focus on learning and free from disruptions is what you want to obtain from the rules and expectations.

Prior to when students first test your rules, you need to think through how you will handle situations. It is always better to practice proactive or preventive discipline rather than find yourself in a reactive or corrective discipline setting. Many discipline situations can be anticipated by teachers who are in tune with students and what is happening in their lives. As class unfolds on a daily basis, be sure to be moving about the classroom in an unpredictable fashion. If you note students talking, sometimes it is sufficient to move in their direction. If that is not sufficient, simply asking them to stop or tone down their conversation, if it is on topic, should be sufficient. Locating yourself to be near a potential problem is a strong preventive technique.

Another important technique is being in position to nip problems in the bud prior to their having a chance to spread. If a student makes a remark that not everyone hears, students will spread the remark across the classroom like ripples on a pond. Being in a position to cancel the ripples or acknowledge the remark yourself stops an undercurrent from developing and shows that you are in tune with what is happening. If you note that something is about to start, you might be in a position to ask a question of the person about to initiate something. In many cases, they may only want some recognition. If you can give it to them in a productive and professional manner, you fulfill their need while preventing the need for a potential corrective disciplinary action.

In dealing with students, consistency is important. Once they know the rules for the classroom, warn once and then act. When students break rules, either they are testing you or the rules are not clear. If it is the former, have the consequences thought through that students will face as a result of their misbehavior. You might establish a particular time before or after school during the week that they will have to serve a detention for you. Your school may have some formal detention program that you could use to serve their time. It is sometimes effective to ask the student to see you after class to discuss the punishment. In each case, they should know why they are being punished and what the punishment is. It is sometimes helpful to have a notebook to have them sign in that such a discussion took place to serve up the importance of it in your eyes and to establish a record of misbehavior should additional individuals have to be brought into the picture later.

In all of your corrective activities with students, see that you use common sense and maintain the dignity of the students. It is their behavior that you are not accepting, not them as individuals. In some cases, they may make you laugh or amuse you with their antics, but even so, you must consistently enforce the classroom rules and expectations. Let them know that you appreciate their humor or wit, but help them find useful ways of expressing their thoughts and feelings.

Some teachers get into infinite sequences of warning students about their behavior. It is best to warn once and then act. Once students see that you are serious and will back up the rules, they will take their misbehavior in other directions. This is especially true when they see that you enforce your rules in a consistent, but fair, fashion. Punishments should never involve academic or homework related punishments. Often the most effective punishments are to sit and do nothing for a period of time.

The most productive forms of discipline are being prepared, being alert to what is going on, positioning yourself in the classroom, stopping situations from getting started, showing enthusiasm for your subject and their learning, and recognizing your students regularly as individuals. One teacher is known for giving out "no-bell" prizes, see Figure 15.3, for outstanding presentations, solutions, or insights in the classroom. Another teacher maintains records for a classroom recognition ceremony held near the end of the year in which student achievements across the year are recognized and the good times of the class are remembered as a group.

Summary

All of these suggestions will not guarantee the start of the year will be a success. You are the person that has to meld them together into a plan that works for you. It is the planning and consistent implementation of the plan that makes classrooms work. Think through what you want your class to look like and then work consistently to achieve that goal.

Exercises 15.1

1. Draw a diagram of your ideal classroom. Show how you would locate desks, six computers, and resource materials to achieve your goals. Explain why you developed this plan.

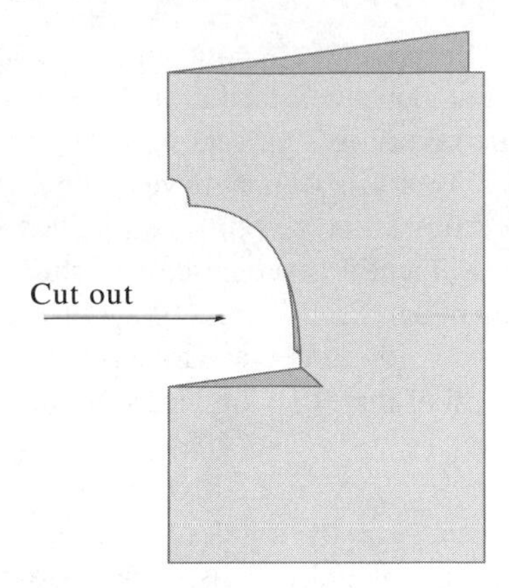

Figure 15.3 One of many ways to reward students in a memorable manner.

2. Write out a list of the behavior goals that you would expect your students to follow. Give a justification for each goal.

3. Create a form that could be given to students on the first day of class to collect the information about the students for your files and as background for getting to know them better.

PROJECTS Find Out for Yourself

1. Interview two teachers about how they start off the year with their students. How do they let students know what their expectations for performance and behavior are?

2. Visit with a dean or disciplinary officer at the school where you will student teach to discuss effective ways of dealing with punishments for classroom incidents at the classroom level.

15.2 Organizing Your Classroom to Improve Focus

Planning and developing instructional materials for your classroom will take more time than you imagine. Developing ways of keeping records of lessons and resources from the start will aid in improving your efficiency and lessening your burden across the years. The development of efficient storage and filing systems, along with a general plan for an orderly classroom will also rub off on the students. A classroom atmosphere where resources are immediately available and references easy to find promotes learning as well.

Efficient Filing and Records Systems

As you begin your classroom organization, it is good to develop a filing system for resources and a filing system for lesson plans. Each of these will help you remember successful approaches and, at the same time, serve to promote integrating new ideas into your teaching from year-to-year.

A filing system for resources might consist of three different, but interrelated, collections of materials. The first would be worksheets, handouts, copies of articles, and other print materials dealing with topics. These materials could provide background as well as structured lesson ideas for your classroom. This filing system might start out arranged by mathematical topics built on the standards: Number and Operations, Algebra, Geometry, Measurement, Data Analysis and Probability, Problem Solving, Reasoning and Proof, Communication, Connections, Representation, Equity, General Curriculum Materials, Teaching Ideas, Learning Theories, Assessment, Technology, and other headings. As your system develops, you might wish to break each area down into finer subdivisions. For example, Algebra might be further subdivided as follows:

Algebra

Concept of Variable
Writing Expressions
Developing and Solving Equations
Developing and Solving Inequalities
Quadratic Equations and Their Solutions
General Polynomial Equations
Systems of Equations
Graphing Equations
Graphing Inequalities
History of Elementary Algebra
Functions in General
Linear Functions
Polynomial Functions
Exponential Functions and Logarithms
Trigonometric Functions
Logistic Function
Matrices

Recursion/Discrete Dynamical Systems
Number Theory Topics
Applications of Algebra
Technology in Algebra
Algebraic Web Resources

The list of subdivisions for Algebra will continue to grow as time progresses. As your system grows, you might find it important to keep a listing of resources in an index file to help locate particular resources. Similar listings of topics will evolve for other areas as your collection grows.

A second area of resources consists of electronic resources. You may want to develop a filing system on your computer of web sites and other materials that can be used to plan and help in conducting classes. A similar filing system in folders on your computer can contain links to various web sites, as well as electronic copies of worksheets and other materials you have developed or acquired to assist in teaching various topics in your curriculum.

A third set of resources consists of texts and other printed materials that you might keep on bookshelves in the classroom. Again these materials should be arranged by topic areas to assist in quickly locating a particular reference when it is needed. You might even wish to develop a filing system for these materials to keep track of them and to be able to index them for easy location as your collection grows. You may also want to keep copies of reviews of new products so that you can lay your hands on them when it is time to order new technology or library materials.

The second major filing system should be for your lesson plans. As you complete a lesson, you may want to have a form on which you can write your reflections on what worked well and what things you would change in teaching that lesson another time. These lesson plans can then be filed, along with any overheads, masters of handouts, and assessments by chapter for use as resources in coming years. In addition to the lesson plans, you may also wish to have a general folder dealing with syllabi, general curriculum papers dealing with that particular course, and other related materials.

The combination of these two general filing systems, one for resources and one for lesson plans grows in importance as your career progresses. They provide you with quick references to ideas, assistance in planning for instruction so that not everything has to start from scratch each year. They provide a foundation for continued growth and a resource of ideas as you, yourself, take on the role of a more senior member of the staff and are called on to help younger teachers with ideas. Such a filing system also provides a basis for finding ideas for students to pursue in project work or for writing papers about topics in mathematics.

Maintaining an Orderly Classroom

In the same manner as you maintain materials for supporting your teaching, you want to maintain your classroom so that it promotes and supports efficient learning of mathematics. Part of this atmosphere is having a classroom that is clean and inviting. This is not just the job of the school's maintenance and janitorial staff. At the end of each period, students should be responsible for cleaning up around their desks and seeing that their desks or tables are in a condition ready for the next class to start. In a like manner you need to police the classroom at the beginning and end of the day to make sure that things are in order and that resources are where they should be for activities coming up in the next few days.

Another part of maintaining an orderly classroom comes from the procedures you set up for distributing papers, for having students put problems on the board, and how you deal with issues like hall passes, students making up assignments, and related matters. If students are seated by some form of grouping, you can establish a mailbox arrangement, as shown in Figure 15.4, at some point in the room where you put papers to be returned to the members of particular groups. Such mailbox systems made of corrugated cardboard are available commercially, or you can easily make one yourself. Just before the beginning of class, specific students can be assigned the responsibility for picking up their group's papers and distributing them before the bell rings. A similar system can be established for distributing manipulative materials or other laboratory equipment for investigations conducted as part of class. The same person can also be responsible, on a rotating basis, for distributing and collecting calculators for their groups if school-owned calculators are used in the classroom.

When students are asked to put work on the board, they should write the problems on the board in order around the classroom. Students should learn to write their name above their work, the number of the problem, and then show all of their work. Having

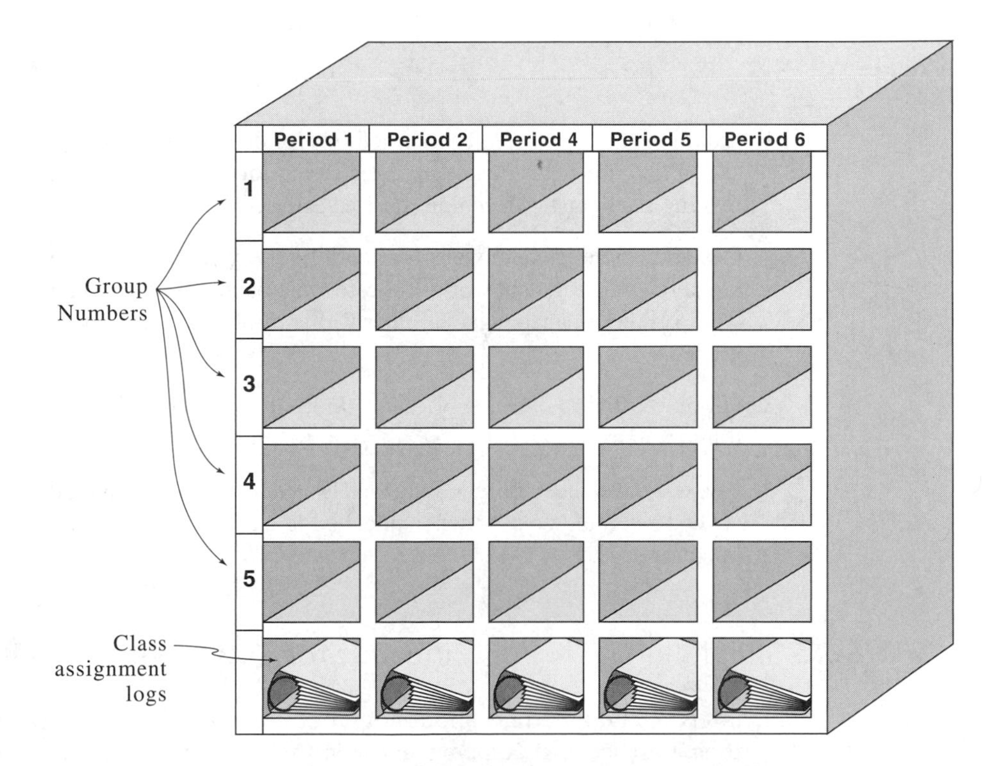

Figure 15.4 A mailbox arrangement.

the problems in order helps in discussing the work. Neither you, nor the students, waste time in looking about to find a specific problem. Having the students' names above the work makes it easy to call on them and lessens your memory load during the coverage of the problems.

Adding these routines to your procedures for running your class increases your efficiency by lessening the mental load of tasks to worry about during instruction and allows you, and your students, to focus on the mathematics being discussed. Combining these techniques with efficient ways of taking the roll, perhaps while students are doing a warm-up at the beginning of class, helps get classes started efficiently and with a minimum of directions having to be given by you. This places more responsibility for monitoring the work of the classroom on the students and frees you to do the things that you have to do.

Another aid to efficiency is developing a classroom notebook of assignments. Each day, either you or an assigned student should take a set of notes, along with the assignment and any handouts for the day and place them in a class notebook. Students who are absent can then get these materials, copy them, and become responsible for what they have to make up. The responsibility for maintaining this notebook can be scheduled for almost the entire year in a short period of time and posted on the bulletin board or on a printed schedule. Each student might serve in this role for a week at a time. The notebooks might be kept in the bottom-row mailbox for each class.

Exercises 15.2

1. Examine some texts on general methods for other suggestions for classroom organization and efficiency. Write up three of the ideas that you found and relate them to making mathematics teaching more effective.

2. Describe some efficient ways for handling the grading of daily papers, and to maintain a record of student work. Indicate what balance you would give to weighing the work on these papers against other forms of assessment in determining a student's grade.

3. Develop a list of subcategories for three of the other areas of the *Standards* for a filing system.

4. Create a coversheet for reflecting on a lesson following teaching it. Indicate how your coversheet could be used when attached to a copy of the lesson plan and filed away for future use.

PROJECTS Find Out for Yourself

1. Interview two teachers about how they maintain their resources and records of their past lessons. Compare and contrast their methods with the method described earlier in this section.

2. Observe and interview a teacher on how they deal with passing back papers, distributing materials, and maintaining oversight on the functioning of their classroom. Which of their methods did you find particularly helpful?

15.3 Motivating Students by Involving Them

Perhaps the largest factor in managing a classroom is keeping students motivated and contributing participants in classroom activities. You cannot count on the materials alone to accomplish this task. Like a conductor of an orchestra, you must continually monitor the class and bring both individuals and small groups of students back into the work of the day. Keeping students involved is not an impossible task, but it does require continual vigilance on your part.

Gaining student involvement is central to getting students into doing mathematics at the bell. The key to motivation is keeping students involved. If they are waiting for you to take roll or read announcements, the message is that the mathematics is secondary. Hence, get them involved before the bell rings and keep them involved over the full period of the class. The warm-up or starter at the beginning class might be a problem covered in the previous class meeting, a recreational puzzle that motivates the topic of the day, a practice review that helps recall skills that will be required in a few days, or a quiz on definitions or theorems. The results produced may be collected, or it may serve as a basis for opening the discussion.

The discussion should involve the teacher asking students questions about what they have done. The questions should be focused questions for which students can give focused answers. Teachers must monitor their actions to see that students really do respond to the questions asked. Johnson (1982; 1986; 1994) suggests that teachers need to

- Pause after asking a question to allow students to reflect on the answer. In doing so, the teacher should refrain from calling on a specific student until all students have had an opportunity to think about the task posed.
- Ask questions that cause students to both give an answer and a justification for why that answer holds.
- Be prepared to follow up with a student if they have difficulty in answering a question. For example, if a student is stumped, have a follow-up question that either asks a related procedural or conceptual question that would help the student move toward a correct answer to the original question. One can also allow the student a "lifeline" to call on another student for help.
- Keep the students from blurting out answers. Maintain your classroom behavioral rules of raising hands to be called upon, or sometimes just tell them to think and you will call on students at random.

- Not ask rhetorical questions or questions that contain their own answers. See that the questions asked cause students to think and become cognitively involved.

In addition to getting students involved in the lesson through questioning and working on problems with their partners, you can increase student motivation through the use of challenging exercises or problems. As students work on these problems, vary whether they work individually or with a partner. Vary the ways in which they work. Also take time occasionally to model a solution for them like a coach. Then give them a similar problem to work to practice the method you just demonstrated. Compare and contrast the two problems and the solution methods used to cement the understanding desired.

You can also take time out to discuss the applications of the material and to link the material to careers using mathematics. When possible, have such material posted on the bulletin board to keep students examining the board for what is new. Information on careers in mathematics can be found at the web sites of a number of professional societies:

American Mathematical Society: **www.ams.org/careers**

American Statistical Association: **www.amstat.com**

Conference Board of the Mathematical Sciences: **www.maa.org/cbms**

Institute for Operations Research and the Management Sciences: **www.informs.org**

Mathematical Association of America: **www.maa.org/students**

National Council of Teachers of Mathematics: **www.nctm.org**

Society of Actuaries: **www.BeAnActuary.org**

Society for Industrial and Applied Mathematics: **www.siam.org/careers/**

Some commercial textbook publishers and corporations produce free poster materials suitable for use on bulletin boards as well. Other suppliers of classroom materials such as Dale Seymour Publications and Creative Publications also have mathematics posters and related materials for sale. Combinations of these can help make your mathematics classroom both an attractive and informative environment for your students.

You might also consider bringing in some individuals from the community across the year to discuss how they make use of mathematics in their career or profession. You would also probably be able to get someone from a local college or university to speak on career opportunities or applications of mathematics for your students. Such days not only provide a change of pace, but also help increase student motivation. Be sure to discuss what your goals are with a speaker prior to their coming to meet with your students. Work to get dynamic and motivating individuals when you replace valuable class time with their presentations.

A final issue in motivating students is to end your classes on a high note. As discussed earlier, you should operate your classes such that you dismiss the students at the end of the hour, not the school bell system. Having this control, you should have the students working right up until the last minutes of class on most days. Johnson (1984;

1986; 1994) offers several suggestions for keeping students involved and motivated right up to the closing moment of class. To do this, you need to develop a number of different "five-minute drills" with which to end class. These can vary from

- Calling students representing different groups within the class to the board to work problems reviewing what has been discussed during the day. They can make use of help from individuals in their group. This is particularly good when the focus of the day has been on procedural topics.
- Focusing students' attention on particular points to consider as they work through the assigned reading and homework for the following day. Such an overview should be used to point out critical features and specific points of major importance.
- Providing students with a challenge problem to consider. Such problems might serve as a springboard into the next topic, extend the present topic, or review a concept or technique students experienced difficulty with on the most recent examination.
- Having students write a summary of the day's work in their journals if journals are used.
- Discussing the classes' progress in the topic over the past period of time. Helping students gain a perspective on their performance and how their progress measures up to your expectations. It is also good to do this individually with each student at least once during each grading period.

Keeping a class flowing smoothly with all of the students involved and participating is a major task for teachers. If you work at it on a daily basis by planning for good openers, healthy questioning during the hour, and solid closings to your classes, it will gradually become another routine that you will continue to perfect. When these issues are not planned, they rarely happen, and you will find yourself dealing more with corrective discipline issues and repeating instructions over and over again. Getting students involved and participating motivated members of the learning community in your classroom is key to maintaining good classroom management.

Exercises 15.3

1. Select a topic of interest from the middle school curriculum. Develop a number of questions that you could ask students about the topic as you develop it. Include a copy of the materials you would expect the students to be working from in the class—pages from the textbook, handouts, and so on.
2. Find a list of additional sources of information on careers in the mathematical sciences on the web or through materials designed for counselors. Select specific materials that appear to be most useful and write a short synopsis of what they have to offer a middle school student and a high school student.

3. Develop outlines for three different openings and closings other than those mentioned in the text. How do they support student involvement and motivation?

PROJECTS Find Out for Yourself

1. Look in a textbook designed for the level of mathematics you want to teach. What resources does it offer in its lesson plans to help you make your classes motivating and informative to your students relative to:
   a. careers in mathematics?
   b. applications of mathematics?
   c. the value of mathematics to society?

2. Visit a number of school mathematics classrooms. What makes some motivating and what makes some uninviting? What features would you like to adopt to make your classroom inviting? Describe these in some detail.

References

Johnson, D. R. (1982). *Every minute counts: Making your math class work*. Palo Alto, CA: Seymour.

——— (1986). *Making minutes count even more: A sequel to every minute counts*. Palo Alto, CA: Seymour.

——— (1994). *Motivation counts: Teaching techniques that work*. Palo Alto, CA: Seymour.

National Council of Teachers of Mathematics (2000). *Principles and standards for school mathematics*. Reston, VA: Author.

Steen, L. A. (ed.) (1992). *Mathematics books: Recommendations for high school and public libraries (MAA)*. Washington, D.C.: Mathematical Association of America.

Thiessen, D., Mathias, M., & Smith, J. (1998). *The wonderful world of mathematics: A critically annotated list of children's books in mathematics* (2nd ed.). Reston, VA: NCTM.

16

Futures—Mathematics Education in the 21st Century

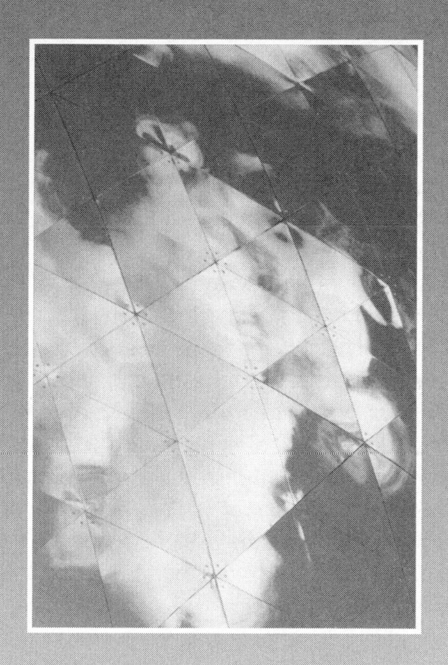

Any vision of school mathematics teaching and learning needs to be subjected to informed criticism. Moreover, it needs to change continually in light of the professions' experience and the better understanding it can achieve through a fair, thorough, and tough-minded debate. If mathematics educators can adopt a more critical stance toward their work, there is good reason to be optimistic that many of the challenges of the next few decades can be met in ways that will lead to more effective professional practice.

Jeremy Kilpatrick and Edward A. Silver (2000)

Predicting the future of mathematics education is a difficult task. However, the current efforts to improve mathematics education provide a set of initial conditions from which we can begin to envision where the teaching and learning of mathematics may be going. The recommendations of the NCTM's *Principles and Standards for School Mathematics* and the CBMS's *Guidelines for the Education of Teachers of Mathematics* call for significant changes in curricular content and instructional approaches at the middle and senior high school levels. At the same time, results of national and international surveys of mathematics education suggest that little has changed in the nation's classrooms. Will this discontinuity between desired states and present states continue?

This is the question that you and your colleagues will have a great say in answering. Change can only take place as teachers and schools begin to reconsider what they want to provide through their curricula and what expectations they will set for their students. The time is right for this change to take place. The original NCTM *Standards* were set in 1989, and instructional materials are now available to support the kinds of programs envisioned by the *Standards* and their sequel, the *Principles and Standards*. Theories of educational change (Fullan 1991) state that change takes place in stages. The stages are those of introduction, initiation, and institutionalization. The *Standards* have been introduced and the stage of initiation is underway. It is during this phase that schools and teachers will seriously begin to modify their curricula and expectations for school mathematics. These recommendations also call for significant changes in instructional delivery and support for mathematics programs.

Part of this transformation in instructional delivery calls for a more constructivist approach in the classroom. For many teachers of mathematics, this is a major change. For most, their models of teaching have been teachers accustomed to standing and delivering lectures or at least running teacher-centered classrooms. It is only in the past few years that a measurable number of teachers have begun to use cooperative learning and associated methods in teaching mathematics. In a like manner, the reforms calling for a more integrated curriculum that builds around a focus on processes and modeling are relatively new. Few teachers have had the opportunities to see the development of modeling techniques and methods afforded by this text. As a result, considering mathematical content and problems from a modeling perspective is both novel and challenging.

As you enter the classroom, you will be outfitted for moving forward on both fronts—curricular and instructional. How all of you choose to integrate your capabilities into developing classrooms that challenge students to learn mathematics in constructive ways will shape the future of mathematics education. Change takes place classroom by classroom, school by school, and district by district. The winds of change are shifting in a positive direction, but will need a consistent source of energy to achieve the shift in instruction envisioned by the recommendations we have considered.

You will have important decisions to make as you begin your career in teaching mathematics. For example,

- How will you provide students with a rich mathematics curriculum built around the precepts set forth in the NCTM's *Principles and Standards for School Mathematics*? What content will you offer students, what expectations will you set for their performance, and what styles of instructional delivery will best suit the situations?

- How will you structure your instructional patterns to best help your students begin to construct their own mathematical knowledge in a reflective manner? How will you learn to step back and ask the questions that will enable your students to take responsibility for monitoring their own learning and for making sense out of mathematics themselves?

- What role will communication and writing play in your classroom? How will your students learn to make representational shifts between verbal, symbolic, graphical, and tabular modes of communicating as they work to solve problems and communicate their results?

- How will you introduce modeling concepts and procedures to your students? What expectations will you hold for them as a group and as individuals to build models for both discrete and continuous situations? What role will probabilistic, or stochastic, models play in your classroom's curriculum?

- What modes of assessment and evaluation will you use in examining your students' progress toward your and your school's goals for their achievement in mathematics? What role will extended student-constructed response items, projects, and journaling play in your classroom?

- How will the mathematics studied be connected with the current events of the local issues, the daily news, and significant applications taking place in your community? How will students of different racial, ethnic, and genders learn that mathematics is for all students, not a chosen few? What will you do to ensure that each of your students feels that they can do mathematics?

- What role will technology, both calculator based and computer based, play in your teaching and learning efforts? How will calculators be integrated in daily instruction and students' homework and project work? What use will be made of dynamic geometry programs and major wordprocessing, spreadsheet, and graphical display systems in your mathematics program?

- How will you deal with individual differences you face in reaching all of your students and serving their varied needs? What will you do to make mathematics a friendly and accessible discipline for all of your students?

Your answers to these questions will say to a great deal about what your classroom becomes. The answers you and your peers give to these questions will determine, to a great deal, what becomes of mathematics education for the next two generations of our nation's students. The mathematics education they receive will enable them to confront problems we cannot even imagine today. The mathematics education they receive will empower them to make decisions that will change our society in ways we cannot imagine. The outcomes associated with decisions made about what you teach, how you teach, and what you expect mathematically of your students will effect changes you cannot imagine. Make those decisions carefully and work hard to implement them for all of your students. The effort is worth it and the journey will be adventurous.

Best wishes from your authors as you embark on an eventful career in mathematics education.

Reference

Fullan, M. G. (1991). *The new meaning of educational change*, 2nd ed. New York: Teachers College Press.

Index

Credits

Photo Credits

Chapter 1. 3: North Wind Picture Archives
Chapter 2. 29: Will Hart/ PhotoEdit
Chapter 3. 65: David Young-Wolff/ PhotoEdit
Chapter 4. 127: Kevin R. Morris/ Corbis
Chapter 5. 159: Bonnie Kamin/ PhotoEdit
Chapter 6. 201: NASA/ Corbis
Chapter 7. 257: ©Duncan Smith/ Corbis
Chapter 8. 317: World Perspectives/ Stone
Chapter 9. 355: ©Lynda Richardson/ Corbis
Chapter 10. 405: ©Joseph Sohm/ Corbis
Chapter 11. 453: Dr. Gary Settles/ Photo Researchers, Inc.
Chapter 12. 485: David Young-Wolff/ PhotoEdit
Chapter 13. 517: Chris Marona/ Photo Researchers
Chapter 14. 549: David Young-Wolff/ Stone
Chapter 15. 589: Bonnie Kamin/ PhotoEdit
Chapter 16. 611: Michael Banks/ Stone

Quotation Credits

Chapter 1: A. B. Chace, 1927.
Chapter 2: Gloria Lappan and Diane Briars (1995). How Should Mathematics Be Taught? *In* I. M. Carl (ed.), *Prospects for School Mathematics*. Reston, VA: NTCM, p. 147.
Chapter 3: William F. Lucas (1983). Problem Solving and Modeling in the First Two Years. *In* A. Ralston and G. S. Young (eds.), *The Future of College Mathematics*. New York: Springer-Verlag, p. 43.
Chapter 4: James T. Fey (1990). Quantity. *In* L. A. Steen (ed.), *On the Shoulders of Giants*. Washington, D.C.: National Academy of Sciences, pp. 91–92.
Chapter 5: August DeMorgan (1837). Preface in *Elements of Algebra*. London, England.

Chapter 6: Marjorie Senechal (1990). Shape. *In* L. A. Sheen (ed.), *On the Shoulders of Giants*. Washington, D.C.: National Academy of Sciences, p. 147.
Chapter 7: David S. Moore (1990). Uncertainty. *In* L. A. Sheen (ed.), *On the Shoulders of Giants*. Washington, D.C.: National Academy of Sciences, pp. 134–135.
Chapter 8: Ross L. Finney (1981). *In* L. A. Sheen (ed.), *Mathematics Tomorrow*. New York: Springer-Verlag, p. 212.
Chapter 9: Maynard Thompson (1981). Application of Undergraduate Mathematics. *In* L. A. Sheen (ed.), *Mathematics Tomorrow*. New York: Springer-Verlag, p. 212.
Chapter 10: SIAM News (1994).
Chapter 11: Robert M. White (1998). *In* L. A. Steen (ed.), *Calculus for a New Century*. Washington, D.C.: Mathematical Association of America, p. 9.
Chapter 12: Thomas Cooney (1994). Teacher Education as an Exercise in Adaptation. *In* D. A. Aichele and A. F. Coxford (eds.), *Professional Development for Teachers of Mathematics*, pp. 10–11.
Chapter 13: James Hiebert and Thomas P. Carpenter (1992). Learning and Teaching with Understanding. *In* D. A. Grouws (ed.), *Handbook of Research on Mathematics Teaching and Learning*. New York: Macmillan, p. 65.
Chapter 14: Jean Kerr Stenmark (ed.) (1991). *Mathematics Assessment*. Reston, VA: NTCM, p. 3.
Chapter 15: Gilbert Cuevas (1998). The Role of Complex Mathematical Tasks in Teacher Education. *In* Mathematical Sciences Education Board (eds.), *High School Mathematics at Work*. Washington, D.C.: National Academy of Science, p. 140.
Chapter 16: Jeremy Kilpatrick and Edward A. Silver (2000). Unfinished Business. *In* M. A. Burke (ed.), *Learning Mathematics for a New Century*. Reston, VA: NCTM, pp. 233–234.

Permissions Credits

11, 14, 16 From Giordano, Weir, & Fox, *A First Course in Mathematical Modeling*, Second Edition. Copyright ©1997 Brooks/Cole Publishing Company. Page 16 data from R. Pearl, ©The Growth of Population, Quart. Rev. Biol. **2**(1927): 532–548.
34 From Dossey, J. A., How Should U.S. Performance in School Mathematics Be Interpreted? *In* Z. Usiskin. (ed.), *Developments in School Mathematics Education Around the World*, 1999, pp. 228–239. Copyright © NCTM 1999. Reprinted by permission of University of Chicago School of Mathematics Project.
36, 38 From Schmidt, W. H., McKnight, C. C., & Raizen, S. A., *A Splintered Vision: An Investigation of U.S. Science and Mathematics Education*, 1997, pp. 70; 99–101. Copyright © 1997 Dordrecht, Netherlands: Kluwer Academic Publishers. Reprinted with kind permission of Kluwer Academic Publishers.
39 From J. Flanders, *Arithmetic Teacher*, **35**(1): 18–23. p. 19. Copyright © 1987 by the National Council of Teachers of Mathematics. Reprinted with permission. All rights reserved.
60–61 From the *NCTM News Bulletin*. Copyright © 1999 by the National Council of Teachers of Mathematics. Reprinted with permission. All rights reserved.

73, 74 From *Mathematics: Modeling Our World, Course 2*, 1998, pp. 6, 7. Copyright © 1998 Lexington, MA: COMAP, Inc. Reprinted by permission.
76 From Boyd, C. J., Burrill, G. F., Cummins, J. J., Kanold, T. D., Malloy, C. *Geometry: Integration, Applications, and Connections*, 1998, p. 533. Copyright © 1998 New York: Glencoe/McGraw-Hill. Reprinted by permission.
86–89, 91–94, 98, 100, 102, 103, 107, 110, 115, 118, 120 From Giordano, Weir, & Fox, *A First Course in Mathematical Modeling*, Second Edition. Copyright © 1997 Brooks/Cole Publishing Company.
151 From Hart, E. W., Algorithmic Problem Solving in Discrete Mathematics. *In* L. J. Morrow (ed.), *The Teaching and Learning of Algorithms in School Mathematics: 1998 Yearbook*, pp. 251–267. Copyright © 1998 by the National Council of Teachers of Mathematics. Reprinted with permission. All rights reserved.
170 From *Addison Wesley-Longman Secondary Math: An Integrated Approach: Focus on Algebra* by Charles, R. I., & Thompson, A., p. 309. © 1998 Addison Wesley Publishing Company. Published by Prentice Hall, Inc. Used by permission of Pearson Education, Inc.
177 Reprinted with permission from *Mathematics, Pedagogy, and Secondary Teacher Education* by Thomas J. Cooney, et al. Copyright © 1996 by Thomas J. Cooney, Erich Wittman, Georg Schrage, John Dossey, and Stephen Brown. Published by Heinemann, a division of Reed Elsevier Inc., Portsmouth, NH.
187 From Dossey, J. A., Jones, C. O., & Klag, P., *Quantitative Literacy and the Middle School Curriculum* (Unpublished Document), May 2000, Copyright © May 2000 New York: The College Board. Reprinted with permission.
188, 189, 191, 192, 212, 214, 215 From Giordano, Weir, & Fox, *A First Course in Mathematical Modeling*, Second Edition. Copyright © 1997 Brooks/Cole Publishing Company.
232 From *UCSMP Precalculus and Discrete Mathematics* by Peressini, A. L. et al., pp. 642–643, © 1992 Scott-Foresman. Published by Prentice Hall, Inc. Adapted by permission of Pearson Education, Inc.
269 From Teaque, D. & Compton, H., The Midge Problem, *Consortium*, 55 Fall, 1995, pp. 8, 12. Copyright © 1995 Lexington, MA: COMAP, Inc. Reprinted by permission.
284, 285, 291–295, 301, 304–306, 309 From Giordano, Weir, & Fox, *A First Course in Mathematical Modeling*, Second Edition. Copyright © 1997 Brooks/Cole Publishing Company.
345 From Crosswhite, F. J., Dossey, J. A., Swafford, J. O., McKnight, C. C., Cooney, T. J., Downs, F. L., Grouws, D. A., & Weinzweig, A. I., *Second International Mathematics Study: Detailed Report for the United States, 1986*, pp. 344–347. Copyright © 1986 Champaign, IL: Stipes Publishing. Reprinted with permission.
347 Reprinted with special permission of King Features Syndicate.
357, 358, 360–364, 369, 373, 375, 378, 378–385, 387, 391, 394–400, 410–412, 414–417, 419, 420, 424, 433–437, 440, 441, 444, 445, 447–450, 455, 457, 458, 462, 464, 465, 471–473 From Giordano, Weir, & Fox, *A First Course in Mathematical Modeling*, Second Edition. Copyright © 1997 Brooks/Cole Publishing Company.
475 From F. Giordano & M. Weir, *Differential Equations: A Modeling Approach*, 1991, p. 51. Copyright © 1991 Menlo Park, CA: Addison-Wesley. Reprinted by permission of Pearson Education.

476 Adapted from UMAP Unit 625, which was written by the authors and David H. Cameron. The adaptation is presented with the permission of COMAP, Inc.
477–482 From Giordano, Weir, & Fox, *A First Course in Mathematical Modeling*, Second Edition. Copyright © 1997 Brooks/Cole Publishing Company.
523 From Larcombe, T., *Mathematical Learning and Difficulties in the Secondary School*, 1985. Copyright © 1985 Milton Keynes, England: Open University Press. Reprinted with permission.
530 Adapted from Larcombe, T., *Mathematical Learning Difficulties in the Secondary School*, 1985, p. 47. Copyright © 1985 Philadelphia, PA: Open University Press. Reprinted with permission.
535 From Secada, W. G., *Race, Ethnicity, Social Class, Language and Achievement in Mathematics*, 1992, pp. 636–637. Copyright © 1992. New York: Macmillian Publishing. Reprinted by permission of Pearson Education.
536 From Zaslavsky, C., Multicultural Mathematics: One Road to the Goal of Mathematics for All. *In* G. Cuevas & M. Driscoll (eds.), *Reaching All Students with Mathematics*, 1993, pp. 45–55. Copyright © 1993 by the National Council of Teachers of Mathematics. Reprinted with permission. All rights reserved.
537 From P. Ron, Spanish-English Language Issues in the Mathematics Classroom. *In* E. Ortiz-Franco, N. G. Hernandez & Y. De La Cruz (eds.), *Changing the Faces of Mathematics: Perspectives on Latinos*, 1999, pp. 22–33. Copyright © 1999 by the National Council of Teachers of Mathematics. Reprinted by permission of the publisher and author. All rights reserved.
555 From Crosswhite, F. J., Dossey, J. A., Swafford, J. O., McKnight, C. C., Cooney, T. J., Downs, F. L., Grouws, D. A., & Weinzweig, A. I., *Second International Mathematics Study: Detailed Report for the United States, 1986*, p. 344. Copyright © 1986 Champaign, IL: Stipes Publishing. Reprinted with permission.